Integrated Watershed Management in Rainfed Agriculture

Integrated Watershed Management in Rainfed Agriculture

Suhas P. Wani

*International Crops Research Institute for the Semi-Arid Tropics (ICRISAT),
Patancheru, Andhra Pradesh, India*

Johan Rockström

Stockholm Environment Institute (SEI), Stockholm, Sweden

K.L. Sahrawat

*International Crops Research Institute for the Semi-Arid Tropics (ICRISAT),
Patancheru, Andhra Pradesh, India*

CRC Press

Taylor & Francis Group
Boca Raton London New York

CRC Press is an imprint of the
Taylor & Francis Group, an **informa** business

CRC Press
Taylor & Francis Group
6000 Broken Sound Parkway NW, Suite 300
Boca Raton, FL 33487-2742

First issued in paperback 2017

*CRC Press/Balkema is an imprint of the Taylor & Francis Group,
an informa business*

© 2011 Taylor & Francis Group, London, UK

Typeset by MPS Limited, a Macmillan Company, Chennai, India

No claim to original U.S. Government works

ISBN-13: 978-0-415-88277-4 (hbk)
ISBN-13: 978-1-138-11779-2 (pbk)

Library of Congress Cataloging-in-Publication Data

Wani, S. P. (Suhas Pralhad), 1952–
  Integrated watershed management in rainfed agriculture /
Suhas P. Wani, Johan Rockström, Kanwar Lal Sahrawat.
    p. cm.
  Includes bibliographical references and index.
  ISBN 978-0-415-88277-4 (hardback : alk. paper)
  1. Dry farming. 2. Watershed management. 3. Dry farming—
Economic aspects. 4. Agricultural productivity. I. Rockström, Johan.
II. Sahrawat, K. L. III. Title.
  SB110.W26 2011
  631.5'86—dc23

                                              2011027889

Published by: CRC Press/Balkema
               P.O. Box 447, 2300 AK Leiden, The Netherlands
               e-mail: Pub.NL@taylorandfrancis.com
               www.crcpress.com – www.taylorandfrancis.co.uk – www.balkema.nl

**Visit the Taylor & Francis Web site at
http://www.taylorandfrancis.com**

**and the CRC Press Web site at
http://www.crcpress.com**

Contents

Preface   xvii
Foreword   xix
List of contributors   xxi

**I Improving livelihoods in rainfed areas through integrated watershed
management: A development perspective**   I

1.1   Introduction and overview   1
1.2   The semi-arid tropics and the importance of rainfed farming   2
1.3   The objectives of watershed management   3
1.4   The main experience   3
1.5   Understanding the nature of rural poverty   4
    1.5.1   Landlessness   4
    1.5.2   Women   5
    1.5.3   Subsistence   5
    1.5.4   Debt and awareness   5
    1.5.5   Health   5
    1.5.6   Human rights   6
    1.5.7   Environment   6
    1.5.8   Trends   6
    1.5.9   Assistance   6
    1.5.10 Equity and inclusion   7
    1.5.11 Migration   7
    1.5.12 Case study   7
    1.5.13 Indicators of progress   8
1.6   Four stages to pull a rural community out of poverty   8
    1.6.1   The Process   9
        1.6.1.1   Stage 1   9
        1.6.1.2   Stage 2   9
        1.6.1.3   Stage 3   10
        1.6.1.4   Stage 4   10
    1.6.2   Recent improvements   10
    1.6.3   Community resource centers   11
    1.6.4   Income-generating activities   11

1.6.5 Second generation interventions: market links 11
1.6.6 Duration of support and coordination 12
1.6.7 Better use of the NGOs as collaborators or implementers 12
1.7 Aspects of the process of watershed implementation 13
1.7.1 Management of watershed 13
1.7.2 Groups and *Jankars* 14
1.7.3 Savings, loans, and credit 15
1.7.4 Literacy and numeracy 15
1.7.5 Expenditure 15
1.7.6 Land ownership 15
1.7.7 Spread and dissemination 16
1.7.8 Lack of spread of locally successful watershed interventions in India 16
1.7.9 Sustainability 16
1.7.10 Subsidies and cost contributions 17
1.8 Technology for the poor 17
1.8.1 General principles 17
1.8.1.1 Participatory varietal selection (PVS) 18
1.8.1.2 Client oriented breeding 18
1.8.1.3 Seed priming 19
1.8.1.4 Technologies for women 19
1.8.1.5 Soil conservation 19
1.8.1.6 Soil fertility 20
1.8.1.7 Crop pest control 20
1.8.1.8 Renewable energy technologies 20
1.8.1.9 Community empowerment 21
1.8.2 Water 21
1.8.3 Trees 23
1.8.4 Livestock and fodder 25
1.8.5 Aquaculture 26
1.9 Future research needs 27
1.10 Concluding remarks and way forward 30
References 30

2 **Watershed development as a growth engine for sustainable development of rainfed areas** **35**

2.1 Introduction 35
2.2 A concept of safe operating space for humanity 37
2.3 Current status of rainfed agriculture 38
2.4 Vast potential to increase crop yields in rainfed areas 39
2.5 Improved water productivity is a key to unlock the potential of rainfed agriculture 42
2.6 Water alone cannot do it 45
2.7 Integrated watershed management is key for sustainable management of land and water resources and improved livelihoods 46
References 48

3 Watershed development for rainfed areas: Concept, principles, and approaches   **53**

3.1 Introduction   53
3.2 Watershed concept   53
3.3 Importance of land use planning in watershed development   56
3.4 Criteria for prioritization of watersheds   57
3.5 Common features of the watershed development model   58
3.6 Evolution of watershed development approach in India   60
3.7 Need for a holistic approach for watershed management   63
3.8 Evolution of the consortium approach   64
3.9 Components of integrated watershed management   65
    3.9.1 Entry Point Activity   65
    3.9.2 Land and water conservation practices   68
    3.9.3 Integrated pest and nutrient management   69
    3.9.4 Farmers' participatory research and development trials   69
    3.9.5 Crop diversification and intensification of crops and systems   70
    3.9.6 Use of multiple resources   71
    3.9.7 Capacity building   72
3.10 Key features of facilitating the consortium approach   74
    3.10.1 Need for a common goal – team building   74
    3.10.2 Building on the strengths   74
    3.10.3 Institutionalization of partnerships   74
    3.10.4 Internal and external institutional arrangements   75
    3.10.5 Dynamic and evolving   75
    3.10.6 Scaling-up/out the approach   75
3.11 Advantages of consortium approach   75
    3.11.1 Sustainability   75
    3.11.2 Cost-effectiveness   76
    3.11.3 Win-win solution through empowerment of partners   76
    3.11.4 Rapid scaling-up   76
    3.11.5 Change in organizational behavior   76
    3.11.6 Public-private partnerships are facilitated (multiplier effect)   77
3.12 Learnings from the experience and triggers for success   77
3.13 Operationalizing community watershed as a growth engine   78
3.14 Watershed as an entry point to improve livelihoods   79
3.15 Convergence in watershed   79
3.16 Multiple benefits from integrated watershed development   80
3.17 Conclusions   81
References   82

4 Equity in watershed development: Imperatives for property rights, resource allocation, and institutions   **87**

4.1 Introduction   87
    4.1.1 The context   87
    4.1.2 Exclusion of landless and women: Role of CPLRs   88

	4.1.3	Groundwater and equity	89
	4.1.4	Project-based equity	90
	4.1.5	Linkages between technology, allocation of funds, and institutions: The issue of mode	90
	4.1.6	Equity in policy and guidelines	91
	4.1.7	Objectives and approach	92
4.2	Equity, property rights, and biophysical characteristics		92
	4.2.1	Equity and property relations in land and water	92
		4.2.1.1 Historically embedded inequalities in access to land and water	92
		4.2.1.2 Land: Domains of ownership	93
		4.2.1.3 Ownership, land use, and CPLR	93
		4.2.1.4 Ownership and landlessness	94
	4.2.2	Water: Availability and increasing water scarcity	94
		4.2.2.1 Property relations in water	95
		4.2.2.2 Water is a local and non-local resource	95
		4.2.2.3 Spatial or location inequities	96
		4.2.2.4 How do the biophysical characteristics actually play out?	96
	4.2.3	Water is a common pool resource and has competing uses	97
		4.2.3.1 The meeting of property relations and the biophysical characteristics	97
		4.2.3.2 Watershed also creates conditions for a positive sum game	97
	4.2.4	Efforts to address equity	98
		4.2.4.1 Equity in coverage	99
		4.2.4.2 Targeted approach	99
		4.2.4.3 Common lands	100
		4.2.4.4 Equitable sharing of increased water	100
		4.2.4.5 Produce sharing arrangements	101
		4.2.4.6 Attempts at risk proofing/pooling and sharing arrangements	102
	4.2.5	Main observations	102
4.3	Funds allocation and subsidies		103
	4.3.1	Nature of watershed treatments among sample villages	104
	4.3.2	Perceived benefits: Sources and beneficiaries	105
	4.3.3	Total number of beneficiaries: Some approximation	106
	4.3.4	Distribution of subsidies and alternative mechanisms	107
	4.3.5	Overall evidence	108
4.4	Equity and institutions		109
	4.4.1	Criticality of institutional process for equity in watershed	109
	4.4.2	Institutional challenges: Learning from the CPR literature	109
	4.4.3	Institutions within WDPs: Provisions in various guidelines	110
	4.4.4	Learning from the past experience: Taking stock	111
		4.4.4.1 Participatory processes: A larger view	112
		4.4.4.2 SHGs and user groups: A tool for equity	112
		4.4.4.3 CPLRs and institutions	113

	4.4.5	Examples of good practices	113
		4.4.5.1 Streamlining equity consideration: Case of CWDP-Orissa	113
		4.4.5.2 Public-private collaboration for forest development in watersheds: A case of IGWDP in Ahmednagar district	114
		4.4.5.3 Sujala watershed and social regulations	115
	4.4.6	Main observations	117
4.5	Gender mainstreaming		117
	4.5.1	Enhancing women's participation and mainstreaming of women SHGs	118
	4.5.2	Promotion of micro-enterprises	118
	4.5.3	Institutional challenges	120
4.6	Policy implications and way forward		121
	4.6.1	Multi-pronged approach	121
	4.6.2	Policy recommendation	121
	4.6.3	Way forward	122
References			124

5 Policies and institutions for increasing benefits of integrated watershed management programs — **129**

5.1	Introduction		129
5.2	Integrated watershed management program in India		130
5.3	Policy endorsement at macro level		131
5.4	Watershed development guidelines		135
	5.4.1	The new common guidelines	136
5.5	Institutional arrangements for watershed development		137
	5.5.1	National Rainfed Area Authority	141
	5.5.2	Central Level Nodal Agency	142
	5.5.3	State Level Nodal Agency	142
	5.5.4	District Watershed Development Unit	142
	5.5.5	Project implementing agency	142
	5.5.6	Watershed Committee	143
	5.5.7	Self-help groups	143
	5.5.8	User groups	143
5.6	Promoting closer institutional links		143
5.7	Dealing with policy and institutional constraints		145
	5.7.1	Collective action	145
	5.7.2	Bottom-up approach	146
	5.7.3	Capacity building	146
	5.7.4	Knowledge-based Entry Point Activity	146
	5.7.5	Empowering women and vulnerable groups	147
5.8	Sustainable watershed management: Role of common guidelines		147
	5.8.1	Institutional responsibilities	148
	5.8.2	Delegation of power to the states	148

	5.8.3 Dedicated institutions	149
	5.8.4 Convergence	149
	5.8.5 Consortium approach	150
	5.8.6 Addressing equity	150
	5.8.7 Project management	151
	5.8.9 Post-project sustainability	151
5.9	Operationalizing policies	151
5.10	Conclusions	153
	References	154

6   Application of new science tools in integrated watershed management for enhancing impacts   **159**

6.1	Introduction	159
6.2	New science tools for watershed management	160
	6.2.1 Geographical information system (GIS)	160
	6.2.2 Remote sensing	161
6.3	Crop-growth simulation modeling	163
6.4	Field sensors and data communication devices	165
	6.4.1 Global Positioning System	166
	6.4.2 Automatic Weather Station	166
	6.4.3 Mobile devices	166
6.5	Data storage and dissemination	167
6.6	Spatial technologies in rainfed agriculture and watershed management	167
	6.6.1 Characterization of production systems in India	167
	6.6.2 Land use mapping for assessing fallows and cropping intensity	168
	6.6.3 Spatial distribution of rainy season fallows in Madhya Pradesh	172
	6.6.4 Spatial distribution and quantification of rice-fallows in South Asia: Potential for legumes	172
	6.6.5 GIS mapping of spatial variability of soil micronutrients at district level	175
	6.6.6 Assessment of seasonal rainfall forecasting and climate risk management options for peninsular India	175
	6.6.7 Baseline studies to delineate watershed	177
	6.6.8 Regional-scale water budgeting for SAT India	180
	6.6.9 Spatial water balance modeling of watersheds	180
6.7	Integrated watershed management for land and water conservation and sustainable agricultural production in Asia	183
	6.7.1 Assessment of agroclimatic potential	183
	6.7.2 Climatic water balance	183
	6.7.3 Climatic water balance of watersheds in China, Thailand, Vietnam, and India	183
	6.7.4 Rainfed length of growing period	184

6.7.5   Drought monitoring at watersheds   185
6.7.6   Weather forecasting for agriculture   185
6.7.7   Watershed monitoring   186
6.7.8   Satellite images for impact assessment   186
6.7.9   Monitoring and evaluation of NWDPRA
        watersheds using remote sensing   187
6.7.10 Monitoring and impact assessment of Adarsha watershed   188
6.8   Technology integration   195
6.8.1   Field data transmission   196
6.8.2   Sensor Web   197
6.8.3   Spatial simulation modeling   197
6.8.4   Use of ICT in watershed management   198
6.8.5   Intelligent watershed information system   198
6.9   Summary and conclusions   200
References   201

7 Soil and water conservation for optimizing productivity and
   improving livelihoods in rainfed areas   **205**

7.1   Introduction   205
7.2   Soil and water conservation practices   207
7.2.1   In-situ soil and water conservation   207
        7.2.1.1   Contour cultivation and conservation furrows   209
        7.2.1.2   Tied ridges   212
        7.2.1.3   Scoops (or pitting)   213
        7.2.1.4   Broad-bed and furrow and related systems   215
7.2.2   Bunding   221
        7.2.2.1   Contour bunding   221
        7.2.2.2   Modified contour bunds   222
        7.2.2.3   Graded bunding   223
        7.2.2.4   Field bunding   224
        7.2.2.5   Compartmental bunding   224
        7.2.2.6   Vegetative barriers   225
7.2.3   Tillage   227
        7.2.3.1   Zero tillage or minimum tillage or
                  conservation tillage   229
7.2.4   Ex-situ soil and water conservation
        (runoff harvesting and supplemental irrigation)   231
        7.2.4.1   Crop responses to supplemental irrigation   235
7.2.5   Indigenous soil and rainwater conservation practices   237
7.3   Enhancing the impacts of soil and water conservation and water
      harvesting interventions through integrated watershed approach 239
7.4   Strategies for improving adoption of soil and water
      conservation practices by farmers   240
7.5   Conclusions   242
References   243

8 **Rainwater harvesting improves returns on investment in smallholder agriculture in Sub-Saharan Africa** **249**

8.1 Introduction 249
8.2 The need to respond to the threat of climate change 251
8.3 Policies and institutional frameworks 251
8.4 Why focus on rainwater harvesting? 253
8.5 Options for rainwater harvesting 254
   8.5.1 RWH from surface runoff and storage in ponds, pans, and tanks (blue water) 255
   8.5.2 Rooftop rainwater harvesting 258
   8.5.3 Small earth dams and weirs 259
   8.5.4 Sand and subsurface dams 259
   8.5.5 Runoff harvesting and storage in soil profile 260
   8.5.6 In-situ water harvesting and conservation 263
   8.5.7 Spateflow diversion and utilization 265
   8.5.8 Conservation agriculture and RWH 266
   8.5.9 Soil fertility management in supporting RWH efforts 268
   8.5.10 Water for livestock 269
   8.5.11 Socioeconomic issues in RWH 270
8.6 Conclusions 272
8.7 Way forward 272
   8.7.1 Support rainwater harvesting 273
      8.7.1.1 Optimizing rainwater harvesting (Integrated watershed management) 273
      8.7.1.2 Runoff harvesting, diversions, and storage in soil profile 273
      8.7.1.3 Small individual water storages in ponds, pans, and tanks 274
      8.7.1.4 Medium-scale storage 274
      8.7.1.5 Rainwater harvesting for underground storages 274
   8.7.2 Provide secure rights to access land and water 274
   8.7.3 Adoption of innovative financing for smallholder farmers 274
   8.7.4 Interactive capacity strengthening for RWH 275
   8.7.5 Enhance policy support 275
   8.7.6 Recommendations 275
References 276

9 **Management of emerging multinutrient deficiencies: A prerequisite for sustainable enhancement of rainfed agricultural productivity** **281**

9.1 Introduction 281
9.2 Soil degradation – organic matter and nutrient status of SAT soils 283
9.3 Balanced nutrient management: Crop productivity and quality 292
9.4 Soil quality and water use efficiency 298

9.5 Strategy for scaling-up the soil test-based approach for
enhancing agricultural productivity 301
9.6 General discussion and conclusions 305
References 307

**10 Increasing crop productivity and water use efficiency in
rainfed agriculture 315**

10.1 Introduction 315
10.2 Water use efficiency: Concepts and definitions 316
10.3 Water balance of crops in different rainfed regions 317
10.4 Gaps in productivity and water use efficiency 318
10.5 Integrated approach to enhance productivity and
water use efficiency 319
10.6 Rainfall management to secure water availability 320
   10.6.1 In-situ soil and water conservation 321
      10.6.1.1 Land surface management 321
      10.6.1.2 Tillage 323
      10.6.1.3 Conservation agriculture 324
   10.6.2 Water harvesting and groundwater recharge 324
10.7 Increasing water use and water use efficiency 326
   10.7.1 Efficient supplemental irrigation 326
      10.7.1.1 Conveyance of water to the field 326
      10.7.1.2 Methods of application of supplemental
water on SAT Vertisols 327
      10.7.1.3 Efficient application of supplemental
water on SAT Alfisols 327
      10.7.1.4 Scheduling of irrigation and deficit irrigation 328
      10.7.1.5 Conjunctive use of rainfall and
limited irrigation water 330
      10.7.1.6 Supplemental irrigation and
crop intensification or diversification 330
   10.7.2 Increasing soil water uptake 331
      10.7.2.1 Improved crop agronomy 331
      10.7.2.2 Balanced plant nutrition 332
      10.7.2.3 Improved crop varieties and
nutrient management 334
      10.7.2.4 Water conservation practices
and nutrient management 334
      10.7.2.5 Crop protection 335
      10.2.7.6 Crop intensification (double cropping) 336
      10.7.2.7 Crop diversification with chickpea in
rice fallows 337
      10.7.2.8 Contingent and dynamic cropping 338
   10.7.3 Reducing soil evaporation 339
      10.7.3.1 Mulches 339
      10.7.3.2 Microclimate modifications 339

10.7.3.3 Land degradation, conservation agriculture, and
water use efficiency 340
10.7.4 Crop breeding for increased water productivity 340
10.8 Promoting adoption of technologies 341
10.8.1 Enabling policies 341
10.8.2 Building institutions 341
10.8.3 Raising awareness and capacity building 342
10.9 Summary and conclusions 342
References 343

11 **Impact of watershed projects in India: Application of various
approaches and methods** **349**

11.1 Introduction 349
11.1.1 An overview of watershed development programs in India 349
11.1.2 Synthesis of past experience of watershed development
in India 353
11.1.3 Need for economic impact assessment of watershed 354
11.1.4 Challenges in impact assessment of watershed
development 355
11.1.4.1 Methods of impact assessment 356
11.1.4.2 Approaches of impact assessment 356
11.1.4.3 Scale or time lags 357
11.1.4.4 Samples for the study 357
11.1.4.5 Selection of indicators 357
11.1.4.6 Choosing the discount rate 358
11.1.5 Indicators for evaluation of watershed development
projects 358
11.2 Approaches 359
11.2.1 Before and after 359
11.2.2 With and without 361
11.2.3 Combination of with and without using
double difference method 361
11.3 Methodologies: Application of watershed evaluation methods 361
11.3.1 Conventional benefit-cost analysis 361
11.3.2 Econometric methods (Economic surplus approach) 362
11.3.2.1 Application of economic surplus method to
watershed evaluation 364
11.3.2.2 Cost of project 365
11.3.2.3 Results of the economic surplus method 365
11.3.3 Bioeconomic modeling approach 370
11.3.3.1 Advantages of bioeconomic modeling in
impact assessment studies 371
11.3.3.2 Application of bioeconomic model for impact
evaluation of watershed development
program in semi-arid tropics of india 371
11.3.3.3 Biophysical and socioeconomic data 372

		11.3.3.4	Bioeconomic modeling	372
		11.3.3.5	Validation of the bioeconomic model	372
		11.3.3.6	Impact of change in yield of dryland crops	373
		11.3.3.7	Impact of change in irrigated area in the watershed	373
	11.3.4	Meta analysis		377
		11.3.4.1	Review of studies on meta analysis	377
		11.3.4.2	Biophysical impacts	378
		11.3.4.3	Socioeconomic impacts	380
		11.3.4.4	Environmental impacts	381
		11.3.4.5	Overall economic impacts	382
	11.3.5	Comparison of the methods		384
11.4	Conclusions and policy recommendations			385
References				386

12 Watershed management through a resilience lens **391**

12.1	Watershed management in smallholder rainfed agroecosystems		391
12.2	Embedding smallholder farming in landscape ecosystem services		393
	12.2.1	Introduction to management successes and failures	393
	12.2.2	Smallholder agroecosystems and ecosystem services	393
	12.2.3	Watershed (landscape) management "successes"	397
	12.2.4	Summary of selected cases	397
	12.2.5	Academic reviews of unmanaged case studies	400
	12.2.6	Additional case studies	401
	12.2.7	Landscape limits and trade-offs between ecosystem services	402
	12.2.8	Long-term sustainability and the hidden impacts of management successes and failures	403
12.3	Impacts on ecosystem services and the relationship to barriers to development of sustainable ecosystem services		405
	12.3.1	Impacts on system stability as obstacles in smallholder rainfed agroecosystems	405
	12.3.2	Understanding barriers as parts of ecosystem processes	405
	12.3.3	Climate change as an over-arching pressure	408
	12.3.4	Barriers for development reinterpreted as management opportunities	411
12.4	Understanding successes and long-term agroecosystem stability		411
	12.3.3	Introduction to agroecosystem stability	411
	12.4.2	Resilience defined	411
	12.4.3	Drivers of system stability	412
	12.4.4	Tipping points and regime shifts	413
	12.4.5	Defining key system components and processes	413
	12.4.6	Interpreting successes in terms of overall agroecosystem stability	414
12.5	Identifying management entry points using a resilience frame		415
	12.5.1	Management entry points – learning from case studies	415

		12.5.2	Recommendations for future research	417
		12.5.3	Resources to assist practitioners	418
	References			418

13 Impacts of climate change on rainfed agriculture and adaptation strategies to improve livelihoods   **421**

13.1	Introduction		421
13.2	Climate change impacts		421
	13.2.1	Crop and livestock production	424
	13.2.2	Water resources	425
13.3	Regional impacts		426
	13.3.1	Sub-Saharan Africa	426
	13.3.2	South Asia	427
13.4	Prices, poverty, and malnutrition		428
13.5	Adaptation		429
	13.5.1	Coping, adaptation, and resilience	430
	13.5.2	Adaptation strategies	434
13.6	Conclusions		436
References			437

Index   441
Colour plates   449

Preface

Feeding teeming global human population, especially in the developing world remains a huge challenge for agricultural scientists and farmers alike. In the developing world, agriculture and related activities still remain a major source of livelihood; and hence the means to alleviate poverty and assure sustenance and rural development. This is more so in sub-Saharan Africa and South Asia, where a large part of the population depends on agriculture to provide food, feed, and fiber, and as a source of livelihoods for the rural poor.

It must however, be emphasized that in the context of overall world food production, rainfed agriculture has played and would continue to play a crucial role in the global food production as a large part of agriculture is rainfed ($\sim$80%), and contributes over 50% to the global food supply. Moreover, rainfed areas are also where poorest of the poor live, and these areas remain hot-spots of poverty and under- and malnutrition. Even a modest increase in productivity of the vast rainfed areas would result not only in a large increase in global food volume, but would also stabilize these areas with enhanced food availability and rural development. Most certainly, agricultural development is at the heart of sustainable development of dryland areas. Increasing the productivity of rainfed areas especially in the semi-arid and arid regions of the world with impoverished soil resource base and water shortages is the greatest challenge of the 21st century.

Water shortage is recognized as the primary constraint in rainfed areas and remains a deterrent to the use of external inputs of plant nutrients even at a modest level, although the soil resource base in these areas are fragile and marginal compared to their irrigated counterpart. Recent research demonstrates that the soils in rainfed areas are not only thirsty (water shortage) but also hungry (nutrient deficiencies). Nevertheless, recent focus on rainfed agriculture along with concomitant enhanced investment in rainfed agriculture, a prerequisite to sustainably increase the productivity in rainfed areas, has helped to intensify research in these areas and develop technology packages that simultaneously address both water shortage and nutrient disorders that are holding back the potential of rainfed areas. These science-based interventions have provided highly encouraging results from extensive on-farm evaluation of the improved technology or through the upgradation of rainfed agriculture. Upgrading rainfed agriculture is the new mantra for an overall sustainable development of rainfed areas. This is based on the principle of addressing the constraints related to water shortage and soil infertility in a sustainable manner.

The book "Integrated Watershed Management in Rainfed Agriculture" is organized in 13 chapters, which address a range of issues involved in the upgradation of rainfed agriculture for improved livelihoods of rural community using integrated watershed management as the means to implement various technologies for improving productivity and livelihoods. Topics cover technical, social, policy, and developmental issues related to integrated watershed development and management. In addition to the issues related to soil and water conservation, rainwater harvesting and efficient use, diagnosis and management of plant nutrients, up-scaling of technology, impacts of climate change on rainfed agriculture and various methods and approaches used for assessing impact of watershed projects on productivity, income and livelihood in Asia and sub-Saharan Africa are also covered. We hope this book will be a useful resource for various stakeholders including students, researchers, development investors and agents, and policy makers.

Suhas P. Wani, Johan Rockström, and
K.L. Sahrawat

Foreword

Rainfed agriculture is important globally as it covers 80% (1.5 billion ha) of arable crop land. While the importance of rainfed agriculture varies regionally, it produces most of the food for poor communities in developing countries. In sub-Saharan Africa, more than 95% of crop land is rainfed, 90% in Latin America, 60% in South Asia, 65% in East Asia, and 75% in Near East and 75% in North Africa.

Since 1960, arable land has expanded in South America (83%), Africa (46%), and Asia (36%) whereas it has decreased in North America (4%) and Europe (25%). Meeting the challenge of improving the livelihood of 950 million poor and malnourished people in the world and achieving food security for the ever growing population with finite and degraded water and land resources is a daunting task. In addition, due to climate change and variability, rainfed production potential would be adversely affected. It is estimated that in developing countries loss of 10–20% of production area may affect 1.3 billion people by 2080.

During the 21st century and in the foreseeable future, agriculture will continue to play an important role in the economies of Africa and South Asia in spite of growing incomes, urbanization, globalization, and declining contribution of agricultural income to gross domestic product. While the quantity of available land and water has increased since 1950, available land and water per capita has declined significantly due to increase in human population. Continuous cultivation with intensified agricultural production has resulted to nutrient depletion and land degradation through erosion. In the future, crop land will continue to expand not only to supply the world with food but also to compensate for land degradation and unsustainable agricultural practices. Current productivity of rainfed agriculture in the tropical arid, semi-arid, and sub-humid regions is oscillating between 1 and 2 t ha^{-1} on farmers' fields, lower by two- to four-folds of achievable potential.

Limitless opportunities exist to unlock the vast potential of rainfed agriculture in achieving food security and reducing poverty in developing and emerging economies in Asia and Africa. However, the Millennium Ecosystem Assessment (MEA) and Comprehensive Assessment of Water for Food and Human Health, Assessment of Biofuels, and Inter-Governmental Panel on Climate Change (IPCC) have raised the key question of how much natural resources can be safely harnessed for human development without declining the sustainability. To explore the vast potential of rainfed agriculture, we need a new paradigm for sustainable intensification of rainfed agriculture through integrated management of water, land, and other resources at watershed scale.

Integrated watershed management is important in rainfed agriculture for reducing poverty through increased agricultural production by enhancing water and other resource-use efficiency. As it is becoming evident, the era of specialized component expertise in addressing the complex and interlinked issues of food security, poverty reduction, sustainable development, and protection of the environment is over. What we need now is a knowledge-intensive integrated approach that is inter-disciplinary and inter-institutional. The new paradigm should include scientific knowledge and recognizing the importance of research, development and extension continuum to achieve the desired impact.

This holistic approach of integrated watershed management can become a growth engine for sustainable intensification of rainfed areas through convergence, capacity building, and collective action with technical back stopping by the consortium of institutions. It is a win-win strategy for a good beginning in the development of rainfed areas.

I am confident that this book on "Integrated Watershed Management in Rainfed Agriculture" will facilitate scientific and policy measures to unlock the vast potentials of and facilitate increased investment needed to operationalize the new paradigm for rainfed agriculture, toward contributing to improved livelihoods and food security while protecting the environment through sustainable intensification measures. It is a very valuable resource material not only for researchers, policy makers, and development workers but also for students and the farmers alike.

William D. Dar
Director General
ICRISAT

List of contributors

Samuel Abraham, Society for Promoting Participative Eco-system Management (SOPPECOM), Pashan, Pune, Maharashtra, India

E.T. Adgo, Bahir Dar University, Bahir Dar, Ethiopia

T.F. Akalu, Amhara Agricultural Research Institute, Bahir Dar, Ethiopia

Amita Shah, Gujarat Institute of Development Research (GIDR), Ahmedabad, Gujarat, India, E-mail: amitagidr@gmail.com

K.H. Anantha, International Crops Research Institute for the Semi-Arid Tropics (ICRISAT), Patancheru, Andhra Pradesh, India

V. Balaji, International Crops Research Institute for the Semi-Arid Tropics (ICRISAT), Patancheru, Andhra Pradesh, India

Jennie Barron, Stockholm Environment Institute (SEI), University of York, York, UK, E-mail: jennie.barron@sei.se

Peter Q. Craufurd, Resilient Dryland Systems, International Crops Research Institute for the Semi-Arid Tropics (ICRISAT), Patancheru, Andhra Pradesh, India, E-mail: p.craufurd@cgiar.org

S.V.K. Jagadish, International Rice Research Institute (IRRI), Metro Manila, Philippines, E-mail: k.jagadish@cgiar.org

P.K. Joshi, International Food Policy Research Institute (IFPRI), MTI Division, CG Block, NASC Complex, New Delhi 110 012, India

K.J. Joy, Society for Promoting Participative Eco-system Management (SOPPECOM), Pashan, Pune, Maharashtra, India

G.J. Kajiru, Lake Zone Agricultural Research and Development Institute, Mwanza, Tanzania

A.V.R. Kesava Rao, International Crops Research Institute for the Semi-Arid Tropics (ICRISAT), Patancheru, Andhra Pradesh, India

B.M. Mati, Jomo Kenyatta University of Agriculture and Technology (JKUAT), Nairobi, Kenya, E-mail: bancym@gmail.com

P.K. Mishra, Central Research Institute for Dryland Agriculture (CRIDA), Hyderabad, Andhra Pradesh, India

W.M. Mulinge, Kenya Agricultural Research Institute (KARI), Nairobi, Kenya

S. Nedumaran, International Crops Research Institute for the Semi-Arid Tropics (ICRISAT), Patancheru, Andhra Pradesh, India

J.M. Nkuba, Maruku Agricultural Research and Development Institute, Bukoba, Tanzania

Jon Padgham, International START Secretariat, Washington DC, USA, E-mail: jpadgham@start.org

K. Palanisami, International Water Management Institute (IWMI), ICRISAT, Patancheru, Andhra Pradesh, India

G. Pardhasaradhi, International Crops Research Institute for the Semi-Arid Tropics (ICRISAT), Patancheru, Andhra Pradesh, India

Prabhakar Pathak, International Crops Research Institute for the Semi-Arid Tropics (ICRISAT), Patancheru, Andhra Pradesh, India, E-mail: p.pathak@cgiar.org

Patrick Keys, Stockholm Resilience Centre, Stockholm University, Stockholm, Sweden; University of Washington, Washington, USA

Piara Singh, International Crops Research Institute for the Semi-Arid Tropics (ICRISAT), Patancheru, Andhra Pradesh, India

K.V. Raju, Institute for Social and Economic Change, Dr. VKRV Rao Avenue, Nagarabhavi, Bangalore, India, (Currently with Govt. of Karnataka, Vidhan Soudha Bengaluru, India) E-mail: kvraju2008@gmail.com

Kaushalya Ramachandran, Central Research Institute for Dryland Agriculture (CRIDA), Santoshnagar, Hyderabad, Andhra Pradesh, India

Johan Rockström, Stockholm Environment Institute (SEI), Stockholm, Sweden

P.S. Roy, Indian Institute of Remote Sensing (IIRS), Dehradun, Uttarakhand, India

K.L. Sahrawat, International Crops Research Institute for the Semi-Arid Tropics (ICRISAT), Patancheru, Andhra Pradesh, India, E-mail: k.sahrawat@cgiar.org

V.N. Sharda, Central Soil Water Conservation Research and Training Institute (CSWCRTI), Dehradun, Uttarakhand, India

Bekele Shiferaw, International Maize and Wheat Improvement Center (CIMMYT), Nairobi, Kenya

A.K. Singh, Indian Council of Agricultural Research (ICAR), New Delhi, India

A. Subba Rao, Indian Institute of Soil Science (IISS), Bhopal, India

R. Sudi, International Crops Research Institute for the Semi-Arid Tropics (ICRISAT), Patancheru, Andhra Pradesh, India

B. Venkateswarlu, Central Research Institute for Dryland Agriculture (CRIDA), Santoshnagar, Hyderabad, Andhra Pradesh, India

Suhas P. Wani, International Crops Research Institute for the Semi-Arid Tropics (ICRISAT), Patancheru, Andhra Pradesh, India, E-mail: s.wani@cgiar.org

M.J. Wilson, Hunmanby, Filey, North Yorkshire, England, E-mail: michaeljwilson@fsmail.net, michaeljwilson42@gmail.com

Chapter 1

Improving livelihoods in rainfed areas through integrated watershed management: A development perspective

M.J. Wilson

1.1 INTRODUCTION AND OVERVIEW

This book "Integrated Watershed Management in Rainfed Agriculture" is about poverty, the environment, agriculture, and rural business in the semi-arid tropics (SAT) of developing countries. It tells about an approach to rural development, the watershed approach, which has proven particularly successful and which deserves wide international application in pursuit of the various Millenium Goals. The approach deploys our current understanding of the social and technical features of the rural SAT and seeks a true partnership with villagers, drawing together ideas and knowledge from the outside to complement their skills and knowledge. The approach gradually builds up awareness and confidence and creates local institutions so as to enable the villagers to change their lives markedly for the better. It recognizes that success requires the poor and disadvantaged to be intimately involved in the development effort and seeks to evolve a diverse economy in the village, albeit with a largely agricultural starting point. It uses the physical boundary of a watershed because this permits the rational harnessing, storage, and use of water, without which there can be but little progress in these predominantly rainfed environments.

From a government and donor perspective, watershed work clearly addresses important objectives related to poverty, productive capacity, environment, and output, and others relating to good governance, women in development, water productivity, and carbon sequestration. Watershed interventions "tick a lot of the right boxes" because they attend to pragmatism rather than rhetoric. Interestingly, whereas previously the soil restoration, land and water management, and drought proofing objectives of watershed work were not necessarily shared by the people being assisted, within the context of the new poor-people-focused type of watershed, they have become so.

This introductory chapter, which is based on experience in India, provides an overview of watershed approaches in the country: the principles, the social, institutional, and management implications, and the technical possibilities. All these aspects are dealt with in depth in the subsequent specialist chapters in which more recent experience in other countries of Asia and Africa is discussed.

1.2   THE SEMI-ARID TROPICS AND THE IMPORTANCE OF RAINFED FARMING

Global success in improving food security and reducing poverty cannot be achieved without success in the SAT, which are home to 38% of the world's poor, 45% of the world's hungry, and 70% of the world's malnourished. Rainfed agriculture in developing countries (much of it in the SAT) is expected to account for 40% of the growth in world cereal production through 2025. Some studies aver that effective investment in agricultural research-and-development in SAT areas may give higher marginal returns, reduce poverty more, and conserve the environment better, than further investments in favored areas (Rosegrant *et al.*, 2002).

Crop yield potential in the SAT is high, due to abundant solar radiation and crop output can be markedly increased with water harvesting, soil and water conservation, soil fertility management, and germplasm selected for the needs of the particular area. The livestock, fish, and tree components of the farming system also have great scope to be improved with better technology. All in all, subsistence can give way to a diverse rural economy based on seizing market opportunities to generate income and commerce. However, developmental success in rural SAT areas has to contend with the following issues:

- **Socioeconomic:** Widespread acute poverty, small-sized holdings, few opportunities for employment and seasonal migration in search of it, poor health and education, lack of capital and problems in accessing information, inputs, credit, and markets, unawareness of entitlements and no empowerment, insecure land tenure, and ineffective collective management of common property.
- **Environmental:** Extreme rainfall variability, soil erosion, forest and biodiversity loss, groundwater depletion, and water scarcity.
- **Political and institutional:** No political standing, separation of research from development, inability to undertake multidisciplinary support or systems-research or to deliver location specific technology.

All these issues together limit agroecosystem productivity. Research at the International Crops Research Institute for the Semi-Arid Tropics (ICRISAT), for example, shows how smallholders only harvest the equivalent of $1.0\,t\,ha^{-1}$ of grain compared to $5-7\,t\,ha^{-1}$ harvested at its research station. But of course, research and technology is only one dimension of broad-based development, the achievement of which may also require the following measures (Wilson *et al.*, 2007):

- Investment in rural infrastructure.
- Stronger public institutions working more closely one with another.
- Improved access to natural resources by the poor.
- More effective and equitable public policies.
- Local adaptation of innovations and closer research–development links.
- More effective risk management strategies.
- Investment in water capture and management for supplementary irrigation.
- Integrated water resource planning at a scale smaller than the river basin.
- More effective collective action by communities.
- Capacity building.

1.3 THE OBJECTIVES OF WATERSHED MANAGEMENT

Well they say it is better to travel in hope than to arrive but hope is not good enough for success in a watershed initiative. Right at the outset of a new venture it must be worth looking at what might realistically be achieved:

* Soil and water conserved, and the environment put back together again in terms of water table height, soil erosion, and forest extent and quality.
* Poverty reduced, its worst aspects removed, and more equity in the community.
* Increased agricultural productivity and more food self-sufficiency.
* A diversified economy with a wider range of bigger incomes as well as with support from, and connections to the broader economy and to markets, and with improved access to and responsiveness of Government schemes.
* A more aware, more informed, and better educated community with more social cohesion and collaboration.
* More community self-reliance, articulation, and assertiveness, and better community leaders.
* A natural resource base and a community more resilient to drought and climate change.

All these would be a good start. But will they be sustainable given inexorable population growth which will reduce holding-size, and increase pressure on the natural and communal resources? How much additional output might be expected? Will some people still be poor and why? Should one aim from the start to get them out of farming and into other industries?

1.4 THE MAIN EXPERIENCE

Much of the SAT lies in India and the whole of SAT India is divided into watersheds for development purposes. What follows is based primarily on experience in India with the Western and Eastern India Rainfed Farming Projects (WIRFP and EIRFP), which were supported by the United Kingdom Department for International Development (DFID). These projects were located in various districts where Gujarat, Rajasthan, and Madhya Pradesh meet and, in the East, where Jharkhand and Orissa meet.

The projects benefited from newly emerging concepts of participation propounded by Chambers *et al.* (1989) from participatory technology work pioneered by Witcombe and Joshi (1995), and from the work of a medley of institutions and non-government organizations (NGOs), including: Myrada, Seva Mandir, Agha Khan Rural Support Programme, and Sadguru Foundation. The work was by no means all new, but the projects drew together trains of thought and modes of practice, and by doing so they became influential. They informed other DFID rural activities worth some £200 million, including: Karnataka Watersheds which in turn was influential; the Rural Livelihoods Projects in Andhra Pradesh, Orissa, and Madhya Pradesh; and the Himachal Pradesh and Karnataka Forestry Projects. These projects in turn influenced State Governments and the World Bank's watershed program. They also influenced ICRISAT's successful watershed work and through ICRISAT the National Watershed

Guidelines and the introduction of watershed approaches in other Asian countries. Various institutions and NGOs currently use the approaches devised and many Government officials, now working throughout India, were associated with the two projects.

WIRFP and EIRFP ran for sixteen years to 2006 and were engaged with some 2,000 villages and two million people, mostly tribal and scheduled caste people living in the most poverty-stricken circumstances. Together with the subsequent projects they helped to move watershed work from a purely land-focused activity to one in which benefits flow to all people living within a watershed, whether landowners or landless. They were important to the evolution of participatory crop selection and breeding approaches, gave rise to a realization of the importance of migration, and found ways to involve women and the poor.

As some of the longest running rural development programs of their type, WIRFP and EIRFP offer an insight into the diversification of a village economy and links to markets. As an example of the best "end result", I refer to Kumpura cluster in southern Rajasthan. From a beginning like that described in the section on poverty below, an initial three villages has expanded to 22 in eight Panchayats. There are now 550 groups and a federation, with a committee of 25 members, which has formed a mini-bank, runs its own tractor and car, and distributes fertilizer in the villages by hired lorry. There are wells, check-dams, a *pucka* road, village electrification, a range of improved crops, milk cooperatives, and a successful Joint Forest Management Scheme. Someone from half of the families still migrates for work, but does so now by choice, to earn good money as a trained mason, carpenter, electrician, or blacksmith. These projects, and others like them, have clearly shown that community-led rural development works. They should now be assessed for sustainability, why some villages do much better than others and, despite best efforts, some families remain poor.

What now follows in this chapter takes us from an examination of poverty through the different stages and aspects of watershed management. The second half of the chapter provides a simple-to-understand overview of the many technologies which can be deployed, and concludes with a look at research needs.

1.5   UNDERSTANDING THE NATURE OF RURAL POVERTY

If rural development is to pay particular attention to the poor, we need to understand the nature of poverty and consider its practical implications. What is the reality for the individuals that make up the large numbers of the poor? Clearly there will be a spectrum of poverty in any rural society varying from the abject through to people who are relatively well-off. However, the "Poverty Line" is not fixed: a drought or family crisis may precipitate people below it. In the Andhra Pradesh scheme, 45% were "Below Poverty Line" and 55–75% were small and marginal farmers on only 20% of the land (Walker 1999).

1.5.1   Landlessness

Substantial numbers of the rural poor are landless. In Orissa, 60–70% of project villagers were marginal farmers or totally landless, all below the poverty line. Some

25% of them were living on less than 40% of poverty line income. WIRFP figures were slightly better with 10–35% landlessness. Most of the landless will have a homestead and, in theory, rights to common land and forest, but encroachment is leading to dwindling common resources in many areas. As families grow, holdings are split leading to more landlessness and land alienation. A study of land tenure in 1988 in Orissa (Taru 1998) found up to 80% of land in some villages in *de facto* possession of other parties, usually moneylenders and large landlords. A small landowner may take a loan and enter a share cropping arrangement until the loan is repaid, so starting on a downward spiral which usually ends in loss of the land.

1.5.2 Women

Many of the rural poor are women, often heading households. Women account for two thirds of the people living in abject poverty. Truly, a poor woman's work is never done and much of it is drudgery. Into bed late yet up betimes at 4 am, light the fire, fetch water, feed the livestock, cook the meals, see to the children, cut and fetch fodder, fuelwood, and water often from some hours' distance away, grind the corn, labor on the farm, victims of domestic violence, lack of credit, ignorance of their reproductive rights and of the laws relating to divorce, property, and violence are important gender issues (Prameela and Bijor 2002).

1.5.3 Subsistence

Those with land are usually subsistence farmers with small-sized holdings, perhaps with insecure land tenure, possibly share croppers. Subsistence can vary from food self-sufficiency throughout the year to food production enough for only a few months. Part time farming is thus normal since those unable to produce sufficient to feed themselves year round, perforce need to gain other income. Many achieve this by seasonal migration to where work is to be found: perhaps cotton picking or sugarcane cutting or building construction.

1.5.4 Debt and awareness

The poor are often persistently indebted and unable to service usurious interest rates on loans taken for food or medical emergencies. They are often without capital, without assets, have difficulty in obtaining credit, whilst being physically isolated and without buying power means commerce is disinterested in supplying services and farming inputs. Lacking confidence and the wherewithal to go out and discover technical information and knowledge of their entitlements, they remain unaware and do without. With little effective cooperation amongst themselves, they have ineffective forms of collective management of common property and no collective bargaining power.

1.5.5 Health

Under- or mal-nourishment is common. If this occurs frequently and severely in the pregnant and the young, then cognitive abilities may be affected and immune systems lowered so that the poor are prone to diseases. Seva Mandir (1999–2003) highlights the following issues: "Poor basic and personal hygiene and eye care, problems caused by tobacco and alcohol, and domestic violence. They list reproductive health needs and concerns as: family planning; safe, aseptic deliveries; child immunization; and supplementary feed for mother and child. Infertility and uterine problems are major issues in some areas, as are sexually transmitted diseases brought back by migrant workers, including HIV. Lack of safe water is commonplace, more so in a drought. Particular water quality concerns lie with ubiquitous industrial pollution, with fluoride and arsenic contamination of groundwater, frequently encountered in eastern India, and with iodine deficiency in the Himalayas.

1.5.6   Human rights

The poor have few basic human rights. With no political standing or local empowerment, they have no voice in decisions that might help matters and are often ignored by Government. With few opportunities for education, especially for women, the poor are often illiterate and innumerate and so less able to fend for themselves outside the village and they fall easy prey to deception and monetary cheating. They are unaware of their legal rights. Without a social support network, sickness and disablement lead to a crisis sales of assets and become at best life-changing and at worst life-threatening.

1.5.7   Environment

The SAT is in any case a risky environment in which to farm. Highly variable rainfall, within and between seasons, and infertile and erosive soils make for secure harvests only five years in ten, with partial failure two years in ten, and complete failure three years in ten. In these circumstances the impact of the poor on their environment is self-destructive. The forest is often gone or degraded, whereas previously they derived 40% of their livelihood from it in the form of timber, fuel, grazing, and medicinal plants; a range of food like frogs, caterpillars, and fungi; and an important transfer of nutrients to cropland (Wilson *et al.*, 1998).

1.5.8   Trends

The water table is commonly dropping and leading to water scarcity. I have seen a river rendered locally useless by industrial pollution. The cropland is often badly eroded. And, as if all this were not enough, there are some significant unfavorable trends: rapid population growth causing landlessness due to holding subdivision, adding yet more pressure on grazing and the forest, and causing encroachment onto marginal land. There is competition for water between agriculture and alternative uses, and there may be the prospect of more frequent extreme weather and greater aridity resulting from climate change.

1.5.9   Assistance

Even when willing to help, government agencies, separated by discipline and function and short of travel costs, have been unable to do the multi-disciplinary work entailed in effective assistance. Over the years, there has been but little systems research, neither consultation nor participatory working, and blanket recommendations have been issued instead of messages adapted to local circumstances. There has been a failure to scale-out successful innovations. Whereas NGOs can commonly help with issues of welfare, social organization, and equity, they are mostly unable to comprehensively address the technical challenges needed for the poor community to break out of its cycle, or even spiral, of poverty (Wilson *et al.*, 1998).

1.5.10   Equity and inclusion

The better-off in the community usually control the best land, and the common assets. They resist attempts by the poor to assert their rights to a share and usually seek to capture new inputs and offers of assistance. Wealthy landowners with private tube wells may draw down water tables below the level of the open wells used by the poor for small-scale irrigation and drinking water. Land-based inputs and subsidies do not of course benefit the landless. Once poor people become the focus for development benefits, then they encounter problems of equity and distribution. Every stage of the development cycle (design, implementation, and maintenance) has its equity and conflict of interest problems. There are difficulties in getting poor women involved early on in the process and the danger of marginal groups being left behind. There is a continual struggle to "operationalize" equity and even "participation" does not necessarily result in equity. Generally the dominant hold sway and the poor fall into line. How can the benefits of improving communal land be distributed equitably when one man has eight cows and another only one? When the best water structure sites are next to the best land, how can the costs and benefits be equitably shared? (K.B. Saxena, ICRISAT, personal communication, 1991)

1.5.11   Migration

It took a while for the Rainfed Farming Projects to realize just how important migration is in terms of: (a) the numbers migrated (before intervention 80% migration was common and whole families migrated); (b) the distinction between migration-by-choice and enforced migration; and (c) the effects of migration on livelihoods, health, and education. When migration is for a long period, it becomes difficult to make and maintain the contact needed to build up a fruitful development relationship, and for migrants to obtain government services.

1.5.12   Case study

Life then is a constant battle against dwindling landholdings, declining soil fertility, indebtedness, recurrent sickness, and disinterestedness. No wonder that reduced drunkenness and wife beating figure largely in women's vision for the future. To complete this overview of the reality of poverty and add even more starkness to it, here are my notes from a recent visit to 150 families of an ultra poor tribal community. The incidents are authentic and formed the basis of interventions by an NGO seeking redress. They record appalling oppression and corruption, cheating, and sexual exploitation.

Only 10% of this community had land title; 90% were seeking title as encroachers; and 70% had borrowed from moneylenders for food and health. When no work was available to enable paying off their debt, some debtors had been bonded for 2–3 years and had their bullocks taken. Illegal cutters of trees in the forest had made false accusations about the community leading to oppressive penalties. Fictitious loans had been alleged and the community had to repay. Where the community was entitled money, an intermediary was appointed and took a hefty cut. Merchants, buying leaves to be sold as plates, bundled up 125 leaves but paid for 100. There had been sexual exploitation by moneylenders when the community was looking for work and rape in the forest when girls were out collecting firewood.

1.5.13   Indicators of progress

The issues catalogued above provide a ready set of indicators for the progress of a developmental intervention. Other indicators can measure the evolving maturity and sustainability of progress and of community institutions that emerge (Mosse 1995).

1.6   FOUR STAGES TO PULL A RURAL COMMUNITY OUT OF POVERTY

The watershed model has evolved from the efforts of many donors, NGOs, and government agencies. Whilst all these efforts have gradually learnt one from another, some practitioners are ahead of the field and others deficient in some aspect or other of understanding or practice, principally in approaches to people participation or in the application of science and technology. But generally nowadays the better models have some or all of the following features in common:

- Participation of villagers as individuals, as groups, or as a whole, increasing their confidence, enabling their empowerment and their ability for self-determination and to plan for the future.
- Capturing the power of group action in the village, between villages, and in even broader collaborations.
- Basic infrastructure construction with contributions in cash or labor from the community.
- Better farming techniques, notably the improved management of soil and water, diversifying the farming system, and integrating the joint management of communal areas and forest.
- The involvement of the landless, often in providing services.
- Arrangements for the provision of basic services and infrastructure.
- The establishment of village institutions and links with the outside world.
- Improved relationships between men and women.
- Employment and income generation by enterprise generation in predominantly but not exclusively agricultural-related activities.
- The fusion of research and development.
- Capturing the extraordinary power of participatory technology development.
- An involvement with enforced migration.

1.6.1 The process

The schema below gives an overview of the process of poverty-focused rural livelihoods development as it now stands. Sodhi *et al.* (2002) give a more detailed version.

1.6.1.1 Stage 1

- Relate the contents of this schema to the village community.
- The worst iniquities and injustices resolved: more income is one immediate consequence.
- People made aware of their rights and entitlements. A determination engendered to do something about the community's plight, resulting in the first plans.
- Groups formed and savings started with self-help groups (SHGs) lending money, mainly for consumption and medical needs.
- Village Specialists and Group Leaders (*Jankars*) identified and trained.
- Problems of migration eased and some reduction in duress migration.
- Some reduction in women's drudgery and a start made with simple income-generating activities.
- Start of the stabilization of soil and some water resource development.
- Start of participatory crop variety selection and introduction of simple technological innovations like vermicompost, and better tools and equipment across the farming system.
- Community more articulate about common concerns and starts to address these.
- "Opportunities and escapes" analysis and planning.
- Start of forest protection.
- A Community Resource Center formed.
- Role of women in decision-making increased.
- Information about government programs and activities in the district obtained.
- Constructive contacts made with government agencies and NGOs.
- Pump-priming activities undertaken in adjacent, then in distant villages.

1.6.1.2 Stage 2

- Agricultural development leading to food self-sufficiency for most of the community.
- Skills and information being acquired and the community exposed to a range of ideas about commercial opportunities, the scope to leave farming, and of the need to stop the division of holdings if a slide back into poverty is to be averted.
- Big reduction in numbers migrating and improvement in the quality of migration.
- Those who are likely to remain below the poverty line identified.
- More agricultural technologies being used and the start of market-oriented agriculture.
- More income-generating activities in agriculture and allied sectors and some service industries set up.
- SHGs lending money for investment.
- Forest protection results in better availability of livestock fodder and other non-timber forest products (NTFPs) and may be extended to joint operation with adjacent villages.

1.6.1.3   Stage 3

- Nutritional security achieved by most of the community.
- More commercial agriculture with higher value items and some value being added.
- More infrastructure development with the completion of soil conservation structures.
- Strong groups branching out into a range of activities.
- Migration is primarily voluntary for better livelihood opportunities.
- Individuals and groups providing farming services like joint marketing of produce, production of natural pesticides, water pump maintenance, etc.
- Financial assets now substantial and financial acumen being sharpened.

1.6.1.4   Stage 4

- The watershed's water resources are now fully harnessed and equitably managed by the community.
- The local generation of renewable energy.
- Assistance with high-value production and marketing. This would embrace seeds and selected vegetables and fruits and perhaps forest products. It would include adding value by partial or full processing and packing.
- Acquisition of the skills and business knowledge sufficient to allow some people to provide services to farming or to move out of farming altogether.
- Special targeting of the remaining poor.
- More information and discussion about climate change and adaptation strategies.
- Working through with the community the implications of the withdrawal of support.

1.6.2   Recent improvements

The schema is only indicative; there will be overlaps and different speeds of progress in different communities. It was welcomed by the NGOs, villagers and professionals to whom I exposed it and seemed to give all a helpful vision and framework of what lies ahead. It reflects a recent realization of the need to:

- Achieve sustained, diverse, and independent economic growth in the village as the final outcome.
- Raise more awareness in the village early on about business opportunities outside of agriculture, commercial farming, family planning, implications of the withdrawal of program support, and the risk of reversion to the poverty-stricken circumstances.
- Establish a Community Resource Center at an early stage; initially to support migration, and also to provide services which reflect the community's present needs.
- Attend to what I have termed a second generation of interventions which more closely links production with the market, fully exploits the natural resources of the village, and deals with remaining poverty (Stage 4 of the schema).

1.6.3 Community resource centers

These centers are now providing migrants with: identification documents; help with the telephone; information about legal rights (helpful in confrontations with police); information about the range of government schemes and benefits and minimum wage rates; registration for National Rural Employment Guarantee Scheme; details of good employers; help with negotiations with difficult employers; and help with messages to and fro migrants and the village. The centers' work has now expanded to the hire of machinery (pumps, winnowers, and implements); sale of seeds and pesticides; and skill training and technical support.

1.6.4 Income-generating activities

In order to become fully food self-sufficient, the landless and those with smaller holdings need to embark on income-generating activities (some of these are farming system based while others are not) chosen according to skills, education, and desire. Generally, such activities will only be attractive if they can yield substantially higher returns than paid labor as they involve risk, but there are plenty of well tried and tested opportunities. A short list of enterprises in which people have already become employed is given:

- Bamboo collection.
- Making leaf plates, broomsticks, bricks, candles, and puffed rice.
- Establishing tree nurseries, small livestock enterprises, and machinery hire and repair businesses.
- Tailoring and weaving.
- Paddy threshing, de-husking, and milling.
- Pump management, operation, and maintenance for irrigation schemes.
- Managing communal grain and seed banks.
- Seed processing.
- Manufacture and sale of the neem oil-based insecticide, HNPV (*Helicoverpa* nuclear polyhedrosis virus), vermicompost, and ayurvedic medicines.
- Expelling niger oil.
- Maize starch processing.
- Manufacturing a nutritious food for the Women and Child Support Program.
- Primary education for children.
- Producing mushrooms, flowers, fruits, and rabbits.

Common property land resources can effectively be regenerated as pasture and biofuel and energy plantations and used to generate income when managed by vulnerable groups. The extensive wastelands are currently being used for oilseed production (Wani *et al.*, 2009) and this offers the poor the opportunity of a direct link to the market. In future work, more efforts are needed to promote fishing rights for the poor in impoundments and to promote the production of fish fingerlings, fodder, and different aspects of renewable energy.

1.6.5 Second generation interventions: market links

With soil fertility improved and some irrigated production enabled, the output of food producing enterprises increases such that some farm families can achieve year-round self-sufficiency and others add two to six months of family food supply to the starting position. Once food self-reliance has been attained, a portion of the rural population will move away from subsistence activity and cereals and towards intensifying and diversifying their farming and livelihood systems to produce fruits, vegetables, legumes, oilseeds, seeds of improved varieties, and high-value fodder, and to engage in commercial livestock production. This is the case especially where water is available and access to urban markets is good. Some WIRFP villages are now specialized in seed production.

Although some progress has been achieved, yet stages 1, 2, and 3 have been insufficiently long to permit a firm move into commercial farming and for the village economy to diversify. So now a "second generation" of interventions is needed. These might comprise both the technology and assistance to make a market link or help to add value by processing, packaging, improving quality, or changing the time of sale with storage. The market might be local, national, or export. It may be direct to an industrial processor. It needs to overcome the problems of small-scale production, isolation, and lack of information. It might look at branding with a local or district name or seek "Organic", "Fair Trade", or a purity guarantee like "Aflatoxin-free".

ICRISAT is ahead of the field in such market-led intensification and has pioneered: soil testing and provision of micronutrients; premium payments made for aflatoxin-free maize and groundnut; bio-diesel production; high-sugar sorghum for ethanol production; and soybean sold for oil production. Other possibilities would include: vegetable and oilseed processing; marketing medicinal and aromatic plants; fish farming; and quality fodder in bulk. In the drier areas, a focus might be expected on: livestock, better bred and managed, and fed on cultivated fodder and better managed rangeland; agroforestry; dryland horticulture and horticulture irrigated from local water resources like collector wells; the strategic irrigation of staple crops; the use of short-duration varieties; and non-agricultural sources of income (Sreedevi and Wani 2009).

1.6.6   Duration of support and coordination

There is now a faster start to the process than was the case hitherto as rural awareness has risen greatly with improved communications and with so much government money being directed towards agriculture and rural poverty. Four to five years period seems not long enough to extricate people from poverty. But if other government resources were coordinated with watersheds, then after eight years of support things might be nearing completion. In India these would include the National Rural Employment Guarantee Scheme and the BAIF/NABARD WADI program. Coordinating government schemes does now seem to be key and getting the Panchayat planning system to function properly and do this is salient. Successful experience of working effectively with the Panchayat needs exposing. Where has the Panchayat Raj Institutions (PRI) system worked and where has it not and why? How have connections with the new rural information hubs helped? A compilation of all government schemes and of rural people's entitlements would also be valuable.

1.6.7   Better use of the NGOs as collaborators or implementers

With some important exceptions, NGOs are generally not fully aware of, nor equipped for, a livelihoods program which uses best practice, rarely getting beyond Stage 1 of the four-stage schema given earlier. Yet their awareness building and social organizational skills are a great asset. With more support, with awareness of the technology outlined in the section below, with a greater sharing amongst themselves of their own collective knowledge, and with the acquisition of additional skills, they could be more powerful agents in the watershed effort.

1.7   ASPECTS OF THE PROCESS OF WATERSHED IMPLEMENTATION

1.7.1   Management of watershed

The management of a watershed works at the nexus between numerous agencies and the primary beneficiaries represented by grassroots institutions like SHGs and Micro-watershed Development Committees. Experience has shown how NGOs with social organization skills, government departments and agencies with technical expertise, elected bodies with statutory powers, donors, and increasingly the private sector can each learn from the others and so modify their approaches and bring new resources to bear (Wani *et al.*, 2002, 2008a, 2008b).

Various forms of project management agency have been tried: a Government Agency with Project Implementation and Support Units; Fertilizer Cooperatives; NGOs; even a Registered Society. The Karnataka project compared different regimes. Management has to oversee participatory village planning and participatory technology development; build local institutions and teach skills which provide better employment opportunities for migrants; resolve conflicts between villages; and promote federations and links with business and industry. The task of watershed work is so great that all the disciplines and agents which can contribute must be deployed and converged to implement the activities and achieve the overall objectives.

Management practice must maintain transparency, be free from corruption and be perceived accountable by the people. This has been successfully achieved, for example, with formal audits and public display boards listing the details of work done, the beneficiaries and their contributions (Mukherjee 2002).

Management needs to look after its own staff too, with a clear management structure and a human resource strategy which addresses future men and women staffing needs, and their career development and skill training. Recruiting women staff needs affirmative action in the recruitment procedures and a project culture in which women can work safely and effectively.

The finance deployed for watershed work must be sufficient to give operational flexibility. This would include: salaries to implementing agencies and NGOs at the sub-watershed level, entry point activities, capacity building, and direct funding of community groups, perhaps with matching grants (Mukherjee 2002).

Watershed implementation is essentially seeking to change the way of life of communities along the lines they desire. Rural communities are suspicious and frightened of any approach that is seeking to change their status quo, no matter how appealing the prospect presented to them. They are fearful of change and of being cheated, or of there being religious objectives behind the approach. The speed of start-up and mobilization is helped by being able to see existing success in another village; by speed in the delivery of something asked for, like a hand pump; by small entry point activities that start to build a relationship; and by income generation or some other clear benefit as from a simple change to existing practice or a simple new initiative.

Community organizers follow a sequence of: selecting villages and village clusters against chosen criteria; rapport building; participatory appraisal to elicit the physical, financial, social, and natural capital of the village; identifying and analyzing community problems; and putting development options in order of priority. This generates a work plan for an initial range of small-scale activities and a broader set of shared development objectives and visions for the community as a whole. These shared visions usually include such things as education for all children; fulfilling basic needs; freedom from debt; migration stopped; improved farming practices; and reversing damage to the local environment (Shahi *et al.*, 2001a).

The first impacts are easy to achieve. There is rapid uptake of new varieties, an enthusiastic application of newly won skills, and immense interest in new equipment, as with a simple bearing that reduces the labor of grinding grain. Interventions that do more than this are longer term and more challenging, but soon, more than just a menu of activities has been produced. Villagers acquire confidence and the ability to interact one with another and to express views freely. They realize that they do have options and choices and can take control of their own destinies. They acquire problem solving and technical skills.

Different methods are used in participatory rural appraisal (PRA): mapping of the village's natural resources and farming system; a time line of historical events; a calendar of seasonal events to reveal livelihood options and priorities; and a social map and wealth ranking. Wealth ranking reveals the extent of social and economic differentiation in villages. It becomes an important tool for monitoring if activities are reaching the whole range of households. It helps to identify which households are particularly vulnerable and informs the choice of specific interventions to assist each group. Villagers can easily undertake data collection. Wealth ranking, work planning, and participatory monitoring and evaluation generate more skills and a virtuous cycle of improved planning and implementation based on regular assessments of the impact of interventions as seen through the community's eyes (Sodhi *et al.*, 2001).

1.7.2 Groups and *Jankars*

The landless and the small landholders form the building blocks of the SHGs with each group based on kinship, common need or interest, caste or creed, and each starting its own savings and credit scheme. Each village and each group selects men and women to be trained to provide a wide range of services. These specialists are motivators and innovators who use their innate leadership skills and build on local knowledge. Called *Jankars*, they provide services, disseminate ideas, and become a powerful force to help expand the project and extend its influence. They undertake: PRA, group formation, rapport building, training, budding and grafting, soil conservation layout, recording of meetings, dealing with banks, village entry, program initiation in new villages and areas, and check-dam construction. Many of these "*Jankars*" subsequently took on important wider roles in the Panchayat and beyond (Shahi *et al.*, 2001b). In health, much has been achieved with village and community health workers as well as home-remedy workers and with the training of traditional childbirth attendants (Seva Mandir 1999–2003).

1.7.3 Savings, loans, and credit

Freeing themselves from the moneylenders is one of the first goals of savings groups, which set their own savings targets and rules of lending. Loans are made for consumption (e.g., medical expenses) and for production (e.g., crop inputs and the purchase of assets). Repayment defaults are usually very low because of peer pressure. Financial skills and acumen are acquired. Groups tend to keep accurate records with their own accountant and have the accounts audited. The interest rates that villagers charge themselves are high (25%) but only half what local moneylenders charge. The Village Development Fund acts as a platform for the community to gather together and discuss things. Some group members purchase equipment which is then lent to other members.

Once sufficient experience and a sufficient sum have been achieved, the group becomes eligible for a bank loan. At later stages, different village groups may federate, create a forum for community issues other than finance, and arrange services like the purchase of transport, collective sales and purchases, maintenance and repair, and specialized production.

1.7.4 Literacy and numeracy

Separate PRAs, and focus-group discussions with women, marginal groups, and those wanting non-formal education identify their literacy and numeracy priorities. Literacy groups are often formed early on. Functional literacy enhances status in the community, and of course confers an ability to read fertilizer packets, bank books, and road signs, and avoid being cheated by moneylenders and merchants.

1.7.5 Expenditure

The path of project expenditure is slow at the start due to the need to achieve satisfactory community development before large expenditure is incurred. But as the process proceeds, communities become more capable and the type, pace, and complexity of the activity expand. This, and the rise in the number of clusters and of villages per cluster create an exponential rise in expenditure. The community may also be successful in attracting additional funding from NGOs and the government, and in building up its own savings.

1.7.6 Land ownership

Watersheds are likely to have Forest Department, Village Revenue, and private lands. From the initial stages of work, management might usefully note the different regimes controlling land use, the ownership and rights to usufruct and benefits, and plan activities so that the fruits of watershed development do not later pass out of control of the communities. Where common property is in short supply, a memorandum of agreement might be entered into between the community and the absentee landlord, and this would include government departments. Groundwater should be regarded as a common property resource (Taru 1998).

1.7.7 Spread and dissemination

In WIRFP, there was an early spread of impact from each initial cluster of two or three villages to another two to six surrounding villages, with ultimately 10 to 14 villages influenced by the two community organizers working in each cluster. Spread was helped by "pump priming" with easily transferable messages and inputs and by field visits. The indirect spread of germplasm and ideas was much wider. Information quickly flows to villages where relatives reside.

1.7.8 Lack of spread of locally successful watershed interventions in India

Donor, NGO, or District Government watershed initiatives are often successful, some eminently so. They commonly influence about twice the number of villages regarded as nucleus by the project. But rarely is there a major impact beyond the project area, except perhaps for the spread of introduced improved germplasm. Some of the reasons for this are:

- Institutional politics and institutional walls.
- Protection of a "patent" approach with undue attention paid to its nuances.
- Lack of understanding of the concept.
- Unwillingness and/or inability to work in participatory, interdisciplinary, and inter-institutional mode with a lack of appropriate skills and incentives.
- Arrogance of officials and unwillingness to treat villagers as equal partners and perhaps fear of the unknown.
- Frustration by government regulations or lack of them.
- Slow or inappropriately timed release of funds.
- No recurrent budget for travel.

1.7.9 Sustainability

The type of rural development discussed here has not been going on for long enough to be able to say with confidence "this or that is sustainable" and the community will not revert to poverty. There is always the fear of division of holdings and consequent increase in the numbers of the poor; also the fear that greed, poverty (and current legislation) will lead to common property being encroached and that a lack of market links and business skills will frustrate progress, or that agencies, which should provide support and services, will be found wanting.

Among the indicators of likely sustainability, one is a diversified village economy based on commerce, and with thriving income-generating activities and good links to the outside world. Another is self-reliance, in part due to the quality of the leaders and SHGs which have emerged during implementation and which can solve problems and take things forward by themselves. In the Rainfed Projects the leaders started as *Jankars* and 25% of them became members of the Panchayat and even the State Legislative Assembly. A community ability to dissolve an institution that has lost its purpose and to create a new one for a new purpose may be important, and one would hope to see the community caring for the least well-off.

Right from the outset, watershed management should have its eye on the exit and gradually evolve an exit strategy with the villagers. Part of this work is to gradually load the village with more responsibilities, so that the balance of power shifts towards the village in the development relationship; this leads to an increasing emphasis on the village monitoring its own progress, and integrating the watershed with government activities. Karnataka created greater autonomy by getting village groups to supervise and pay para-workers (Mukherjee 2001). Development helps the villagers to achieve their vision of education for all children; early attention to discussing the desirability of family planning and limiting family size and the acquisition of skills, will also help. To set a range of tasks which the community has to do on its own, but with watershed support in the background, would be good preparation too.

1.7.10 Subsidies and cost contributions

Extensive subsidies pose the danger of promoting dependence and imposing supply-driven technologies and solutions pushed by the implementing agencies. Villagers must conceive some cost to themselves of being involved and village institutions perceive an ability to do things on their own without subsidy. These things are vitally important for long-term sustainability and of course also to lower the total cost of the program, making it easier to replicate. Nevertheless, financial support, especially in the early stages, can give paid employment to the community and simultaneously create productive assets like forest, water supplies, and more stable and fertile land. Farmers who benefit from a particular intervention have commonly contributed 25–30% of the cost, the community setting such rates in discussion (Mukherjee 2001).

1.8 TECHNOLOGY FOR THE POOR

This entire section is an expanded version of Wilson (2008). The role of science and technology has been central to watershed development efforts and can start by providing an income-generating, knowledge-based entry point (Wani *et al.*, 2003). An immense amount of effective technology has now been devised for rural India. Much of it is simple and inexpensive. In brief, some examples are: improved varieties of field crops, vegetables, fruit trees, and forages, all selected and bred by participatory methods; seed coatings and seed priming; organic fertilizers; application methods for inorganic fertilizers and micronutrient supplements; low-cost water structures and inexpensive drip irrigation systems; aquaculture systems, including those suited to seasonal rather than permanent ponds; caterpillars (*Helicoverpa*) containing NPV for pest control; improved tools devised with women in mind; the smokeless stove; machinery appropriate to the village and household; and energy producing or saving devices like the hydraulic rams, biogas plants, and solar lighting, which the Government is seeking to promote with subsidies. An overview of the field is given, followed, because of their importance and potential, by a little more detail about water, trees, fodder, and aquaculture.

1.8.1 General principles

Technology by itself, whilst fundamentally important, is only a partial solution to development, and is merely one tool. It must be set within a context of infrastructure, social cohesiveness, institution formation, financial resources and financial acumen, market links and the availability of support and service. It should deal with all aspects of rural economy and, as much as there is a spectrum of poverty, so there should be a spectrum of technical responses.

Technology must be appropriate to the community's needs and inclinations of the moment. Clearly, these change over time with economic and social growth. To ensure appropriateness, technology introduction and development should be participatory. This means an equal partnership between the agent and the community. Not the "top-down" passing on of a standard package of seed, fertilizer, and pesticide as was the early case with the Green Revolution in irrigated areas, but rather the careful analysis of the need, an examination of local technology and a joint development and search for solutions.

Generating locally appropriate technology by involving the end-user has proven highly successful. As an example: in seeking to introduce aquaculture to poor communities in Eastern India, the standard package of small fingerlings and management of a permanent pond was entirely inappropriate as poor people only have access to seasonal ponds. A new strategy involving stocking seasonal ponds with larger fingerlings and selling a smaller fish, either for consumption or to someone who had the resources to grow it on further, proved entirely successful (G. Haylor, personal communication, 1999).

Technology has to be devised within terms of the "farming system" and the "livelihood system" rather than be approached from a single discipline. Within the farming system, technology is needed for crops, livestock, fish, trees, soil conservation, soil fertility enhancement, water conservation, and water harvesting. Technology is needed within the livelihood system in order to reduce drudgery, to produce renewable energy, and to generate income. In this way, issues can be dealt with across the spectrum of disciplines and in order of the priority of the moment. Of course, there will be many perceptions of priority: those of the men and those of the women, and others pertaining to particular interest groups.

1.8.1.1 Participatory varietal selection (PVS)

Germplasm offers a quick way to make an impact if it is location specific, provided it is suitable for the particular soil and climate and appropriate to the need. This can be achieved by identifying with men and women in the community those characters most desired. These might be earliness of harvest, high yield, an ability to smother weeds or grow in an intercrop, pest tolerance, or cooking quality. These characters are set in order of desirability and then sought in the released crop varieties and germplasm collections. A selection of the most likely varieties is given to the farmer for planting and assessing the traits. Those of no use are rejected, and those which offer advantages are quickly adopted and can be locally multiplied in an income-generating venture (Witcombe *et al.*, 1996).

1.8.1.2 Client oriented breeding

This technology is used after exploring PVS. Broadly speaking, modern crop breeding offers yield increases of 1.5% per annum. Thus, it should be possible to replace a forty-year-old landrace with something "off the shelf" that offers 60% more yield.

And so it often proves, as such work has given impressive results of 50 to 100% yield increases in maize, rice, and various legumes in a short period (Witcombe *et al.*, 1996; Bourai *et al.*, 2003). Again, targeted at traits identified as important by farm families (not necessarily yield potential) and with an ordering of priorities, the gene-banks are searched to identify material, which may be crossed with the varieties identified by PVS and the farmer is intimately involved in the testing and selection of the resulting new material. Of course such breeding is aimed specifically at the infertile soil and irregular rainfall often associated with poor rural societies, since crop breeders have largely concentrated hitherto on irrigated regimes and on better soils and sophisticated management (Ryan *et al.*, 2000).

Whilst PVS and breeding work is beyond the capacity of watershed managers or NGOs, it should be possible to commission it for the main crops of the particular agro-environment. It would be wise too to check that the local landraces are preserved.

1.8.1.3 Seed priming

On-farm seed priming is a simple, robust, low risk, and widely applicable technique of overnight pre-soaking seed, surface drying it, and then sowing in the usual manner. About 30 to 50% yield increases are obtained in a range of cereal and leguminous crops and germination is more even and crop duration shortened (Harris 2001).

1.8.1.4 Technologies for women

The more successful technologies developed for women have involved improving conditions in the home, reducing drudgery and injury, and generating income (APICOL 1998; Ashok *et al.*, 2005). They include:

- Transparent plastic roof sheets to lighten the kitchen.
- Improved wood-burning kitchen stoves (*chulas*) and biogas cookers.
- Solar lights.
- Indoor toilets.
- Ball bearing fittings for the hand grain-mill and a hand maize-sheller.
- Farm implements of lighter construction than those used by men.
- Accessible fodder and fuel-wood plots.
- Well water close at hand.
- Small livestock and a host of income-generating technologies.

1.8.1.5 Soil conservation

Farming cannot be improved much without a stable and fertile soil. With neither, farmers are unwilling to risk their scarce resources on improved material and methods: hence the importance of watershed initiatives. Soil conservation has to be mostly a communal rather than an individual initiative and may only be feasible after about one year of establishing community cohesion, although it can provide early employment. As always with physical soil conservation structures, adequate care is needed not to concentrate runoff, so the situation is made worse. Diversion ditches above the arable area with safe disposal, sufficient contour bunds to cope with the slope, and carefully designed waterways are the crux of the issue. Responsibilities and arrangements for maintenance need to be quite clear and agreed.

1.8.1.6   Soil fertility

Soil fertility rapidly improves once the soil is stabilized and organic material and compost added. For example: bunds and gulley plugs rapidly lead to fertile strips of soil behind the bund and eventually to terraces and new fields and a change to more valuable crops.

In some areas, especially in eastern India, there are acute deficiencies of zinc, sulfur, and boron. These micronutrients can be readily ameliorated with inexpensive material and may result in immediate yield increases of 50 to 120% (Dangarwala 1983; Ali 1992; Rego *et al.*, 2005). The cost of application of the major nutrients, nitrogen, phosphorus, and potassium, is effectively reduced by seed pelleting, by subsurface application, and by micro dosing rather than broadcasting. Routine soil testing as part of the watershed approach may pay handsome dividends (Sahrawat 2005).

1.8.1.7   Crop pest control

There are many environmentally neutral methods of pest control:

- Disease-free seeds and seedlings produced by tissue culture.
- Choice of variety: Plant breeding has produced varieties of various crops such as groundnut, pigeonpea, chickpea, and sorghum, with resistance to specific diseases, superior vigor, and shorter duration, which all help the plant to resist pests and diseases (Gowda *et al.*, 2009).
- Scouting and hand-picking or shaking caterpillars off the plant.
- Paper bags over inflorescence of fruits.
- Inter-row cropping with insect-repellent crops.
- Sex-pheromone insect traps.
- Trap crops, e.g., sunflower.
- Bird perches.
- Tobacco and neem extract washes and NPVs produced in the village.
- Release of parasitic wasps, bees, mites, and sterile males.
- Aflatoxin contamination of maize and groundnut can be reduced by combining variety with the avoidance of stress, crop husbandry, postharvest management, and industrial involvement (ICRISAT 2007).
- Circulating information about outbreaks.

1.8.1.8   Renewable energy technologies

Smokeless *chulas*, biogas (minimum two or three cows), and micro-hydels have been well tried and tested. Solar lanterns are now widely sold. Small-scale windmills, watermills, and gasifiers are not yet commercialized and hydraulic rams are only just being promoted, despite having been around for two generations. *Jatropha* and *Pongamia* plantations for bio-diesel are becoming popular, but there is work to be done on selecting the best germplasm. Renewable energy technology may need more work on local adaptation but there is clearly an opportunity to do more in this field (Wani *et al.*, 2008a).

1.8.1.9 Community empowerment

Providing technical services to the community provides employment for some and a profitable new livelihood for other individuals or groups. Village specialists can be trained and then paid for their acquired skills, for example, in contour bund layout, budding and grafting, seed priming and pelleting, seedling nurseries, managing water extraction and distribution, servicing pumps, or providing a machinery contracting service. The preparation of NPVs provides important employment for women and reduces the cost of pest control immensely.

The development of community land provides a wonderful opportunity to reverse environmental degradation, generate income, and provide food, fodder, and non-timber forest products. Use of the land should be planned by the community and poor people given special roles, or title to the produce. Improved pasture grasses and legumes may be used, plantations of useful trees established, a bio-diesel enterprise started, or the forest regenerated from natural re-growth by rest and respite from cutting and grazing (D'Silva *et al.*, 2004; Dixit *et al.*, 2005).

There would be great benefit in collating the technologies available, perhaps on a CD. This is especially so for equipment and machinery. Single discipline approaches in India have frustrated a broad farming and livelihood systems perspective, but watershed staff and NGO partners and communities could be easily made aware of what is available. An NGO, once suitably appraised and trained, would then be a much more effective vehicle. Technologies which help a community to contend with smaller holdings and with climate change will undoubtedly become increasingly important.

1.8.2 Water

The first priority in watershed management is often water. Just as estate agents in London who say there are three important issues when selling property: location, location, and location, in rainfed areas the three important things are firstly water, secondly, water, and thirdly water. Water is used for human and livestock drinking, for irrigation and supplementary irrigation, for domestic use, village industry, sanitation and fish. The water component of watershed programs has tended to be supply-led when what is needed is better management and more efficient use of rainwater, and avoidance or reduction of losses to the system. Water development should concentrate on recharge, harvesting, and conservation of surface water and not on deep tube wells. Using gravity flow wherever possible will minimize costs and energy use. Water technology embraces:

- Capture, i.e., soil and water conservation are intertwined.
- Harvest, storage, and groundwater recharge.
- Avoidance of waste and loss: reduction of evaporative losses from the soil and from stored water surfaces; use of unavoidable waste and of low quality water; recycling and multiple-use, as with culturing fish in stored water bodies.
- Purification for drinking. Domestic sanitation.
- Improving the productivity of use.
- Audit of water use and cost-benefit analysis of the different options for use with trade-offs between competing demands.

- Drought tolerant varieties.
- Predictive meteorology and its practical application.
- The resilience of ecosystems and livelihoods to the consequences of climate change.
- Communal management of water resources supply and demand.

Maximizing water resources and their equitable use are commonly seen by the community, at the outset and in retrospect, as one of the most important aspects of watershed work. Water resource audits have been successfully introduced. These combine participatory approaches with high technology to assess the status of the water resource, current access and entitlement to water for productive and domestic uses, the demand for change and new sources, and to inform decision-making and policy formation. Particularly in areas of low natural rainfall and runoff, the choices made for water capture and use involve profound trade-offs. For example, afforestation and surface water bodies reduce the annual runoff from the watershed and so there will be less water to fill tanks and for downstream users (Bachelor 2001).

Communal management of water is important to seek equitable distribution and access. This may dictate small, dispersed capture and storage structures rather than one large structure and the strict monitoring and control of groundwater extraction to ensure drinking water supplies are not compromised. Simple technology and instruments can assist in this work. There is much effective traditional water technology, which can be built upon. Low-cost structures of stone and earth and a clay subsurface barrier to harvest and store water in the arid and semi-arid areas are now well proven, cost only one-fifth of masonry per litre stored, and give greater equity in access to water. Another simple approach is to direct road runoff into pits where it recharges the groundwater. There is a legion of simple, well-known approaches such as crop choice, reduced conveyance losses, field leveling for more uniform in-field distribution, mulches, drip irrigation, contours and ridges with alternate ridges planted, and the use of stand pipe waste (Gischler and Jáuregui 1984).

Lift irrigation from open wells, check-dams, and weirs is well proven and practiced. Hydraulic rams are in use and offer virtually free lift of a constant flow of water for whatever purpose. Certainly the availability of hydram sites should be part and parcel of the initial watershed survey.

Climate change may destine parts of the SAT towards aridity, most parts to more frequent fluctuations of rainfall within the average and an increased frequency of extreme events (Wilson *et al.*, 2007). Hence, working towards ecosystem and livelihood resilience to the changes in store would seem prudent. Watershed managers need to position themselves for this.

There is promising agro-meteorological work on crops and cropping patterns, and work to ally satellite and historical information. In due course, this may enable farmers to receive information through the season, thus allowing them to change crop or tactic if they are likely to receive more rain or less than the initial prediction. The success rate of predictions, the speed of information flow and the channels used for it, all need further research as do the implications for seed supply of the meteorological predictions provided and the optimal size of the recommendation zone. Wherever meteorological predictions are known to be valuable they could be associated with information hubs and watershed work (Wilson *et al.*, 2007).

1.8.3 Trees

There are timber trees, fruit trees, nut trees, fodder trees; trees, which give shade, fuel-wood, browse, medicines, mulch, or oil-bearing seeds; trees, which help to curtail soil erosion; trees, which provide the host for the lac insect and the tassar silkworm; trees, which give nitrogenous rich organic matter to help restore soil fertility; multi-purpose trees; and trees, creating an environment giving pleasure and solace (an environment with a biodiversity of animals, birds, insects, and game).

Some trees and bamboo are grown for quick returns of produce or income, or planted as a long-term investment. Trees are grown in plantations or in forests, as a private venture by individuals or small groups; around the homestead, perhaps with drip irrigation, or on farm bunds and boundaries, or in niches of small rocky areas and on "wasteland". Trees are also grown in larger units by farmers with suitable contiguous landholdings, or on common land or in Joint Forest Management (JFM) Schemes with the Forest Department.

Trees can be left to grow to their own devices or felled, lopped, browsed, or managed as coppice or as pollards. They may be of improved genetic material grafted and budded onto mature native stock. Trees can be grown in orchards of mixed species, sometimes in tiers or on terraces; or as a single species, as with a fuel plantation near at hand to reduce women's labor. They may be grown densely in pure stand and gradually thinned, or intercropped in the early years of establishment with cereals, oilseeds, legumes, fruits, vegetables, grass, or fodder.

What a contrast is the common reality of brown, barren, denuded landscapes and villages devoid of shade so frequently seen in poverty-stricken semi-arid rural areas. Lines of women head-loading 35 kg loads of fuel-wood to the village, or to the town for sale, are a common sight and provide one obvious cause for the constantly receding forest boundary, though corrupt contractors are a commonly cited other reason. There are also "fossil forests", especially in the uplands. Here, intensive grazing has removed all the natural replacement saplings so the forest is not regenerating itself.

In many communities, the older generation will describe quite vividly what the environment used to be like in "olden days". The lowering of water tables, the loss of forest grazing and of dozens of traditional medicinal plants, all these are ascribed to the loss of forest. Further, the decline of fertility of cropped areas is due, at least in part, to a reduced transfer of nutrients from the forest as bedding or animal dung.

Clearly, given such a variety of options and possibilities to improve matters, tree and tree germplasm introduction needs to be based upon very clearly articulated needs and desires by the village community at large and by the focus groups and individuals. Men and women, young villagers and old, the better-off and the poor often have different perceptions of what is needed together with some shared visions. There are often two or three different schemes in one village. Ventures on common land and JFM schemes may involve adjacent villages.

Where common land is to be planted, equity concerns should be at the forefront of discussion to ensure access and a role for marginal groups and women and the poor. Non-timber produce is likely to be of particular and immediate concern to them. Protection from cattle and goats and theft, management techniques, and rules and responsibilities need to be quite clear at the outset as does how the short- and long-term benefits and products are to be shared.

There is a wealth of varieties of trees and shrubs, fruit trees, fodder grasses and pasture legumes, which could be planted simultaneously and which offers impressive improvements on those commonly used by poor rural communities. Much of the work done on this material will have been with experimental station inputs and management so it needs to be clear from the experience of others that it is best suited to village circumstances.

Trees are a long-term investment and expensive to plant, so it is most important to take particular care to obtain the best planting stock or seed using established provenances and selected mother trees for seed, or budding and grafting material and to ensure that nursery practice is optimal.

Tree nurseries can provide a ready source of income for poor people if they are instructed in basic principles, but the type and size of nursery and its operation and management ultimately depends on the nature of the planting envisaged.

Where planting is subsidized, there is always a danger of holes being dug and paid for and then the planting grazed off so that a repeat becomes necessary. Villagers need to contribute sufficiently to completely own what is going on. A good indication of how well things are going is given by the survival percentage of new plantings: 70% should be the minimum for non-fruit trees and 90% for fruit trees.

As long as the root stock of a degraded forest has been left in-situ then recovery and regeneration merely requires protection from grazing, burning, and human activity, and a little replanting to introduce new or favored species. The speed of regeneration in these cases can be astonishing. One degraded site at Kumpura village in Rajasthan, India has recovered in eighteen years to the stage where timber is being cut and sold at good profit to the community.

Joint Forest Management offers the opportunity to address four of Government and Donor goals simultaneously: *Pulling people out of poverty* by helping their livelihoods in various ways: *climate change* by carbon sequestration; *biodiversity conservation* by maintaining the integrity of the forest and by the sustainable use of non-timber forest resources; and the *prevention of land degradation* consequent on managing the forest.

If the JFM is associated with a broader livelihoods intervention, then the introduction of renewable energy methodology adds to the climate change benefits; soil and water conservation augments the prevention of land degradation; and, to a lesser extent, there can be an impact on the phasing out of persistent organic pollutants by minimizing the use of chemical inputs with natural pesticides and fertilizers.

The enormous extent of degraded forest in India provides the scope for immense impact. In appropriate areas, eco-tourism ventures may be possible, generating new income for the community and thus giving it greater reason to continue managing the forest sustainably. It would be possible to select areas suited to JFM, using remote sensing and other data, the sites being either for the protection of intact forest or for the restitution of degraded forest. Clearly, within the boundaries of what the regulations permit, forest department staff must be completely sympathetic to the desires and actions of the villagers, seeing people as equally important to their other objectives in a JFM Scheme.

1.8.4   Livestock and fodder

Livestock play an important role in food and income generation, in maintaining soil fertility, as animals of draft and transport, and as an asset which reduces vulnerability. Of the world's 1.3 billion poor who survive on less than US$1 per day, almost half own livestock. In many rural societies poor women derive their income from livestock, especially small ruminants and dairy. Some 70% of livestock in India is reputedly owned by landless people.

India's currently low per capita consumption of livestock products could change rapidly as the economy and the population grow and as incomes rise. The projected demand for meat, eggs, milk, and milk products was expected to double between 2000 and 2020 presenting a significant market opportunity (Pinstrup-Anderson *et al.*, 1999).

The most widespread constraint to improving livestock productivity is the inability of producers to feed their animals adequately throughout the year, due to both a shortage of feed and the poor nutritional quality of what is available. The main resources used as livestock feed are unimproved pasture and hay made from it; crop residues and weeds; fodder trees or shrubs; and household wastes, cultivated forages, and concentrate feeds. The relative importance of all these varies with the farming system and with the season and the rainfall.

Natural pastures are the major source of livestock feed for poor smallholders but grazing land is decreasing as cropping, urbanization, and industry expand. The consequent increase in the pressure of uncontrolled grazing is causing a decline in the natural legume content of swards, soil erosion, conflicts among different users, and yet more grazing pressure on the remaining forest. Villagers know these issues well, so once given the skills to resolve conflicts and act in unison, it has proved relatively easy to get them to agree to set aside a part of their communal grazing, to cut, cart, and share the resulting hay crop, and even sell that surplus. As with cropped land, some areas may give good returns to the application of micronutrients, along with major nutrients.

Food crop residues are the most important of fodders. Until perhaps the 1990s, crop breeding had generally emphasized grain yield rather than the livestock feed value of the vegetative parts, and so varieties and hybrids were produced with less edible residue than unimproved varieties. This has now changed and much research effort is devoted to breed dual purpose cereals (e.g., sorghum and millet) and legumes (e.g., chickpea, pigeonpea, cowpea, and common beans). Both the volume and the nutritional qualities of the vegetative material are important. So-called "stay green" types do not senesce upon grain production and so retain nutritional quality. All this material can now of course be included in PVS.

Many of the native trees and shrubs in semi-arid areas are used as fodder and browse. A great many shrubs have been selected for their suitability to the drier areas and to salinity (Mohammad 1989). As with communal forested areas, an agreed system of management and protection would need to be in place before introductions were attempted. In some places individuals hold rights (titles or *pattas*) to individual trees.

Cultivated forages would seem to offer the most likely and promising method to address the shortage and quality of feed. The range of material available should permit finding something appropriate to the differing mixes of animal, land, available labor, and finance found in all village communities. The widow with a few goats and limited land will have a different requirement to a larger family owning a few cattle.

In the 1980s and 1990s forages were widely evaluated for adaptation and yield in many sites throughout the tropics, but this early work focused on biophysical adaptation without due regard for socioeconomic and policy issues, neither with farmer participation nor the identification of niches within the farming system. There was but little awareness among farmers of what was available, a lack of evidence of the economic profitability, and inadequate seed availability and technical support. Subsequently, species have been evaluated and superior accessions identified for many farming systems and niches. Whilst the germplasm available has not yet been widely used, there are now a number of success stories, based on participatory evaluation, which convert forages into milk, meat, and money (Hill and Roothaert 2002a). Various efforts are afoot to design expert systems which use regional soil and climate data to allow farmers and watershed managers to choose the best forages to suit their needs (Hanson 2002; Roothaert 2002).

Improved forages contribute to soil fertility through their leaf litter and microbial nitrogen fixation, and reduce erosion by providing dense ground cover. Fodders and forages have been shown to be important feed supplements for ducks, fish, and chickens as well, and producing seed or vegetative planting material generates a good income (Hill and Roothaert 2002b). In short, all the pieces are now in position to make another significant advance within watersheds using a range of cultivated fodders, which meets the different needs of different types of livestock owners.

Animal health and improved breed are secondary to the dominant problem of feed. Isolated rural areas rarely receive regular veterinary attention, but in some watershed schemes it has proved possible to arrange for veterinary camps to which large numbers of livestock arrive at a central point for vaccination against major prevailing diseases. Trained "bare-foot" veterinarians can administer basic attention and undertake artificial insemination.

Improved breeds of goats and chickens are commonly available; also the *Murrah* buffalo for those who can afford it. For cattle, until the feed issue is resolved, it is folly to attempt to upgrade with higher producing breeds, which will not withstand malnutrition. Also, the adaptation to disease of native stock should not be discarded lightly. This adaptation to disease has been put to good use by researchers at the International Livestock Research Institute (ILRI), who have selected breeds of sheep more resistant to local endoparasites in the Philippines and associated them with improved housing slatted floors, controlled grazing, stall feeding cut and carry forages, chemical dewormers, and controlled breeding (ILRI 2002).

Again, working with communities in the Philippines, ILRI has achieved 37% higher live-weight gains with multi-nutrient block licks. These improve dry matter intake in tethered cattle during the dry season when the quality of available forages is the poorest. Farmers are able to produce these licks on their own (ILRI 2002).

1.8.5   Aquaculture

The wetter parts of the SAT offer attractive options for profitable fish production including fish culture in water storage bodies and fish integrated with rice – the rice fields are slightly modified to contain a deeper fish refuge. In the drier areas, opportunities are limited to water storage bodies. Both systems require that pesticide use on crops is minimized and integrated pest management systems dominate. There is plenty of practical experience and documentation about a variety of integrated systems (fish-horticulture, fish-mushroom, fish-chicken, fish-duck, and fish-sericulture) (Edwards and Demaine 1970). Single species or poly-cultures are possible. A big fish is not necessarily the most desirable end product; it may be fry, or fingerlings to sell-on and this means that permanent water is not essential. Catfish can be produced in very limited water near the homestead and fed on household waste. Communal management requires clarity at the outset on ownership and rights. In water bodies dedicated to aquaculture, cow dung to stimulate food organisms can provide the majority of feed for filtering and omnivorous fish like silver carp and *catla*; cultivated fodders can be used for herbivorous species like common carp and grass carp (Little *et al.*, 1996).

The genetic improvement of Nile Tilapia was undertaken in the mid 1990s and the resulting, highly productive germplasm was widely distributed across Asia. The breeding methods devised are now being applied to various carp species (Mair *et al.*, 1994). There are three constraints: (1) aquaculture is not usually considered in rural development initiatives; (2) the cost of the water body, if this is dedicated to fish production only; and (3) the supply of fry.

1.9 FUTURE RESEARCH NEEDS

Ultimately, without the application of science and technology, there will be no improvement in the lives of the rural poor and in the environment surrounding them; nor will needed increases in agricultural productivity be met. Whereas some outcomes of science and technology are taken up rapidly, for example, locally appropriate crop varieties and telephony, other technologies cannot be effectively used until a community is sufficiently empowered to accept them, financially, socially, and institutionally. This may well be why many successful watershed initiatives have had impressive, but only local impact. Yet there is so much area to cover with watershed programs that if those successful aspects and technologies which can be readily adopted are widely promoted in areas awaiting a full watershed intervention, then this would be immensely valuable.

How best to spread the impact of watershed interventions locally and further afield? Research is needed to establish what has been the spontaneous sideways spread of impact of the major watershed programs undertaken by various agencies: Where does the spread of influence stop and why? What spreads by itself? What will spread with modest support? What requires the total activity and investment of a full watershed program? With what minimum intervention can a substantial benefit be achieved? What institutions are needed to promote spread of a concept and with what organizational arrangements and scientific and financial support? How much can be achieved by local institutions, rather than by Government? How best can one overcome the human and institutional barriers to working in participatory and inter-institutional mode?

There must also be lessons to be learned from the variability of impact achieved among villages and clusters of villages within one watershed. Is the variability due, for example to deficiencies of execution, to the convergence of other schemes, or to particular strengths within a community? The lessons of failure would be just as important as those of success. There are sustainability issues too. What happens when a program is left alone? What are the needs of a community post-intervention?

The Comprehensive Assessment of Watershed Programs in India (Wani *et al.*, 2009) concludes that the best impacts have been achieved in the 700 mm to 1100 mm annual rainfall zone for which the best technologies are available; also that watersheds of 2,000 to 3,000 ha would be the best operational unit offering, *inter alia*, economies of scale and greater hydrological efficiency. This conclusion might usefully be tested at and beyond the drier end of this annual rainfall range as different assemblies of technology are devised, not least those relating to hydrology and to the optimal size of the domains for weather forecasting.

There may be more work to be done too on the most strategic order of priority of sites for watershed intervention. Would expansion plans best be based on poverty, agroecology, socioeconomic similarities, market opportunities, existing infrastructure, information hubs, or previous but inadequate watershed treatment? Clarity about strategic objectives and cost-benefits will be important here.

The "Comprehensive Assessment" also gives a range of "Best-bet Technical Options". Some of these may well be promoted to good effect outside of a watershed intervention.

As we have seen, PVS commonly gives 30% to 100% yield increases in a short period and further increases can then be achieved by client oriented crop breeding. The remediation of micronutrient deficiencies gives an enormous and immediate impact on crop yield of 50% to 120%. As such work can be done without a broader intervention, and its potential impact is so great, there must be value in researching how far and how fast the principles can be applied to the main crops of those predominantly rainfed areas largely bypassed by the Green Revolution and in recommendation zones not yet addressed by watershed initiatives. How best can the two technologies be combined and with what resulting synergy?

So far, this section has looked at research which might benefit areas awaiting watershed treatment, or indeed, which have been treated but would benefit from being revisited. What about research needed to improve performance in watershed areas themselves?

Deficient rainfall or outright drought is the constant bane of the rainfed farmer. Arid zone farmers and researchers are even more familiar with it, so it would be well to take full advantage of what arid zone research has to offer the semi-arid areas. This would be both in terms of "mining the existing knowledge resource" and building on existing research collaborations. The stated need for more technologies for the zone with less than 700 mm of annual rainfall, and the likelihood that as climate change advances, fluctuations from the norm will increase in frequency and intensity, underscores this strategy. Vulnerability to drought differs across the poverty spectrum, so different technologies and policies are needed to buffer all in a community against drought and to manage aridity.

Predictive agro-meteorology is providing exciting prospects to reduce farmers' seasonal risks. It compares current observations against statistical probabilities derived from historical records. Its use should become the norm, with long range and pre-season forecasts to permit farmers to plan, choose, and change tactics, and medium and short range forecasts for crop management.

Wilson *et al.* (2007) give the following list of work needed or ongoing relating to climatic risk:

- Research on seasonal rainfall forecasts and decision support systems to guide farmer investment decisions on fertilizer and other inputs.
- Downscaling climate forecasts and adapting cropping systems models to develop decision support tools that help manage climatic risks.
- The acceptability to farmers of forecasting error.
- The extent of the most appropriate recommendation domain within each region.
- Mapping for varietal adaptation, in part based on cultivar responses to photoperiods leading to improved extrapolation domains.
- Modeling the interactions of predictive meteorology, soil analysis and fertilizer recommendations.
- More research on water harvesting and conservation agriculture to improve crop water productivity in rainfed agro-ecosystems.
- Exploring the interactions of seed supply with meteorological predictions.

Access to water and its equitable allocation and management are of course key parts of a response to aridity. Bachelor (2001) has shown how in the drier areas, major choices have to be made about how water is used; e.g., water transpired by a plantation established on a denuded hill cannot be stored in the watercourse. Taking forward such work requires more clarity about the social and institutional problems of its application, as does also preventing abuses like the lowering of groundwater tables and helping the poor and disadvantaged to make best use of what water they are able to access.

Where water rather than land is the most limiting factor, maximizing water productivity is more important than maximizing yields. The International Center for Agricultural Research in the Dry Areas (ICARDA) reports two key findings which may have resounding implications: (1) supplemental irrigation done conjunctively with rain is far more efficient than when applied alone; and (2) applying only 50% of the full-irrigation requirement led to 10% yield loss only. Thus 90% of a full-irrigation yield could be obtained on twice the rainfed area previously irrigated. ICARDA is also using satellite mapping and geographical information system (GIS) to design the most efficient water harvesting arrangements (ICARDA 2002).

Information flows to rural communities are being examined through village internet hubs, virtual expert systems, and community radio. These clearly present immense opportunities for the rapid transmission of urgent information and the presentation of research recommendations, perhaps synthesized into bundles. Much work needs to be done on the acceptability of different formats of such work and in selecting the technical content (Wilson *et al.*, 2007).

All aspects of renewable energy technology need more mainstreaming in watershed work: hydro-electric, hydraulic, wind, solar, and biofuels. Fodders too deserve much more prominence, including fodder shrubs and trees in dry areas (ICARDA 2002).

Aquaculture has shown remarkable benefits where it has been used but has not yet been widely promoted because of its institutional separation from land-based research and development, and in that most aquaculture research has pertained to the wetter end of the rainfall spectrum and to the more wealthy in the community. None of this is insuperable.

Whilst some larger NGOs are already engaged in watershed work as agents of Government or in providing specialist services, many smaller NGOs could be but are not. Empowering communities is often their *raison d'être* but many are unable to progress much beyond Stage 1 of the four stages of village improvement outlined earlier in this chapter. How best then to support and augment their knowledge and skills so they may better contribute to major watershed initiatives seems a worthy topic for investigation.

1.10   CONCLUDING REMARKS AND WAY FORWARD

The experiences described above clearly show how well community-based rural development works and how watersheds provide an appropriate framework in which agencies concerned with human and renewable natural resources conservation and development can come together. The watershed approach is now undoubtedly the most effective rural development method for the rainfed semi-arid regions, certainly in Asia, and possibly in Africa.

A compendium of best practice and numerous ideas as to how to build on previous watershed efforts (Wani *et al.*, 2007). Work is underway, or identified to research key technical, social, and institutional issues to increase impact. More recent experience in various other countries in Asia and in Africa is beginning to throw up new insights.

Thus, enough is known now about the impact, relevance, and fundamentally important aspects of watershed approaches to warrant a concerted, widely applied international effort to use them to make a substantial contribution to the various international poverty and environmental goals: removing poverty, increasing economic output, protecting the environment, and building human and natural resource resilience to cope with a range of future challenges.

How might this happen? Clearly a lead country and institution needs to present a case in an international forum and to interested donors. A planning exercise, the appointment of lead agents and institutions, and sourcing of finance would follow, succeeded by logical sequence of preparatory and implementing events. Some international agricultural research centers such as ICRISAT, ICARDA, ILRI, International Centre for Research in Agroforestry (ICRAF) and Centro International Agricultura Tropical (CIAT) should be involved. The constraints would undoubtedly revolve around two issues: the fatigue of international agencies with too many targets and initiatives, and the financial wherewithal in an era of retrenchment in the West. Nevertheless, there is a strong and attractive case to be made.

REFERENCES

Ali, M.H. 1992. *Studies on micronutrients*. An overview of the work done by the Indo-British Fertilizer Education Project. (In two parts.) India: Hindustan Fertilizer Corporation.

APICOL. 1998. *Identification of various non-land based income generation activities and market opportunities in Bolangir and Nuapada districts of Orissa*. Bhubaneswar, Orissa, India: Asian Information Marketing and Social Research.

Ashok, M.S., T. Jafry, and J.S. Gangwar. 2005. *An appraisal of technologies promoted by EIRFP*. Gramin Vikas Trust, Silsoe Research Institute, and Cirrus.

Bachelor, C.H. 2001. A fine balance: managing Karnataka's scare water resources. Karnataka, India: DFID Watershed Development Project.

Bourai, V.A., A. Choudhary, M. Misra, D.S. Virk, and J.R. Witcombe. 2003. *Participatory crop improvement in Eastern India: A first impact assessment*. New Delhi, India: Gramin Vikas Trust; and Bangor: Centre for Arid Zone Studies.

Chambers, R., A. Pacey, and L.A. Thrupp. (ed.) 1989. Farmer first. London, UK: Intermediate Technology Publications.

Dangarwala, R.T. 1983. *Micronutrient research in Gujarat*. Gujarat, India: Gujarat Agricultural University.

Dixit, S., J.C. Tewari, S.P. Wani, C. Vineela, A.K. Chourasia, and H.B. Panchal. 2005. Participatory biodiversity assessment: Enabling rural poor for better natural resource management. *Global Theme on Agroecosystems Report No. 18*. Patancheru, Andhra Pradesh, India: International Crops Research Institute for the Semi-Arid Tropics.

D'Silva, E., S.P. Wani, and B. Nagnath. 2004. The making of new Powerguda: community empowerment and new technologies transform a problem village in Andhra Pradesh. *Global Theme on Agroecosystems Report No. 11*. Patancheru, Andhra Pradesh, India: International Crops Research Institute for the Semi-Arid Tropics.

Edwards, P., and H. Demaine. 1970. *Rural aquaculture: overview and framework for country reviews*. FAO RAP Publ. 1997/36. Bangkok, Thailand: FAO.

Gischler, C., and C.F. Jáuregui. 1984. Low-cost techniques for water conservation and management in Latin America. *Nature and Resources* 20(3):11–18.

Gowda, C.L.L., R. Serraj, G. Srinivasan *et al*. 2009. Opportunities for improving crop water productivity through genetic enhancement of dryland crops. In *Rainfed Agriculture: unlocking the potential*, ed. S.P. Wani, J. Rockström, and T. Oweis, 133–163. Comprehensive Assessment of Water Management in Agriculture Series. Wallingford, UK: CAB International.

Hanson, J. 2002. *Development of a database and retrieval system for the selection of tropical forages for farming systems in the tropics and sub-tropics*. Internal document. Addis Ababa: International Livestock Research Institute (ILRI).

Harris, D. 2001. *On-farm seed priming: a key technology to improve the livelihoods of resource-poor farmers in marginal environments*. Bangor: Centre for Arid Zone Studies.

Hill, R., and R. Roothaert. 2002a. *Farming success story in Northern Vietnam*. Forages for Smallholders Project. Metro Manila: CIAT.

Hill, R., and R. Roothaert. 2002b. *When farmers are the scientists. Exotic forages boost livestock production and reduce soil erosion in poor upland communities*. Forages for Smallholders Project. Metro Manila: CIAT.

ICARDA. 2002. *Annual Report 2001–2002*. Aleppo, Syria: ICARDA.

ICRISAT. 2007. *Food security through food safety – Concerted efforts to overcome aflatoxin contamination*. (A bulletin listing partnerships and research projects.) Patancheru, Andhra Pradesh, India: ICRISAT.

ILRI. 2002. In Compendium Report of Future Harvest Centres to Annual Meeting of the CGIAR in Manila.

Little, D.C., P. Surintaraseree and N. Innes-Taylor. 1996. Fish culture in the rice fields of northeast Thailand. *Aquaculture* 140: 295–321.

Mair, G.C., J.S. Abucay, J.A. Beardmore *et al*. 1994. *Genetic manipulation for improved tilapia (GMIT) technology*. Swansea University; and Philippines: Freshwater Aquaculture Centre of Central Luzon State University.

Mohammad, N. 1989. *Rangeland management in Pakistan*. Kathmandu, Nepal: International Centre for Integrated Mountain Development (ICIMOD).

Mosse, D. 1995. *Karnataka Watershed Development Project, monitoring indicators and gender strategy*. Swansea: University of Swansea; and Wales: Center for Development Studies.

Mukherjee, K. 2001. People's participation in watershed development schemes in Karnataka – changing perspectives. (Unpublished)

Mukherjee, K. 2002. Revisiting the KAWAD Project – random thoughts of an insider. (Unpublished)

Pinstrup-Anderson, P., R. Pandya-Lorch, and M.W. Rosegrant. 1999. *World food prospects: Critical issues for the early twenty-first century*. 20:20 Vision Report. Washington, DC, USA: International Food Policy Research Institute.

Prameela, V., and N. Bijor. 2002. *Improving the lot of women in agriculture*. Bangalore, India: Cirrus Management; and Silsoe Research Institute.

Rego, T.J., S.P. Wani, K.L. Sahrawat, and G. Pardhasaradhi. 2005. Macro benefits from boron, zinc and sulfur application in Indian SAT. *Global Theme of Agro ecosystems Report No. 16*. Patancheru, Andhra Pradesh, India: ICRISAT.

Roothaert, R. 2002. *Guide to select the best forages for farmer evaluation*. Forages for Smallholders Project. Metro Manila: CIAT.

Rosegrant, Mark, Ximing Cai, Sarah Cline, and Naoko Nakagawa. 2002. *The role of rainfed agriculture in the future of global food production*. Discussion Paper No. 90. Washington, DC, USA: IFPRI.

Ryan, J.G., J.S. Kanwar, and N.G.P. Rao. 2000. Review of rainfed agro-ecosystem research. India: National Agricultural Technology Project.

Sahrawat, K.L. 2005. *Standardized, economically and statistically sound method for representative soil sampling of a micro-watershed (500–1,000 ha)*. Patancheru, Andhra Pradesh, India: ICRISAT.

Seva Mandir. 1999–2003. Annual Reports. India: Seva Mandir.

Shahi, M., S. Jones, P.S. Sodhi *et al.* 2001a. Participatory planning approach for livelihoods enhancement. *WIRFP Guidance Bulletin No. 1*. New Delhi, India: Gramin Vikas Trust.

Shahi, M., S. Jones, P.S. Sodhi *et al.* 2001b. Empowering communities – Jankar System. *WIRFP Guidance Bulletin No. 4*. New Delhi, India: Gramin Vikas Trust.

Sodhi, P.S., S. Jones, D. Mosse *et al.* 2002. Project livelihoods model for primary stakeholders. *WIRFP Guidance Bulletin No. 14*. New Delhi, India: Gramin Vikas Trust.

Sodhi, P.S., M. Shahi, and A. Parey. 2001. *Participatory monitoring and evaluation*. A guidance bulletin of the Western India Rainfed Farming Project. New Delhi, India: Gramin Vikas Trust.

Sreedevi, T.K., and S.P. Wani. 2009. Integrated farm management practices and upscaling the impact for increased productivity of rainfed systems. In *Rainfed agriculture: unlocking the potential*, ed. S.P. Wani, J. Rockström, and T. Oweis, 222–257. Comprehensive Assessment of Water Management in Agriculture Series. Wallingford, UK: CAB International.

Taru, L.E. 1998. A study of land tenure in Orissa, Balagir and Buapada. Scoping of issues for the DFID Western Orissa Rural Livelihoods Project.

Walker, J. 1999. Social Development Annex to submission No. 14 of 1999 to the DFID Project Evaluation committee (PEC) for Andhra Pradesh Rural Livelihoods Project. London, UK: DFID.

Wani, S.P., P.K. Joshi, K.V. Raju *et al.* 2008a. Community watershed as a growth engine for development of dryland areas. A Comprehensive Assessment of Watershed Programs in India. *Global Theme on Agroecosystems Report No. 47*. Patancheru, Andhra Pradesh, India: International Crops Research Institute for the Semi-Arid Tropics.

Wani, S.P., P. Pathak, H.M. Tam, A. Ramakrishna, P. Singh, and T.K. Sreedevi. 2002. Integrated watershed management for minimizing land degradation and sustaining productivity in Asia. In *Integrated Land Management in the Dry Areas. Proceedings of a Joint UNU-CAS International Workshop*, 8–13 September 200, Beijing, China, ed. Zafar Adeel, 207–230. Tokyo, Japan: United Nations University.

Wani, S.P., K.L. Sahrawat, T.K. Sreeedevi, G. Pardhasaradhi, and S. Dixit. 2009. Knowledge-based entry point for enhancing community participation in integrated watershed management. In *Proceedings of the Comprehensive Assessment of Watershed Programs in India*, 25–27 July 2007, ICRISAT, Patancheru, 53–68. Patancheru, Andhra Pradesh, India: ICRISAT.

Wani, S.P., H.P. Singh, T.K. Sreedevi *et al.* 2003. Farmer-participatory integrated watershed management: Adarsha watershed, Kothapally India, An innovative and up-scalable approach. *A case study.* In *Research towards integrated natural resources management: examples of research problems, approaches and partnerships in action in the CGIAR*, ed. R.R. Harwood and A.H. Kassam, 123–147. Washington, DC, USA: Interim Science Council, Consultative Group on International Agricultural Research.

Wani, S.P., T.K. Sreedevi, T.S. Vamsidhar Reddy, B. Venkateshvarlu, and C. Shambhu Prasad. 2008b. Community watersheds for improved livelihoods through consortium approach in drought prone rainfed areas. *Journal of Hydrological Research and Development* 23:55–77.

Wilson, M.J. 2008. Annex to Report to GEF New Delhi: First Impressions of the GEF Small Grants Programme. New Delhi, India: Global Environment Facility (GEF), UNDP.

Wilson M.J., L. Harrington, and M. Wopereis. 2007. Report of the Centre Commissioned External Review on the Global Theme – Land, Water and Agro-diversity Management. Patancheru, Andhra Pradesh, India: ICRISAT.

Wilson, M.J., S. Jones, A. Ghauri *et al.* 1998. Submission No. 27 of 1998 to the DFID Project Evaluation Committee (PEC). Western India Rainfed Farming Project. London, UK: DFID.

Witcombe, J., and A. Joshi. 1995. *Farmer participatory approaches for varietal breeding and selection and linkages to the formal seed sector.* Centre for Arid Zone Studies, University of Wales; and India: Krishak Bharati Cooperative.

Witcombe, J., A. Joshi, K.D. Joshi, and B.R. Sthapit. 1996. Farmer participatory crop improvement, varietal selection and breeding methods and their impact on bio-diversity. (In three parts.) Experimental Agriculture 32:445–496.

Chapter 2

Watershed development as a growth engine for sustainable development of rainfed areas

Suhas P. Wani and Johan Rockström

2.1 INTRODUCTION

Globally humankind is facing a great challenge of achieving food security for the ever growing population and growing per capita incomes particularly in the emerging giant economies like Brazil, Russia, India, China, and South Africa. The urgent need to decrease poverty and undernourishment while protecting the environment means delicately balancing development and sustainability resulting in increased additional pressure on the global food production system. In the foreseeable future agriculture will continue to be the backbone of economies in Africa and South Asia in spite of growing incomes and urbanization, globalization, and the declining contribution of agricultural income to gross domestic product (GDP) of the developing (in sub-Saharan Africa with 35% contribution to GDP agriculture employs 70% population) as well as emerging economies like India (with 18% contribution to GDP agriculture employs 65% population). In the past 40 years, increased crop productivity with adoption of new technologies and increased agricultural inputs along with expansion of agricultural land by about 20–25% has enabled an extraordinary progress in food security and nutrition level (FAO 2002). In 2009, more than one billion people went undernourished not because there is not enough food, but people are too poor to buy food. The percentage of hungry people in the developing world had been dropping for decades (Figure 2.1) even though the number of hungry worldwide barely dipped. The food price crises in 2008 reversed these decades of gains made in the area of food production and security (Nature 2010).

Increased food production has to come from the available and limited land and water resources which are finite. Neither the quantity of available land or water has increased since 1950, but the availability of land and water per head has declined significantly due to increase in global human population. Distribution of land and water varies differently in different countries and regions in the world and also the current population as well as expected growth which is estimated to grow rapidly in developing countries. The world is facing a severe water and land scarcity which is already complicating the national and global efforts to achieve food security in several parts. As estimated by 2025, areas where one-third of the global population resides will be facing physical and economic scarcity of water. The land expansion for agriculture is at the cost of grasslands, savannahs, and forests and land expansion for urban and infrastructure areas at the expense of agricultural land (Holmgren 2006).

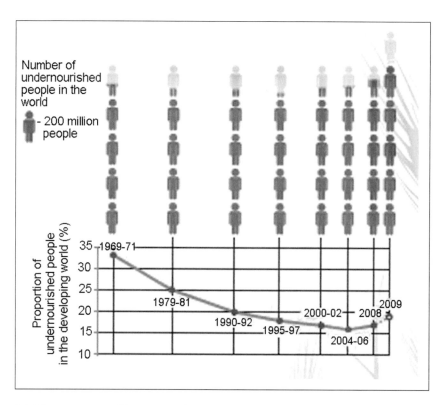

Figure 2.1 The percentage of hungry people in the developing world had been dropping for decades (bottom) even though the number of hungry worldwide barely dipped (top). But the food price crisis in 2008 reversed these decades of gains (Source: Nature 2010)

Crop land currently comprises nearly 12% (1.5 billion ha) of the world land area. In the future, crop land will not only be expanding to supply the world with food, but also to compensate for land degradation and unsustainable agricultural practices. Between 1981 and 2003 an absolute decline in net primary productivity (NPP) across 24% of the global land areas largely in southern Africa, Southeast Asia, and Southern China indicated not only the loss of farms and forests but also overall loss of 950 million tons of carbon (Bai *et al.*, 2008). The complexities of issues and multiple challenges for humankind in the 21st century such as growing population and urbanization, increasing incomes, changing diets, wastages of food, increasing land and environmental degradation, and the millennium development goal to reduce the number of poor people to half by 2015 are interlinked and call for inter-disciplinary thinking and new ways of doing agriculture. In this chapter we analyze the interlinked factors associated with food production by zooming in rainfed agriculture which covers 80% of cultivated land globally (Table 2.1) and assessing the evidence-based available options for unlocking the potential of rainfed agriculture through sustainable intensification with integrated watershed management approach.

Table 2.1 Global and continent-wise rainfed area and percentage of total arable land[a]

Continent/ Regions	Total arable land (million ha)	Rainfed area (million ha)	% of rainfed area
World	1551.0	1250.0	80.6
Africa	247.0	234.0	94.5
Northern Africa	28.0	21.5	77.1
Sub-Saharan Africa	218.0	211.0	96.7
Americas	391.0	342.0	87.5
North America	253.5	218.0	86
Central America and Caribbean	15.0	13.5	87.7
South America	126.0	114.0	90.8
Asia	574.0	362.0	63.1
Middle East	64.0	41.0	63.4
Central Asia	40.0	25.5	63.5
South and East Asia	502.0	328.0	65.4
Europe	295.0	272.0	92.3
Western and Central Europe	125.0	107.5	85.8
Eastern Europe	169.0	164.0	97.1
Oceania	46.5	42.5	91.4
Australia and New Zealand	46.0	42.0	91.3
Other Pacific Islands	0.57	0.56	99.3

[a] FAO (2010); FAOSTAT (2010).

2.2   A CONCEPT OF SAFE OPERATING SPACE FOR HUMANITY

The key question arising from various assessments such as the Millennium Assessment (MA), the Comprehensive Assessment on water management in agriculture, and Inter-governmental Panel on Climate Change (IPCC) is how much natural resources can be safely harnessed for human development without jeopardizing sustainability (MA 2005; IPCC 2007; Molden *et al.*, 2007).

Recently, Rockström *et al.* (2009) have proposed a novel concept of a safe oper-ating space for humanity suggesting acceptable levels of nine biophysical processes linked to the ability of the Earth system to remain in the current stable state. These describe a corridor for human development where risks of irreversible and significant damage seem tolerably low. For seven key parameters they suggested a quantifica-tions of safe boundary levels, of which three (climate change, loss of biodiversity, and atmospheric nitrogen fixation) have already been exceeded. Agriculture is the world's second largest consumer of water after forestry. The second most important factor controlling food production globally is the soil quality which is severely affected due to growing problem of land degradation which results in crop land expansion. Over the last decade deforestation has occurred globally at a rate of about 13 million ha per year, whereas plantations have increased in few countries such as Indonesia; since 1990, Europe, South America, Southeast Asia, and Africa continue to see high rates of net forest loss (Cossalter and Pye-Smith 2003; Bringezu *et al.*, 2009; UNEP/SEI 2009). The growing need to produce more food, feed as well as biofuel for energy means increasing pressure on scarce water and land resources. These interlinked and multiple challenges suggest strongly that business as usual will not be able to achieve the goal of sustainable development. The challenge is how to enhance the water, land, and other natural resource use efficiencies to meet the goals of food, feed, energy, and water security within the safe operating space for humankind.

2.3   CURRENT STATUS OF RAINFED AGRICULTURE

Out of 1.55 billion ha arable crop land globally, 1.25 billion ha is rainfed with varying importance regionally (95% in sub-Saharan Africa, 90% in Latin America, 60% in South Asia, 65% in East Asia, and 75% in Near East and 75% in North Africa) (FAOSTAT 2010) (Table 2.1), but produces most food for poor communities in developing countries. These challenges are exacerbated by climatic variability, the risk of climate change, population growth, health pandemics (AIDS, malaria), degrading natural resource base, poor infrastructure and changing patterns of demand and production (Ryan and Spencer 2001; Wani *et al.*, 2009; Rockström *et al.*, 2010; Walker 2010). There is a correlation between poverty, hunger, and water stress (Falkenmark 1986). The UN Millennium Development Project has identified the hot-spot countries in the world suffering from the largest prevalence of malnourishment that coincide closely with the countries hosted in semi-arid and dry subhumid hydroclimates in the world (Figure 2.2), i.e., savannahs and steppe ecosystems, where rainfed agriculture is the dominating source of food, and where water constitutes a key limiting factor to crop growth (SEI 2005).

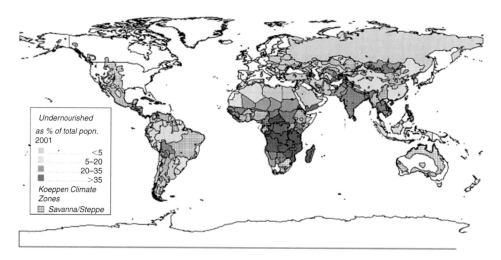

Figure 2.2 The prevalence of undernourished in developing countries (as percentage of population 2001) together with the distribution of semi-arid and dry subhumid hydroclimates in the world, i.e., savannah and steppe agroecosystems. These regions are dominated by sedentary farming subject to the world's highest rainfall variability and occurrence of dry spells and droughts (Source: SEI 2005) (See color plate section)

Even with growing urbanization, globalization, and better governance in Africa and Asia, hunger, poverty, and vulnerability of livelihoods to natural and other disasters will continue to be greatest in the rural tropical areas. The importance of rainfed sources of food weighs disproportionately on women, given that approximately 70% of the world's poor are women (WHO 2000). As most of the poor are farmers and landless laborers (Sanchez *et al.*, 2005), strategies for reducing poverty, hunger, and malnutrition should be driven primarily by the needs of the rural poor and should aim to build and diversify their livelihood sources. Substantial gains in land, water, and labor productivity as well as better management of natural resources are essential to reverse the downward spiral of poverty and environmental degradation, apart from the problems of equity, poverty, and sustainability – and hence, the need for greater investment in the semi-arid tropics (World Bank 2005; Rockström *et al.*, 2007; Wani *et al.*, 2008a, 2009).

Evidence is emerging that climate change is making the variability more intense with increased frequency of extreme events such as drought, floods, and hurricanes (IPCC 2007). A recent study assessing rainfed cereal potential under different climate change scenarios, with varying total rainfall amounts, concluded that it is difficult to estimate the exact degree of regional impact. But most scenarios resulted in losses of rainfed production potential in the most vulnerable developing countries. In these countries, the loss of production area was estimated at 10–20%, with an approximate potential of 1.3 billion people affected by 2080 (IIASA 2002). In particular, sub-Saharan Africa is estimated to lose 12% of the cultivation potential mostly projected in the Sudan-Sahelian zone, which is already subject to high climatic variability and adverse crop conditions.

2.4   VAST POTENTIAL TO INCREASE CROP YIELDS IN RAINFED AREAS

In the past 40 years, 30% of the overall grain production growth is due to expansion of agricultural areas and the remaining 70% growth originated from intensification through yield increases per unit land area. However, the regional variation is large, as is the difference between irrigated and rainfed agriculture. In developing countries rainfed grain yields are on an average $1.5\,t\,ha^{-1}$ compared to $3.1\,t\,ha^{-1}$ for irrigated yields (Rosegrant *et al.*, 2002), and increase in production from rainfed agriculture has mainly originated from land expansion.

In sub-Saharan Africa, with 99% rainfed production of main cereals such as maize, millet, and sorghum, the cultivated cereal area has doubled since 1960 while the yield per unit land has nearly been stagnant for these staple crops (FAOSTAT 2010). In South Asia, farmers shifted away from more drought tolerant low-yielding crops such as sorghum and millet, whilst wheat and maize have approximately doubled in area since 1961 (FAOSTAT 2010). During the same period, the yield per unit land for maize and wheat has more than doubled. For predominantly rainfed systems, maize yields per unit land have nearly tripled and wheat more than doubled during the same time period.

In the temperate regions, rainfed agriculture generates among the world's highest yields with relatively reliable rainfall and inherently productive soils. Even in tropical

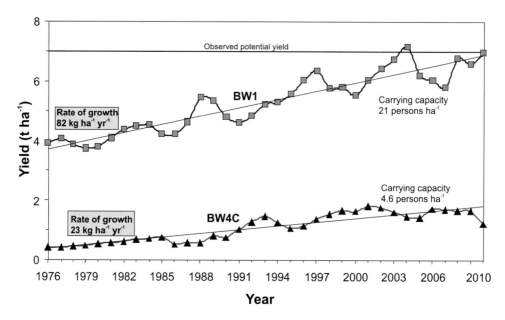

Figure 2.3 Three-year moving average of crop yields in improved (BW1) and traditional (BW4C) management systems during 1976–2009 at ICRISAT, Patancheru, India

regions, particularly in the subhumid and humid zones, agricultural yields in commercial rainfed agriculture exceed 5–6 t ha^{-1} (Rockström and Falkenmark 2000; Wani *et al.*, 2003b, 2003c) (Figure 2.3). At the same time, the dry subhumid and semi-arid regions have experienced the lowest yields and the weakest yield improvements per unit land. Here, yields oscillate between 0.5 and 2 t ha^{-1}, with an average of 1 t ha^{-1} in sub-Saharan Africa, and 1–1.5 t ha^{-1} in South Asia, Central Asia, and West Asia and North Africa (WANA) for rainfed agriculture (Rockström and Falkenmark 2000; Wani *et al.*, 2003b, 2003c, 2009, 2011; Rockström *et al.*, 2010).

Yield gap analyses carried out for Comprehensive Assessment for major rainfed crops in semi-arid regions in Asia and Africa and rainfed wheat in WANA, revealed large yield gaps with farmers' yields being a factor 2 to 4 times lower than achievable yields for major rainfed crops (Figures 2.4 and 2.5) (Agarwal 2000; Rockström *et al.*, 2007; Fisher *et al.*, 2009; Singh *et al.*, 2009; Wani *et al.*, 2011). In countries in Eastern and Southern Africa the yield gap is very large (Figure 2.5). Similarly, in many countries in West Asia, farmers' yields are threefold lower than achievable yields, while in some Asian countries the figure is closer to twofold. Historic trends present a growing yield gap between farmers' practices and farming systems that benefit from management advances (Wani *et al.*, 2003c, 2009, 2011) and vast scope exists to unlock the potential of rainfed agriculture through sustainable management of natural resources through scaling-out the experiences from the islands of success spread sporadically throughout the globe (Kijne *et al.*, 2009; Wani *et al.*, 2009).

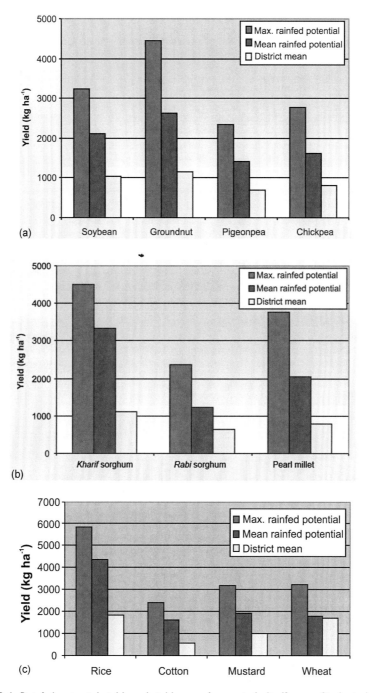

Figure 2.4 Rainfed potential yields and yield gaps of crops in India (Source: Singh *et al.*, 2009)

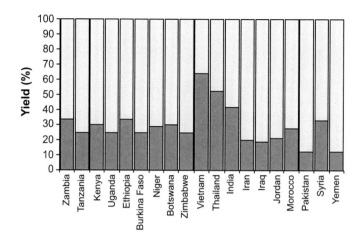

Figure 2.5 Examples of observed yield gap (for major grains) between farmers' yields and achievable yields (100% denotes achievable yield level, and columns actual observed yield levels) (Source: Rockström *et al.*, 2007)

2.5 IMPROVED WATER PRODUCTIVITY IS A KEY TO UNLOCK THE POTENTIAL OF RAINFED AGRICULTURE

An adequate human diet takes about 4000 liters of water per day to produce, which is over 90% of the daily human water requirement. If farmers continue to use the current methods for producing food then to feed the world's growing population we will need a great deal more water to keep everyone fed: another $1600 \, km^3 yr^{-1}$ just to achieve the UN Millennium Development Goal of halving hunger by 2015 (SEI 2005), and another $4500 \, km^3 yr^{-1}$ with current water productivity levels in agriculture to feed the world in 2050 (Falkenmark *et al.*, 2009; Rockström *et al.*, 2009) This is more than twice the current consumptive water use in irrigation, which already contributes to depleting several large rivers before they reach the ocean. It is becoming increasingly difficult, on social, economic, and environmental grounds, to supply more water to farmers.

Water scarcity is a relative concept and as explained by Rockström *et al.* (2009) current estimates of water scarcity are using the conventional approach and assessing the amount of renewable surface and groundwater per capita (i.e., so called blue water) without taking into consideration the full resource of rainfall, and notably "green water", i.e., soil moisture used in rainfed cropping and natural vegetation. As Figure 2.6a illustrates, South Asia, East Asia, and the Middle East North Africa (MENA) regions are the worst affected in terms of blue water scarcity. However, according to a recent assessment that included both green and blue water resources, the level of water scarcity changed significantly for many countries (Figure 2.6b). Large parts of China, India, and sub-Saharan Africa are conventionally water scarce but still have sufficient green and blue water to meet the water demand for food production. If green water (on current agricultural land) for food production is included, per capita water availability is doubled or tripled in countries such as Uganda, Ethiopia,

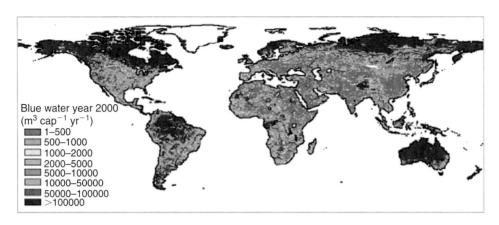

Figure 2.6a Renewable liquid freshwater (blue water) stress per capita using LPJ dynamic modeling year 2000 (after Rockström *et al.*, 2009) (See color plate section)

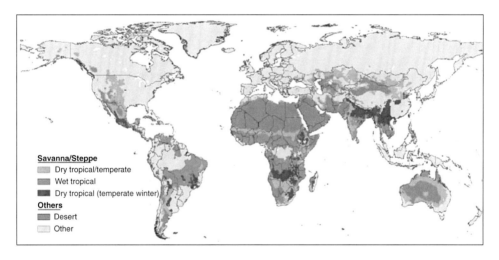

Figure 2.6b Renewable rainfall (green and blue water) stress per capita using LPJ dynamic modeling year 2000 (after Rockström *et al.*, 2009) (See color plate section)

Eritrea, Morocco, and Algeria. Moreover, low ratios of transpiration to evapotranspiration (T/ET) in countries such as Bangladesh, Pakistan, India, and China indicate high potential for increasing water productivity through vapor shift (Rockström *et al.*, 2009).

Absolute water stress is found most notably in arid and semi-arid regions with high population densities such as parts of India, China, and the MENA region. The MENA region is increasingly unable to produce the food required locally due to increasing water stress from a combination of population increase, economic development, and

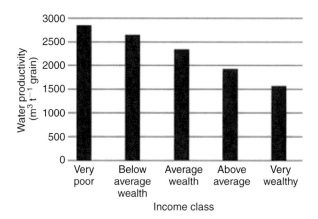

Figure 2.7 Water productivity for maize yields in smallholder farming systems in sub-Saharan Africa (based on Holmen 2004)

climate change, and will have to rely more and more on food (and virtual water) imports. For the greater part of the world the global assessment of green and blue water suggested that water stress is primarily a blue water issue, and large opportunities are still possible in the management of rainfed areas, i.e., the green water resources in the landscape (Rockström *et al.*, 2009). The current global population that has blue water stress is estimated to be 3.17 billion, expected to reach 6.5 billion in 2050. If both green and blue water are considered, the number currently experiencing absolute water stress is a fraction of this (0.27 billion), and will only marginally exceed today's blue water stressed in 2050.

Given the increasing pressures on water resources and the increasing demands for food and fiber, the world must succeed in producing more food with less water. Hence, it is essential to increase water productivity in both humid and arid regions. Some describe the goal as increasing the "crop per drop" or the "dollars per drop" produced in agriculture. Regardless of the metric, it is essential to increase the productivity of water and other inputs in agriculture. Success will generate greater agricultural output, while also enabling greater use of water in other sectors and in efforts to enhance the environment.

Water productivity can vary with household income, as farmers' yields vary as a result of local input and management styles. In a household level study of 300 farmers in eight sub-Saharan countries, the more wealthy farmers had generally higher yield levels (Holmen, 2004), and subsequently better water productivity (Figure 2.7). The differences were significant between the wealthier classes and poorest classes. More than $1,000 \, m^3$ additional water was required per ton of maize grain produced by the poorest farmers compared to the wealthiest farmers. Data suggest that yield improvements for the purpose of poverty alleviation can also significantly improve water productivity, especially in current low yielding rainfed (green water) agriculture, in sub-Saharan Africa and parts of South Asia. Improved water use efficiency and productivity can improve food security.

2.6 WATER ALONE CANNOT DO IT

Successful efforts to increase crop yields through improved management practices other than water help in bridging the yield gaps by increasing water use efficiency. Hence, it provides good opportunities to examine other critical constraints which are holding back the potential of increasing crop yields. Soil health is severely affected due to land degradation and is in need of urgent attention. Often, soil fertility is the limiting factor to increased yields in rainfed agriculture (Stoorvogel and Smaling 1990; Rego et al., 2005). Soil degradation, through nutrient depletion and loss of organic matter, causes serious yield decline closely related to water determinants, as it affects water availability for crops, due to poor rainfall infiltration, and plant water uptake, due to weak roots. Nutrient mining is a serious problem in small-holder rainfed agriculture. In sub-Saharan Africa soil nutrient mining is particularly severe. It is estimated that approximately 85% of African farm land in 2002–04 experienced a loss of more than 30 kg ha^{-1} of nutrients (nitrogen, phosphorus, and potassium) per year (IFDC 2006). There is a need to manage soil fertility through integrated fertility management options through optimum use of biological, organic, and chemical sources of required nutrients.

On-farm diagnostic work of the International Crops Research Institue for the Semi-Arid Tropics (ICRISAT) in different community watersheds in different states of India as well as in Southern China, North Vietnam, and Northeast Thailand showed severe mining of soils for essential plant nutrients including secondary and micronutrients along with macronutrients (Rego et al., 2007; Sahrawat et al., 2007). Evidence from on-farm participatory trials in different rainfed areas in India clearly indicated that investments in soil fertility improvement directly improved water management resulting in increased rainwater productivity and crop yields by 70 to 120% when both micronutrients and adequate nitrogen and phosphorus were applied (Rego et al., 2005; Sahrawat et al., 2007; Srinivasarao et al., 2010). Similarly, integrated land, nutrient, and water management options as well as use of improved cultivars in semi-arid regions increased significantly rainwater productivity (Wani et al., 2003a, 2011; Sreedevi and Wani 2009). Gains in rainwater use efficiency with improved land, nutrient, and water management options were far higher in low rainfall years (Figure 2.8).

In addition, soil organic matter, an important driving force for supporting biological activity and crop productivity in soil, is very much in short supply particularly in tropical countries (Lee and Wani 1989; Syers et al., 1996; Katyal and Rattan 2003). Management practices that augment soil organic matter and maintain at a threshold level are needed. Improved agricultural management practices in the tropics such as intercropping with legumes, application of balanced plant nutrients, suitable land and water management and use of stress-tolerant high-yielding cultivars improved soil organic carbon content and also increased crop productivity (Lee and Wani 1989; Wani et al., 1995, 2003b, 2005, 2007; ICRISAT 2005; Gowda et al., 2009; Srinivasarao et al., 2009). Good quality organic materials in fields using farm bunds and degraded common lands in the villages could be produced by growing nitrogen-fixing shrubs and trees to generate nitrogen-rich loppings. Also, large quantities of farm residues and other organic wastes could be converted into valuable source of plant nutrients and organic matter through vermicomposting (Nagavallemma et al., 2005; Sreedevi et al., 2007; Wani et al., 2008b; Sreedevi and Wani 2009).

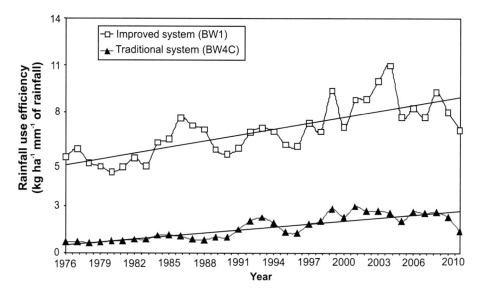

Figure 2.8 Increased rainwater use efficiency in low rainfall years at ICRISAT, Patancheru, India

2.7 INTEGRATED WATERSHED MANAGEMENT IS KEY FOR SUSTAINABLE MANAGEMENT OF LAND AND WATER RESOURCES AND IMPROVED LIVELIHOODS

Of the $100,000\,km^3$ of water that falls on the land each year, 65% becomes green water, i.e., water contained in the root zone of the soil and 35% results in blue water, i.e., runoff water into lakes, rivers, reservoirs, and aquifers. Integrated watershed management provides a good opportunity to manage land and water resources in an integrated manner for sustainable livelihoods while protecting the environment (Wani *et al.*, 2002, 2008a, 2009, 2011). Both green and blue water is generated in the landscape and integrated water resource management is the key for sustainable development and management of water at small catchment scale that is recommended for enhancing the efficiency of water in rainfed areas (Wani *et al.*, 2002, 2009, 2011; Molden *et al.*, 2007; Rockström *et al.*, 2007, 2010). Rockström *et al.* (2009) have indicated that green water dominates food production as consumptive use of green water is four times larger than that of blue water (fresh water in rivers, lakes, reservoirs, and aquifers).

Evidence from water balance analyses on farmers' fields around the world shows that only a small fraction, less than 30% of rainfall, is used as productive green water flow (plant transpiration) supporting plant growth (Rockström 2003). In arid areas, typically as little as 10% of the rainfall is consumed as productive green water flow (transpiration) and 90% flows as non-productive evaporation flow, i.e., no or very limited blue water generation (Oweis and Hachum 2001). In temperate arid regions, such as WANA, a large portion of the rainfall is generally consumed in the farmers' fields as productive green water flow (45–55%) that resulted in higher yield levels (3–4 t ha^{-1} as compared to 1–2 t ha^{-1}) and 25–35% of the rainfall flows as non-productive green water flow, and remaining 15–20% generate blue water flow. These indicate a large scope of opportunity. Low agricultural yields in rainfed agriculture, often blamed as rainfall deficits, are in fact caused by other factors than rainfall.

This suggests great scope and opportunity for improvement of green water productivity, as it entails shifting non-productive evaporation to productive transpiration, with no downstream water trade-off. This vapor shift (or transfer) through improved management options is a particular opportunity in arid, semi-arid, and dry subhumid regions (Rockström *et al.*, 2007).

Field measurements of rainfed grain yields and actual green water flows indicate that by doubling yields from 1 to 2 t ha^{-1} in semi-arid tropical agroecosystems, green water productivity may improve from approximately 3500 m^3 t^{-1} to less than 2000 m^3 t^{-1}. This is a result of the dynamic nature of water productivity improvements when moving from very low yields to higher yields. At low yields, crop water uptake is low and evaporative losses are high, as the leaf area coverage of the soil is low. This results in high losses of rainwater as evaporation from soil. When yield levels increase, shading of soil improves.

Business as usual to manage rainfed agriculture as subsistence agriculture with low resource use efficiency cannot sustain the economic growth and needed food security. There is an urgent need to develop a new paradigm for upgrading rainfed agriculture. The conventional sectoral approach to water management produced low water use efficiencies resulting in increased demand for water to produce food. We need to have a holistic approach based on converging all the necessary aspects of natural resource conservation, their efficient use, production functions, and income enhancement avenues through value chain and enabling policies and much needed investments in rainfed areas (Wani *et al.*, 2003c, 2009; Rockström *et al.*, 2007).

Furthermore, the current focus on water resource planning at the river basin scale is not appropriate for water management in rainfed agriculture, which overwhelmingly occurs on farms of <5 ha at the scale of small catchments, below the river basin scale. Therefore, focus should be to manage water at the catchment scale (or small tributary scale of a river basin), opening for much needed investments in water resource management also in rainfed agriculture (Rockström *et al.*, 2007).

Evidence collected during the Comprehensive Assessment of water for food and water for life revealed that business as usual in global agriculture would not be able to meet the goal of food security and reducing poverty. If the situation continued it will lead to crises in many parts of the world (Molden *et al.*, 2007). However, the world's available land and water resources can satisfy future demands by taking the following steps:

- Upgrade rainfed agriculture by investing more in rainfed agriculture to enhance agricultural productivity (rainfed scenario).
- Discard the artificial divide between rainfed and irrigated agriculture and adopt integrated water resource management approach for enhancing resource efficiency and agricultural productivity.
- Invest in irrigation for expanding irrigation where scope exists and improving efficiency of the existing irrigation systems (irrigation scenario).

- Conduct agricultural trade within and between countries (trade scenario).
- Reduce gross food demand by influencing diets and reducing postharvest losses, including industrial and household waste.

To upgrade rainfed agriculture in the developing countries, community based participatory and integrated watershed management approaches are recommended and are found effective through a number of islands of success in Asia and Africa (Wani *et al.*, 2002, 2003a, 2008b; Rockström *et al.*, 2007). In the rainfed areas of the tropics, water scarcity and growing land degradation cannot be tackled through farm-level interventions alone and community-based management of natural resources for enhancing productivity and improving rural livelihoods are urgently needed (Wani *et al.*, 2003a, 2009; Rockström *et al.*, 2007). A holistic approach is needed that integrates all the necessary aspects of natural resource conservation, their efficient use, production functions, and income enhancement avenues through value chain and enabling policies and much needed investments in rainfed areas. A major research and development challenge to upgrade rainfed agriculture is to integrated knowledge systems amongst different stakeholders and scientific disciplines by coming out of disciplinary compartments and translate available blue prints into operational plans and implement them (Wani *et al.*, 2003a, 2006, 2009; Rockström *et al.*, 2007). We know what to do but the challenge is how to do it.

The community-based management of natural resources calls for new approaches (technical, institutional, and social), which are knowledge-intensive and need strong capacity building measures for all the stakeholders including policy makers, researchers, development agents, and farmers. The small and marginal farmers are deprived of new knowledge and materials produced by the researchers. There is a disconnect between the farmers and the researchers as the extension systems in most developing countries are not functioning to the desired level. There is an urgent need for a holistic approach to address issues of rainfed agriculture to achieve food security and alleviate poverty to meet the Millennium Development Goals.

REFERENCES

Agarwal, A. 2000. Drought? Try capturing the rain. Briefing paper for members of parliament and state legislatures – An occasional paper from the Centre for Science and Environment (CSE), New Delhi, India.

Bai, Z.G., D.L. Dent, L. Olsson, and M.E. Schaepman. 2008. *Global assessment of land degradation and improvement: 1. Identification by remote sensing.* GLADA Report 5. Wageningen, The Netherlands: LADA, World Soil Information; and Rome, Italy: FAO.

Bringezu, S., H. Schütz, K. Arnold *et al.* 2009. Global implications of biomass and biofuel use in Germany – Recent trends and future scenarios for domestic and foreign agricultural land use and resulting GHG emissions. *Journal of Cleaner Production* 17:57–68.

Cossalter, C., and C. Pye-Smith. 2003. *Fast-wood forestry: myths and realities.* Forest Perspectives No. 1. Center for International Forestry Research.

Falkenmark, M. 1986. Fresh water – Time for a modified approach. *Ambio* 15(4):192–200.

Falkenmark, M., L. Karlberg, and J. Rockström. 2009. Present and future water requirements for feeding humanity. *Food Security* 1:59–69.

FAO. 2002. Agriculture: Towards 2015/30. *Technical Interim Report.* (http://www.fao.org/es/esd/at2015/toc-e.htm)

FAO. 2010. AQUASTAT database (http://www.fao.org/nr/aquastat).

FAOSTAT. 2010. http://faostat.fao.org/site/567/default.asp.ancor. (Accessed in April 2010.)

Fisher, G., V.V. Harrij, G. Eva *et al.* 2009. Potentially obtainable yields in the semi-arid tropics. *Global Theme on Agroecosystems Report No. 54.* Patancheru, Andhra Pradesh, India: International Crops Research Institute for the Semi-Arid Tropics.

Gowda, C.L.L., R. Serraj, G. Srinivasan *et al.* 2009. Opportunities for improving crop water productivity through genetic enhancement of dryland crops. In *Rainfed agriculture: unlocking the potential,* ed. S.P. Wani, J. Rockström, and T. Owies, 133–163. Wallingford, UK: CAB International.

Holmen, H. 2004. Why no green revolution in Africa? *Currents* 34:12–16. (www.slu.se)

Holmgren, P. 2006. Global land use area change matrix: Input to GEO-4, Rome, Italy: FAO.

ICRISAT. 2005. Identifying systems for carbon sequestration and increased productivity in semi-arid tropical environments. *A Project Completion Report* funded by National Agricultural Technology Project, Indian Council of Agricultural Research, New Delhi, India. Patancheru, Andhra Pradesh, India: International Crops Research Institute for the Semi-Arid Tropics.

IFDC. 2006. Agricultural production and soil nutrient mining in Africa: Implications for resource conservation and policy development. *IFDC Technical Bulletin.* Alabama, USA: IFDC.

IIASA. 2002. *Climate change and agricultural vulnerability.* Special report for the UN World Summit on Sustainable Development, Johhanesburg, 2002. IIASA. (http://www.iiasa.ac.at/Research/LUC/JB-Report.pdf)

IPCC. 2007. Climate change impacts, adaptation and vulnerability. Technical summary of Working Group II to Fourth Assessment Report of Inter-governmental Panel on Climate Change, ed. M.L. Parry, O.F. Canziani, J.P. Paultikof, P.J. van der Linden, and C.E. Hanon, 23–78. Cambridge, UK: Cambridge University Press.

Katyal, J.C., and R.K. Rattan. 2003. Secondary and micronutrients: Research gaps and future needs. *Feriliser News* 48(4):9–14, 17–20.

Kijne, J., J. Barron, H. Hoff *et al.* 2009. *Opportunities to increase water productivity in agriculture with special reference to Africa and South Asia.* Stockholm, Sweden: Stockholm Environment Institute.

Lee, K.K., and S.P. Wani. 1989. Significance of biological nitrogen fixation and organic manures in soil fertility management. In *Soil fertility and fertility management in semi-arid tropical India,* ed. C.B. Christianson, 89–108. Alabama, USA: IFDC.

MA. 2005. Millennium ecosystem assessment, ecosystems and human well-being: current status and trends. In *Fresh water,* Vol. 1, ed. R. Hassan, R. Scholes, and N. Ash. Washington, USA: Isaland Press. (http://www.millenniumassessment.org/en/Products.Global.Condition.aspx)

Molden, D., Karen Frenken, Randolph Barker *et al.* 2007. Trends in water and agricultural development. In *Water for food, water for life: a comprehensive assessment of water management in agriculture,* ed. D. Moden, 57–89. London, UK: Earthscan; and Colombo, Sri Lanka: International Water Management Institute (IWMI).

Nagavallemma, K.P., S.P. Wani, S. Lacroix *et al.* 2005. Vermicomposting: recycling wastes into valuable organic fertilizer. *Journal of Agriculture and Environment for International Development* 99(3/4):188–204.

Nature. 2010. The growing problem. World hunger remains a major problem, but not for the reasons many suspect. Nature analyses the trends and the challenges of feeding 9 billion by 2050. News Feature Food 466:546–547.

Oweis, T., and A. Hachum. 2001. Reducing peak supplemental irrigation demand by extending sowing dates. *Agricultural Water Management* 50(2):109–123.

Rego, T.J., K.L. Sahrawat, S.P. Wani *et al.* 2007. Widespread deficiencies of sulfur, boron and zinc in Indian semi-arid tropical soils: on-farm crop responses. *Journal of Plant Nutrition* 30:1569–1583.

Rego, T.J., S.P. Wani, K.L. Sahrawat *et al.* 2005. Macro-benefits from boron, zinc, and sulphur application in Indian SAT: A step for grey to green revolution in agriculture. *Global Theme on Agroecosystems Report No. 16.* Patancheru, Andhra Pradesh, India: International Crops Research Institute for the Semi-Arid Tropics (ICRISAT).

Rockström, J. 2003. Water for food and nature in drought-prone tropics: vapour shift in rainfed agriculture. *Philosophical Transactions of the Royal Society of London Series B – Biological Sciences* 358(1440):1997–2009.

Rockström, J., and M. Falkenmark. 2000. Semiarid crop production from a hydrological perspective: Gap between potential and actual yields. *Critical Reviews in Plant Sciences* 19(4):319–346.

Rockström, J., M. Falkenmark, L. Karlberg *et al.* 2009. Future water availability for global food production: the potential of green water for increasing resilience to global change. *Water Resources Research* 45. (W00A12, doi:10.1029/2007WR006767)

Rockström, J., N. Hatibu., T. Oweis *et al.* 2007. Managing water in rainfed agriculture. In *Water for food, Water for life: a comprehensive assessment of water management in agriculture*, ed. D. Molden, 315–348. London, UK: Earthscan and Colombo, Sri Lanka: International Water Management Institute (IWMI)

Rockström, J., L. Karlberg, S.P. Wani *et al.* 2010. Managing water in rainfed agriculture – The need for a paradigm shift. *Agricultural Water Management* 97:543–550

Rosegrant, M., C. Ximing, S. Cline *et al.* 2002. The role of rainfed agriculture in the future of global food production. *EPTD Discussion Paper No. 90.* Washington, DC, USA: Environment and Production Technology Division, IFPRI.

Ryan, J.G., and D.C. Spencer. 2001. *Challenges and opportunities shaping the future of the semi-arid tropics and their implications.* Patancheru, Andhra Pradesh, India: International Crops Research Institute for the Semi-Arid Tropics (ICRISAT).

Sahrawat, K.L., S.P. Wani, T.J. Rego *et al.* 2007. Widespread deficiencies of sulphur, boron and zinc in dryland soils of the Indian semi-arid tropics. *Current Science* 93(10):1–6.

Sanchez, P., M.S. Swaminathan, P. Dobie, and N. Yuksel. 2005. Halving hunger: It can be done. Summary version of the report of the Task Force on Hunger, New York, USA. USA: The Earth Institute at Columbia University.

SEI. 2005. Sustainable pathways to attain the millennium development goals – assessing the role of water, energy and sanitation. Document prepared for the UN World Summit, Sep 14, 2005, New York, USA. Stockholm, Sweden: Stockholm Environment Institute. (http://www.sei.se/mdg.htm)

Singh, P., P.K. Aggarwal, V.S. Bhatia *et al.* 2009. Yield gap analysis: modelling of achievable yields at farm level. In *Rainfed agriculture: unlocking the potential*, ed. S.P. Wani, J. Rockström, and T. Oweis, 81–123. Comprehensive Assessment of Water Management in Agriculture Series. Walingford, UK: CAB International.

Sreedevi, T.K., and S.P. Wani. 2009. Integrated farm management practices and up-scaling the impact for increased productivity of rainfed systems. In *Rainfed agriculture: unlocking the potential*, ed. S.P. Wani, J. Rockström, and T. Oweis, 222–257. Comprehensive Assessment of Water Management in Agriculture Series. Wallingford, UK: CAB International.

Sreedevi, T.K., S.P. Wani, and P. Pathak. 2007. Harnessing gender power and collective action through integrated watershed management for minimizing land degradation and sustainable development. *Journal of Financing Agriculture* 36:23–32.

Srinivasa Rao, Ch., K.P.R. Vittal, B. Venkateswarlu *et al.* 2009. Carbon stocks in different soil types under diverse rainfed production systems in tropical India. *Communications in Soil Science and Plant Analysis* 40(15):2338–2356.

Srinivasa Rao, Ch., S.P. Wani, K.L. Sahrawat *et al.* 2010. Effect of balanced nutrition on yield and economics of vegetable crop in participatory watersheds in Karnataka. *Indian Journal of Fertilizers* 6(3):39–42.

Stoorvogel, J.J., and E.M.A. Smaling. 1990. *Assessment of soil nutrient depletion in sub-Saharan Africa*: 1983–2000. Vol. 1, Main Report. Report 28. Wageningen, Netherlands: Winand Staring Centre.

Syers, J.K., J. Lingard, J. Pieri *et al.* 1996. Sustainable land management for the semiarid and subhumid tropics. *Ambio* 25:484–491.

UNEP/SEI. 2009. *Rainwater harvesting: a lifeline for human well-being*. Nairobi, Kenya: United Nations Environment Programme (UNEP); and Stockholm, Sweden: Stockholm Environment Institute (SEI).

Walker, T. 2010. Challenges and opportunities for agricultural R & D in the semi-arid tropics. Internal document for Strategic Planning. Patancheru, Andhra Pradesh, India: International Crops Research Institute for the Semi-Arid Tropics (ICRISAT).

Wani, S.P., K.H. Anantha, T.K. Sreedevi, R. Sudi, S.N. Singh, and M. D'Souza. 2011. Assessing the environmental benefits of watershed development: evidence from the Indian semi-arid tropics. *Journal of Sustainable Watershed Science & Management* 1(1):10–20.

Wani, S.P., P.K. Joshi, K.V. Raju *et al.* 2008a. Community watershed as a growth engine for development of dryland areas. A comprehensive assessment of watershed programs in India. *Global Theme on Agroecosystems Report No. 47*. Patancheru, Andhra Pradesh, India: International Crops Research Institute for the Semi-Arid Tropics.

Wani, S.P., P.K. Joshi, Y.S. Ramakrishna, T.K. Sreedevi, P. Singh, and P. Pathak. 2008b. A new paradigm in watershed management: A must for development of rainfed areas for inclusive growth. In *Conservation farming: enhancing productivity and profitability of rainfed areas*, ed. Anand Swarup, Suraj Bhan, and J.S. Bali, 163–178. New Delhi, India: Soil Conservation Society of India.

Wani, S.P., A.R. Maglinao, A. Ramakrishna, and T.J. Rego. 2003a. *Integrated watershed management for land and water conservation and sustainable agricultural production in Asia*. Proceedings of the ADB-ICRISAT-IWMI Annual Project Review and Planning Meeting, Hanoi, Vietnam, 10–14 December 2001. Patancheru, Andhra Pradesh, India: International Crops Research Institute for the Semi-Arid Tropics.

Wani, S.P., P. Pathak, L.S. Jangawad, H. Eswaran, and P. Singh. 2003b. Improved management of Vertisols in the semi-arid tropics for increased productivity and soil carbon sequestration. *Soil Use and Management* 19:217–222.

Wani, S.P., P. Pathak, T.K. Sreedevi, H.P. Singh, and P. Singh. 2003c. Efficient management of rainwater for increased crop productivity and groundwater recharge in Asia. In *Water productivity in agriculture: limits and opportunities for improvement*, ed. J.W. Kijne, R. Barker, and D. Molden, 199–215. Wallingford, UK: CAB International; and Colombo, Sri Lanka: International Water Management Institute (IWMI).

Wani, S.P., P. Pathak, H.M. Tam, A. Ramakrishna, P. Singh, and T.K. Sreedevi. 2002. Integrated watershed management for minimizing land degradation and sustaining productivity in Asia. In *Integrated land management in dry area*s. Proceedings of a Joint UNU-CAS International Workshop, 8–13 September 2001, Beijing, China, ed. Z. Adeel, 207–230. Tokyo, Japan: United Nations University.

Wani, S.P., Y.S. Ramakrishna, T.K. Sreedevi *et al.* 2006. Issues, concepts, approaches and practices in integrated watershed management: Experience and lessons from Asia. In *Integrated management of watersheds for agricultural diversification and sustainable livelihoods in Eastern and Central Africa: lessons and experiences from semi-arid South Asia, 17–36*. Proceedings of the International Workshop held at ICRISAT, Nairobi, 6–7 December 2004. Patancheru, Andhra Pradesh, India: International Crops Research Institute for the Semi-Arid Tropics (ICRISAT).

Wani, S.P., T.J. Rego, S. Rajeswari *et al.* 1995. Effect of legume-based cropping systems on nitrogen mineralisation potential of Vertisol. *Plant and Soil* 175:265–274.

Wani, S.P., K.L. Sahrawat, T.K. Sreedevi *et al.* 2007. Carbon sequestration in the semi-arid tropics for improving livelihoods. *International Journal of Environmental Studies*, 64(6): 719–727.

Wani, S.P., P. Singh, R.S. Dwivedi, R.R. Navalgund, and A. Ramakrishna. 2005. Bio-physical indicators of agro-ecosystem services and methods for monitoring the impacts of NRM technologies at different scale. In *Natural resource management in agriculture: methods for assessing economic and environmental impacts,* ed. B. Shiferaw, H.A. Freemen, and S.M. Swinton, 97–123. Wallingford, UK: CAB International.

Wani, S.P. T.K. Sreedevi, J. Rockström, and Y.S. Ramakrishna. 2009. Rainfed agriculture–Past trends and future prospects. In *Rainfed agriculture: unlocking the potential,* ed. S.P. Wani., J. Rockström, and T. Oweis, 1–35. Comprehensive Assessment of Water Management in Agriculture Series. Wallingford, UK: CAB International.

WHO. 2000. Gender, health and poverty. *Factsheet No. 25.* World Health Organization. (http://www.who.int/mediacenter/factsheets/fs251/en/)

World Bank. 2005. *Agricultural growth for the poor: An agenda for development.* Washington, DC, USA: The International Bank for Reconstruction and Development/The World Bank.

Chapter 3

Watershed development for rainfed areas: Concept, principles, and approaches

Suhas P. Wani, B. Venkateswarlu, and V.N. Sharda

3.1 INTRODUCTION

Land, water, and vegetation are the natural resources, which provide food, feed, fiber, and fuel needs for the survival of human beings. However, the growing biotic pressure and overexploitation of the natural resources are leading to their accelerated degradation, resulting in reduced productivity. The sustainable management of natural resources is the key for the sustenance and well-being of human beings. Water is a finite resource and an elixir of life; however, water is becoming scarce due its overexploitation to meet the demands of the ever increasing demographic pressure. Agriculture is a major consumer (75–80%) of water for food production globally. For meeting the food demand of the growing global population by 2025, it is estimated that additional $2000\,km^3$ water will be required with the current practices of food production (Falkenmark 1986). An integrated approach to rainwater management is necessary, where the links are addressed between investments and risk reduction, between land, water, and crop, and between rainwater management and multiple livelihood strategies. The conservation linked development of vital natural resources on a sustained basis without impairing its productivity for the future generation is the need of the hour. In this context, watershed scale becomes very effective and handy to manage water and land resources effectively, particularly in the drought-prone rainfed areas, which are the hot-spots of poverty, malnutrition, and water scarcity and are prone to severe land degradation (Wani *et al.*, 2003a, 2009; Rockström *et al.*, 2007, 2009, 2010). For sustainable development of rainfed agriculture in tropical Asia and Africa, small catchment or watershed management approach is recommended for the sustainable development and to achieve food security through enhanced green water (rainwater stored as soil moisture) and blue water (runoff water harvested in tanks and groundwater) use efficiency (Wani *et al.*, 2002, 2009; Rockström *et al.*, 2007, 2010).

3.2 WATERSHED CONCEPT

A watershed is a catchment area from which all water drains into a common point, making it an attractive hydrological unit to manage water and soil resources through science-based technical management options. Along with water and soil, biodiversity also can be effectively managed at the watershed scale to harness the agroecological potential on a sustainable basis (Wani and Garg 2009). Watershed is a spatial unit that includes diverse natural resources (soil, water, trees, biodiversity, etc.) that are unevenly distributed within a given geographical area (Knox and Gupta 2000; Johnson *et al.*, 2002). The water flowing in a watershed interconnects upstream and downstream areas, and provides life support to rural people holding unequal use rights, making people and animals an integral part of watersheds (Wani *et al.*, 2010). Activities of people/animals affect the health and sustainability of watersheds and vice versa. Clearly, watersheds are geologically, ecologically, and socially complex geographical units characterized by temporal and spatial interdependence among resources as well as resource users. Watersheds are also inhabited by socially and economically heterogeneous groups of people located at different points along the terrain, creating potential conflicts in the resource use among those on the upper, middle, and lower reaches of the catchment. This implies that the effectiveness of the watershed interventions will depend on the ability to treat the entire hydrological landscape, following the ridge to valley approach and not just a part of it. In a watershed, along with the on-site impacts, there are off-site impacts also. The quality and status of land, water, and vegetation vary as per the toposequence position; and appropriate management practices are needed for their development and sustainable use as per their capability. For practical purposes, a systematic scientific and rational approach would be to use watershed as the unit of planning and development, and to achieve this objective a framework for the watershed is a prerequisite.

The terms catchment, sub-catchment, and watershed are often synonymously employed, defined by a single river system and further grouped into macro, meso, and micro level in a hierarchical system for management using a codification system linking different levels. It is thus essential to have a hierarchical system of delineating bigger hydrological units into watersheds, and also a codification system needs to be developed so that each watershed could be identified as an individual entity without losing linkage with the bigger units, i.e., catchments, sub-catchment, etc., to which it belongs. The size of the smallest hydrological unit while delineating a bigger system into watersheds/sub-watersheds/micro-watersheds could be restricted to viable size dictated by the working feasibility. In general, the area under different categories covered is 30–50 km^2 for sub-watershed, 10–30 km^2 for mini-watershed, 5–10 km^2 for micro-watershed, and 500–5000 ha for the implementation unit.

The concept of stream order is often followed in the geomorphic analysis of the natural drainage system. Every stream, tributary, or river has an associated watershed, and small watersheds aggregate together to become a larger watershed. Water travels from headwater to the downward location and meets with similar strength of stream, then it forms one order higher stream as shown in Figure 3.1.

The stream order is a measure of the degree of stream branching within a watershed. Each length of stream is indicated by its order (for example, first-order, second-order, etc.). The start or headwaters of a stream, with no other streams flowing into it, is called the first-order stream. First-order streams flow together to form a second-order stream. Second-order streams flow into a third-order stream and so on. Stream order describes the relative location of the reach in the watershed. Identification of stream order is useful to assess the amount of water availability in reach and its quality; stream orders are also used as criteria to divide a larger watershed into smaller units. Moreover, the criteria for selecting the watershed size also depend on the objectives

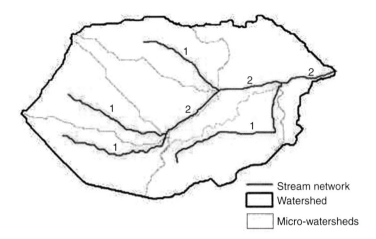

Figure 3.1 A large watershed divided into six micro-watersheds based on stream order (Note: Numbers on the stream network show the stream order of respective stream) (Source: Wani and Garg 2009)

for development and terrain slope. A large watershed can be managed in plain valley areas or where forest or pasture development is the main objective (Singh 2000). In hilly areas or where intensive agriculture development is planned, the size of watershed relatively preferred is small.

Moreover, because of the lateral and downhill movement of soil and water resources unilateral action taken by any single resource user may impose positive or negative consequences (externalities) on any other resource user. The ability to exclude or prevent these externalities is determined by the nature of property rights held by the resource users. When negative externalities are difficult to exclude or prevent at a low cost, some of the production and resource use decisions for certain resources may fall under the control of other agents. When the externalities are negative, the production or resource use levels may be socially supra optimal. The reverse is true for desirable externalities for which the individual resource users are not fully compensated. The ability to internalize these kinds of mutual spillover effects resulting from spatial and temporal interdependence among resource users requires interventions mediated through targeted policies and institutional incentives that encourage cooperation and collective action. Fragmented land ownership and the settlement patterns coupled with unequal access and use rights create conflict and diverging interests. This reduces the incentives for cooperation and increases the transaction costs involved in organizing the resource users for collective action (Shiferaw *et al.*, 2009).

However, a participatory framework of watershed development calls for a different approach, indicative of macro- and micro-level of delineation encompassing different communities and administrative units, avoiding social conflicts. Considering the role of communities and importance of their participation for sustainable development of the watersheds in India, watersheds of 500 ha were used for development in India as community watersheds covering one village or a cluster of inhabitations.

However, through the meta analysis of available data, it was found that small watersheds were not as effective in terms of economic, environmental, and social impacts as were the larger watersheds of >1200 ha (Joshi *et al.*, 2005, 2008; Wani *et al.*, 2008a). The Common Watershed Guidelines released by the Government of India (2008) adopted larger size (1000–5000 ha) of watersheds by developing them in clusters. Each of the big drainage system is divided and sub-divided through stages using different codes to indicate various stages starting with macro-level and going down to micro-level.

In recent years, watershed management has become the focal point of agricultural and rural development in rainfed areas with the objective of livelihood enhancement. Watershed development is fundamentally the creation of new opportunities both in an institutional sense and ecologically feasible manner. The ultimate indicator of success is the ability of communities to take advantage of new opportunities and the extent to which benefits are sustained in the long-term. Today, the concept of integrated watershed management (IWM) is recognized to go beyond traditional technical interventions for soil and water conservation, to include multiple crop-livestock-tree, and market related innovations that support and diversify livelihoods to build the resilience against changes in the future, including those by globalization and climate change (Wani *et al.*, 2008b). The concept ties together the biophysical notion of a watershed as a hydrological unit with that of the community and institutional factors that regulate the demand and determine the viability and sustainability of such interventions. The hydrological approach helps to identify the appropriate technical interventions on the supply side, while the village or community-based planning and implementation is fundamental for creating institutions for community empowerment and sustainability on the demand side (Shiferaw *et al.*, 2009). The landscape level, but community-based IWM interventions create synergies among targeted technologies, policies, and institutions that improve productivity, resource use sustainability, and market access for the resource users.

Integrated watershed management has become an approach integrating sustainable management of natural resources through collective action of the resource users for improving livelihoods of people in harmony with nature rather than a mere unit for managing natural resources and has shown the potential for scaling-out the benefits ensuring community participation, due largely as a result of the tangible economic benefits as well as capacity development through knowledge sharing (Farrington and Lobo 1997; Wani *et al.*, 2000, 2003a).

3.3 IMPORTANCE OF LAND USE PLANNING INWATERSHED DEVELOPMENT

The unevenly distributed, diverse, and interconnected natural resources and interdependence of human beings and animals for their living and sustainability calls for proper planning for the development, management, and use of land resources. Adinarayana (2008) employed Watershed Management Information System (WATMIS) to evaluate the agroecological characteristics using primary data, soil erosion assessment, and aspects of conservation management. Data from various sources such as National Bureau of Soil Survey and Land Use Planning (NBSSLUP), remote sensing, groundwater, agriculture, forestry, rural development departments, and markets can be effectively used with the help of geographical information systems (GIS), simulation models (crop, water, soil loss, runoff), and bioeconometric models for the sustainable development and management of watersheds (Wani *et al.*, 2008a, 2008b, 2009).

3.4 CRITERIA FOR PRIORITIZATION OF WATERSHEDS

One of the conventional approaches for prioritization of the watershed was based on the silt yield index (SYI) method developed by the All India Soil and Land Use Survey (AISLUS), which consumed a lot of time and sizable human and financial resources. Sidhu *et al.* (1998) used these approaches and prioritized the development of detailed work plan for watersheds. To provide efficient framework of watersheds in the country, AISLUS first developed Watershed Atlas of India comprising 17 sheets at a 1:1 million scale. The country was hydrologically demarcated into six major water resource regions, 35 river basins, 112 catchments, 500 sub-catchments, and 3237 watersheds. Subsequently, Digital Watershed Atlas of India was developed by the organization for a GIS-based Web service on watershed, soil, and land information.

Andhra Pradesh Rural Livelihoods Programme (APRLP) adopted integrated watershed development approach to improve rural livelihoods in Andhra Pradesh, India and devised a nine-point selection criteria (Sreedevi and Wani 2009) for watersheds, integrating the natural resource degradation criteria with multiple deprivation criteria (social and material deprivation) to arrive at reliable indicators for both technical and social features (Table 3.1). Micro- and macro-watersheds were identified and prioritized based on the SYI indicators of land degradation due to erosion and the dependability of rainfall and evapotranspiration, which depend on the variability and deviation of rainfall. Multiple deprivation criteria include the indices of poverty, considering the multiple dimensions of poverty as reflected in deprivation of income, accessibility to services, and social status. Since APRLP took a holistic view of people towards their livelihoods and opportunities, it integrated the indices of natural resource degradation and multiple deprivations, and a matrix was drawn up where each parameter was given equal importance while selecting the watersheds. A probation period of up to 18 months was made mandatory for the capacity building plans for the primary and secondary stakeholders and the preparation of strategic (perspective plan for 5 years) and annual action plans. Thus, it is a farmer-friendly Participatory Net Planning (PNP) approach (Sreedevi and Wani 2009).

The IWM program has adopted similar criteria for prioritizing the watersheds in different states in India as well as for prioritizing and allocating the financial resources for the program. Higher priority and weightage is given to the extent of rainfed area in the state, level of poverty, drinking water and groundwater status, low crop yields, poverty index [people in the categories of below poverty line (BPL), scheduled caste (SC)/scheduled tribe (ST), etc.], area owned by small and marginal farmers, SC/ST and BPL people, contiguity to the already treated/ongoing watersheds, the extent of treatable common property resources, and willingness of the villagers to participate, contribute, and support the program.

Table 3.1 Nine-point criteria for the selection of watersheds under APRLP in Andhra Pradesh, India[a]

Parameters	Range	Mark	Weightage
% of small and marginal farmers	<25	5	
	>25 to 50	10	
	>50	15	15
% of SC/ST holdings	<10	3	
	>10 to 25	10	10
% of women organized in self-help groups and participating in the program	<20	3	
	>20 to 50	5	
	>50	10	10
Status of groundwater (m)	<10	2	
	>10 to 15	3	
	>15	5	5
Andhra Pradesh Remote-sensing Application Centre (APSRAC) prioritization	Very low	6	
	Low	12	
	Medium	18	
	High	24	
	Very high	30	30
Livestock population	<1000	2	
	>1000 to <2000	3	
	>2000	5	5
Number of families affected/involved in migration	<50	3	
	>50 to <100	5	
	>100	10	10
Contiguity	Yes	5	
	No	0	5
Availability of fallow/wasteland and common property resources for the poor to utilize usufruct (%)	<10	3	
	>10 to <20	5	
	>20	10	10
Total			100

[a]Source: Sreedevi and Wani (2009).

3.5 COMMON FEATURES OF THE WATERSHED DEVELOPMENT MODEL

Government agencies, development thinkers, donors, researchers and non-government organizations (NGOs) have gradually learnt from each other and evolved the watershed management approach (though some are ahead in the field and others deficient in some aspect or other, principally in people participation or in the science). But in general nowadays, the improved models have some or all of the following features in common (Wani *et al.*, 2008a):

- Participation of the villagers as individuals, as groups or as a whole, improving their confidence level, enabling their empowerment and ability to plan for the future with self-determination;
- Capturing the power of group action in the village, among villages, and from the federations, e.g., capturing economies of scale by collective marketing;

- Construction of basic infrastructure with contributions in cash or as labor from the community;
- Better farming techniques, notably the improved practices for the management of soil, water, diversifying the farming systems, and integrating the joint management of communal areas and forest;
- Involvement of the landless, often in providing services;
- Arrangements for the provision of basic services and infrastructure;
- Establishment of village institutions and their linking with the outside world;
- Improved relationships between men and women;
- Employment and income generation through enterprise development predominantly in but not exclusively agricultural-related activities.

And in some instances:

- The fusion of research and development (R&D) by capturing the extraordinary power of participatory technology development, including varietal selection with direct links to germplasm collections as has happened in the case of the establishment of model watersheds as well as some of the internationally funded watershed programs, for example, the APRLP project supported by Department for International Develoment (DFID), UK, the World Bank-funded Sujala Watershed Development Program in Karnataka, India, and the Integrated Watershed Development Program supported by Asian Development Bank (ADB), Manila, the Philippines in Thailand, Vietnam, China, and India implemented by International Crops Research Institute for the Semi-Arid Tropics (ICRISAT);
- A complete avoidance of corruption so that the trust is built and all the benefits are passed on to the community;
- Reduction in distress migration from the rural areas to towns in search of a livelihood.

Recent additions to the watershed model:

- A pragmatic use of the scientific knowledge as the entry point rather than money, leading to tangible economic benefit from low-cost interventions that generate rapid and substantial returns for large number of people at low level of risk. Among these are novel interventions focusing on seeds of improved cultivars, the assessment of soil fertility status, integrated pest management (IPM), micronutrients, and soil conservation and water table recharge structures (Wani *et al.*, 2003a, 2008b; Dixit *et al.*, 2007).
- A broad-based approach to income generation, involving private sector associated with scientific advances and markets; for instance, in the remediation of micronutrient deficiencies; in marketing of the products from medicinal and aromatic plants; with premium payments paid by industrial processors for aflatoxin-free maize and groundnut; with high sugar sorghum, and selected crops such as *Jatropha* and *Pongamia* sold to industry for ethanol and bio-diesel production; with the production for sale of commercial seed, hybrid varieties, and biopesticides (Wani *et al.*, 2003a; Sreedevi and Wani 2009).

- Employing remote sensing, GIS, and simulation modeling for planning, execution, and monitoring and to provide feedback to farmers; yield gap analysis and rapid assessment of the fertility status of watersheds to improve productivity (Wani *et al.*, 2009).
- Building the capacity of the formal and informal rural institutions through bottom-up approach with emphasis on owning of the program from the beginning for sustainability. The consortium partners have come up with an innovative model of "Advisory Council" (*Salaha Samithi*) in addition to the watershed committee, at the watershed level, for transparency, equity, gender related issues, and conflict resolving.
- Building productive partnership and alliance in a consortium model for conducting research and technical backstopping and for this the members of the consortium work together right from the planning stage (Wani *et al.*, 2003a, 2008b, 2009).
- Focus to build resilience in the watershed and its community against climate change and events of the post-program intervention (Wani *et al.*, 2008a).

Where implemented properly, the model has led to profound changes in the farming systems including improved food self-sufficiency, enhanced employment, commerce and income. Where indifferently executed, the approach has not brought desirable results to the community. Thus, yield gap is observed between research station and farmers' yields. Much of the difference can be captured by the implementing agencies by adopting the best practice. The more recent linking of natural resource science with private sector markets and with peoples' broader livelihoods in consultation with them, has been transforming the dynamic and success rate of the developmental efforts (Wani *et al.*, 2008a).

3.6 EVOLUTION OF WATERSHED DEVELOPMENT APPROACH IN INDIA

In the beginning, watershed development in the rainfed areas was synonymous to soil and water conservation and was achieved by constructing field bunds and structures to harvest runoff water (Singh *et al.*, 1998; Wani *et al.*, 2002). In these activities, the techno-centric, compartmental, and target-oriented approaches were followed by involving one or two departments of the Government without much coordination. It was a top-down target-based approach with hardly any involvement of the stakeholders in the planning, implementation, and maintenance of the structures and bunds. Hence, such efforts did not make headway in impacting livelihoods of the rural poor in the rainfed areas (Farrington and Lobo 1997; Joshi *et al.*, 2000; Dixit *et al.*, 2001; Kerr 2001; Wani *et al.*, 2002; Kerr and Chung 2005; Shah 2007).

Rainfall pattern in the tropical and subtropical rainfed areas is highly variable both in terms of total amount and distribution. This leads to moisture stress during critical stages of crop production and makes agriculture production vulnerable to pre- and post-production risks. Watershed development projects in India have been sponsored and implemented by the Government from early 1970s onwards. The phases in the journey through the evolution of watershed approach are shown in Figure 3.2 (Wani *et al.*, 2005, 2006a). Various watershed development programs like Drought Prone

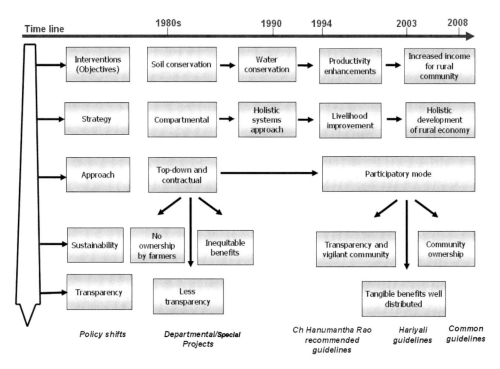

Figure 3.2 Journey through watershed approach in India (Source: Wani *et al.*, 2005, 2006a)

Area Programme (DPAP), Desert Development Programme (DDP), River Valley Project (RVP), National Watershed Development Project for Rainfed Areas (NWDPRA), and Integrated Wasteland Development Programme (IWDP) were launched subsequently in various hydro-ecological regions consistently affected by water stress and drought like situations. The entire watershed development program was primarily focused on structure-driven compartmental approach of soil conservation and rainwater harvesting during 1980s and benefited a few well to do farmers through increased groundwater availability. In spite of putting efforts for maintaining soil conservation practices (contour bunding, pits excavation, grassed waterways, etc.), the farmers used to plow through the structures from their fields. It showed that a straightjacket top-down approach cannot bring desired impact in watersheds and a mix of the individual- and community-based interventions are essential along with the productivity enhancement measures to ensure tangible benefits to small farm holders.

The integrated watershed development program with participatory approach was emphasized during the mid 1980s and in the early 1990s. This approach had focused on raising crop productivity and livelihood improvement in watersheds (Wani *et al.*, 2006b) along with the implementation of soil and water conservation measures. The Government of India appointed a committee in 1994 under the chairmanship of Prof. C.H. Hanumantha Rao, which strongly felt a need to move away from the conventional government department approach based on bureaucratic planning without involving the local communities (Raju *et al.*, 2008). The new guidelines were recommended in 1995, which emphasized collective action and community participation, including participation of the primary stakeholders through community-based organizations (CBOs), NGOs, and Panchayat Raj Institutions (PRIs) (GOI 1994, 2008; Hanumantha Rao *et al.*, 2000; DOLR 2003; Joshi *et al.*, 2008). The watershed development guidelines were again revised in 2001 (called Hariyali guidelines) to further simplify and facilitate the involvement of PRIs more meaningfully in planning, implementation, and evaluation, and in the community empowerment (Raju *et al.*, 2008). The guidelines were issued in 2003 (DOLR 2003). Subsequently, "Neeranchal Committee" (2005) evaluated the entire government-sponsored, NGO and donor-implemented watershed development programs in India and suggested a shift in focus "away from a purely engineering and structural focus to a deeper concern with livelihood issues" (Raju *et al.*, 2008).

Appreciating the fact that to reduce poverty in the rainfed areas holistic watershed development is a must (Wani *et al.*, 2003a, 2009; Rockström *et al.*, 2007, 2010), the new generation of watershed development programs are implemented with a larger aim to address the problems such as food security, equity, poverty, gender, severe land degradation, and water scarcity in the dryland areas. Hence, in the new approach, watershed, a land unit to manage water resources has been adopted as a planning unit to manage total natural resources of the area. Improving livelihoods of local communities is highlighted by realizing the fact that in the absence of them, sustainable natural resource management (NRM) would be elusive. With these considerations, the watershed programs have been looking beyond soil and water conservation into a range of activities from productivity enhancement through interventions in agriculture, horticulture, animal husbandry, livelihoods, community organization, and gender equity (Wani *et al.*, 2002; APRLP 2007; GOI 2008). This holistic approach required optimal contribution from different disciplinary backgrounds, creating a demand for multi-stakeholder agenda in the watershed development programs (Wani *et al.*, 2002, 2003b).

During 1990s, there has been a paradigm shift in the thinking of policy makers based on the learnings from the earlier programs. In India, the watershed programs are silently revolutionizing agriculture in the rainfed areas (Joshi *et al.*, 2005; Wani *et al.*, 2006b); and by 2006 (up to 10th Five Year Plan) about US\$7 billion have been invested by Government of India and other donor agencies treating 38 million ha in the country (Wani *et al.*, 2008b). During a detailed evaluation of the on-farm watershed programs implemented in the country, the ICRISAT team observed that once the project team withdrew from the villages the farmers reverted back to the earlier practices and very few components of the improved soil, water, and nutrient management options were adopted and followed. Although, the economic benefits of improved technologies were observed in the on-farm experiments, the adoption rates were quite low. Individual component technologies such as summer plowing, improved crop varieties, and intercropping were continued by the farmers. However, the soil and water conservation technologies were not much favored (Wani *et al.*, 2002).

The importance of the local community participation in the watershed programs to enhance their efficiency and sustainability has been widely acknowledged (Kerr *et al.*, 2000; Samra and Eswaran 2000; Wani *et al.*, 2002; Joshi *et al.*, 2004). As a result, through a series of policies and guidelines the responsibilities of managing the watershed programs have shifted towards the local communities. But achieving participation of the primary stakeholders has not been easy. One of the major learnings over a period of time has been that unless there is some tangible economic benefit to the community, peoples' participation does not come forth (Olson 1971; Wani *et al.*, 2002). To enhance community participation, it is necessary to achieve tangible impact of the watershed development activities. To achieve a tangible impact, it is necessary that different agencies such as research centers, development line departments of the Government, training institutions, CBOs, and NGOs come together and share their expertise in a complementary way through the convergence of approaches, actors, and actions (Wani *et al.*, 2002).

3.7 NEED FOR A HOLISTIC APPROACH FOR WATERSHED MANAGEMENT

The watershed implementing agencies have inbuilt strengths in community organization but majority of these agencies lack technical competencies in the development and management of natural resources. They depend heavily on the technical resource agencies for building capacities of their staff and the community members involved in NRM. Since different resource agencies have their compartmental specializations in specific areas, there is a lack of holistic approach in providing technical support to the NGOs, government line departments, and other implementing agencies, thereby affecting their performance in implementing the watershed programs. On the other hand, the research organizations are usually mandated to work at the individual farm level. Biophysical scientists often have limited experience in the dynamics of forming the collective action groups that is essential for watershed-based activities. However, with the approach of ultra disciplinary specialization (reductionist approach) and the lack of professional reward mechanisms in the research institutions, and disciplinary hierarchy, scientists are more comfortable to work in their own area of specialization rather than working in the multidisciplinary teams (Wani *et al.*, 2009). In projects that have been led by research centers, researchers seem to document results and findings mainly for the scientific sector (Gündel *et al.*, 2001). Focus on social organization is less in these programs, reducing their effectiveness. Government departments have their strengths in specific technical competencies and wider reach, but they lack skills in social organization. Traditionally, the watershed programs implemented by government departments have been supply-driven and target based. The Central and State governments allocated resources for watershed development. Subsequently, the officials used to identify locations and decide various activities for implementation. Such an approach did not match the needs of stakeholders in the watershed (Kerr *et al.*, 2000; Joshi *et al.*, 2004). Since these departments operate mostly in a compartmental way, integrated approach was lacking in such programs and thus desired success could not be achieved. Due to such deficiencies in capacities of the implementing agencies, most of the watershed programs failed to achieve optimal benefits (Farrington and Lobo 1997; Kerr and Chung 2005). This situation has strongly supported the idea of different agencies coming together to support watershed programs. But bringing together organizations with different strengths, weaknesses, and styles of functioning on a common platform to work together for a common cause is challenging. ICRISAT has successfully evolved a scalable model based on the 'Consortium Approach' through

ADB-supported watershed development program at Kothapally in Rangareddy district of Andhra Pradesh (Wani *et al.*, 2003a).

3.8 EVOLUTION OF THE CONSORTIUM APPROACH

ICRISAT was one of the earliest CGIAR (Consultative Group on International Agricultural Research) centers to give formal recognition in its mandate to supplement research on individual crops with research in farming systems. Watershed-based research was an example of interdisciplinary research even before the term assumed significance (Shambu Prasad *et al.*, 2005, 2006). This interdisciplinary research, over the years, has shaped up into an Integrated Genetic Natural Resources Management (IGNRM) approach at ICRISAT (Twomlow *et al.*, 2006). But in the beginning, ICRISAT also faced the problems of hierarchy of disciplines among scientists who were working together. After realizing the importance and potential of combining disciplinary expertise in a complementary way, such issues were sorted out which gave rise to the idea of the Consortium Approach based on the success of multidisciplinary approach at the research station (Wani *et al.*, 2003a).

The Consortium is a convergence of agencies/actors/stakeholders who have a significant role to play in the watershed development project. Facilitated by a leader/leading organization, member-organizations prepare common plans and work towards achieving the agreed common objectives. After witnessing ICRISAT's pioneering and quality work and its results, many agencies approached ICRISAT for sharing of the knowledge/approach/technology in their areas. It was decided to work along with the reputed NGOs in the on-farm watersheds, which much helped in strengthening the idea of the Consortium Approach.

The ADB-supported project enabled ICRISAT to test the integrated watershed model at Kothapally village of Rangareddy district in Andhra Pradesh, to attempt to minimize the gap between research findings and on-farm development. The purpose of the work was also to adopt the learning loop in the planning of strategic research based on the participatory model in the research for development. There was also a request from the Government of Andhra Pradesh to demonstrate the benefits by increasing crop productivity through watershed approach in the rainfed areas under farmers' conditions. The beginning itself of this work was highly encouraging as there was a demand for demonstrating the success of the new approach of watershed development on farmers' fields and space was created for innovations in the ongoing DPAP under the watershed program by the Government of Andhra Pradesh.

For this model, as opposed to single institution based approach, relevant organizations were identified and brought into the network to form a consortium of institutions for technical backstopping of the project. ICRISAT, M Venkatarangiah Foundation (MVF), an NGO, Central Research Institute for Dryland Agriculture (CRIDA), National Remote Sensing Agency (NRSA), and DPAP, now called the District Water Management Agency (DWMA), Rangareddy district administration of the Government of Andhra Pradesh along with farmers of the watershed formed the consortium (Figure 3.3) (Wani *et al.*, 2003a).

The first success of the new approach was evident when more number of farmers came forward to undertake the participatory evaluation of the technologies for which

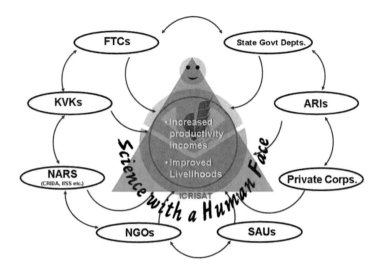

Figure 3.3 A pictorial representation of different partners in the Adarsha Watershed Consortium (See color plate section)

except for the knowledge, farmers had to pay for inputs in cash or kind. Farmers obtained threefold increase in the yields of pigeonpea and other crops in the first rainy season itself (Table 3.2). During the second year, people from four villages surrounding Kothapally came to ICRISAT and asked for technical help, promising that they would show similar or better results than shown by Kothapally farmers, in a shorter period. This indicated to the Consortium team members that the approach was self-replicating as people from the surrounding villages saw tangible benefits from the approach. ICRISAT and DWMA of Rangareddy district decided to provide technical support and necessary inputs on a cost basis to these four villages. True to their words, the villagers demonstrated the benefits in terms of doubling their crop productivity (Table 3.3). The model has become a success story and henceforth the model has been suitably adapted and scaled-up/out in many other locations.

3.9   COMPONENTS OF INTEGRATED WATERSHED MANAGEMENT

3.9.1   Entry Point Activity

Entry Point Activity (EPA) is the first formal project intervention, which is undertaken after the transect walk, selection, and finalization of the watershed. It is highly recommended to use the knowledge-based EPA to build the rapport with the community (Wani *et al.*, 2002, 2008a). Direct cash-based EPA must be avoided as such activities give a wrong signal to the community at the beginning of the implementation of various interventions. A detailed discussion on the knowledge-based EPA to build rapport with the community ensuring tangible economic benefits to the community members is provided by Dixit *et al.* (2008).

Table 3.2 Average crop yields with different practices at Adarsha watershed, Kothapally, 1999–2009[a]

Cropping system	Yield (kg ha[-1])													
	Before 1998	1999–2000	2000–01	2001–02	2002–03	2003–04	2004–05	2005–06	2006–07	2007–08	2008–09	Mean	CV (%)	SE ±
Improved system														
Sole maize	–	3250	3760	3300	3480	3920	3420	3920	3630	4680	4810	3820	17.8	80
Maize/Pigeonpea intercrop system	–	5260	6480	5600	5650	6290	4990	6390	6170	6120	6680	5960	16.7	116
Sorghum/Pigeonpea intercrop system	–	5010	6520	5830	–	5780	4790	5290	5310	–	–	5500	13.4	154
Sole sorghum	–	4360	4590	3570	2960	2740	3020	2860	2500	–	–	3330	23.9	141
Farmers' practice														
Sole maize	1500	1700	1600	1600	1800	2040	1950	2250	2150	–	–	1890	17.2	53
Sorghum/Pigeonpea intercrop system	1980	2330	2170	2750	3190	3310	3000	3360	3120	–	–	2900	19.2	110
Hybrid cotton	–	2295	7050	6600	6490	6950	–	–	–	–	–	5880	37.0	511
BT cotton	–	–	–	–	–	–	–	6210	5590	7310	9380	7120	26.1	315
Mean		3477	4970	3833	4018	4814	3651	4584	4320	6268	7396			
CV (%)		11.9	31.4	10.7	8.0	14.5	20.3	10.8	12.2	16.7	16.2			
SE ±		415	1559	410	323	698	742	495	525	1049	1201			

[a]Source: Updated from Wani et al. (2008b).
In Sorghum/Pigeonpea intercrop system with farmers' practice, the improved pigeonpea variety ICPL 87119 was grown along with local sorghum variety Pacha Jonna from 2001; pigeonpea variety used earlier was discontinued as it was highly susceptible to fusarium wilt.

Table 3.3 Crop response to the improved management practices in 98 farmers' fields in four villages around Kothapally watershed, Andhra Pradesh, India in 2001 rainy season and 2001–02 postrainy season[a]

			Maize yield (kg ha^{-1})	
Cropping system	Farmers' practice (kg ha^{-1})	Improved practice (improved seed + management) (kg ha^{-1})	Farmers' practice	Improved practice
Maize/Pigeonpea				
Maize	1900	4365	1900	4365
Pigeonpea	350	1130	1240	3995
Sorghum/Pigeonpea				
Sorghum	1200	2725	1755	3990
Pigeonpea	330	1185	1170	4190
Maize/Chickpea				
Maize	2200	4800	2200	4800
Chickpea	650	1085	2380	3870

[a] Derived from Wani *et al.* (2009).

Table 3.4 Farmers' fields deficient in plant nutrients in various states of India[a]

			Farmers' fields (%) deficient[b]					
State	No. of districts	No. of farmers	OC	Av P (ppm)	Av K (ppm)	Av S (ppm)	Av B (ppm)	Av Zn (ppm)
Andhra Pradesh	11	3650	76	38	12	79	85	69
Gujarat	1	82	12	60	10	46	100	85
Jharkhand	2	115	42	65	50	77	97	71
Karnataka	10	27500	70	46	21	84	67	55
Kerala	3	28	11	21	7	96	100	18
Madhya Pradesh	12	341	22	74	1	74	79	66
Rajasthan	9	421	38	45	15	71	56	46
Tamil Nadu	5	119	57	51	24	71	89	61
Total	53	32256	69	45	19	83	70	58

[a] OC = Organic carbon; Av P = Available phosphorus, Av K = Available potassium, Av S = Available sulfur, Av B = Available boron, Av Zn = Available zinc.
[b] Below critical limit for a particular nutrient.

During the process of scaling-out of the consortium model for APRLP of the Government of Andhra Pradesh supported by DFID, UK, in the states of Rajasthan, Madhya Pradesh, Gujarat and Karnataka, supported by Sir Dorabji Tata Trust, Mumbai, ADB, Sujala Watershed (program of Government of Karnataka supported by World Bank), the baseline characterization of soils in the watersheds was used as a knowledge-based EPA along with other EPAs. The analysis of a large number of soil samples from farmers' fields in various states of India revealed that soil resource base in the tropics is not only thirsty, but also hungry, especially for micronutrients like zinc (Zn) and boron (B) and secondary nutrients like sulfur (S), along with macronutrients including nitrogen (N) and phosphorus (P) (Table 3.4). About 80–100% farmers' fields

Table 3.5 Increase in crop productivity with micronutrient (MN) amendments in 50 watersheds in three districts of Andhra Pradesh, 2002

| Crop | Average grain yield (kg ha^{-1}) | | Yield increase (%) over control |
	Control	MN treatment[a]	
Maize	2800	4560	79
Mung bean	770	1110	51
Castor	470	760	61
Groundnut (pod)	1430	1825	28

[a]Micronutrients applied: boron (0.5 kg ha^{-1}), sulfur (30 kg ha^{-1}), and zinc (10 kg ha^{-1}).

Table 3.6 Increase in crop productivity with micronutrient amendments and recommended dose of macronutrients in 50 nucleus watersheds in Andhra Pradesh, 2003[a]

| Crop | Yield (kg ha^{-1}) | | | | | |
	Control (C)	Sulfur (S)	Boron (B)	Zinc (Zn)	C + SBZn	C + NP + SBZn
Maize	2790	3510 (26)	3710 (33)	3710 (33)	4140 (49)	4890 (75)
Groundnut	830	930 (12)	1000 (20)	1060 (27)	1230 (48)	1490 (78)
Mung bean	900	1210 (33)	1130 (24)	1320 (46)	1390 (54)	1540 (70)
Sorghum	900	1190 (32)	1160 (29)	1330 (47)	1460 (62)	1970 (119)

[a]Source: Rego *et al.* (2007).
Figures in parentheses indicate yield increase (%) over control.

in several states of India were found critically deficient in Zn, B, and S (Sahrawat *et al.*, 2007). Subsequent follow-up participatory research and development (PR&D) trials in 50 micro-watersheds in Andhra Pradesh with amendments of Zn, B, and S increased yields by 30–174% for maize, 35–270% for sorghum, 28–179% for groundnut, 72–242% for pearl millet, and 97–204% for pigeonpea (Tables 3.5 and 3.6) (Rego *et al.*, 2007).

3.9.2 Land and water conservation practices

The implementation of soil and water conservation practices is basic in the watershed management program. These practices can be divided into two main categories: (1) In-situ, and (2) Ex-situ. Land and water conservation practices made within an agricultural field like construction of contour bunds, graded bunds, field bunds, terrace scoops, broad-bed and furrow (BBF) system and other soil moisture conservation practices are known as in-situ water management interventions. These practices minimize land degradation, improve soil health, and increase soil moisture availability and groundwater recharge. The construction of check-dams, farm ponds, and gully control structures, and pits excavation across the stream channel are known as ex-situ management interventions (Figure 3.4). Ex-situ watershed management practices reduce the peak discharge in order to reclaim gully formation and harvest substantial amount of runoff, which increases groundwater recharge and irrigation

Figure 3.4 Ex-situ water management in Andhra Pradesh, India. (left) A dugout farm pond at Guntimadugu watershed in Kadapa; (right) Mini percolation tank in Kothapally (See color plate section)

potential in the watersheds. Soil and water conservation measures for sustainable watershed management are dealt in detail in this volume by Pathak *et al.*

3.9.3   Integrated pest and nutrient management

Water alone cannot increase crop productivity to its potential level without other interventions. A balanced nutrient supply along with adequate moisture availability and pest and disease-free environment can increase agricultural production extremely compared to unmanaged land. Integrated nutrient management (INM) involves the integral use of organic manure, biological inputs such as biofertilizers, crop straw, and other plant and tree biomass materials along with the application of chemical fertilizer (both macro- and micronutrients). Integrated pest management involves the use of different crop pest control practices like cultural, biological, and chemical methods in a combined and compatible way to suppress pest infestations. Thus, the main goals of INM and IPM are to maintain soil fertility, manage pests and the environment so as to balance cost, benefit, public health, and environmental quality.

3.9.4   Farmers' participatory research and development trials

The PR&D approach helps in empowering and capacity building of the farmers should be an integral part of any watershed development program. Farmers' participatory selection of improved crop cultivars in 150 micro-watersheds of APRLP in five districts of Andhra Pradesh resulted in identification of improved cultivars of sorghum, pearl millet, maize, castor, green gram (mung bean), groundnut, pigeonpea, and chickpea (Table 3.7). Through this approach farmers are able to identify location specific material considering the traits, which researchers would not have considered while selecting the cultivars. Further, to ensure the availability of the seeds of improved cultivars, self-help groups (SHGs) in the watershed villages are trained to handle village seed banks (Dixit *et al.*, 2005; Ravinder Reddy *et al.*, 2007). Trained farmers undertook seed production using breeders' seeds for sowing, and with the help of consortium

Table 3.7 Farmers' participatory evaluations for productivity enhancement in watersheds under APRLP in Andhra Pradesh, India during 2002–04

District	Watershed villages	Crop	No. of trials	Cultivars	Yield (kg ha⁻¹) Farmers' practice	Best-bet	Yield gain (%)
Kurnool, Nalgonda, Mahabubnagar	17	Castor	41	Kranthi	780	1240	59
Mahabubnagar, Nalgonda	22	Maize	40	Ratna 2232	2770	4510	63
Kurnool	13	Groundnut/ Pigeonpea	53	ICGS 76, ICGV 86590	775	1320	70
Kurnool	19	Sole groundnut	52	ICGS 76, ICGV 86590	1075	1605	49
Kurnool	2	Chickpea	34	ICCV 37	1370	1930	41
Anantapur	19	Sole groundnut	35	ICGS 76, ICGV 86590	770	1100	43

partner farmers maintained their purity. The village seed banks are very effective in overcoming the bottleneck of availability of good quality seed in the villages particularly of improved varieties of cereals such as pearl millet and sorghum, and legumes such as groundnut, chickpea, and pigeonpea, which the private seed companies do not handle.

The Government of Andhra Pradesh has scaled-up this initiative by providing ₹100,000 (US$2200) as a revolving fund to each SHG and organizing breeder or foundation seeds for the SHGs. In all, 200 village seed banks are operating in the state (Shanti Kumari 2007). In all, 255 farmers' participatory evaluation trials with improved cultivars of castor, maize, groundnut, sorghum, and chickpea along with improved nutrient management showed 41–70% increase in crop yields over farmers' management practice (Table 3.7).

In 208 watersheds in Asia, yields of several crops increased by 30 to 242% over the baseline yields varying from 500 to 1500 kg ha⁻¹. Recently under the World Bank aided Sujala-ICRISAT initiative in 22 villages in five districts of Karnataka, the results from 232 on-farm PR&D trials showed 56–198% increase in productivity of groundnut, maize, finger millet, sunflower, and other crops (Table 3.8).

3.9.5 Crop diversification and intensification of crops and systems

Crop diversification refers to bringing about a desirable change in the existing cropping patterns towards a more balanced cropping system to reduce the risk of crop failure; and crop intensification is the enhancement of cropping intensity and production to meet the ever increasing demand for food in a given landscape. Watershed management puts emphasis on crop diversification and sustainable intensification through the use of advanced technologies, especially good variety of seeds, balanced fertilizer application, and supplemental irrigation.

Table 3.8 Farmers' participatory evaluations for productivity enhancement in watersheds under ICRISAT-Sujala project in Karnataka, India during 2005–06

District	Watershed villages	Crop	No. of trials	Cultivars	Yield (kg ha⁻¹) Farmers' practice	Best-bet	Yield gain (%)
Kolar and Tumkur	7	Groundnut	63	JL 24, ICGV 91114, K1375, K6	915	2260	146
Kolar and Tumkur	9	Finger millet	62	MR 1, L 5, GPU 28	1154	1934	67
Chitradurga	2	Sunflower	30	KBSH-41, KBSH-44, GK 2002	760	2265	198
Chitradurga and Haveri	4	Maize	49	PA 4642, GK 3014	3450	5870	70
Haveri	4	Sole groundnut	16	ICGV 91114	1100	1716	56
Dharwad	4	Soybean	12	JS 335, JS 9305	1350	2470	83

Farmers in watersheds in northern Vietnam diversified maize-based systems by including groundnut and vegetable crops, obtained increased productivity as well as income (Table 3.9). The inclusion of groundnut, a legume, reduced inorganic N fertilizer requirement for maize and also increased yield by 18%. Soil and water conservation measures such as staggered contour trenching, planting of *Gliricidia*, or pineapple vegetative border, rainwater harvesting pits and loose boulder gully control structures on the sloping lands, improved water availability in open wells (Figure 3.5) and enabled the farmers to grow high-value watermelon crop with the highest benefit-cost ratio amongst the cropping systems (Table 3.10). In the Tad Fa and Wang Chai watersheds of Thailand, the farm incomes increased by 45% within three years (US$1,195 per cropping season). The Lucheba watershed in Guizhou, China transformed its economy through crop-livestock integration with buckwheat as an alley crop that controlled soil erosion, provided fodder, and increased per capita income from US$200 to US$325 in two years. The implementation of improved soil, water, nutrient, and crop management practices reduced runoff and soil loss in the nucleus micro-watersheds in Vietnam, China, Thailand and India (Table 3.11).

3.9.6   Use of multiple resources

Farmers solely dependent on agriculture hold high uncertainty and risk of failure due to various extreme events, pest and disease attack, and market shocks. Therefore, integration of agricultural (on-farm) and non-agricultural (off-farm) activities is required at various scales for generating consistent source of income and support for farmers' livelihood. For example, agriculture, livestock production, and dairy farming together can make a more resilient and sustainable system compared to adopting agricultural production alone. The products or by-products of one system could be utilized for the other and vice-versa. In this example, biomass production (crop straw) after crop harvesting could be utilized for livestock feeding and dung/excreta obtained from livestock could be used for energy production through biogas plants, and the plant nutrient rich

Table 3.9 Crop yields as influenced by best-bet options in Andhra Pradesh and Karnataka[a]

District	Watershed	Grain yield (t ha^{-1})		
		Improved practice	Traditional practice	Yield advantage (%)
Andhra Pradesh				
Nalgonda	Kacharam	4.40	1.68	162
	D. Gudem	2.96	2.25	32
	K. Gudem	3.83	2.34	64
	Sadhuvelli	4.02	2.84	42
	Gouraipalli	3.85	1.91	102
Mean		3.81	2.20	73
Mahabubnagar	Sripuram	5.76	4.44	30
	Uyyalawada	3.90	2.02	93
	Aloor	4.37	2.40	82
	Nallavelli	5.81	4.27	36
	Vanapatla	5.92	4.31	37
	Naganool	5.64	4.20	34
	Malleboinpally	3.89	1.62	140
	Sripuram	8.32	3.04	174
	Naganool	8.00	3.12	156
	Vanapatla	8.39	5.52	52
	Gollapally	4.73	3.56	33
Mean		5.88	3.50	68
Grand mean		5.24	3.10	69
Karnataka				
Kolar and Tumkur (Groundnut)		2260	915	247
Kolar and Tumkur (Finger millet)		1934	1154	167
Chitradurga (Sunflower)		2265	760	298
Chitradurga (Maize)		5870	3450	170
Haveri (Sole groundnut)		1720	1100	156
Dharwad (Soybean)		2470	1350	183
Mean		2753	1454	203

[a]Source: Derived from Sreedevi and Wani (2009).

slurry from biogas plants can be applied to agricultural plots to maintain soil fertility. The integrated system includes horticulture plantation, aquaculture, and animal husbandry at an individual farm, household, or the community scale. In all the community watersheds, the equity issues are addressed through productivity enhancement and income-generating activities in addition to the normal soil and water conservation measures.

3.9.7 Capacity building

Watershed development requires multiple interventions that jointly enhance the resource base and livelihoods of the rural people. This requires capacity building of all the stakeholders from the farmer to the policy makers. Capacity building is a process

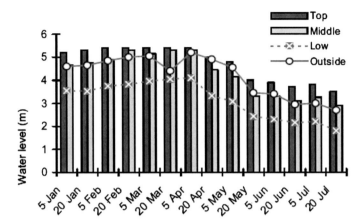

Figure 3.5 Groundwater levels in the open wells in Thanh Ha watershed in Vietnam during 2004

Table 3.10 Economics of crops grown in Thanh Ha Commune, Ho Binh Province, Vietnam

Crop	Area (%)	Yield (t ha^{-1}) Average	Range	Income (US$)	Benefit-cost ratio
Maize	83	3.4	0.9–7.0	421	1.41
Watermelon	6	17.8	10.0–36.0	2015	1.73
Sugarcane	8	58.3	20.0–83.0	1270	1.06

Table 3.11 Seasonal rainfall, runoff, and soil loss from different benchmark watersheds in Thailand and India

Watershed	Seasonal rainfall (mm)	Runoff (mm) Treated	Untreated	Soil loss (t ha^{-1}) Treated	Untreated
Tad Fa, Khon Kaen, NE Thailand	1284	169	364	4.21	31.2
Kothapally, Andhra Pradesh, India	743	44	67	0.82	1.90
Ringnodia, Madhya Pradesh, India	764	21	66	0.75	2.2
Lalatora, Madhya Pradesh, India	1046	70	273	0.63	3.2

to strengthen the abilities of people to make effective and efficient use of resources in order to achieve their own goals on a sustained basis (Wani *et al.*, 2008b). Unawareness or ignorance of the stakeholders about the objectives, approaches, and activities is one of the reasons that affects the performance of the watersheds (Joshi *et al.*, 2008). Capacity building programs focus on the construction of low-cost soil and water conservation methods, production and use of biofertilizers and biopesticides, income-generating activities, livestock-based activities, wasteland development, organizing groups and promoting collective action, participatory monitoring, social auditing, leadership skills, and market linkage for primary stakeholders. Clear understanding of strategic planning, monitoring, and evaluation mechanism and other expertise in the field of science and management is essential for the government officials and policy makers. The stakeholders should be aware about the importance of various activities and their benefits in terms of economics, social, and environmental factors. Therefore, organizing various training programs at different scales is important for watershed development.

3.10 KEY FEATURES OF FACILITATING THE CONSORTIUM APPROACH

The key features of the consortium approach for watershed management are described by Wani *et al.* (2009).

3.10.1 Need for a common goal – team building

Working in partnership is successful only if all the members share a common goal. For the consortium approach, ICRISAT attempted to achieve this by identifying important institutions whose objective is to enhance agricultural productivity and income and reduce rural poverty, and which are working in the area of watersheds. A series of team building workshops addressed the following objectives:

- Develop a common vision of the watershed development program among consortium partners.
- Inculcate a team spirit among the members to achieve the goal of sustainable NRM for improved rural livelihoods.
- Develop an understanding of and appreciation for the efforts and initiatives taken up by various teams.
- Discuss and develop action plans for the desired impacts.
- Develop a combined strategy to up-scale the impact to neighboring watersheds.

3.10.2 Building on the strengths

The consortium's main principle is to harness the strengths of the partners and overcome the weaknesses. This principle must be ingrained amongst all the partners and the strengths of each partner's valuable inputs need to be highlighted to ensure the feeling of importance.

3.10.3 Institutionalization of partnerships

The process of institutionalizing partnership began with the identification of suitable scientists with required expertise and necessary institutions for the project to achieve its goal. This approach was found to be more effective than that based on identifying organizations first and then trying to find people within those organizations who can get represented in the consortium. While being part of the consortium, participating organizations appreciated strengths of each other and rapport was built. This collaborative spirit has been shared in many other projects that followed.

3.10.4   Internal and external institutional arrangements

For facilitating the functioning of the consortium, there is a need to put in place an institutional mechanism, both internal and external, to review the progress of the project from time to time and to take necessary action (Wani *et al.*, 2009).

3.10.5   Dynamic and evolving

The consortium approach is not a static model but should be adapted based on field situation and requirements. It provides the philosophy and framework, while specific components need to be added to make it relevant as per the situation. In addition to the critical stakeholders such as NGOs, national agricultural research systems (NARS), State and Central Government line departments, and farmers' organizations, based on the need, relevant private industries can also be brought into the consortium.

3.10.6   Scaling-up/out the approach

Following the success of the model, the Consortium Approach has been scaled-up to several locations. Starting with different districts in Andhra Pradesh through APRLP, other states in India with financial support from the Sir Dorabji Tata Trust, Sir Ratan Tata Trust, and Government of India have established watershed sites as sites of learning. Also with the support from ADB, Bureau of Agriculture, Government of the Philippines watersheds have been established in India, China, Vietnam, and Thailand. There has been spill over effects of the learnings of this approach in Africa, particularly in Eastern Africa through ICRISAT's association with the Association for Strengthening Agricultural Research in Eastern and Central Africa (ASARECA). The South-South collaboration between Indian Council of Agricultural Research (ICAR) and ASARECA in the area of integrated watershed management is facilitated by ICRISAT and International Water Management Institute (IWMI).

3.11   ADVANTAGES OF CONSORTIUM APPROACH

The main advantages of the approach include synergy and creativeness in the tackling of NRM challenges for which solutions are rarely found with a single discipline expertise; for example, in the management of livestock-fisheries-agricultural systems along with credit-markets and institutions. In the Consortium Approach where a multidisciplinary team addresses the problem situation, there is a potential for creative thinking and new ideas, which benefit the farmers as well as researchers and developmental workers (Wani *et al.*, 2008b).

3.11.1   Sustainability

The Consortium Approach facilitates members of the network to have ownership of the objectives of the program. This leads to optimal contribution from diverse disciplinary backgrounds providing a holistic systems approach. As a result solutions for problems are effective. Activities are planned and are demand driven, and implementation is in a participatory manner leading to effective solutions with good prospects of their being sustainable initiatives.

3.11.2 Cost-effectiveness

At the time of project implementation, working linkages are established among various actors in the consortium. This ensures quick access to relevant people when primary stakeholders encounter a situation, leading to timely solutions. One of the main issues in NRM work is the involvement of various departments independently and this in many cases results in the duplication of work. In the Consortium Approach, each of the actors knows what other departments are doing. So there is less chance for duplication of the work.

3.11.3 Win-win solution through empowerment of partners

The Consortium Approach allows members to learn from one another. It spreads interdisciplinary knowledge among partners. Strengths of each of the partners are harnessed and help is provided mutually by partners to get over their weaknesses. When there is an effort to build upon strengths of each of the partners, weaknesses get covered with strengths of other partners. In the team, biophysical scientists not only started offering solutions for issues related with other related disciplines but also got sensitized with socioeconomic, gender, and institutional issues. One team became more cohesive overcoming conventional disciplinary hegemony (Wani *et al.*, 2008c).

3.11.4 Rapid scaling-up

Many studies on NRM indicated that it is important to work with different partners to facilitate scaling-up. The Consortium Approach ensures intensity and closeness in which communication and collaboration takes place among partners, which contributes to an effective scaling-up. Impact could be further enhanced through new innovative partnerships. Since different partners are involved, necessary enabling institutions and policies are put in place in a shorter time. For example, when the need to undertake research on *Jatropha* arose in Andhra Pradesh, the watershed consortium partners came together and formed the consortium in a short period and included new members as needed (Wani *et al.*, 2008c) while working in a watershed consortium model to benefit landless people through bio-diesel plantations in common property resources.

3.11.5 Change in organizational behavior

The general tendency of a researcher is to develop technology in the laboratory/research station and transfer it to the field through extension agencies. This tendency got reengineered into working closely with primary stakeholders and developing technology in a participatory way. Governmental and non-governmental extension agencies also find it worthwhile to play a role in developing the technology by listening to farmers carefully and contributing through feedback and sharing indigenous knowledge options with researchers. Different researchers within ICRISAT and other partner institutions also got sensitized in social, gender, equity, and other disciplines and this helped to overcome the disciplinary bias. Good research and management practices got internalized amongst the partners (Wani *et al.*, 2008c).

3.11.6   Public–private partnerships are facilitated (multiplier effect)

Backward and forward linkages are important to enhance income and agricultural production in rural areas. Private entrepreneurs came forward to join the consortium for harnessing the opportunity. For example, during baseline characterization, widespread deficiency of not only B, Zn, and S but also N and P was observed in 80–100% of farmers' fields (Sahrawat *et al.*, 2007). Farmers' participatory trials with amendments of deficient nutrients showed substantial yield increases and enhanced incomes (Rego *et al.*, 2007). However, the availability of B and other micronutrients in remote villages became an issue. Borax Morarji Ltd., producer of B fertilizers in India, came forward to join the consortium to ensure the availability of B fertilizers in the villages through SHGs. Similarly, for handling the marketing of produce and processing, various industries came forward to join the consortium; for example, in the case of bio-diesel initiative, a public–private partnership amongst GTZ-Southern Online Bio-Technology (SBT) and ICRISAT is ongoing under which SBT is operating $40\,kl\,d^{-1}$ bio-diesel plant in Nalgonda district, Andhra Pradesh with German technology provided by Lurgi and ICRISAT is providing technical support to the farmers for cultivating bio-diesel plantations and facilitating buy-back arrangements between the farmers and SBT (Kashyap 2007).

3.12   LEARNINGS FROM THE EXPERIENCE AND TRIGGERS FOR SUCCESS

The most crucial issue that determines the success of a consortium is capable leading or facilitating partners. Partnerships need to be nurtured by the lead partner. As mentioned earlier, the Consortium Approach is dynamic and continually evolves. Following the framework and philosophy, the lead partner should be innovative enough to facilitate adaptation and evolution of the model to suit the local needs. Quite often there would be conflicting ways of working among partners. The consortium leader needs to understand this fact and ensure flexibility and transparency among partners to accommodate opinions of the members without causing damage to the overall objectives.

Each member of the team should know that he/she can influence the team agenda. There should be a feeling of trust and equal influence among team members that facilitates open and honest communication. This allows each member to provide their technical knowledge and skills in helping to solve the problems, complete the project, and develop new programs.

The consortium leader, where possible, should help select or influence the composition of consortium members. Selection of members should be based on their willingness to work in a team and share their resources, both technical skills and financial contributions. The selection of a right set of partners determines the success to a major extent. Learning behavior among the partners is essential for the Consortium Approach. More importantly, there should be predisposition to work collectively for the community development. It is essential to achieve shared understanding of objectives by the members. They should be able to identify themselves with the common objectives. The lead organization should facilitate this process. Once the objectives are evolved, it is again the responsibility of the lead partner to always bring members' attention to the objectives and help in ensuring focused work in the correct direction (Wani *et al.*, 2009).

There is a need to develop, understand, and accept a set of principles by the members, which include norms for operating within the team. Team building measures go a long way in developing stronger partnerships and internalizing the operating guidelines. Sharing of the credit for the impact, publications, and policy guidelines amongst the partners is very critical. The leader has to ensure that in all communications about the consortium activities, all partners are recognized, acknowledged, and rewarded. Such measures go a long way to build a trust amongst the consortium partners. Similarly, open communication and conflict resolving mechanisms must be in place.

Tangible economic benefits to individual primary stakeholders are essential for participation in the consortium. Integration of new science tools such as GIS and remote sensing enhanced the efficiency of recommendations and resulted in higher benefits to the community. Knowledge-based EPA is another reason for enhanced sustainable community participation. Motivation of the farmers was sustained due to the fact that there is continuous learning which is directly relevant to their fields. Capacity building of partners and sensitization of policy makers helped in building partnerships. Transactions costs (time and money) are higher for partnership building but higher benefits call for partnerships (Shambhu Prasad *et al.*, 2006; Wani *et al.*, 2009).

3.13   OPERATIONALIZING COMMUNITY WATERSHED AS A GROWTH ENGINE

For community watershed development program to become the growth engine for sustainable development of rainfed areas, the major challenge is the scaling-up to large areas as the successful watersheds remained a few and unreplicated (Kerr *et al.*, 2002; Joshi *et al.*, 2005). An integrated consortium approach for the sustainable development of community watersheds with technical backstopping and convergence is developed and evaluated in Asia (Wani *et al.*, 2002, 2003a) and adopted by Government of India (GOI 2008). The systems approach looks at various components of the rural economy – traditional food grains, new potential cash crops, livestock, and fodder production as well as the socioeconomic factors such as alternative sources of employment and income. The adoption of this new paradigm in rainfed agriculture has shown that with proper management of the natural resources, the systems productivity can be enhanced and poverty can be reduced without causing further degradation of the natural resource base (Rockström *et al.*, 2007, 2010; Wani *et al.*, 2008a).

3.14 WATERSHED AS AN ENTRY POINT TO IMPROVE LIVELIHOODS

Watershed as an entry point should lead to exploring the multiple livelihood interventions (Wani et al., 2006a, 2006b, 2007, 2008b). The overall objective of the whole approach being poverty elimination through sustainable development, the new community watershed management provides an envelope that fits into the framework as a tool to assist in providing sustainable rural livelihoods. The task is to sustainably intensify complex agricultural production systems, while preventing damage to the natural resources and biodiversity, and to improve the welfare of the farmers through value addition and market linkages.

ICRISAT's consortium model of community watershed management espouses the principles of collective action, convergence, cooperation and capacity building (four Cs) with technical backstopping by a consortium of institutions to address the issues of equity, efficiency, economics, and environment (four Es) (Wani et al., 2006b). The new integrated community watershed model provides technological options for the management of runoff, water harvesting, in-situ conservation of rainwater for groundwater recharging and supplemental irrigation, appropriate nutrient and soil management practices, waterway system, crop production technology, and appropriate farming systems with income-generating micro-enterprises for improving livelihoods, while protecting the environment. The current model of watershed management as adopted by ICRISAT watershed consortium team involves environment-friendly options and the use of new science tools (Wani et al., 2000, 2002, 2008a; Sreedevi et al., 2004).

Adarsha watershed (in Kothapally, Ranga Reddy district in Andhra Pradesh) led by ICRISAT consortium, has clearly demonstrated increased crop productivity from rainfed systems through integrated watershed management approach (Table 3.3). Similar benefits are recorded by several other watersheds in Asia (Wani et al., 2008c; Pathak et al., 2009; Sreedevi and Wani 2009).

3.15 CONVERGENCE IN WATERSHED

Convergence in watersheds evolved with the community watershed management model, which apart from the IGNRM strategy, encompasses several other entities. The holistic community watershed is used as an entry point to converge and to explicitly link watershed development with rural livelihoods and effective poverty eradication and in the process identify policy interventions at the micro-, meso-, and macro-levels. Convergence takes place at different levels; at the village level it requires facilitation of the processes that bring about synergy in all the watershed-related activities. The scope for the issues relate to the processes for the change in micro-practices, macro-policies, convergence, and information and management systems (Wani et al., 2008b, 2009).

Government of India has come up with an innovative approach of converging the various schemes and programs at the watershed platform to avoid duplication and for efficient use of available funds. There are several schemes that are in operation and one of them is Mahatma Gandhi National Rural Employment Guarantee Act (MGNREGA) that provides employment for the wage seekers. In most of the land-based developmental activities, the landless were left out, but the MGNREGA is helping the poor by providing employment at the doorstep. Under this program, most of the activities are land-based involving earth work. Besides, the landless, marginal and small farmers can take up in-situ and ex-situ soil and water conservation works on their own land and the cost met by the program. Some of the states in India like Gujarat and Madhya Pradesh have already initiated MGNREGA planning on a watershed basis even in the non-IWM program areas; and in Andhra Pradesh, the activities of MGNREGA have been converged with the IWM program.

The activities in IWM approach which involves convergence include: rainwater conservation and harvesting, productivity enhancement through improved crop and management options, soil test based integrated nutrient management options including the application of micro- and secondary nutrients that are deficient in the production systems, soil conservation, crop diversification using high-value crops, establishing the village seed banks through SHGs, processing for value addition (seed material, poultry feed, animal feed, grading and marketability, quality compost preparation), rehabilitation of degraded common lands with suitable soil, water, and nutrient management options using grass and plantation including bio-diesel plant systems, livestock-based livelihood activities through improvement of breed, health, and feed quality, poultry rearing for egg and meat production and local hatching to provide chicks, and vermicomposting with cow dung, fodder waste, and weeds to provide quality compost locally.

3.16 MULTIPLE BENEFITS FROM INTEGRATED WATERSHED DEVELOPMENT

The adoption of IWM effected remarkable multiple impacts on resource-poor farm households in the semi-arid tropics (SAT).

Reducing rural poverty in the watershed communities was evident in the transformation of farmers' economies. The improved productivity with the adoption of cost-efficient water harvesting structures as an entry point improved livelihoods through crop intensification and diversification with high-value crops (Wani *et al.*, 2008b, 2009; Sreedevi and Wani 2009). It also benefited the women, landless, and vulnerable members through income-generating activities.

Building on social capital made a huge difference in addressing rural poverty in watershed communities. Crop–livestock integration is another facet harnessed for poverty reduction. The Lucheba watershed, Guizhou province of southern China has transformed its economy through modest injection of capital-allied contributions of labor and finance to create basic infrastructure like access to roads and drinking water supply. In the Tad Fa and Wang Chai watersheds in Thailand, there was 45% increase in farm income within three years. Farmers earned an average net income of US$1195 per cropping season.

Increasing crop productivity is a common objective in all the watershed programs. Enhanced crop productivity is achieved after implementation of soil and water conservation practices along with appropriate crop and nutrient management. Overall, in the 65 community watersheds in Andhra Pradesh and 30 watersheds in Karnataka (Tables 3.5 to 3.9) (each measuring approximately 500 ha), implementation of best-bet practices resulted in significant yield advantages in sorghum (35–270%), maize (30–174%), pearl millet (72–242%), groundnut (28–179%), sole pigeonpea (97–204%), and intercropped pigeonpea (40–110%). In Thanh Ha watershed of Vietnam, yields of soybean, groundnut, and mung bean increased three- to fourfold (2.8–3.5 t ha^{-1}) as compared with baseline yields (0.5–1.0 t ha^{-1}), reducing the yield gap between potential farmers' yields. A reduction in N fertilizer (90–120 kg urea t ha^{-1}) by 38% increased maize yield by 18% in Thanh Ha watershed in Vietnam. In Tad Fa watershed in northeastern Thailand, maize yield increased by 27–34% with improved crop management (Wani et al., 2008b; Sreedevi and Wani 2009).

Improving water availability in the watersheds was attributed to the efficient management of rainwater and in-situ conservation, establishment of water harvesting structures, and improved groundwater levels. Even after the rainy season, the water level in wells close to the water harvesting structures sustained good groundwater yield and benefited the village women through drinking water availability as well as men with increased irrigation (Wani et al., 2006a, Pathak et al., 2009; Sreedevi and Wani 2009). Supplemental irrigation played a very important role in reducing the risk of crop failures and in optimizing the productivity in the SAT region.

Sustaining development and protecting the environment are the two-pronged achievements of the watersheds for reducing soil loss and runoff loss. Introduction of IPM in cotton and pigeonpea substantially reduced the number of chemical insecticidal sprays in Kothapally, Andhra Pradesh during the season and thus reduced the pollution of water bodies with harmful chemicals. The introduction of IPM and improved cropping systems decreased the use of pesticides worth US$44 to 66 ha^{-1} (Ranga Rao et al., 2007). Increased carbon sequestration of 7.4 t ha^{-1} in 24 years was observed with improved management options in a long-term watershed experiment conducted at ICRISAT, Patancheru, India. By adopting fuel-switch for carbon, the women SHGs in Powerguda (a remote village of Andhra Pradesh) have pioneered the sale of carbon units (147 t CO$_2$ C) to the World Bank from their 4,500 *Pongamia* trees, seeds of which are collected for producing saplings for distribution/promotion of bio-diesel plantation (Wani et al., 2009). Normalized difference vegetation index (NDVI) estimation from satellite images showed that within four years, the vegetation cover could increase by 35% in Kothapally (Wani et al., 2005).

Conserving biodiversity in the watersheds was engendered through participatory NRM. Pronounced agrobiodiversity impacts were observed in Kothapally watershed where farmers now grow 22 crops in a season with a remarkable shift in cropping pattern from cotton (200 ha in 1998 to 100 ha in 2002) to maize/pigeonpea intercrop system (40 ha in 1998 to 180 ha in 2002), thereby changing the crop agrobiodiversity factor (CAF) from 0.41 in 1998 to 0.73 in 2002. In Thanh Ha, Vietnam the CAF changed from 0.25 in 1998 to 0.6 in 2002 with the introduction of legumes (Wani et al., 2005). Building resilience is an important facet of IWM approach. The resilience framework for watersheds using case studies is described in detail by Barron and Keys in this volume.

3.17 CONCLUSIONS

The growing biotic pressure and over-exploitation along with inappropriate management of natural resources are causing their degradation and reducing agricultural productivity. Rainfed areas in the tropics are fragile ecosystems that are prone to severe land degradation and are the hot-spots of poverty and water scarcity.

Management of a small catchment or watershed through participatory approach is recommended for sustainable use of natural resources to increase agricultural productivity and reduce rural poverty. Integrated watershed management is an approach integrating sustainable management of natural resources through collective action of resource users for improving livelihoods of people in harmony with nature rather than a mere hydrological unit. Interdependence of human beings and animals for their living through sustainable use of scientific land use planning on interconnected natural resources need codification up to national level and scientific criteria to prioritize development of watersheds in the country. Improved models of watershed development include some or all of the features such as community participation, collective action, consortium of soil and rainwater conservation structures, better farming practices, involvement of women and landless people through income-generating activities, fusion of research and development, transparency, science-based productivity enhancement, monitoring and evaluation measures, building capacity of the formal and informal rural institutions, building productive partnerships and alliances in a consortium model, and building resilience of the communities and the natural resources. The current IWM program in India has evolved over the past thirty years from top-down target oriented approach to conserve soil and water to community participatory integrated holistic livelihood approach to improve rural livelihoods through sustainable management of natural resources. The holistic, participatory, and consortium approach integrates biophysical interventions with socioeconomic and institutional innovations to sustainably develop and manage natural and human resources for reducing poverty and provisioning ecosystem services. When implemented the IWM programs produced multiple benefits in terms of conserving soil, water, and biodiversity, increased productivity and family incomes, building social capital and resilience to cope with impacts of changes in future including those due to climate change and globalization. Integrated watershed management approach could become the growth engine for sustainable development of dryland areas in the tropics.

REFERENCES

Adinarayana, J. 2008. *A systems-approach model for conservation planning of a hilly watershed centre.* Bombay Powai, Mumbai, India: Studies in Resources Engineering, Indian Institute of Technology.

APRLP. 2007. *Andhra Pradesh Rural Livelihoods Project: an overview.* Government of Andhra Pradesh, India: Andhra Pradesh Rural Livelihoods Programme.

Dixit, S., C. Ramachandran, G. Sastri *et al.* 2001. Natural resources management vs. social resources management: The watershed paradox. In *Management issues in rainfed agriculture in India,* ed. K.H. Vedini, 69–78. Hyderabad, India: National Institute for Agricultural Extension Management (MANAGE).

Dixit, S., J.C. Tewari, S.P. Wani *et al.* 2005. Participatory biodiversity assessment: Enabling rural poor for better natural resource management. *Global Theme on Agroecosystems Report No. 18.* Patancheru, Andhra Pradesh, India: International Crops Research Institute for the Semi-Arid Tropics.

Dixit, S., J.C. Tewari, S.P. Wani et al. 2007. Participatory and conventional biodiversity assessments: creating awareness for better natural resource management. *Annals of Arid Zone* 46(2):197–204.

Dixit, S., S.P. Wani, T.J. Rego et al. 2008. Knowledge-based entry point and innovative up-scaling strategy for watershed development projects. *Indian Journal of Dryland Agriculture and Development* 22(1):22–31.

DOLR. 2003. *Guidelines for Hariyali*. New Delhi, India: Department of Land Resources, Ministry of Rural Development, Government of India. (http://dolr.nic.in/HariyaliGuidelines.htm)

Falkenmark, M. 1986. Fresh water – time for a modified approach. *Ambio* 15(4):192–200.

Farrington, J., and C. Lobo. 1997. Scaling up participatory watershed development in India: Lessons from the Indo-German Watershed Development Programme. *Natural Resource Perspective*, Number 17, February 1997. London, UK: Overseas Development Institute.

GOI. 1994. *Guidelines for watershed development*. New Delhi, India: Ministry of Rural Development, Government of India.

GOI. 2008. *Common guidelines for watershed development projects*. India: National Rainfed Area Authority (NRAA), Ministry of Land Resources, Government of India.

Gündel, S., J. Hancock, and S. Anderson. 2001. *Scaling-up strategies for research in natural resource management – A comparative review*. Chatham, UK: Natural Resources Institute.

Hanumantha Rao, C.H. 2000. Watershed development in India: recent experience and emerging issues. *Economic and Political Weekly* 35(45):3943–3947.

Johnson, N., H.M. Ravnborg, O. Westermann et al. 2002. User participation in watershed management and research. *Water Policy* 3(6):507–520.

Joshi, P.K., A.K. Jha, S.P. Wani et al. 2005. Meta-analysis to assess impact of watershed program and people's participation. *Comprehensive Assessment Research Report 8*. Colombo, Sri Lanka: Comprehensive Assessment Secretariat, IWMI.

Joshi, P.K., A.K. Jha, S.P. Wani et al. 2008. Impact of watershed program and conditions for success: A meta-analysis approach. *Global Theme on Agroecosystems Report No. 46*. Patancheru, Andhra Pradesh, India: International Crops Research Institute for the Semi-Arid Tropics.

Joshi, P.K., L. Tewari, A.K. Jha et al. 2000. Meta analysis to assess impact of watershed. *In* Workshop on Institutions for Greater Impact of Technologies. National Centre for Agriculture Economics and Policy Research, New Delhi, India.

Joshi, P.K., P. Vasudha, B. Shfieraw et al. 2004. Socioeconomic and policy research on watershed management in India: Synthesis of past experiences and needs for future research. *Global Theme on Agroecosystems Report No. 7*. Patancheru, Andhra Pradesh, India: International Crops Research Institute for the Semi-Arid Tropics.

Kashyap, Divya. 2007. PPP on farmers' participation in the value chain of biodiesel production: German Development Cooperation's experience. Presented at GTZ-ICRISAT Meeting on Public Private Partnership (PPP), 14 September 2007, ICRISAT, Patancheru, India.

Kerr, J. 2001. Watershed project performance in India: conservation, productivity and equity. *American Journal of Agricultural Economics* 83:1223–1230.

Kerr, J.M., and K.R. Chung. 2005. Evaluating the impact of watershed management projects: A practical econometric approach. In *Natural resource management in agriculture: methods for assessing economic and environmental impacts*, ed. B. Shiferaw, H.A. Freeman, and S.M. Winton, 223–243. Wallingford, UK: CAB International.

Kerr, J., G. Pangare, and V. Lokur. 2002. *Watershed development projects in India. Research Report 127*. Washington, DC, USA: IFPRI.

Kerr, J., G. Pangare, L.V. Pangare et al. 2000. An evaluation of dryland watershed development projects in India. *EPTD Discussion Paper 68*. Washington, DC, USA: Environment and Policy Production Technology Division, International Food Policy Research Institute (IFPRI).

Knox, A., and S. Gupta, 2000. Collective action and technical workshop on watershed management institutions: A summary paper. *Collective Action PRI Working Paper No. 8.* Washington, DC, USA: IFPRI.

Olson, M. 1971. *The logic of collective action: public goods and theory of groups.* Revised edition. New York, USA: Schocken Books.

Pathak, P., K.L. Sahrawat, S.P. Wani, R.C. Sachan, and R. Sudi. 2009. Opportunities for water harvesting and supplemental irrigation for improving rainfed agriculture in semi-arid areas. In *Rainfed agriculture: unlocking the potential,* ed. S.P. Wani, J. Rockström, and T. Oweis, 197–221. Comprehensive Assessment of Water Management in Agriculture Series, Wallingford, UK: CAB International.

Raju, K.V., A. Aziz, M.S.S. Sundaram *et al.* 2008. Guildelines for planning and implementation of watershed development program in India: A review. *Global Theme on Agroecosystems Report No. 48.* Patancheru, Andhra Pradesh, India: International Crops Research Institute for the Semi-Arid Tropics.

Ranga Rao, G.V., O.P. Rupela, S.P. Wani *et al.* 2007. Bio-intensive pest management reduces pesticide use in India. *Journal of Integrated Pest Management, Pesticides News* 76, June 2007, pp. 16–17.

Ravinder Reddy, Ch., K. Gurava Reddy, G. Thirupathi Reddy *et al.* 2007. Increased adoption of seed treatment for groundnut disease management through farmer participatory evaluation: A micro study in Kurnool district of Andhra Pradesh. *MANAGE Extension Research Review* VII(1):101–109.

Rego, T.J., K.L. Sahrawat, S.P. Wani *et al.* 2007. Widespread deficiencies of sulfur, boron and zinc in Indian semi-arid tropical soils: on-farm crop responses. *Journal of Plant Nutrition* 30:1569–1583.

Rockström, J., M. Falkenmark, L. Karlberg *et al.* 2009. Future water availability for global food production: the potential of green water for increasing resilience to global change. *Water Resources Research* 45, W00A12, doi:10.1029/2007WR006767.

Rockström, J., L. Karlberg, S.P. Wani *et al.* 2010. Managing water in rainfed agriculture – The need for a paradigm shift. *Agricultural Water Management* 97: 543–550.

Rockström, J., Nuhu Hatibu, Theib Oweis *et al.* 2007. Managing water in rainfed agriculture. In *Water for food, water for life: A comprehensive assessment of water management in agriculture,* ed. D. Molden, 315–348 London, UK: Earthscan; and Colombo, Sri Lanka: International Water Management Institute (IWMI).

Sahrawat, K.L., S.P. Wani, T.J. Rego *et al.* 2007. Widespread deficiencies of sulphur, boron and zinc in dryland soils of the Indian semi-arid tropics. *Current Science* 93(10):1–6.

Samra, J.S., and H. Eswaran. 2000. Challenges in ecosystem management in a watershed context in Asia. In *Integrated watershed management in the global ecosystem,* ed. Ratan Lal, 19–33. Boca Raton, USA: CRC Press.

Shah, Amita. 2007. Benchmark survey for impact assessment of participatory watershed development projects in India. Draft report submitted to Planning Commission, Government of India, New Delhi, India.

Shambu Prasad, C., A.J. Hall, and S.P. Wani 2005. Institutional history of watershed research: the evolution of ICRISAT's work on natural resources in India. *Global Theme on Agroecosystems Report No. 12.* Patancheru, Andhra Pradesh, India: International Crops Research Institute for the Semi-Arid Tropics.

Shambu Prasad, C., T. Laxmi, and S.P. Wani. 2006. Institutional learning and change (ILAC) at ICRISAT: a case study of the Tata-ICRISAT Project. *Global Theme on Agroecosystems Report No. 19.* Patancheru, Andhra Pradesh, India: International Crops Research Institute for the Semi-Arid Tropics.

Shanti Kumari. 2007. Building watershed based livelihoods: APRLP experience. Presented at the Review Meeting of A Comprehensive Assessment of Watershed Programs in India, 23–27 July 2007, ICRISAT, Patancheru, India.

Shiferaw, B., V. Ratna Reddy, and S.P. Wani. 2009. Watershed externalities, shifting cropping patterns and groundwater depletion in Indian semi-arid villages: The effect of alternative water pricing policies. *Ecological Economics* (Special Section: Biodiversity) 67(2):327–340.

Sidhu, G.S., T.H. Das, R.S. Singh, R.K. Sharma, and T. Ravishankar. 1998. Remote sensing and GIS techniques for prioritization of watersheds: a case study in upper Machkund Watershed, Andhra Pradesh. *Indian Journal of Soil Conversation* 26(2):71–75.

Singh, A.K., Ashok Kumar, and K.D. Singh. 1998. Impact of rainwater harvesting and recycling on crop productivity in semi-arid areas – a review. *Agriculture Review* 19(1):6–10.

Singh, Raj Vir. 2000. *Watershed planning and management.* Bikaner, India: Yash Publishing House.

Sreedevi, T.K., B. Shiferaw, and S.P. Wani. 2004. Adarsha watershed in Kothapally: Understanding the drivers of higher impact. *Global Theme on Agroecosystems Report No. 10.* Patancheru, Andhra Pradesh, India: International Crops Research Institute for the Semi-Arid Tropics.

Sreedevi, T.K., and S.P. Wani. 2009. Integrated farm management practices and up-scaling the impact for increased productivity of rainfed systems. In *Rainfed agriculture: unlocking the potential,* ed. S.P. Wani, J. Rockström and T. Oweis, 222–257. Comprehensive Assessment of Water Management in Agriculture Series. Wallingford, UK: CAB International.

Twomlow, S., B. Shiferaw, Peter Cooper *et al.* 2006. *Integrating genetics and natural resource management for technology targeting and greater impact of agricultural research in the semi-arid tropics.* Patancheru, Andhra Pradesh, India: International Crops Research Institute for the Semi-Arid Tropics.

Wani, S.P., and K.K. Garg. 2009. Watershed management concept and principles. In *Proceedings of the Comprehensive Assessment of Watershed Programs in India,* 25–27 July 2007, ICRISAT, Patancheru, India, 1–11. Patancheru, Andhra Pradesh, India: International Crops Research Institute for the Semi-Arid Tropics.

Wani, S.P., P.K. Joshi, K.V. Raju *et al.* 2008a. Community watershed as a growth engine for development of dryland areas. A comprehensive assessment of watershed programs in India. *Global Theme on Agroecosystems Report No. 47.* Patancheru, Andhra Pradesh, India: International Crops Research Institute for the Semi-Arid Tropics.

Wani, S.P., P.K. Joshi, Y.S. Ramakrishna, T.K. Sreedevi, P. Singh, and P. Pathak. 2008b. A new paradigm in watershed management: A must for development of rain-fed areas for inclusive growth. In *Conservation farming: enhancing productivity and profitability of rain-fed areas,* ed. Anand Swarup, Suraj Bhan, and J.S. Bali, 163–178. New Delhi, India: Soil Conservation Society of India.

Wani, S.P., A.R. Maglinao, A. Ramakrishna, and T.J.Rego. 2003a. *Integrated watershed management for land and water conservation and sustainable agricultural production in Asia.* Proceedings of the ADB-ICRISAT-IWMI Annual Project Review and Planning Meeting, Hanoi, Vietnam, 10–14 December 2001. Patancheru, Andhra Pradesh, India. International Crops Research Institute for the Semi-Arid Tropics.

Wani, S.P., Meera Reddy, T.K. Sreedevi *et al.* 2006a. *Corporate science and technology institutions – partnerships for inclusive and sustainable development and economic growth*: Proceedings of the CII-ICRISAT Workshop, 27 February 2006 at ICRISAT, Patancheru, India: Patancheru, Andhra Pradesh, India: International Crops Research Institute for the Semi-Arid Tropics.

Wani, S.P., P. Pathak, H.M. Tam, A. Ramakrishna, P. Singh, and T.K. Sreedevi. 2002. Integrated watershed management for minimizing land degradation and sustaining productivity in Asia. In *Integrated land management in dry areas.* Proceedings of a Joint UNU-CAS International Workshop, 8–13 September, Beijing, China, ed. Z. Adeel, 207–230. Tokyo, Japan: United Nations University.

Wani, S.P, Y.S. Ramakrishna, T.K. Sreedevi *et al.* 2006b. Issues, concepts, approaches and practices in integrated watershed management: Experience and lessons from Asia. In *Integrated management of watersheds for agricultural diversification and sustainable livelihoods in Eastern and Central Africa: lessons and experiences from semi-arid South Asia.* Proceedings of the International Workshop held at ICRISAT, Nairobi, 6–7 December 2004, 17–36. Patancheru, Andhra Pradesh, India: International Crops Research Institute for the Semi-Arid Tropics.

Wani, S.P., H.P. Singh, T.K. Sreedevi *et al.* 2003b. Farmer-participatory integrated watershed management: Adarsha Watershed, Kothapally, India. An innovative and up-scalable approach. A case study. In *Research towards integrated natural resources management: examples of research problems, approaches and partnerships in action in the CGIAR,* ed. R.R. Harwood and A.H. Kassam, 123–147. Washington, DC, USA: Interim Science Council, Consultative Group on International Agricultural Research.

Wani, S.P., P. Singh, R.S. Dwivedi, R.R. Navalgund, and Y.S. Ramakrishna. 2005. Biophysical indicators of agro-ecosystem services and methods for monitoring the impacts of NRM technologies at different scale. In *Natural resource management in agriculture: methods for assessing economic and environmental impacts,* ed. B. Shiferaw, H.A. Freeman and S.M. Swinton, 97–123. Wallingford, UK: CAB International.

Wani, S.P., P. Singh, P. Pathak *et al.* 2000. Sustainable management of land resources for achieving food security in the semi arid tropics. In *Advances in land resources management for 21st Century,* ed. S.P. Gavande, J.S. Bali, D.C. Das, T.K. Sarkar, D.K. Das, and G. Narayanamurthy, 290–306. New Delhi, India: Angrokor Publishers.

Wani, S.P., T.K. Sreedevi, J. Rockström *et al.* 2007. Improved livelihoods and food security through unlocking the potential of rainfed agriculture. In *Food and water security,* ed. U. Aswathanarayana, 89–106. UK: Routledge.

Wani, S.P., T.K. Sreedevi, J. Rockström, and Y.S. Ramakrishna. 2009. Rainfed agriculture – Past trends and future prospects. In *Rainfed agriculture: unlocking the potential,* ed. S.P. Wani, J. Rockström, and T. Oweis, 1–35. Comprehensive Assessment of Water Management in Agriculture Series. Wallingford, UK: CAB International.

Wani, S.P., T.K. Sreedevi, R. Sudi *et al.* 2010. Groundwater management an important driver for sustainable development and management of watersheds in dryland areas. In *2nd National Ground Water Congress, 22 March 2010, New Delhi, India,* 195–209. New Delhi, India: Ministry of Resources, Government of India.

Wani, S.P., T.K. Sreedevi, T.S. Vamsidhar Reddy, B. Venkateshvarlu, and C. Shambhu Prasad. 2008c. Community watersheds for improved livelihoods through consortium approach in drought prone rainfed areas. *Journal of Hydrological Research and Development* 23:55–77.

Chapter 4

Equity in watershed development: Imperatives for property rights, resource allocation, and institutions

Amita Shah, Samuel Abraham, and K.J. Joy

4.1 INTRODUCTION

4.1.1 The context

Juxtaposed with the conventional thrust on irrigated agriculture, the watershed development program with their central focus on dryland farming, is a major policy initiative in India for striking a balance between the two broad categories of farming systems, viz., irrigated and dryland. In that sense, the watershed development project (WDP) is an important initiative to attain equity across regions, farms, and farmers operating under different agroclimatic conditions in India. However, there are natural, agronomic, and fiscal constraints that may limit the potential equity in the WDPs. Also, the equity concerns in WDPs, till now, have focused mainly on the intra-community and intra-households, especially in a micro setting rather than dealing with the inter-regional inequities and/or upstream-downstream issues in the context of a larger watershed.

It is imperative thus to bear in mind that WDPs, with their central thrust on land and water resources management, are deemed to benefit mainly those having access to the land, and/or have an effective say in the decision-making process within the project. Prima facie landless, voiceless, and women are likely to be excluded from the benefits from the WDPs, unless special care is taken to ensure their equitable inclusion right from the initial phase of the project cycle[1]. The central question therefore being asked in the context of watershed development is: Who gains and who loses in the process of participatory WDPs in a micro setting?

However, the issue of equity has gained increasing attention in the light of the fact that much of the benefits, during the initial phase of WDPs during the late 1990s, were found to be limited only to a small sub-set of the community (Shah 2001; Wani *et al.*, 2002) as the emphasis was on groundwater augmentation and controlling soil erosion and not on productivity enhancement and income-generating activities. Unfortunately, the objective of equity in the WDPs has remained more or less unattended. This could be mainly due to the complexity involved in planning, execution, and benefit-sharing

---

[1] Even though the project guidelines encouraged greater participation of women in watershed groups, they are not being seen as the primary stakeholders in WDPs. In fact women often tend to lose in WDPs because of the restricted access to the common property land resources (CPLRs) (Ruth *et al.*, 2004). This may imply that women pay the cost of watershed development.

mechanisms of WDPs. The complexities arise primarily due to the various trade-offs involved in the project such as: (a) private vs social benefits; (b) short-term vs long-term gains; and (c) scientific (i.e., 'ridge to valley' and integrated) approach vs crop productivity centric approach for resource management[2].

Since a large proportion of the investment in the WDPs is allocated to land-based activities[3], and the access to augmented water for irrigation is also linked to the ownership of the land, the project benefits are found to be tilted in favor of the landed and the male head of the households who generally own the land. Development of common property resources (land, water, forest) and formation of self-help groups (SHGs) for promoting income-generating activities thus, assume the central stage for addressing the issues of landless women and other marginalized segments of the community. A particular aspect that needs special attention is that income-generating activities, designed for helping the marginalized segments of the community, are often non-land based; this may imply further marginalization of the landless and women from attaining equitable access/control over productive resources and also sharing of the project benefits as discussed subsequently in the chapter (Iyengar *et al.*, 2001). The limited impact has raised doubts about the potential of the micro-watershed-based projects in terms of extending the livelihood support to the poor. The evidence, by and large, suggests that watershed development if properly designed and implemented could serve as necessary but not sufficient condition for reducing rural poverty. Also, it is alleged that if not properly designed and implemented, these projects may worsen the equity scenario in the events where increased availability of groundwater, for the want of regulatory mechanisms, may lead to further concentration and depletion of the resource (Kerr 2002b).

It is thus, important to examine the issue of equity in sharing of the benefits from WDPs, focusing on the common property land as well as water resources. It is, however, important to recognize that there is a limit beyond which the inherent inequality in ownership of land, which in turn determines the access to groundwater, could be overcome even if regulatory mechanisms for the benefit sharing are in place. Also, the issue of equity goes beyond the watershed boundaries, as treatment and water use in the upper catchments of a larger watershed can create negative externality for the availability of water in the downstream micro-watersheds. Understanding the limits set up by the existing biophysical and property regimes is, thus, an essential precondition for understanding the issues of equity and benefit sharing within the context of WDPs (Shah 2007; Joy and Paranjape 2009).

4.1.2 Exclusion of landless and women: Role of CPLRs

Contrary to the irrigation-centric approach, the commons play two distinct roles in upholding the special features noted above. First, development of CPLRs, including

---

[2] In a broader context, equity concerns in a project are influenced by a number of factors such as the differing conceptualization among various agents, limits to the radical agenda that could be taken up within a given time and space, macro-level policies, and the revealed preference of the society for the kind of development approach to be followed (Sangameshwaran 2006). Much of these is beyond the control of the local community in general and the marginalized people in particular.

[3] An analysis of the expenditure pattern shows that more than 70% of the funds are used for land and water management interventions that predominantly benefit larger farmers. Only 7.5% are being used to support the livelihoods of poor and landless families (Reddy and Soussan 2003).

forest, is essential for extending direct benefits to the landless (Farrington *et al.*, 1999). Second, and perhaps more important, is the sharing of benefits from the newly augmented water resources resulting mainly from watershed treatments on public or common lands. An important feature about watershed development, unlike irrigation, is that private benefits are often smaller than the social benefits. The experience from WDPs has brought mixed outcomes. Whereas the projects have helped augmenting rainwater, thereby enhancing productivity of land, these benefits have often remained limited and selective covering only a sub-set of the landed households. On the other hand, most of the watershed projects have overlooked development of CPLRs owing to a number of constraints (legal, procedural, socioeconomic, and institutional). And, in the cases where CPLRs have been treated, and developed with suitable institutional arrangements to ensure the benefits for landless and vulnerable members, it has helped to improve the benefits for the vulnerable groups in the society (Dixit *et al.*, 2005; Wani *et al.*, 2009a) in terms of improved availability of fodder and income. However, such projects are very few in the country. The same holds true in the case of provision for drinking water, which otherwise would have helped women in terms of mitigating the time as well as drudgery in collecting it. The larger reality therefore, is the exclusion of landless and often voiceless as in the case of women, whose interests are often overlooked at the stage of design and implementation. Hence, it appears that more than complete exclusion of small and marginal farmers, the critical issue is that of limited and selective benefits from the project among the marginalized communities.

It is therefore, imperative that the design of the watershed treatment should include equity and sustainability aspects while planning for productivity enhancement. To the extent equity is constrained by the structural aspects like biophysical and property regime, the onus is on ensuring that the expected benefits are actually realized and later on shared equitably. It is in this context that participatory institutions and processes assume special significance. This, in fact, has been the central thrust of the various policy guidelines put in place since the mid 1990s.

4.1.3   Groundwater and equity

Growing inequity in access to groundwater is also fueling a process of social differentiation, which impacts directly on the livelihoods of some groups and contributes to the consolidation of power relations within communities (Batchelor *et al.*, 2003). Groundwater being under private ownership/access represents a classic situation – the 'tragedy of commons'. The negative externalities of over-extraction of groundwater may convert a temporary meteorological process into a long-term ecological one of prolonged water scarcity (Bandopadhyay 1989)[4]. Competitive extraction also leads to inequity in access at the intra-watershed and inter-watershed level. Watershed development activities are directed at augmenting water supply as opposed to managing demand and, in many semi-arid areas, demand is fast outstripping supply. Often the locations for the most beneficial structures in a watershed like check-dams are captured by the powerful.

---

[4] For details of the different dimensions of ownership and legal aspects of water – both the modern laws and the customary laws, see Iyer (2009).

4.1.4 Project-based equity

Recognizing that much of the inequity in the benefit of WDPs emanate from natural and/or socioeconomic-political hierarchies, the discourse on equity within WDPs has remained mainly within the limits of what can be called 'project-based equity', which by and large refers to attaining equity in sharing of additional resources/benefits resulting from the project. This of course, is fairly limiting. Nevertheless, the merit of an approach such as this lies in the fact that the project-based equity could pave a way for breaking the structural inequities across class, caste, and gender, provided the issue of equity is brought to the center right from the initial phase of watershed development.

The above perspective on equity within watershed projects is derived from the fact that much of the enhancement of ecosystem services and productive assets takes place with the help of public funds and through community efforts. It is therefore imperative, as a minimalist formula that the additional resource that has been created be assured equitably to everyone in the watershed, even as prior right to previously existing resources are recognized and left largely undisturbed (Joy et al., 2006). This would encompass the issues of what kind of technologies/activities to be undertaken, how much would be the flow of benefits in short- and long-term, who will share the benefits and what type of institutional mechanisms are required to ensure equitable sharing of the additional benefits amongst all households. Attainment of the project-based equity is fairly complex as it necessitates the adoption of equity centric approach right for the planning stage and addressing the various trade-offs as noted above.

It is pertinent that the formulation of the XI plan has taken special cognizance of the issue of equity in benefits flowing from WDPs. While the concern is valid, it is essential to place the issue in a proper context of relatively adverse agroclimatic, economic, and financial setting within which these projects are being implemented. Addressing the equity issue through WDPs requires substantial increase in time change to the process oriented implementation and monitoring along with institutional support with the corresponding increase in funding which could be through convergence of various development schemes in the area (Wani et al., 2008). What is equally important is the building up of the consensus on the issue of equity across various developmental interventions and convergence among them. Watershed projects may create a basis for a progressive move towards equity and empowerment for the other interventions and processes to build further.

4.1.5 Linkages between technology, allocation of funds, and institutions: The issue of mode

What follows from the above discussion, therefore, is that the issue of equity cannot be addressed as an add-on to the 'mode' in which WDPs operate, i.e., the design, technical specification of the treatments, processes involved, and institutional mechanisms for sharing of costs and benefits within the project. In fact, the same applies to the concern of sustainability. Two important messages emerge in the light of the discussion on equity and sustainability: first, addressing these concerns necessitates dealing with the issues of technology, processes, and institutions; and second, these issues need to be dealt with by adopting a systems approach where decisions pertaining to these aspects are closely interconnected and mutually reinforcing. This implies that a meaningful understanding of equity (and sustainability) aspect necessitate a deeper inquiry into the mode of WDPs.

For instance, a large proportion of WDPs are being implemented in dryland regions with low and uncertain rainfall conditions with predominance of water centric approach for watershed treatments. This, in turn, influences not only the allocation of funds but also the nature of use as well as users of water within the project. Further, this may also influence the involvement of a segment of the communities who often feel left out from the project design. This eventually influences the nature of the institutional norms for sharing of costs and benefits among households within the community. Since the investment in WDPs are largely funded by the Government or donor agency, choice of treatments, thereby dissymmetry in the allocation of funds across households, gives rise to the issue of unequal access to benefits from WDPs (Shah 2000). Invariably, the choice of treatments and access to the project benefits across households within the community is shaped by the processes through institutions formed and also their operating. This suggests inter-linkages between technology, funds allocation, and access to the benefits within the project.

4.1.6   Equity in policy and guidelines

Recognizing the problem of marginalization, watershed guidelines especially since the mid 1990s have made special provisions for inclusion of the landless and women into various stages of planning and implementation (Arya 2007). For instance, the stated objectives of the MoRD (Ministry of Rural Development) guidelines following the Hanumantha Rao Committee Report had placed special emphasis on 'improving economic and social conditions of the resource-poor and the disadvantaged sections of the watershed community such as the assetless and women'. Subsequently, the revised Guideline (2001) made it mandatory to form user groups (UGs) and SHGs, which included women and also recommended special groups consisting only of women. The promotion of income-generating activities through training, credit, and marketing support was envisaged as the main plank for addressing the issue of equity in these guidelines. Besides these, membership of women and landless was made mandatory in the village watershed committee. Similarly, special emphasis was laid on technology for reducing drudgery and extending support mechanisms through Mahila Mitra Kisan, in the Guidelines for National Watershed Development Project for Rainfed Areas (NWDPRA) prepared by the Ministry of Agriculture. Targeted budgetary allocations were made for promotion of income-generating activities, focusing specifically on women and the asset-poor.

Although useful, these measures are not sufficient for addressing the constraints faced by women and the landless to become direct beneficiaries of the 'core' activities of the WDPs. Given this context, stakes of women and landless remained confined mainly to on-site employment gains, development of activities allied to agriculture (such as the rearing of small livestock, backyard plantation, nursery raising, inland fishery, etc.), thrift groups, and non-land based activities (such as food processing, traditional crafts, tailoring, etc.) (Shah 2000). It is not clear as to what extent these interventions per se may pave the way for empowerment of the poor, unless the perspective on equity is clearly articulated and shared among all the stakeholders within the project.

The new common guidelines effective from April 2008 lay special emphasis on the equity aspect of watershed development. In fact, equity and gender sensitivity has been recognized as an important guiding principle for project implementation (GOI 2008). The Common Guidelines envisage that "Watershed Development Projects should be considered as the levers of inclusiveness. Project Implementing Agencies must facilitate the equity processes such as (a) enhanced livelihood opportunities for the poor through investment in their assets and improvement in productivity and income, (b) improving access of the poor, especially women to the benefits, (c) enhancing the role of women in decision making, and (d) ensuring access to usufruct rights from the common property resources for the resource poor." The other guiding principles such as decentralization, community participation, and capacity building are dovetailed with the larger objectives of equity. Taking a stock of what has been achieved so far and what are the important lessons emerging from a large number of innovative WDPs from different parts of the country, is critical before launching the next phase for enhancing the impact of WDPs during the XI Plan with new common watershed guidelines of GOI (2008), which have emphasized equity in watershed development.

4.1.7 Objectives and approach

The analysis in this chapter seeks to examine the issue of equity in the light of the systemic approach for watershed development discussed above. More specifically the analysis addresses the following questions:

- To what extent the existing property rights regimes impinge on the planning and access to benefits from WDPs?
- How far is it possible to attain equity through cost-sharing and redistribution of subsidies across households?
- What are the critical processes, spaces, and imperatives for creating equitable institutions and their actual functioning?
- What do we learn from the examples of good practices for attaining equity and mainstreaming of gender concerns?
- What are the major policy implications?

The analysis, based mainly on the existing literature, synthesizes the existing evidence and field experiences of the authors as to distill out important policy guidelines for the future.

4.2 EQUITY, PROPERTY RIGHTS, AND BIOPHYSICAL CHARACTERISTICS

4.2.1 Equity and property relations in land and water

4.2.1.1 Historically embedded inequalities in access to land and water

Watershed based development is increasingly seen as a strategy to stabilize livelihoods in the rural areas, especially in the dryland regions. The fulfillment of livelihood needs depends crucially on who has access to how much and what kind of productive resources. Thus, the issue of livelihoods brings in its wake the issue of whose livelihoods – in other words, the question of equity. One of the important dimensions of equity is "the concern about the intra-generational distribution of human well-being across typical barriers of class, ethnicity, and gender, etc., including concerns about fairness of outcome as well as processes" (Lele 2004). This dimension of equity is related to historically embedded inequalities. Class, caste (or community), and gender are the three major dimensions in which inequality manifests itself in India. Of course, there are other forms of inequality also: for example, the division between tribal and non-tribal. In fact access to land and water very often follow the contours of these historically embedded inequalities.

It is therefore essential to understand the present property relations with regard to land and water as well as the biophysical and social characteristics of these two resources. In this section we try to analyze both these issues as they have a bearing on the equity concerns in watershed development. We also show that the resources which are not constrained by private property relations (or resources that are available under the public and common property domain) are indeed substantial and with proper institutional arrangements and policy support, they could become the main vehicle to further equity in the watershed development program. Of course this does not imply that restructuring (or transforming) the existing property relations is not important. In fact it is important in its own right and also has to be pursued accordingly.

4.2.1.2   Land: Domains of ownership

The issue of property rights in land is rather complex in India (Moench 1998) with different layers of "public" and "private" ownership and the coexistence of formal legal forms with the customary ones. For example, the public or common land could be under the category of revenue department, forest department, or under the *Gram Panchayat*. The land which is categorized as forest could be a reserve forest, protected or non-categorized with different degrees of access and management regimes. In states like Orissa and Uttarakhand, there are also community forests. Often these uses are governed by customary rules and regulations. In the domain of private ownership of the land share, cultivation is also widely practiced. In the case of water too there are many state, region, and community specific customary laws that govern access, use, and operation and maintenance (Vani 2009).

4.2.1.3   Ownership, land use, and CPLR

The ownership pattern of land varies with the different land use class. Kadekodi (2004) made an attempt to gauge the ownership pattern of land under different land use categories in India and make an assessment of what could be considered as CPLR and open access system. According to the classification by Kadekodi, as much as 59% of the land is by and large under the private ownership, whereas 38% is under government or public ownership, including forest. A part of the private or government/public land might have the features of a CPLR especially in terms of conventional user rights and access, considering the conventional user rights. According to this study by Kadekodi, nearly 14% of the geographical area might have the characteristics of common property resources, whereas another 12% of the geographical area is partially/possibly under the common property resources regime. Together they account for nearly 25%[5] (which is about 82 million ha of the total geographical area of 328.7 million ha), which is fairly substantial and with proper institutional and financial arrangements this could be an important resource in the hands of the landless and marginal farmers. However, most of the WDPs have overlooked development of CPLRs owing to a number of constraints (legal, procedural, socioeconomic, and institutional) despite the fact that much of the CPLRs are highly degraded (Joy *et al.*, 2006).

4.2.1.4 Ownership and landlessness

The ownership of land, irrespective of land reforms and land re-distribution of various types is skewed. According to the India Rural Development Report of 1992, 43% of the country's rural population was absolutely or near landless (http://www. empowerpoor.org/backgrounder.asp?report=162). Though the data are dated, the situation would have only worsened with the displacement and dispossession taking place with the reforms started in the 1990s. While we do not intend to get into the details of these compelling realities, it might suffice to state that landlessness and semi-landlessness is a predominant feature and much of the land is devoid of clear record and title (Shah 2009).

4.2.2 Water: Availability and increasing water scarcity

The annual utilizable water in India is estimated to be 1,122 bcm (billion cubic meters). Of this, surface water accounts for 690 bcm (61%) and the groundwater accounts for the remaining 432 bcm (39%). The actual use in terms of utilization of the irrigation potential reveals a somewhat different scenario: groundwater has surpassed the surface irrigation and accounts for nearly 60% area under irrigation (Shah 2009). In the case of drinking water too, groundwater is the main provider.

    The estimated per capita availability of water per year has been also decreasing: in 1951 the per capita availability was $5,137 \text{ m}^3$, in 2000 it was $1,865 \text{ m}^3$, and by 2005 it decreased to $1,342 \text{ m}^3$ (Paranjape and Joy 2004). This is the average value for India as a whole and averages always hide extremes. So, there are areas where the per capita water availability would be very much lower than the average value. The study by the International Water Management Institute (IWMI) attempted to project the scenario of water availability by dividing a country into four categories as per the relative availability of water to meet its needs in 2025. The study notes that while India will not have major water problems on an average, there will be massive regional variations in the water availability. By 2025, large areas in India will face physical scarcity of water (Seckler *et al.*, 1998).

---

[5] Quoting the World Bank estimates, the PACS study on "Land use and ownership in India" assessed as roughly 20% of the total land area is 'commons', which includes both cultivable and uncultivable wasteland and some forestland (http://www.empowerpoor.org/backgrounder.asp?report=162).

4.2.2.1 Property relations in water

In India, water is a state subject except in the case of inter-state water disputes where the Union of India can intervene and also set up a water dispute tribunal to allocate water amongst the contending states. The state owns all water resources in the state. The surface water – basically water flowing through rivers and streams and stored in dams – operates under the public domain in the sense that the governance and management of these resources are by state agencies. In many of the states, the major and medium irrigation projects are handled by the state water resources department or irrigation departments. In the case of minor irrigation, arrangements differ. For example in Karnataka, the minor irrigation is transferred to Panchayat Raj Institutions (PRIs), and in Maharashtra it is with the water conservation department. As per the 74th constitutional amendment, the village water bodies are with the *Gram Panchayat* (GOI 1992). Groundwater operates under the private property regime and the land owner has unlimited right over the water under his/her land and is free to decide how much to extract, for what use and so on (Joy and Paranjape 2009).

4.2.2.2 Water is a local and non-local resource

Water is both a local and non-local resource. It is present at many scales – from a small watershed to a basin – and at nested and interacting scale and boundaries. Modifying water regime in a watershed, however small it may be, ultimately, has the basin-wide implications. And the way water development is planned, used, and managed causes externalities – both positive and negative – many of which are uni-directional and asymmetric (Joy and Paranjape 2004; Joy *et al.*, 2008). The emphasis on converting all water into groundwater results in users in the valley part having access to most of the benefits. In short, in the case of water, the location in the watershed often determines one's access – people who own land in the valley portion benefit most from the augmented resource[6].

Recognizing that the impact of watershed development extends beyond the treated watershed (what is called externalities), a commitment to equity means ensuring inter-watershed or basin-level equity as well. Every community has a right to water as part of its right to assured livelihood. This implies that the local communities should be assured of adequate access to the water necessary for their livelihood – from local as well as non-local or exogenous sources together (if livelihood needs cannot be met through local water). From this perspective, everybody in the watershed has a right to a basic quantum of water (which also includes the aspect of quality in the case of the drinking water component) as part of his/her right to livelihood. Only after meeting the basic needs or service of all should 'surplus' water be provided to people as extra, economic service for commercial production, whether agricultural or industrial (Paranjape and Joy 1995; Joy and Paranjape 2004). Unfortunately, the mainstream viewpoint within watershed development discourse looks at the water from a local viewpoint and there

---

[6] Therefore, while slogans like '*gaonka pani gaonme*' (basically meaning the rain that falls in a village is for that village) may help conserve water, they go against the grain of collective regulation and control of water resources. While we can argue in the case of many other local resources (except water) that local communities should have full right over the resources in their areas, the same cannot be said about water.

is hardly any sensitivity to the issue of scale or externalities or impacts on downstream water bodies.

4.2.2.3 Spatial or location inequities

The nature and quantum of benefits from watershed development depends on where one's land is located within the watershed. This leads to the second dimension of equity that emanates from spatial or location inequalities (the first one relates to historically embedded inequalities). Primarily because of the biophysical characteristics of the watershed (like slope, depth, and structure of soil, underlying geology, and a host of other factors), benefits accrue unevenly across the different parts of the watershed. Those in the valley portion (lower reach) are likely to benefit more, especially in terms of water resources, as compared to those located on the upper or the middle reaches within a watershed. This is because no matter what measures are taken in the upper reaches of the watershed, the effects of water percolation are greater in the valley portion, which is the lower part of the watershed. Watershed development is asymmetrical also because the people in the upper reaches have no real control over this process. In the case of irrigation command areas, the asymmetry works in the reverse. The head reach farmers can control the flow going to the tail-end portion of the command and the tail-enders do not have any control over this process. The reverse is true for watershed because the watershed hydrology changes as a result of the watershed interventions (Joy and Paranjape 2004). This issue of upstream-downstream difference is not limited to these differences within the watershed, it is also an issue between adjoining watersheds, between upstream and downstream communities, right up to those differences within the entire river basin itself[7].

4.2.2.4 How do the biophysical characteristics actually play out?

Most of the water-holding structures such as check-dams, storage tanks, terminal structures on streams, and larger *nallas* are generally located in the valley portion of the watershed area. Plots closer and downstream to these structures and water sources get more water as compared to those plots that are located farther away and upstream from the structures. One of the important physical attributes of water, which causes negative externalities, is the unidirectional water flow by gravity. Since watershed interventions are carried out mostly in the upper reaches of the watershed, this can impose greater costs on families in the upper reaches. The upper reaches of the watersheds contain a larger proportion of uncultivated common land that is often denuded. Protecting such land against erosion requires vegetation in the landscape, which, in turn, places restrictions on grazing and firewood collection. This imposes costs on the poor. The landless and women, who make use of the commons the most, are likely to be most affected. Moreover, the benefits of water harvesting are mostly downstream where wealthy farmers invariably have larger area of the land. Inhabitants of the upper reaches, therefore, are providing an unpaid environmental service to the lower reaches (Kerr 2002b). Generally, resource-poor farmers like Dalits and small and marginal

---

[7] For a detailed discussion of the asymmetries in watershed and other ecosystem processes see Lele (2004) and Kerr *et al.* (2002).

farmers tend to have inferior quality land mostly in the upper reaches of the watershed while rich farmers are concentrated in the valley portion.

4.2.3 Water is a common pool resource and has competing uses

Irrespective of the property regime if water is to be used in a sustainable and equitable manner, water must be seen as a common pool resource. This partially emanates from the biophysical and social characteristics of water that distinguish it from other resources. Water is also not a public good that can be used in common, but a common pool resource – available commonly but used individually. Many people may use the same streetlight, but if someone uses water, some other person is denied the use of that water. However, though there is an increasing awareness of the common pool character of water, it is clearly accepted only in the case of surface water; and the groundwater is treated as private property (Joy and Paranjape 2004, 2009).

4.2.3.1 The meeting of property relations and the biophysical characteristics

Watershed is assumed to be a natural hydrological unit for the management of water. This is only partially true. It is true so far as surface flow is concerned: groundwater flows do not necessarily follow the watershed boundaries. Also, surface and groundwater flows are not separate entities and are governed by complex interactions and conversion-reconversion phenomena.

Most watershed development activities aim at tilting the balance in the favor of groundwater by converting as much of surface or subsurface water flows into groundwater as possible. The paradox is that watershed development converts water flows from a form that is most suited for handling to a form that is not conducive for handling. Related to this is yet another paradox. In India, as things stand today, surface flows are generally considered a common/state/collective resource, whereas groundwater is virtually considered a private resource. Watershed development then may be seen as a process that transfers a resource from the public domain to virtually private domain. Sufficient attention has not been paid to these 'paradoxes' in the implementation of WDPs (Joy and Paranjape 2004). Probably keeping a balance between surface water storage and groundwater storage, making certain critical irrigation to all as part of the design of the WDP, and evolving institutions' norms for access and use of water could be some of the ways to solve this paradox. Of course there are good examples, though few in number, where attempts have been made to solve this paradox and Hivre Bazar in Maharashtra is one such case.

4.2.3.2 Watershed also creates conditions for a positive sum game

While livelihood and cash income needs are met through one form or other of the biomass, to produce biomass access to land and water is needed. The equity in watershed development could be brought out by creating access to land and water to all through a land redistribution and equitable water distribution program; and by creating access or a fair distribution of the increased productivities of land, and increased water and biomass. Both these are important from the point of equity in watershed development, although often it is the latter that forms the core of the discourse – both in terms of policy and practice.

Watershed development results in the enhancement of ecosystem resources and productive potential. Moreover, this enhancement takes place through the use of public funds and collective community effort. Thus, it can be argued that the additional resource that has been created be assured equitably to everyone in the watershed, even as prior right to previously existing resources are recognized and left largely undisturbed. Thus, without greatly disturbing prior rights and use, potential access to productive resources for the rural poor could be created by watershed development. It creates the possibility of providing equitable access within a positive sum game framework. This, in fact, represents the most important aspect of the potential that watershed development creates to go beyond the constraints imposed by both the property relations and biophysical characteristics.

4.2.4   Efforts to address equity

Kerr (2002b) worked out a detailed typology of approaches used by different projects in India to address the issue of equity. They include: (i) working in particularly poor areas; (ii) employing poor people to construct watershed works; (iii) counting on trickle down benefits to reach poor people; (iv) being sensitive to poor people's needs during implementation; (v) undertaking non-land based activities that support poor people's livelihoods; (vi) giving poor people the decision-making power; (vii) using subsidies selectively; and (viii) guaranteeing poor people usufruct rights to the resources, whose productivity the project enhances (Kerr 2002b). In what follows we present a few examples where efforts to go beyond the constraints imposed by the property relations or the biophysical characteristics of watersheds by the implementing agencies. The examples however, might not suggest an actual success on the ground.

In Adarsha watershed, Kothapally where the consortium led by ICRISAT (International Crops Research Institute for the Semi-Arid Tropics) developed integrated watershed management through convergence, collective action, capacity building and consortium for technical backstopping, equity issue for water is addressed through low-cost water harvesting structures throughout the toposequence in the watershed (Wani *et al.*, 2002) for women income-generating activities through SHGs and for small farmers, productivity enhancement measures through in-situ moisture conservation, and integrated genetic and natural resource management (IGNRM) ensured flow of benefits which enhanced community participation (Wani *et al.*, 2002, 2008; Sreedevi *et al.*, 2009). In the Adihalli-Myllanhalli watershed in Karnataka, BIRD-K consciously decided to spend as much money as possible on the upper reaches of the watershed so that the poorer farmers who have land in the upper reaches benefited. A conscious decision was taken to spread the expenditure evenly across the watershed (Joy and Paranjape 2004). Similarly, Indo-Swiss Participatory Watershed Development-Karnataka (ISPWDK) has adopted an innovative strategy by focusing on the regeneration of fallow lands at the foothills or close to the ridge on a priority basis as these lands belonged to the poor. These lands are of poor quality and very difficult to reclaim. Another strategy has been to bring the non-cultivated, inferior quality land of the poor, under cultivation.

4.2.4.1 Equity in coverage

There are many who see the ridge-to-valley approach, which has been adopted in many watershed programs, as an attempt to correct the imbalance in spending. This approach gives preference to small and marginal farmers who are located on the degraded slopes of the higher reaches of the watershed. This is commonly known as equity in coverage (Reddy and Soussan 2003).

Another related mechanism to ensure equity is through coverage and saturation of the watershed area and spreading the investments more or less equitably, which could be achieved through treating all the lands through soil conservation and moisture harvesting and by taking up small low-cost water harvesting measures in different locations spread over the watersheds and private lands of the marginal sections. The WDPs of ICRISAT, IGWDP (Indo-German Watershed Development Project), DANIDA (Danish International Development Assistance), KAWAD (Karnataka Watershed Development Project), Sujala, etc., which operated under different modes did better in this regard in comparison to the mainstream projects of the Ministry of Agriculture and MoRD and had an inclusive strategy for the planinng and implementation of the conservation measures, especially land-based treatments.

Such attempts to address inequities within the watershed through land-based activities however, had several shortcomings. First, while the approach helped spread expenditure more equitably across the watershed, it did not guarantee the resource-poor any share in the improved resources like water, which is generally appropriated by farmers who are located in the valley portion. Second, it tended to help those having land as the benefits are often in terms of irrigation. Third, there are short-term costs of the interventions in the upper reaches. For instance, the protection of common lands implies loss of access to grazing land especially among the landless. Similarly, the redistribution of land to the landless, by itself does not ensure effective use of the land in the absence of the financial support required for putting the land under cultivation. Though there are government schemes to benefit the resource-poor[8], people are not aware of them.

4.2.4.2 Targeted approach

The other important variant is the targeted approach to equity. Action for Agricultural Renewal in Maharashtra (AFARM) has experimented with this approach in Maharashtra. After many years of work, AFARM has come to the realization that the farmers who have lands in the valley portion are the ones who siphon off the benefits of watershed development. AFARM has already initiated a program in the Sugirpada village, Nandurbar district, Maharashtra covering about 60 marginal households with an area of 100 acres. Though the program does not use a fully-fledged watershed approach (for example, they do not follow the ridge-to-valley approach here), they do contour and soil surveys, and use several soil and water conservation measures as well as dryland agricultural techniques such as furrow and ridge method, contour plowing, and interculturing (loosening soil). Crop-based training in dryland agriculture is also part of the program. AFARM has also started a grain bank and an implement

---

[8] An example is the Ganga Kalyan scheme in Karnataka, wherein people from the Dalit community can get a well dug with electric connection free of cost.

bank. The implement bank, together with sharing of work, helps in timely agricultural operations. Attempts have also been made to initiate preparation of various types of composts, by National Agribusiness Development Programme (NADEP), etc., so that more organic matter is returned to the soil. Farm ponds for protective irrigation have also been initiated[9].

4.2.4.3   Common lands

There are many who see that the development of the commons offers the best bet for the resource-poor, especially landless, to get any tangible benefits from the watershed program. At the least, the rejuvenation of the commons improves availability of fodder and fuel and to the extent the poor benefit more from this, inequity will be addressed. Although most watershed projects aim at regenerating the commons through soil and water conservation measures, planting, and protection, we find that by and large, the performance has not been good, except in some of the successful early generation projects like Sukhomajri and Ralegaon Siddhi. There are two reasons for this: (1) as mentioned above, certain interventions hurt the poor in the short run; and (2) there are implementation problems. Grazing ban is a case in point. In many villages of Maharashtra where watershed development has taken place, Dalits and agricultural laborers who have small ruminants have been affected by the grazing bans (Kerr *et al.*, 1998). There are of course, exceptions to the rule where CPLR-management have significantly helped the poor (Dixit *et al.*, 2005).

4.2.4.4   Equitable sharing of increased water

Despite the fact that compared to other measures, applied water makes the most vital difference to productivity enhancement in the context of watershed development (Shah 1998), equitable access to water is one of the most overlooked aspects of watershed development in the country. Watershed programs rarely aim to guarantee a certain minimum access to water for irrigation and productive purposes. By and large, there is neither prioritization in water use nor any norms for water distribution; in most cases the institution arrangement for sharing of water is non-existent as borne out by the ForWaRD-study in three states in India. Since water rights are tied to land rights, both the location as well as the size of one's holding generally determines who gets how much water.

There are, however, a few examples where equity is addressed in a more redistributive manner. Some of the oft-cited examples in this regard in the watershed context are Sukhomajri, Ralegaon Siddhi, and Hivre Bazar. In Sukhomajri, the money collected from water charges was equally distributed amongst all the households (Kerr 2002a). In Ralegaon Siddhi, collective wells were promoted and informal groups of farmers were organized to manage these wells (Paranjape *et al.*, 1998). In the case of Hivre Bazar bore wells were banned for irrigation and also the growing of the water intensive crops (Menon *et al.*, 2007). Though Pani Panchayat in Maharashtra has been

---

[9] Similar programs have now been started in few villages (like Budhehal and Chopdi) in Sangola taluka and Kini village in Akhalkot taluka – both in Solapur district of Maharashtra (Joy and Paranjape 2004).

pioneer in equitable water distribution with its per capita water allocation (Pangare 1996; Paranjape *et al.*, 1998), it falls outside the watershed experience.

There have also been other interesting experiments with regard to water distribution in watershed programs. In Adihalli-Myllanhalli in Karnataka, BIRD-K took the initiative in constructing farm ponds for farmers near the ridgeline as part of the watershed program. Each farmer had a pond constructed on his/her plot. It was also decided that the neighboring farmers could manually take water for their trees and horticultural crops. In the Manjanahalli watershed in Karnataka, all the people were allowed to take water manually from the ponds.

Under the integrated watershed management program, ICRISAT-led consortium has developed degraded CPLRs through collective action in Bundi watershed, Rajasthan and Velchal in Andhra Pradesh. The degraded CPLRs were developed by forming SHGs of landless and women to undertake soil and water conservation measures accompanied with plantation of *Jatropha* and grass. Men and women SHGs received wages to work at the project sites. The social fencing and mechanical fencing in Rajasthan ensured that no stray animals came in and within one year degraded CPLRs started producing grass. The committee of the SHGs allowed all the villagers to cut and carry the grass for their animals but 50% grass had to be left for the committee. Through this cut and carry system and no open grazing in three years time, the demand for grass for the whole village was fulfilled and the surplus grass was sold to the surrounding villagers earning ₹75,000 per annum for the committee (Dixit *et al.*, 2005). Panchayat gives the maintenance rights to the committee of the SHGs and this system is working well in Rajasthan (Dixit *et al.*, 2005).

At Velchal, degraded CPLRs were rehabilitated with soil and water conservation measures and plantation with *Jatropha* and *Pongamia* trees. Eight SHGs with 80 members have been awarded the usufruct rights by the Collector for harvesting fruits from 18 ha of land for each of the SHGs. In August 2008, the SHG representatives and the state representatives formed Velchal Village Bio-energy Committee to generate electricity using vegetable oil through value-chain approach (Sreedevi *et al.*, 2009; Wani *et al.*, 2009a) to sell the energy to the commercial entrepreneurs in the village.

About 15 farmers could provide protective irrigation manually from the ponds for horticulture and forestry plants. This was one form of sharing practiced by the farmers. The owner of the pond was allowed to pump the water. This water sharing arrangement was decided in one of the meetings by the Sangha (group of farmers in the watershed) (Joy and Paranjape 2004).

4.2.4.5 *Produce sharing arrangements*

In the mid 1990s, in a small experiment in the Khudawadi village in Maharashtra landless women took 10 ha of wasteland on a 15-year produce sharing arrangement from the owners to produce grass and other tree biomass. The Water Users Association (WUA) in the village also gave them some water from a medium irrigation project. They also undertook water and soil conservation work in the 10 ha land. Since the women had assured fodder, they also took up collective goat rearing with Integrated Rural Development Project (IRDP) support (Kulkarni and Rao 2008). There were problems in the lease period, mainly because there was no social agency locally to hold it. However, this showed an important way of bringing private degraded land under productive use and that with institutional and financial arrangements it could be an important avenue for the resource-poor. As we had seen in the earlier section on land use and ownership, there is private substantial area of wasteland/degraded land in the country.

In the Chikamatti village in Chinnahagari watershed of Chitradurga district in Karnataka (KAWAD project), efforts were made to bring revenue land (10 acres) under plantation with a produce sharing arrangement between the Micro Watershed Development Committee (MWSDC) and the village Panchayat. MYRADA as the implementing agency had a lead in the project. People voluntarily came forward and also performed the maintenance work (Joy and Paranjape 2004).

4.2.4.6 *Attempts at risk proofing/pooling and sharing arrangements*

As we had mentioned earlier one of the ways of achieving equity is by sharing the biomass as ultimately the increased productivity of land and the increased water show up in increased biomass production. Thus, sharing biomass produce is also another way of sharing the increased water resources or the increased land productivity; for example, grain banks, fodder banks, and other forms of biomass banks. In a couple of places, like Ralegaon Siddhi and Dornali (both in Maharashtra), grain banks were started. Anyone in need of grain could become a member of the grain bank. The banks distributed grain to the members in summer months or during the drought periods. It is reiterated that all these efforts might not have been fully successful or might offer only partial answers, nonetheless, they seem creative interventions and all of them put together would show us the pathway to move forward.

4.2.5 Main observations

In the context of equity and watershed development, the main divide is between the landless and the marginal farmers on the one hand and the better off farmers on the other, and between those with sloping lands in the upper reaches of the watershed and those with relatively flat lands in the lower reaches of the watershed. The most pressing issue is however regarding the access to common pool resources, especially water, to all sections of the community. It is important to assert that at least the augmented resources that are generated through watershed development through public funds and collective effort must count as common pool resources, subject to a collective decision for their fair allocation. This would involve de-linking water rights from land rights fostered with clearly defined property rights on water. While this requires an appropriate legal framework and effective institutional arrangements (Reddy and Soussan 2003), asserting the principle of treating augmented water as common property resource is a critical minimum requirement (Joy and Paranjape 2004). The following measures could be taken up to provide adequate access for the rural poor to the augmented resources:

- A certain proportion of public as well as private wastelands may be leased to the rural poor on a produce sharing arrangement and additional support for managing them could be provided through targeted assistance from the project funds or through convergence of other programs (like Mahatma Gandhi National Rural Employment Guarantee Scheme) or as post-watershed intervention. The recent initiative of the Ministry of Environment and Forests (MoEF), the National Mission for a Green India, envisages bringing nearly 4 million ha of the total

28.84 million ha open forests under eco-restoration (MoEF 2010). Synergy of this with the watershed development program might open major opportunity for the resource-poor section of the community.

- A certain share of the augmented water resources should be earmarked for the rural poor and they should be encouraged to use it productively by extending targeted support to them.
- All planting and plantation management activity within the watershed could be pooled together and handed over to the tree growers' groups formed with the participation of disadvantaged section of the community.
- The bulk biomass could be pooled and made accessible to the resource-poor sections at concessional terms with the necessary processing facilities for value addition.

The underlying principle in all these is that the rural poor should be entitled to a share of the augmented resource in a proportion that should not be dependent on their property holding. This would mean that the share vests not in the land but in the individual, who, as an individual, has a right to the augmented resources. This would imply that every member of the watershed community has equal share.

However, it should be recognized that the concepts of equity vary, and so long as the main principle is accepted and its spirit is not flagrantly defied in application, communities should be allowed sufficient space to work out their own ways of realizing equity. Once a community accepts a principle, it may find ways to work out creative and adaptive mechanisms. The precondition for the success of all of the above is that the watershed development program planning must be such that no resource is generated or augmented until the social arrangements for its sharing are decided upon and the institutional arrangements for that sharing are in place.

4.3   FUNDS ALLOCATION AND SUBSIDIES

Though watershed development aims at developing the entire set of natural resources, viz., land, water, and vegetation, the treatment is often incomplete and/or asymmetric. This may impinge on fully realizing the potential benefits from the project. At the same time, the project involves choices in terms of sequence, intensity, nature of treatments, and supplementary agronomic practices for both private as well as public land within the watershed. This obviously has significant bearing on the size and distribution of private benefits resulting from the project intervention.

Together these factors lead to lower than the potential flow of benefits on the one hand, and at times, iniquitous sharing of benefits among the different categories of stakeholders – landed with access to irrigation; landed without irrigation; and landless. Within each of the three categories, there is a problem of iniquitous distribution depending on the location of the land and, also on the socio-political space of the household essential for influencing the technology choice as well as mechanisms for the benefit sharing among the stakeholders[10].

---

[10] There are three main factors creating biases in benefits from WDPs. These pertain to patterns and processes determining investment, technology, and capital formation. For details see Farrington *et al.* (1999).

Table 4.1 Coverage of villages under major watershed treatments

(a) Drainage line treatments

Description	Check-dams (Pucca)	Check-dams (Kachcha)	Percolation tanks	Nala/Gulley plugs, reinforced masonry structures (Public)	Gabion/ Bori/ Sanchi structures	Continuous contour trenches (Public)
No. of villages	217	78	47	284	28	189
% of the total villages	62.72	22.54	13.58	82.08	8.09	54.62

(b) Regeneration of CPR

Description	Plantation (Public)	Village tank	Deepening of village tank	Farm ponds (Public)	Development of village pastures
No. of villages	253	209	74	41	19
% of the total villages	73.12	60.40	21.39	11.85	5.49

(c) On-farm treatments

Description	Field bunds (Private)	Farm ponds (Private)	Land leveling (Private)	Plantation (Private)	Small distribution of input kits	Dykes (Private)	All villages
No. of villages	261	100	16	265	9	15	346
% of the total villages	75.43	28.90	4.62	76.59	4.91	4.33	–

4.3.1   Nature of watershed treatments among sample villages

A recent study of a large number of WDPs in the states of Madhya Pradesh, Maharashtra, and Karnataka revealed that the direct benefits from the project reached the minority households within the community[11]. This section discusses the issue of distribution of benefits in the light of the primary data collected from the study in Madhya Pradesh. In doing so the analysis highlights the fact that even when the direct benefits from WDPs are moderate to good, the coverage of households is fairly limited, i.e., about 35–40%.

Table 4.1 provides details of the different treatments carried out under the 346 sample micro-watersheds covered by the study. As expected, water harvesting structures such as check-dams and plantation, besides drainage line treatments, are fairly widespread among the sample watersheds. Also, multiple water harvesting structures have been created in each micro-watershed. These structures constitute a major source

---

[11] For further details visit www.forward.org.in.

Table 4.2 Benefits from watershed projects: initial responses

Details of the benefits	Villages[a] (%)
Increase in water table	81.79 (283)
Reduction in soil erosion	77.46 (268)
Employment opportunity on the work sites	56.65 (196)
Promotion of allied activities	2.89 (10)
Increased tree cover	8.38 (29)
Drinking water for livestock	11.56 (40)
Reduced mgration	0.86 (3)
Improved availability of drinking water	1.45 (5)
No benefits	2.89 (10)
All	346

[a]The Chi-square test is significant. The number of villages is given in parenthesis.

Table 4.3 Main benefits by schemes

Increase in resource availability	Villages[a] (%)
Irrigated area	86.7 (300)
Cultivated land	71.7 (248)
Fuel wood	17.1 (59)
Fodder	56.9 (197)
Soil moisture	90.5 (313)

[a]The Chi-square test is significant. The number of villages is given in parenthesis.

of direct benefits to the farmers. Similarly on the private farms, field bunds and plantation are found to have fairly large coverage among the micro-watersheds evaluated in the study. Compared to these, land leveling and construction of farm ponds seem to have limited coverage in the sample micro-watersheds.

4.3.2 Perceived benefits: Sources and beneficiaries

A number of benefits were reported by the village group prior to the physical verification (Tables 4.2 and 4.3). These include increase in the water table (81.8%); reduced soil erosion (77.5%); employment on the work sites (56.6%); increased drinking water for livestock (11.6%); and increased tree cover as well as vegetation (8.4%). Only 1.4% of the villages reported improvement in the availability of drinking water, whereas reduced migration was reported by only 3 villages, i.e., less than 1% of the sample villages.

A significantly large proportion of the villages reported benefits in terms of increase in soil moisture (90.5%), irrigation (87%) as well as in the net area under cultivation (72%). But benefits in terms of increased fodder and fuel-wood was limited to only 57% and 17% of the villages, respectively. While these observations are largely in tune with the existing understanding on the impact of WDPs, what needs to be highlighted is the fact that most of the village groups perceived the benefits in terms of increased soil moisture. To what extent it may help improve soil productivity however, is a complex issue as noted by a number of earlier studies (Reddy and Ravindra 2004).

Table 4.4 Coverage of beneficiaries: tentative estimates

Treatment	Total no. of structures	No. of structures in good condition	Average no. of beneficiaries	Total no. of beneficiaries
Pucca check-dams	885	465	7	3255
Kachcha check-dams	1048	401	3	1203
Village tanks	733	645	5	3225
Deepening of village tanks	135	100	NA[a]	NA
Percolation tanks	283	255	6	1530
Farm ponds (public)	362	276	2.5	690
Farm ponds (private)	1317	1167	1	1167
Total	–	–	–	11070

[a]NA = Data not available.

An important observation that emerged from the study is that the direct benefits in terms of enhanced availability of drinking water and also of fuel as well as fodder was reported in a small minority of the sample watersheds (Tables 4.2 and 4.3). While about 57% of the sample watersheds were found to have increased availability of fodder, much of this was from the private land that had received direct benefits in terms of irrigation. Similar observations were also made by the study in Maharashtra (Samuel and Joy 2009).

We tried to probe further into the perceived benefits in terms of irrigation, which generally gets translated into enhanced productivity as well as income. Although 87% of the households reported irrigation benefits, the coverage of households deriving such benefits varies significantly. It was observed that 50% of the villages had less than 20% of their households benefiting from the project, while nearly 17% villages had beneficiaries in the range of 20–40%; the remaining 20% of the villages reported more than 40% of the households receiving irrigation benefits from the project. Of course, a higher proportion of the households covered by irrigation benefits was also influenced by the total number of households within each village, besides the number as well as the size of the structures and the present conditions thereof.

4.3.3   Total number of beneficiaries: Some approximation

We worked out the number of beneficiaries using the estimates provided during physical verification (Table 4.4). These estimates, of course, are only approximation so as to get broad contours of the coverage of beneficiaries through major watershed treatment. According to the rough estimates, 346 watershed projects may have covered about 11,000 beneficiaries, especially from the various water harvesting structures as indicated in Table 4.4. The average number of beneficiaries (without considering any overlap) was about 32% of the village. This of course, leaves out the number of households that received direct benefits from the activities like drainage line treatments, plantation, and the construction of field bunds. In the absence of database for generating approximate number of beneficiaries from the remaining activities/treatments, it is difficult to gauge the expanse of the project in terms of beneficiaries within the project villages. In any case the issues pertaining to the intensity, equity, and sustainability remains un-addressed.

Table 4.5 Ranks of watershed treatments by investment[a]

Type of Work	All states	Andhra Pradesh	Gujarat	Madhya Pradesh	Rajasthan
Check-dam	33.5	42.4	55.1	31.4	5.7
Percolation tank	11.0	24.1	2.3	16.4	0.9
Farm pond	15.4	14.1	16.0	23.6	8.5
Land leveling	1.9	0.3	2.6	3.7	1.3
Khadin	1.2	0	0	0	4.5
Farm bunding	7.7	7.0	15.0	7.2	1.9
Contour bunding	4.7	6.1	0.8	9.1	2.9
Diversion drain	0.1	0.3	0.0	0.0	0.2
Plantation	6.2	5.6	8.2	8.6	2.7
Drinking water tank	18.3	0	0	0	71.3
Total	100.0	100.0	100.0	100.0	100.0

[a]Source: Sen *et al.* (2007).

4.3.4   Distribution of subsidies and alternative mechanisms

Asymmetry in the coverage and beneficiaries essentially indicate lopsided flow of public investment/subsidies. Given that water harvesting structures assume central place in watershed treatments and the direct benefits derived thereof, a large proportion of the subsidies is channelized for benefiting a relatively smaller segment of the village community. While we do not have information on the treatment-wise investments among the sample watersheds, we draw evidence from another study covering about 200 micro-watersheds in four states of Madhya Pradesh, Rajasthan, Gujarat, and Andhra Pradesh. The study looked into the issue of investment and maintenance in the post-project phase[12]. Check-dams, farm ponds, and percolation tanks were among the top three treatments on which large amounts of investment have been made through WDPs in Madhya Pradesh. The same pattern is more or less found in the case of other three states except Rajasthan where construction of tanks for drinking water assumed the top priority (Table 4.5).

A general scenario often observed across a large number of micro-watersheds suggest that these three treatments account for nearly two-thirds of the total expenditure incurred for watershed treatments across the states. Since those deriving benefits from such structures happen to be those whose land as well as sources of irrigation are located in the proximity of the regenerated groundwater aquifers, this raises two important issues with respect to equity: first relates to the privatization of benefits from the treatments that often take place on the public land/resources; and second relates to the disproportionate share of subsidies actualized by a selected sub-set of farmers.

Institutional mechanism, which ensures that landless are not paying the local contribution from a reduced wage for watershed work is an important aspect during the implementation phase of the watershed. This also reflects the sensitivity and commitment of both the community and facilitating organizations for equity issues. But field experiences show that often it is the wage labor that contributes to the beneficiary

[12] For details see DSC (2010).

share of the investments. Often the justification is that even after the reduction they get a better deal due to 'schedule of rates' and there is no need to travel outside in search of work[13]. An impact assessment mentioned that 36.7% of watersheds from Gujarat and 6.7% of watersheds from Madhya Pradesh reported the practice of local contribution as per the norms. The remaining project either had full or partial wage cut to meet the requirements.

Off-farm asset creation and economic activities supported through project measures or through availing other subsidies is another institutional mechanism adopted to address the issue of equity. However, quite a few of the projects do not have budget provision for such activities or have very little resources in the form of a revolving fund for the SHGs. Hivre Bazar, a flagship watershed village, reported other subsidy programs such as housing, livestock, etc. in favor of the landless even though the marginal farmers also benefited from the same (Sangameshwaran 2006). Of course, a scenario such as this is not confined only to WDPs. Since natural resources on which such access could be explored are within the realm of the common property resources, the strategy available is developing and creating access to benefits arising from that. Other options available are more 'soft' often refereed as watershed 'plus' through budget allocation and formation of savings and credits and so on. However, quality and intensity of the required strategies and institutional mechanisms vary in both the cases.

It is therefore, essential to evolve an institutional mechanism whereby some kind of parity is attained with respect to the two issues raised above. Provisioning of water rights to all; cross subsidization by those who derive direct benefits from such structures; allocation of water for regeneration of CPLRs, including plantation on public land; and creation of seed money from the contribution collected from the direct beneficiaries of irrigation, for promoting income-generating activities among the poor are some of the possible mechanisms that could be explored in this context.

4.3.5 Overall evidence

The above summary of some of the existing case studies suggests that there is very little evidence on 'Who benefits' and 'How much' from the WDPs. A study by Chandrudu (2006) provides important insights into the issue of processes. For instance, the study indicated that of the 55 micro-watershed projects, 25 had made some efforts in the identification of the poor households at the initial stage of the project implementation. The number however, got reduced to 11 at the subsequent stages of the evolving institutions for the poor and planning; only 7 projects had sustained the focus on poor households at the time of execution. It is further noted that whereas the inclusion of members of the weaker sections and women was an important criterion while formulating the watershed committee, the focus got diluted at the time of planning. Moreover, consultation if any was limited to the dominant section of the village. Participation was low in 40% of the sample watersheds. What is more striking is that while setting the priorities or activities to be included in the action plan, strong bias is usually towards the rich families of the village or the convenience of implementation in terms of the availability of funds, labor, etc. Use of labor is often given lower priority under the guise of the non-availability of local labor.

---

[13] For details see Smith (1998) and Samuel and Joy (2009).

The above observations clearly suggest that the economic benefits are not only limited in terms of coverage of beneficiaries, but also heavily influenced by the decision-making processes at various stages of implementation. This brings us back to the central importance of institutional mechanisms which ensure choice of appropriate treatments and at the same time distribution of benefits flowing thereof.

4.4 EQUITY AND INSTITUTIONS

4.4.1 Criticality of institutional process for equity in watershed

Watershed is a complex resource unit in comparison to certain mono resources common pool regimes such as a grazing land or water body. It involves multiplicity of resources with differing values to different stakeholders under varying property regimes but interconnected through a hydrology cycle. The primacy of institutions in watershed development emerged mainly with the concerns of stakeholder participation and sustainability issues. Problems related to sustainability of assets (and the regenerated resources and impacts) created as part of the intervention led to the conclusion that the institutional mechanisms are necessary for the maintenance and management of the assets. It followed from the fact that the participatory mechanisms improve the efficiency of the intervention besides building stakes.

Unfortunately, the participatory institutions – both in watershed policies and practice – have accorded primacy to the organizational aspects, leaving the crucial issues such as the formation of norms, rules, conventions to the local dynamics. Similarly, it is important to move away from the erstwhile top-down approach; whether this has helped mobilizing collective actions is not quite clear. Realizing this, the Parthasarathy Committee Report observes "any development program based on the local initiative needs to be necessarily accompanied by effective social mobilisation in favor of these socially and economically disadvantaged groups. Detailed agreements on sharing of water and other benefits need to be worked out well before any construction activity is started. The interests of the landless have to be specially borne in mind. Otherwise, all the water harvested will be cornered by the dominant elite. And this is what has happened in most watershed programs in India. The bottom-line is that benefits from any resource created through the project must be equitably shared".

Macro policies such as earmarked funds, usufructs on common property resources, and integration of various vulnerability reduction programs with the watershed agenda certainly create favorable conditions. But the systems for equity or fair sharing of the benefits have to be sorted out through negotiation at the local level and cannot be legislated in a uniform manner (Vaidyanathan 2006). Appropriate institutions for participation, conflict resolution, and usufruct sharing and monitoring, thus assume special significance for ensuring that the benefits are also reaching the vulnerable sections.

4.4.2 Institutional challenges: Learning from the CPR literature

From a sociological perspective, institutions could be defined as the regulatory agencies or systems or standardized patterns of established and rule governed behavior. Institutions have a normative structure and from a social functionalist perspective

limit or guide the behavior. In natural resource management, the role of the institutions is acknowledged as the key factor not only in generating collective actions
for its governance and management, but also ensuring a winsome proposition for
the resource appropriators/users. Most of the common property analysts especially
Ostrom (1990) focus on the ways in which institutions could be crafted purposively
to solve the dilemma of collective action and to create mechanism for governing the
commons. Ostrom argues that collective action for the common property resource
management would be long enduring and successful under conditions of well defined
boundaries, congruence between appropriation and provision rules, effective monitoring, graduated sanctions, efficient conflict-resolution mechanisms, and minimal
recognition of the rights to organize (Ostrom 1990).

Most of these analysis, however takes a fairly romanticized view of the complex
reality where the community is assumed to be homogenous and cohesive. Even though
recent work on common property resource acknowledges the issues of differences
and heterogeneity in the people's capabilities, preferences, and knowledge, they still
are open to critique for neglecting the socio-cultural dimensions of the beliefs and
information as well as power asymmetries[14]. At the same time though there is some
field-based research, which concludes that participation and access to benefits resulting from common property resource management is tilted in favor of the economically
and socially better off, arguing that the other factors like the social and economic
positions and cultural assets like education, access to bureaucracy, and networks play
a crucial role in the participation and resource access (Aggarwal and Gupta 2005).
Even in the case of resource system managed within traditions and customary practices, equity may not be of overriding concern as the traditions and customs are not
necessary egalitarian. However, majority of the critiques of the institutional approach
to the common property resource agree that it is not the institution as such that limits
the opportunity for the disadvantaged, but the nature, composition, and content of
the institution as it often gets built on the existing inequalities and power asymmetry.

4.4.3   Institutions within WDPs: Provisions in various guidelines

As noted earlier, equity at least till recently, has not been a primary concern in the institutional arrangements. In all the guidelines released, institutional arrangements deal
with project administration and to ensure decentralized participation in project implementation. All the guidelines deal with different organizational arrangements required
at various levels from the center to the village/watershed. Participatory processes are
the first step in ensuring inclusiveness and equity; and with the advent of participatory
watersheds, there is an increasing stress on the participation of the local community.

Participation is understood and operationlized in a number of ways in watershed
development from implementation of participatory rural appraisal (PRA) methods to
regular and detailed consultation with the stakeholders. It is considered as the first and
important step in the sustainable management of resources in the watershed context.
However, the participation is not a neutral concept and it involves a set of political
issues such as: Who has the decision-making power? Who has the access to resources?

---

[14] For a detailed critique of institutional perspective of common property resources see Mehta
et al. (1999).

Who benefits from resources? (Farrington *et al.*, 1999). The type of participation in watershed activities varies from contractual, cooperating consultative to collegiate mode of participation (Wani *et al.*, 2009b) and the leader of participation is associated with the efforts put in to build the capacity of the community and empowerment. Hence to address the concerns of equity, it is important to demystify the often-repeated terms such as 'consensus', 'community', 'people' and institutionalize 'equity of space' in negotiation and decision and 'equity in rules' with respect to specific development and distributional outcome. However, the experience in the field shows that equity oriented institutional mechanism is hardly being practiced in a project, except for a few NGO (non-government organization) implemented projects. However, with the lack of disaggregated information on 'participation', it is very difficult to conclude empirically on who actually participated.

The first comprehensive guidelines of 1994 and its revised version in 2001 are silent on the equity aspect of institutional arrangements, except suggesting that SHGs of the scheduled caste/tribe, agriculture labor, and landless persons are organized as a part of the intervention. User groups are suggested for those who are affected by each work/activity and shall include persons having landholding within the watershed (GOI 2001). However, in the Common Guidelines of 2008, there is an explicit statement that 'the Watershed Committee with the help of Watershed Development Team (WDT) shall facilitate resource use agreement among the user groups based on the principles of equity and sustainability. These agreements must be worked out before the concerned work is undertaken' (GOI 2008).

Various guidelines suggest that all types of land, especially degraded forestland, government and community lands can be brought under conservation. There is however, hardly any mention of the usufruct rights to the resource-poor. The Common Guidelines also talk about the treatment of forestland but mainly from the point of view of environment services such treatments will provide for the downstream. However, under the section 'management of developed natural resources' there is mention of the issue of formal allocation of user's right over common property resources. Similarly property or usufructs over water resources is seldom mentioned but the reality is that except for farm ponds and in-situ moisture conservation measures through bunding in private lands most of the water sources are developed on the common lands or drainage course. Off-farm income-generation activities are envisaged and 10% of the developmental cost is earmarked to take up such activity as part of the common guidelines – an improvement from the earlier allocation where only a small amount as Revolving Fund was entrusted with SHGs for such activities. The non-government supported watersheds also follow more or less the same institutional mechanisms such as community organizations, participatory mechanisms, and development of all types of natural resources under various property regimes. The main difference in such a program is with respect to cost and the way the intervention is facilitated and implemented.

4.4.4 Learning from the past experience: Taking stock

This section examines how the provisions made in the various guidelines have been put in actual practice. While there are promising cases, which are far and few between, we deal with a sample that spread across different programs and regions to build an average scenario of how provisions are practiced. Some of the promising ones will be highlighted as cases for appreciating the potential and evolving lessons for replication.

4.4.4.1   Participatory processes: A larger view

Chandrudu (2006) presented a broad canvas of the policy approach for addressing the issue of equity in the major WDPs and examined the actual performance in the light of three sets of parameters, viz., the processes of implementation at critical stages, institutional space, and allocation of funds. Based on an empirical assessment of 55 WDPs spread across seven states in the country, the author brought out a fairly realistic depiction of practicing equity through the three parameters noted above, highlighting the substantial gaps between the policy guidelines and the practice.

On a larger plane, the studies undertaken by ForWaRD in 2009 covering about 1000 micro-watersheds in three states (Madhya Pradesh, Maharashtra, Karnataka) suggest that the participation has been very bleak. Other studies also point towards this anomaly (Chandrudu 2006). Often, the inclusion of the weaker section is to meet the procedural norms and a mechanical operationalization of norms and guidelines are observed. Of 405 watershed sites from the ForWaRD survey in Maharashtra, only 119 watersheds report inclusion of the landless, schedule caste/tribe members or women in the Watershed Committee. The Watershed Committee is basically a negotiated space of the powerful in the community and their interest.

These findings have been corroborated by many reviews and studies. Even where 'consensus approach' was used, the landless felt that they lost out and were unable to stand against the will of a powerful majority (Kerr *et al.*, 2002). There are also instances when the disadvantaged do not get adequate space in the local institutions to address their concerns, approaching other platforms and alliances such as the unit of local political party and other extended networks. Farrington and Lobo (1997), albeit in a different tone, narrate an instance of local politics collaborating with the shepherd community to destabilize a 'consensus' cultivated among landholders and shepherds or grazing community in favor of ban on grazing.

4.4.4.2   SHGs and user groups: A tool for equity

The SHGs are implemented in the watersheds to ensure the participation of women and other weaker sections. The reality however, is quite different. In a large number of projects, the SHGs are either not promoted or are generally a couple of them in each project. An analysis of 12 watersheds of a very successful watershed program like IGWDP shows that while 50–70% of households were members in the SHGs during project implementation, the predominance of the relatively better off women and women from locally dominant caste/sections were quite noticeable. Even accessing loans and other project supported SHG programs generally flow in favor of this section (Vaze and Samuel 2004). Also, the SHGs seize to exist after the completion of the project as shown by the study undertaken by ForWaRD in the three states of India.

User Groups are expected to be directly involved in the management of resources in terms of its maintenance and assigning benefits. However, none of the studies drawing on a wider canvass could establish the case of UGs or its supposed functions in the watershed context and conclude that it is the least facilitated institutional mechanism. Similar observations about non-functioning of UGs in all the watershed programs in Andhra Pradesh, Maharashtra, and Karnataka was noted (Sreedevi *et al.*, 2009). However, in Sujala program watersheds, the group concept of Area Groups (AGs) worked well as each AG representative was a Watershed Committee member and involved in decision-making for all the watershed activities for the AG controlled area (Sreedevi *et al.*, 2009). Conclusions that it is nonfunctional could be drawn from the absence of any effective strategy of maintaining the assets created as part of the watershed. In a large survey of 75 watersheds from 11 districts in Madhya Pradesh and 60 watersheds from 8 districts in Gujarat, only 7 watersheds, that too very recently completed, reported continuity of all institutions. No single watershed from Madhya Pradesh reported continuity (NIRD 2010). Review studies that draw on secondary materials also substantiate the limited reach and influence of the village institutions in ensuring equity in the participation and deriving benefits from the interventions in the WDPs (Joy and Paranjape 2004; Sen *et al.* 2007; Samuel and Joy 2009).

4.4.4.3 CPLRs and institutions

Very few watersheds have developed a successful CPLR model from the point of environmental services and equity aspects. There are many reasons cited for the CPLR problems ranging from administrative apathy to existing claims and encroachments to dwindling of CPLRs due to change in the institutional arrangements including the change in the legal status of the property in question. Though common pool resources of land and water could be leveraged for equity if proper institutional arrangements are evolved, the field realities show that often WDPs sideline CPLRs due to several reasons. There are a few exceptions where attempts have been made to develop CPLRs such as in IGWDP, Comprehensive Watershed Development Project (CWDP), Orissa, DANIDA supported project and ICRISAT watersheds in Rajasthan, Karnataka, and other states in India (Wani *et al.*, 2009b). A few cases where CPLR is successfully restored suggest that the institutional collaboration, utilizing the existing legal and policy provisions like Joint Forest Management (JFM), including devolution together with tackling the issue of management and usufruct rights, helped.

Designing equity indicators, activities, and outcomes, beyond the routine aspects related to the number of meetings, attendance, etc. and integrating it in the overall project design/detailed project report is required for facilitating the concern as suggested by the following case study.

4.4.5 Examples of good practices

4.4.5.1 Streamlining equity consideration: Case of CWDP-Orissa

The CWDP, implemented in Koraput and Malkangiri districts of Orissa, made considerable effort in building stakes for the poor and monitoring the distribution of benefits through an innovative and participatory system of monitoring. Efforts made to address the issues of equity in participation, project processes and implementation, and benefits sharing systems are worth looking into to understand the strategies and institutional mechanisms and policy implications. The project was supported by DANIDA (Danida Watershed Development Programme – DANWADEP), implemented in the four states of Orissa, Madhya Pradesh, Karnataka, and Tamil Nadu. The area of each watershed ranged from 2000 to 5000 ha and the approach was to simultaneously develop a cluster of linked micro-watersheds to enhance the hydrological impacts. Eighty percent of the population belonged to scheduled tribe and scheduled caste. The project adopted environmental regeneration of the watersheds with emphasis on earthen and vegetative measures together with extensive biomass development. Public and private lands were given equal importance. The project had introduced a number of innovative measures. Some of these are listed below:

- Water harvesting structures on common sources were assigned to resource-poor and were supported with pedal pumps to lift water for small plot cultivation.
- Landless were given specific targeted support for income generation and capacity building.
- Landless and resource-poor families were organized into UGs to access usufructs on cashew and other plantations in revenue and village common lands.
- Other Government subsidy programs like housing and tribal sub-plan programs were facilitated through the NGO support.

The most striking feature of the project was the targeted approach towards the 'poorest of the poor'. In order to ensure livelihoods to such section, they were organized into SHGs and various income-generation activities were undertaken with them. At the time of the project closure, about 2700 people were involved in these SHGs (men and women) in which 850 families belonged to the poorest as identified through a participatory wealth ranking. They were mainly landless (but some of them were involved in '*podu*' cultivation), single parent households, physically challenged, and very marginal farmers. They were assisted with specific individual and group schemes to the tune of ₹4,066,577. Poor families were also given priority in the physical work on land and water resources and local low-cost technologies were used making easy maintenance and local replication.

A regular and innovative system of monitoring the impact of the intervention on these 850 resource-poor families were put in place known as 'Assessment of Poverty Reduction Through Monitoring of the Poorest in the Village' (see Lobo and Samuel 2005). Thus, the project showed that even with low budgetary allocation, proper system, procedure, and participatory institutional mechanism could pave the way for equitable benefit sharing in a watershed to some extent. The collaboration with government organizations or NGOs also helped in building the processes. Socioeconomic and cultural specificity of the tribal and the specific provisions in the Panchayat Raj Extension to Scheduled Areas (PESA) helped the project locally and at the policy level, notwithstanding the problems related to fully accessing formal user rights on revenue and forestlands.

4.4.5.2 Public–private collaboration for forest development in watersheds: A case of IGWDP in Ahmednagar District

The IGWDP was initiated in 1993 in Maharashtra and the project is supported by KfW and GTZ from Germany and the Indian counterparts are Watershed Organisation Trust (WOTR) and National Bank for Agriculture and Rural Development (NABARD). The project at the village level is implemented by community-based organizations (CBOs) supported by an NGO. The project had two phases, a Capacity Building Phase

(CBP) and a Full Implementation Phase (FIP). The watershed area ranged from 800 to 1500 ha.

Though in the initial stages of the project it was realized that 15–20% area in the upper reaches of the watershed was mostly forestland, there were a number of administrative bottlenecks in treating this land. There was one window of opportunity and that was the Government Resolution (GR dated 16-3-1992 'Forest Management Through Involvement of Rural People') issued by the Government of Maharshtra. However, the provisions granted as part of a resolution was not always implemented in the field. Therefore, efforts were made to work closely with the District Forest Officials under small grants program. The forest department also contributed through its own planned funds. This helped in creating goodwill between the facilitating agency and the community. This relationship got formalized in the form of special permission given by the forest department to treat forestlands under IGWDP. In July 1994, the forest department officially granted the permission through an official order. In July 1995, Conservators of Forest were officially directed to the registration of forest committees in all the villages under the program.

Another landmark order was the exclusion of such villages from the Forest Conservation Act of 1980 and granting usufructs to the community. A forest official was also deputed to the project to oversee the implementation and also help the forest protection committee (FPC) to streamline the process. As part of this collaboration, the project earmarked a part of the cost of conservation and afforestation to the FPC and the department made their own investment. Work was implemented according to a plan worked out at the village level within a time frame of the watershed work. This is important as the forest works often get spread over for 10–12 years within their respective range plan.

The arrangement got implemented in all the watersheds implemented in Ahmednagar under IGWDP and also in other projects funded by WOTR. Altogether in 40 locations such collaborative JFM was undertaken and it resulted in good green cover, fodder availability, and water recharge. 'Maswandi' watershed in the northern part of the district having a higher percentage of forest, common land, and wasteland worked out a system by which the Adivasi 'Takker' community which has around 25 houses get a privileged access to cut and use or sell grass from assigned plots. The village Watershed Committee also auctions grass and earns a few thousand rupees. A detailed impact assessment of Shekta watershed in Ahmednagar district of Maharashtra implemented under IGWDP by WOTR has documented the benefits of net planning as well as targeted income-generating activities for vulnerable groups (Sreedevi *et al.*, 2008).

4.4.5.3 *Sujala watershed and social regulations*

Based on the above approach, the Sujala Watershed in Karnataka has taken the initiative to provide the user rights to different stakeholders in the proportion of 20:40:40. The 20% share of income (annual or one time income) from common property resources is to be given to *Gram Panchayat* for the benefit of the larger community; 40% share is to be retained by the management committee of multiple UGs towards repair, maintenance, watch and ward, further development of common property resources, etc. and the remaining 40% share is to be shared among the eligible

UG members. The above proportion may however, vary depending upon the type of product under common property resources, i.e., biomass from common land, fish from water pond, irrigated crops from cultivated land, etc. At this stage, a distinction however, is made between mandatory rights to be given to all eligible UGs and actual rights to be availed by those who have taken entrepreneurial risk (during a particular year) by using the common property resources through competitive bidding/auction. In such cases, it would be desirable that bidding/auctioning is restricted to those entrepreneurs who belong to the local community rather than opening it to external persons who could usually be the contractors.

It is however, essential to simultaneously adopt legislative approach to provide legal euthenics to the above mechanism. This may be done through an Act as being currently attempted in Karnataka for the creation of Tank Users Panchayat.

Advance commitment from the community about social regulation before finalization of watershed site: A number of experiences are available regarding social regulation on use of community oriented surface water resource. However, such experiences are very rare with regard to groundwater resource. Nevertheless, each of these experiences have clearly brought out that advance commitment from the community is crucial if social regulations are to be facilitated after the development of water resource under the project. The proposed commitment from the community may be facilitated after exposing them to completed watershed where such arrangements have been made in order to have a lasting impact. Needless to mention that the commitment may be taken in an open meeting of *Gram Samba* before finalizing the watershed site. A copy of above commitment may be sent to block/district authorities, besides keeping it in *Gram Panchayat*/Watershed Committee office. The commitment may consist of the following aspects:

- Social regulation on digging of new wells in the watershed area.
- Promotion of community oriented wells exclusively for resource-poor families and for only low water requiring crops.
- Ban on pumping of surface water collected at the water harvesting structures designed for recharging of groundwater.
- Discouraging conversion of traditional irrigation tanks into percolation tanks unless adequate provision of water has been made for those families who do not own wells in the command area, but have riparian rights over the irrigation water.
- Sharing of groundwater in such a way that the owner of the well uses a part of the water (as per community agreed allocation by the community) and the remaining quantity is shared (on nominal payment basis) with other families whose wells have dried up.
- Improving the efficiency of water use by moving towards critical irrigation (to rainfed crops) from normal irrigation (to high water requiring crops). The efficiency may further be enhanced through the adoption of efficient methods of irrigation and by banning the inefficient use of groundwater.

The above approach however, has remained limited only to one district and the institutional mechanisms for ensuring equity are yet to be evolved on a wider scale.

4.4.6 Main observations

The general scenario based on the outcome from the watershed work suggests that participatory mechanisms and institutional arrangements often follow the contours of inequality as seen in the wider society. Majority of experiences show that the institutional mechanisms to tackle equity in the project, both in terms of inclusion and outcomes are poor, with a few exceptions of highly facilitated projects. The reasons include a mechanical application of provisions in the guidelines without taking the local situations into consideration, a lack of understanding and sensitivity among the agencies involved on participation and equity, compulsions on the deliverables in a stipulated time and the fear of attempts to negatively reverse the existing dynamics that are affecting the project. The performance anxiety and the fear of failure is also often cited by many workers as a reason for not trying hard to push for changing the rules. Lack of assessment of structures and agencies with respect to prevailing inequity in the local context also contribute. We also need to look critically at 'supply of institutions' and search for building institutions forging practices and everyday negotiating of people as anthropologists argue. This may be a time consuming effort.

Equity in participation and empowerment of the disadvantaged go hand in hand and it needs consistent facilitation starting from the initial stages. This is true for all other major project decisions such as prioritization of conservation, payment of wage as granted in the project, and forming of the SHGs of the resource-poor for monitoring the poverty impact of the intervention. Integrating these concerns in the project process and cycle is very important for the facilitating agency.

4.5 GENDER MAINSTREAMING

Most of the policy space within various guidelines could be viewed as instrumentalist approach to gender equity, where the focus is mainly on the practical, as against strategic, concerns of women, viz., food, fodder, fuel, and water for domestic use. These indeed converge with the core concerns of watershed development. What is however, missing is that the present gender perspective within watershed development may not go beyond the primary concerns of women, by proactively trying to correct some of the inherent gender inequities with respect to rights to natural resources, adequate information for making informed choices, and public space for negotiating their preferences as equal partner in the productive spheres along with men.

With regard to gender equity, the issues are further complicated as the discrimination is often more deep rooted, goes beyond the community to household and individual levels. The constraints faced specifically in the context of gender equity are: productivity gains are often limited to only a sub-set of the households thereby limiting the percolation effect to cover the resource-poor; limited access to credit in the absence of ownership of land; administrative difficulties in developing and managing CPLRs especially in the absence of clearly delineated user rights; non-sustenance of gains in productive employment; increased work burden among women in the absence of simultaneous changes in gender division of work and requisite amenities at work; and lack of new skills to be able to benefit from the emerging market opportunities (Shah 2007).

Also, the policy based approach has not addressed the issues of equity and gender as equitable distribution of the benefits, at least from the newly regenerated resources, and has not assumed central importance in the guidelines. Hence, equity did not appear in the list of critical factors in assessing the success of a project. Similarly, gender analysis with respect to gendered priorities in resource management and use were seldom given a central place in the project design and implementation.

4.5.1   Enhancing women's participation and mainstreaming of women SHGs

The participation of women under public sector watershed program is very low despite sufficient evidence regarding their deep interest and heavy dependence on natural resources. Due to increasing migration, men are not readily available in the villages to actively participate in the program. On the other hand, the participation by women in the WDP does not take place properly unless they are organized in sustainable groups. Hence in future, there is a need to mainstream women SHGs and their federations not only for addressing women related agenda but also for an overall management of the WDP through participatory approach.

A possible scenario for enhancing women's participation in the WDPs was developed by Pangare (1996) during the early phase of the participatory approach. Apparently, most of the features suggested a decade ago, appear to be valid even now (Table 4.6). Enhancing women's participation may necessitate integrating within the framework of the watershed program some important aspects such as: (i) organizing all willing adult women in SHGs and their federations; (ii) allocation of separate fund for women specific agenda; (iii) preferential development of land and water resources owned by women headed households/widows or single women; (iv) payment of equal wages to women in development works; (v) adequate representation of women into management committees; (vi) management of watershed program by all women committee (having members from only women SHGs) and carrying out rest of the developmental activities through women SHGs; (vii) preferential allocation of usufruct rights as well as bidding rights over common property resources to women SHGs and their federations; and (viii) focus on development of water resource for drinking water purpose.

In order to facilitate the above approach, the following three steps may be considered at the village level: (i) adoption of a participatory hydrological monitoring system to assess the quantity of water recharged in a particular year; (ii) regulated extraction of groundwater from wells as per the annual recharge and also as per only users right over the recharged water; and (iii) adoption of group action for regulation against over-extraction of groundwater by owners of wells, which are interconnected through the base flow.

4.5.2   Promotion of micro-enterprises

The policies till recently, seem to be mainly on making women (or landless) participate in the process towards empowerment of the marginalized. Thus the equity concern in the watershed projects is an afterthought rather than prior planning. This implies

Table 4.6 Perceptions of the role and contribution of women in watershed development

Stage and activity	Present role	Possible future role
Planning and decision-making		
– Formation of watershed committees	Limited representation on committees	Equal representation on committees
Preparing project proposals		
– Survey	None	Assist in survey
– Identifying soil and water conservation structures	None	Share knowledge and experience
– Contributing land for selected sites	None	Assist in decision-making related to contribution of land
– Preparing budgets	None	Participate in decisions related to labor and other costs
Implementation		
– Constructing conservation structures	Provide labor	Supervise, maintain ledgers, musters, make payments
– Maintaining conservation structures	Provide labor	Initiate decisions on maintenance, supervision, maintaining ledgers, etc.
– Developing agriculture	Provide labor, limited decision-making on family lands	Receive training inputs, contribute to community-level decisions
– Forestry and tree planting	Provide labor, limited decision on selection of species, maintain nurseries	Participate equally in decision-making on selection of species and areas for plantation, provide technical inputs, supervise plantation sites
– Fodder and grazing lands	None	Participate equally in decisions related to development of grasslands, rotational grazing, stall feeding
– Horticulture	Provide labor	Select fruit, process, market
– Dairy and animal husbandry	Provide labor	Technical inputs, process milk, market
Management and delivery systems	None	Participate equally in decision-making, provide technical inputs, share knowledge and information, ensure just distribution of resources
Social and welfare aspects	Have limited access to health and education	Identify community needs, ensure availability of welfare
Alternative energy programs	Provide labor	Identify alternate resources for energy; plan and implement

that the equity issues, could at best, be addressed only when the productivity aspect is satisfactorily addressed.

The Common Guidelines, as noted earlier, have created significant space for addressing the issues of equity and gender empowerment. This is reflected in the budget allocation where nearly one-fourth of the budget is to be spent on (a) livelihood activities for the asset poor (10%); and (b) production systems and micro enterprises (13%). This is a fairly significant step. Assuming a unit cost of ₹12,000 per ha, this amounts to about ₹1.362 million per micro-watershed of 1000 ha over a period of

7 years. How should one make the best use of these opportunities? This question could be examined in the light of some of the important lessons learnt in the past.

The past experiences suggest that most of the micro-watersheds have refrained from addressing some of the important issues that have special relevance to women farmers and other producers. These are: (a) fodder and livestock development, (b) provisioning of drinking water for human and livestock population, (c) water for developing marginalized land including homestead cultivation and promotion of labor/skill intensive agronomic practices, (d) adoption of technologies that may reduce drudgery for the laborers including women's domestic work, and (e) creation of basic amenities. On the other hand, several of the activities, especially those promoted through the SHGs attempt to focus mainly on traditional craft/artisan skills based activities. Most of these suffer from the inter-related problems of: (a) technology, quality, and marketing and market saturation, (b) scale of operation, (c) credit flows, (d) forward-backward linkages, and (e) entrepreneurial trait (knowledge, skills, and attitudes).

It is also often found that the SHG-driven enterprises are in the realm of 'women alone'. It is not clear how far this helps in enhancing the economic viability of such activities on the one hand and gender empowerment on the other. It is likely that these two may operate in a mutually conflicting manner at least in the short run. These are some of the issues that need special attention in the light of the possible avenues for attaining convergence between equity, gender empowerment, and economic viability in the development of micro-enterprises.

The next phase of WDPs therefore, needs to address some of the important missing links in livelihood generation and micro-enterprise development. It is imperative that if women (and other marginalized segments) do not have access and control over the productive resources/assets, then they should be compensated by putting them in the forefront of skill development, technology adoption, and dissemination. An important precondition for attaining this is to shift farm production from input intensive to knowledge intensive technologies that would play critical role in enhancing economic viability and sustainability of the sector. This in turn would imply going beyond the primary concerns and traditional role of the poor and the women in the production spheres and elevating their status as skilled and knowledgeable producers rather than merely as low/unpaid laborers within and outside the domestic arena.

This approach is somewhat different from the one that is currently adopted where women's primary concerns are seen as independent of the overall economic goals of productivity and livelihood enhancement within the land-based primary production in watershed development or as an add on to the mainstream watershed activities. Such an approach may perhaps provide a more legitimate space to the poor and the women, along with their increased roles in the decision-making processes as envisaged in the guidelines.

4.5.3 Institutional challenges

At present, most of the participatory processes are carried out in a mechanistic manner, with bias in favor of the rich and the powerful and often the men. The rest of the community and women do not find this 'greatly unjust or objectionable' because of the two important features characterizing the highly stratified society in India. Under this scenario, addressing the equity issues may require fundamental changes in the way local institutions are evolved. The acid test for an institution such as this could be that it should at least bring on board the conflicting or differential interests of communities – landed and landless; and men and women. It is critical that the conflicting interests are brought to the surface, even if not resolved completely.

Given the fact that social transformation in terms of narrowing the class and gender differentials is a complex process, mere legislative enactments or statutory provisions within watershed guidelines may not be effective. It is here that the role of social movements and civil society organizations (CSOs) working towards the larger goals of equitable development with a transformative agenda may assume special role. It is encouraging that a number of initiatives have come from NGOs working with greater flexibility – procedural, financial, and temporal. What is more heartening is that some of the learning emerging from the NGO or donor agency supported projects are internalized in the state-supported watershed projects. Now, we attempt to summarize the major features of the approach and the salient lessons emerging from each of these experiences.

4.6 POLICY IMPLICATIONS AND WAY FORWARD

4.6.1 Multi-pronged approach

A number of operational devices have been suggested for attaining the equity objective in watershed projects (Kerr 2002b). These include: (a) give priority to poorer areas; (b) use local labor; (c) consolidate the impact of water/irrigation to increased demand for labor and better wages; (d) being sensitive to poors' needs; (e) promote non-farm activities for the landless; (f) involve in the decision-making process; (g) use the subsidies selectively; and (h) ensure user rights to the poor.

While all these are useful tips, the actual result may depend on the location specific situations and the quality of the processes followed and institutions built. It may be recognized that the livelihoods of the rural resource-poor cannot be met entirely through primary production. It needs to be supplemented by non-farm incomes through different value addition avenues. Here the biomass produced as part of the watershed development programs (especially in areas which are not suitable for shallow-rooted crops for bulk biomass production like small dimension timber, bamboo, fiber, medicinal plants, etc.) and make it available to the resource-poor on favorable terms so that the resource-poor could take up value addition activities.

4.6.2 Policy recommendation

The suggestions made above may necessitate certain steps, especially at the macro-level planning and in the initial stage of project implementation; and the following points deserve special attention.

- Increased emphasis on tribal dominated forest-based economies with high incidence of poverty and at the same time, better potential for economic benefits under relatively favorable rainfall and growing conditions, and a large proportion of households cultivating marginal lands.

- Dovetailing MG NREGA with watershed development should ensure systematic treatment rather than haphazard activities pertaining to land and water resources in the region. The NREGA-WDP Convergence Act in Madhya Pradesh should be assessed and suitably modified for adoption in other states.
- Resolving legal complications in treating CPLRs, both under revenue and forest departments, and also for accessing the benefits from regeneration of such land in a sustainable manner.
- Introducing special package for the communities who received land under the distribution of surplus land. Since the land distributed under the scheme is highly degraded, development of such land may deserve special support under watershed projects.
- Ensuring water rights to all by distributing the harvested water under the project.
- Treatments like land leveling, farm pond digging, and farm forestry wherever feasible, may be undertaken irrespective of the poor farmers' ability to pay for the cost-contribution. Ban the deepening of well and the incentives for adoption of water saving devices/crops should be introduced. At the same time, encourage bore well scheme on group-basis. This may be of special significance to tribal areas as demonstrated by experience of Aga Khan Rural Support Programme (India) (AKRSP) in south Gujarat.
- Special support for adoption of sustainable agricultural practices that are knowledge intensive rather than input intensive, among small and marginal farmers. Promotion of the biomass production-based enterprises.
- SHGs consisting of landless households or women's group may be provided with additional seed money from the Watershed Development Fund. This should be based on recognizing the poors' stake on land (especially CPLRs) and water (incremental water harvested through the project).
- Provision for fodder bank in order to ensure smooth supply of fodder during the initial phase when CPLRs is under protection. Special emphasis should be on livestock development especially among landless and small and marginal farmers.
- Availability of water for the life-saving irrigation of plantation on community land should be treated as priority.
- Revival of the *Gram Sabhas* and the continued involvement of project implementing agencies in the post-project period for ensuring at least the first round of repair and maintenance.
- Involvement of NGOs should be encouraged by implementing NREGA on watershed basis. The model adopted by Rajiv Gandhi Mission for Watershed Management (RGMWM) in Madhya Pradesh provides a useful example of involving NGOs as partner NGOs.
- Last, but not the least, the emphasis should be in developing state specific guidelines as demonstrated in the case of Andhra Pradesh and Madhya Pradesh.

4.6.3 Way forward

While it is difficult to make a complete shift in the approach for planning and implementation of the WDPs, special efforts should be made by the state/district level agencies to ensure the minimum achievement in the equity related aspects listed above. In this context, the Report by the Parthasarathy Committee (MoRD 2006) clearly mentioned that benefits of public investment must be seen as public goods, to be shared with equity among all the sections. The concern for equity should be in all the stages, viz., beneficiary selection, benefits sharing, conflict resolution, and monitoring and evaluation. Similarly, gender-equity may be addressed by adopting a comprehensive approach for increasing the representation of women through a separate women's watershed council, which would focus on providing equal wages for equal work, reduction of drudgery, and income enhancement.

Despite development of common property resources (land, water, and forest) under a number of WDPs, there is no clarity about user rights, which assume critical importance for the sustainability of the benefits, especially as it takes several years before the full impact of the development of such resources are realized. While the guidelines lay the responsibilities for the management of common resources on UGs, they do not make clear provisions for the devolution of rights that these groups should in turn enjoy. This, as discussed above, would lead to unsustainable development and ineffective devolution. It is therefore, essential that a comprehensive policy is evolved at the national and state levels for devolving and decentralizing the governance and administration of natural resources with special focus on common property resources.

A clear national policy accompanied by a Model Bill on Common Property Resources may be evolved to crystallize the notion of such resources and create a set of clearly identified rights for the local community. The Bill may clearly state, in a graded manner, different kinds of rights and entitlements of the community (the three categories of rights and powers and functions as indicated in the following recommendations) and the legal nature of relationship of the state, line departments, PRIs over the resources by making the resource-dependent community as the primary stakeholders, entrusted with the rights and responsibilities of maintaining, managing, and improving the quality of the resources while deriving benefits from them. The details regarding the collection of user charges and modality of sharing the benefits between different stakeholders may be spelled out in such a way that major benefits from the common property resources goes to UGs. Likewise, the modalities for sustainable utilization and management of resources may be spelled out where the major responsibilities rest with respective UGs and/or management committee of multiple UGs. The following aspects deserve special attention:

- An integrated policy for land and water use across different agroecological zones with special emphasis on water use efficiency should precede macro-level planning for watershed-based development of natural resources. This should take care of the spatial prioritization and also the compensation mechanism within an upstream-downstream context.
- Promotion of the equitable use of augmented water through the project by providing incentives for the adoption of water-use regulation and water-saving crops/technologies so as to facilitate the resource-poor to gain from the project.
- Legislative and administrative mechanisms for facilitating the poor's access to CPLRs; their intensive management through enhanced availability of water, and development of livestock, and other high valued farming.
- Some of the processes essential for bringing women or the poor's practical as well as strategic concerns and representation of their SHGs into watershed committees should be treated as the non-negotiable right from the initial phase.

- Need to invest in creating local institutions for governing the use of scarce resources, and increasing the size of the economic surplus through productivity enhancement such that the poor benefit from both direct intervention for income generation and the trickle down effect as well as market development.

It is imperative to note that the social transformation such as equity across class and gender necessitates continued convergence among various developmental initiatives; WDPs could play a key role in bringing such convergence at the micro, meso and macro levels.

REFERENCES

Aggarwal, A., and Gupta, K. 2005. Decentralization and participation: the governance of common pool resources in Nepal's Terai. *World Development*, Vol. 33, No. 7.

Arya, S.L. 2007. Women and watershed development in India: issues and strategies. *Indian Journal of Gender Studies* 14(2):199–230.

Bandopadhyay, J. 1989. Riskful confusion of droughts and man induced water scarcity. *AMBIO* 8(s):284–292.

Batchelor, C.H., M.S. Rama Mohan Rao, and S. Manohar Rao. 2003. Watershed development: a solution to water shortages in semi-arid India or part of the problem? *Land Use and Water Resources Research* 3:1–10. (http://www.luwrr.com)

Chandrudu, M.V.R. 2006. Understanding processes of watershed development in India. *Birds eye view of processes: status across states, facilitators and donors*. Secunderabad, Andhra Pradesh, India: Watershed Support Services and Activities Network (WASSAN).

Dixit, S., J.C. Tewari, S.P. Wani, C. Vineela, A.K. Chourasia, and H.B. Panchal. 2005. Participatory biodiversity assessment: enabling rural poor for better natural resource management. *Global Theme on Agroecosystems Report No.18*. Patancheru, Andhra Pradesh, India: International Crops Research Institute for the Semi-Arid Tropics.

DSC. 2010. Post-project management and use of watershed development fund in four states of India. *Policy Brief*. Development Support Centre.

Farrington, J., and Lobo, C. 1997. Scaling up participatory watershed development in India: lessons from the Indo-German Watershed Development Programme. *ODI Discussion Paper No. 17*. London, UK: Overseas Development Institute.

Farrington, J., C. Turton, and A.J. James. 1999. *Participatory watershed development: challenges for the twenty-first century*. New Delhi, India: Oxford University Press.

GOI. 1992. The Constitution (Seventy-Fourth Amendment) Act, 1992. New Delhi, India: Government of India. (http://indiacode.nic.in/coiweb/amend/amend74.htm)

GOI. 2001. Guidelines for Watershed Development (revised 2001). New Delhi, India: Department of Land Resources, Ministry of Rural Development.

GOI. 2008. *Common Guidelines for Watershed Development Projects*. New Delhi, India: Government of India.

Iyengar, S, A. Pastakia, and Shah, Amita. 2001. Karnataka watershed development project (KAWAD): Mid term evaluation report. Ahmedabad, Gujarat, India: GIDR.

Iyer, R. 2009. *Water and the laws in India*. New Delhi, India: Sage Publications.

Joy, K.J., B. Gujja, S. Paranjape, V. Goud, and S. Vispute. 2008. A million revolts in the making: Understanding water conflicts in India. In *Water conflicts in India: A million revolts in the making*, ed. K.J. Joy, Bikshan Gujja, Suhas Paranjape, Vinod Goud, and Shruti Vispate. Routledge.

Joy, K.J., and S. Paranjape. 2004. *Watershed development review: issues and prospects.* Technical Report. Bangalore, India: Centre for Interdisciplinary Studies in Environment and Development (CISED).

Joy, K.J., and S. Paranjape. 2009. Water use: legal and institutional framework. In *Water and the laws in India*, ed. Ramaswamy Iyer. New Delhi, India: Sage Publications.

Joy, K.J., A. Shah, S. Paranjape, S. Badiger, and S. Lele. 2006. Issues in restructuring. *Economic and Political Weekly*, July 8–15, Vol. 41, No. 27 & 28.

Kadekodi, G.K. 2004. *Common property resource management, reflections on theory and the Indian experience.* New Delhi, India: Oxford University Press.

Kerr, J. 2002a. Watershed development, environmental services, and poverty alleviation in India. *World Development* 30(8):1387–1400.

Kerr, J. 2002b. Sharing the benefits of watershed management in Sukhomajri, India. In *Selling forest environmental services: market-based mechanisms for conservation and development*, ed. S. Pagiola, J. Bishop, and N. Landell-Mills, 53–63. London, UK: Earthscan.

Kerr, J., G. Pangare, and V.L. Pangare. 2002. Watershed development projects in India: an evaluation. *Research Report 127.* Washington, DC, USA: International Food Policy Research Institute (IFPRI).

Kerr, J., G. Pangare, L.V. Pangare, and P.J. George. 1998. The role of watershed projects in developing rainfed agriculture in semi-arid tropics. Draft. New Delhi, India: ICAR/World Bank Research Project on Rainfed Agricultural Development.

Kulkarni, S., and N. Rao. 2008. Gender and drought in South Asia: dominant constructions and alternative propositions. In *Droughts and integrated water resource management in South Asia: issues, alternatives and futures*, Jasveen Jairath and Vishua Ballabh, 70–97. New Delhi, India: Sage Publications.

Lele, S. 2004. State-community and bogus "Joint"ness: crafting institutional solutions for resource management. In *Globalisation, poverty and conflict: a critical "Development" reader*, ed. Mar Spoor, 283–303. Dordrecht: Kluwer Academic Publishers.

Lobo, C., and A. Samuel. 2005. *Participatory monitoring systems in watershed development: case studies of applied tools.* Maharashtra, India: WOTR.

Mehta Lyla, Leach, Melissa *et al.* 1999. Exploring understanding of institutions and uncertainty: new directions in national resource management. *IDS Discussion Paper 372.* UK: Environment Group, Institute of Development Studies, University of Suzzez.

Menon, A., P. Kumar, E. Shah, S. Lele, and K.J. Joy. 2007. Community-based natural resource management: issues and cases from South Asia. New Delhi, India: Sage Publications.

MoEF 2010. National Mission for a Green India (Under The National Action Plan On Climate Change). Draft submitted to Prime Minster's Council on Climate Change. (http://moef.nic.in/downloads/public-information/green-india-mission.pdf, accessed 8 February 2011.)

Moench, M. 1998. Allocating the common heritage: debates over water rights and structure, *Economic and Political Weekly* 33(26):A-46–A-54.

MoRD. 2006. From Hariyali to Neeranchal. Report of the Technical Committee on Watershed Programmes in India. New Delhi, India: Ministry of Rural Development, Government of India.

NIRD. 2010. Impact assessment of watershed development in Madhya Pradesh and Gujarat. Report of the study conducted by SOPPECOM and GIDR. India: National Institute of Rural Development.

Ostrom, E. 1990. *Governing the common evolution of institutions for collective action.* UK: Cambridge University.

Pangare, G. 1996. *The good society: The Pani Panchayat Model of sustainable water management.* New Delhi, India: Indian National Trust for Art and Cultural Heritage.

Paranjape, S., and K.J. Joy. 1995. *Sustainable technology: making the Sardar Sarovar Project viable – A comprehensive proposal to modify the project for greater equity and ecological sustainability.* Ahmedabad, India: Centre for Environment Education.

Paranjape, S., and K.J. Joy. 2004. *Water: sustainable and efficient use.* Ahmedabad, India: Centre for Environment Education.

Paranjape, S., K.J. Joy, Terry Machado, Ajaykumar Varma, and S. Swaminathan. 1998. *Watershed based development: a source book.* New Delhi, India: Bharat Gyan Vigyan Samiti.

Reddy, C.B., and A. Ravindra. 2004. Watershed development programme – understanding investments and impacts: Report of study of impacts in five watershed projects of Andhra Pradesh. Secunderabad, Andhra Pradesh, India: Mission Support Unit, AP Water Conservation Unit, Watershed Support Services and Activities Network (WASSAN).

Reddy, V.R., and J. Soussan. 2003. Assessing the impact of participatory watershed development: a sustainable rural livelihoods approach. In *Method for assessing the impacts of natural resource management research,* 84–93. Patancheru, Andhra Pradesh, India: International Crops Research Institute for the Semi-Arid Tropics.

Ruth, M., S. Suseela, and C. Lisa. 2004. *Increasing the effective participation of women in food and nutrition security in Africa.* Issue Briefs 20. Washington, DC, USA: International Food Policy Research Institute.

Samuel, A., and K.J. Joy. 2009. *Watershed development in Maharashtra: a large scale rapid assessment.* Pune, India: SOPPECOM.

Sangameshwaran, P. 2006. Equity in watershed development: a case study in Western Maharashtra. *Economic and Political Weekly,* May 27, 2006, 41(21).

Seckler, D., U. Amarasinghe, D. Molden, R. de Silva, and R. Barker. 1998. World water demand and supply, 1990 to 2025: scenarios and issues. *Research Report No. 19.* Colombo, Sri Lanka: International Water Management Institute.

Sen, S., A. Shah, and A. Kumar 2007. Watershed development programs in Madhya Pradesh: present scenario and issues for convergence. *Technical Report, Forum for Watershed Research and Policy Dialogue.* Pune, India: SOPPECOM.

Shah, A. 1998. Watershed development programs in India: reflections on environment-development perspectives. *Economic and Political Weekly,* Vol. 33, No. 26, pp. A66–A79.

Shah, A. 2000. Watershed programs: a long way to go. *Economic and Political Weekly* 25(35 and 36):3155–3164.

Shah, A. 2001. Who benefits from watershed development? Evidence and issues. *IIED Gatekeeper Series No. 97.* London, UK: International Institute of Environment and Development.

Shah, A. 2007. Equity in the impact of watershed development: class, gender and regions. In *Impact of watershed management on women and vulnerable groups,* 21–36. Patancheru, Andhra Pradesh, India: International Crops Research Institute for the Semi-Arid Tropics.

Shah, A. 2009. *Revisting watershed development in Madhya Pradesh: evidence from a large survey.* Ahmedabad, Gujarat, India: GIDR.

Smith, P. 1998. The use of subsidies for soil and water conservation: a case study from western India. *Agreen Network Paper No. 87.* UK: ODI.

Sreedevi, T.K., S.P. Wani, Raghavendra Sudi, Harshavardhan K. Deshmukh, S.N. Singh, and Marcella D'Souza. 2008. Impact of watershed development in low rainfall region of Maharashtra – a case study of Shekta Watershed. *Global Theme on Agroecosystems Report No. 49.* Patancheru, Andhra Pradesh, India: International Crops Research Institute for the Semi-Arid Tropics.

Sreedevi, T.K., S.P. Wani, Ch. Srinivasa Rao, Raghu Chaliganti, and Ramalinga Reddy. 2009. Jatropha and Pongamia rainfed plantations on wastelands in India for improved livelihoods and protecting environment. In *Proceedings of 6th International Biofuels Conference,* March 4–5, 102–122. New Delhi, India: Winrock International.

Vaidyanathan, A. 2006. Farmers' suicides on the agrarian crisis. (Commentary.) *Economic and Political Weekly* 41(34):4009–4013.

Vani, M.S. 2009. Community engagement in water governance. In *Water and the laws in India,* ed. Ramaswamy Iyer, 167–212. New Delhi, India: Sage Publications.

Vaze, V., and Samuel, A. 2004. *Women and watershed development: an evaluation study of Indo-German Watershed Development.* Pune, India: SOPPECOM.

Wani, S.P., P.K. Joshi, K.V. Raju *et al.* 2008. Community watershed as a growth engine for development of dryland areas. A Comprehensive Assessment of Watershed Programs in India. *Global Theme on Agroecosystems Report No. 47.* Patancheru, Andhra Pradesh, India: International Crops Research Institute for the Semi-Arid Tropics.

Wani, S.P., P. Pathak, H.M. Tam, A. Ramakrishna, P. Singh, and T.K. Sreeedevi. 2002. Integrated watershed management for minimizing land degradation and sustaining productivity in Asia. In *Integrated land management in the dry areas.* Proceedings of a Joint UNU-CAS International Workshop, 8–13 September 2001, Beijing, China, ed. A. Zafar, 207–230. Tokyo, Japan: United Nations University.

Wani, S.P., J. Rockström, and T. Oweis. (ed.) 2009b. *Rainfed agriculture: unlocking the potential.* Comprehensive Assessment of Water Management in Agriculture Series. Wallingford, UK: CAB International.

Wani, S.P., T.K. Sreedevi, S. Marimuthu, A.V.R. Kesava Rao, and C. Vineela. 2009a. Harnessing potential of *Jatropha* and *Pongamia* plantations for improving livelihoods and rehabilitating degraded lands. In *Proceedings of 6th International Biofuels Conference,* March 4–5, 256–272. New Delhi, India: Winrock International.

Chapter 5

Policies and institutions for increasing benefits of integrated watershed management programs

K.V. Raju, K.H. Anantha, and Suhas P. Wani

5.1 INTRODUCTION

One of the most challenging policy issues for a long time has been conservation and management of land and water resources for sustainable agriculture and poverty reduction, specifically in rainfed areas. Rainfed agriculture contributes 60% of world's staple food and is being practiced on 80% of the world's agricultural area (FAOSTAT 2005). Water is a limiting factor in achieving food production (crop growth) in semi-arid and dry subhumid zones (SEI 2005). Nearly two-thirds of India's agriculture is based on rainfed areas and contributes about 9% of Gross Domestic Product (GDP)[1]. As the source of growth in irrigated areas declines, rainfed agriculture must increase to fill the gap. The recent Comprehensive Assessment of Water for Food and Water for Life showed that challenges of poverty and food security with looming water scarcity cannot be met by irrigated agriculture alone, and major gains have to come through upgrading rainfed agriculture (Molden 2007) and recent forecasts warn of aggravated global water scarcity unless effective water resource management at all levels is done (Seckler *et al.*, 1998; Seckler and Amarsinghe 2000; Shiklomanov 2000; Rosegrant *et al.*, 2002, 2006; Falkenmark and Rockström 2004; SEI 2005). In most rainfed areas, water availability is not a problem but rainfall distribution and poor management creates water scarcity for crops, resulting in low rainwater use efficiency and low crop production (Wani *et al.*, 2003a)[2].

In addition, it is estimated that the ownership of land is highly skewed, nearly 65% of the rural households owning less than one ha (GOI 2007). The landless population

---

[1] These areas are fraught with soil erosion, land degradation, and loss of productivity. These have serious equity implications as they affect the subsistence of poor marginalized people. In addition, burgeoning population, poverty, lack of awareness of improved farm technologies and lack of knowledge and skills to use them, low income levels, and resource-poor farmers constitute major threat to the sustainable development in these areas. These rainfed areas have scarce water resources and are prone to severe land degradation (Wani *et al.*, 2002, 2003d, 2008a, 2009).

[2] On the other hand, the working group on watershed development, rainfed farming, and natural resource management for the Tenth Five Year Plan constituted by the planning commission had assessed that 88.5 million ha degraded wasteland including rainfed areas would need development. The working group report envisaged to cover the entire area in four successive Five Year Plans, commencing from the Tenth Plan up to Thirteenth Plan at an estimated cost of ₹72,750 crores (1994 prices) (GOI 2001b).

covers 12% of rural households. Fragmentation of farm holdings continues unabated owing to the burgeoning population and land acquisition for industrialization and urbanization. Per capita land availability has also dropped from 0.48 ha in 1951 to 0.16 ha in 1991 and is expected to drop to 0.08 ha in 2035 (GOI 2007). Thus, enhancing and sustaining productivity and income of small farms through crop-livestock integration and multiple opportunities through agro-processing, value addition, and biomass utilization must be a high priority.

In recognition of these challenges, governments, donors, and development partners have devoted substantial resources to develop and promote rainfed areas at a catchment/watershed scale for sustainable intensification of agriculture and rural livelihoods. This approach produces multiple benefits in terms of increasing food production, improving livelihoods, protecting the environment and addressing gender and equity issues along with biodiversity concerns (Wani et al., 2003b, 2003c, 2009; Rockström et al., 2007, 2010). However, the evidence from a large number of studies clearly suggests that the economic benefits are not only limited in terms of coverage of beneficiaries but also heavily influenced by the decision-making processes at various stages of implementation. It is in this context, participatory institutions have special significance. It is therefore, imperative that the design of the watershed treatment should include equity and sustainability aspects while planning for productivity enhancement.

This chapter focuses on policy and institutional aspects of watershed approaches thereby seeking to complement recent studies which have concentrated on the overall impact of watershed projects (Wani et al., 2008a) and those focusing on the institutional aspects of watershed management (Joshi et al., 2004; Raju et al., 2008).

5.2 INTEGRATED WATERSHED MANAGEMENT PROGRAM IN INDIA

In tropical rainfed areas, infrequent distribution of rainfall results in long dry spells as well as severe runoff and soil erosion during the crop growing period. Since soil and water are critical natural resources for production activities, watershed development aims at optimum and prudent use of these resources in a sustainable and cost-effective manner. Augmentation of water resources and minimizing soil degradation are the main activities of watershed development programs (Wani et al., 2010, 2011).

In India, since the beginning, watershed program went through the structure driven approach for soil conservation and rainwater harvesting, aiming at only some productivity enhancements. Soil conservation program became synonymous with contour bunding and water conservation with check-dams. This was a compartmental and top-down contractual approach. The watershed development approach in India has seen many changes since 1980s. The objective of the program has undergone substantial modifications to include and address several components of rural livelihoods aspects. Therefore, the approach shifted from traditional top-down approach to more holistic participatory approach to address sustainability and transparency through community participation (Wani et al., 2006a). However, at the present time, watershed models are being developed giving priority to the empowerment of the community and the stakeholders so that the projects operate not as a supply driven project but as a demand driven project. Multidisciplinary teams are involved to provide all the technical expertise to solve the problems at the community level. As a result, the level of participation has improved. This approach ensured participation of stakeholders and the watershed is considered as an entry point for improving the livelihoods of the people (Wani *et al.*, 2008a, 2008b; Sreedevi and Wani 2009).

Watershed approach has shown great promise for increasing groundwater recharge and crop yields since the Seventh Five Year Plan (Sharma 2002; Wani *et al.*, 2003b, 2003c; Joshi *et al.*, 2005). The Government of India, therefore, accorded high priority to the holistic and sustainable development of rainfed areas through the integrated watershed development program (Wani *et al.*, 2008a). The range of other government initiatives and incentives also has an influence on watershed development. Some serve a supporting role in improving the benefits to be derived from watershed resources and include sectoral policies on markets and prices, policies and legislation on land, resources and water rights, and the reorientation of extension and research services in the agricultural, livestock, forestry, and wildlife sectors (Turton *et al.*, 1998).

Currently, the emphasis is on the augmentation of water resources by implementing small watershed projects. The majority of watershed development projects in the country are sponsored and implemented by the Government of India with the help of various state departments, non-government organizations (NGOs)[3], self-gelp groups (SHGs), etc. The Drought Prone Area Programme (DPAP), the Desert Development Programme (DDP), the National Watershed Development Project for Rainfed Areas (NWDPRA), Watershed Development in Shifting Cultivation Areas (WDSCA), and the Integrated Wasteland Development Project are a few of the important development programs that plan, fund, and implement watershed development projects. A total sum of US$6 billion has been invested in the country in various watershed development projects from the inception (early 1980s) of the projects until 2006 (Wani *et al.*, 2008a).

Increasing support to watershed development is being extended by a number of international donors. The Department for International Development (DFID), the Deutsche Gesellschaft for Technishe Zusammenarbeit (GTZ), the Swiss Agency for Development and Cooperation (SDC), the World Bank, and the International Fund for Agricultural Development (IFAD) also sponsor and implement watershed development projects, but a significant proportion (about 70%) of the investment in these projects is being made by the Government of India.

5.3   POLICY ENDORSEMENT AT MACRO LEVEL

The watershed program produces multiple tangible and intangible benefits for individuals as well as for the community as a whole. The present watershed development program follows a holistic approach in building resilience of natural resources and human resource to cope with future challenges (Wani *et al.*, 2008c). Therefore, watershed management has been a key component of development planning of rainfed areas

---

[3] The 1994 guidelines paved the way for participation of public bodies such as NGOs, educational institutions, corporate houses and banks in the form of project implementing agencies (PIAs), leading to a massive growth in the number of NGOs (Turton *et al.*, 1998).

since the early 1980s and got good policy support from the central and state governments in the country. Several programs were launched to target watershed development with a focus to improve food security, alleviate poverty, and sustain the quality of the natural resource base. Several important policies have been launched by the Government of India that affects the success of the watershed development programs. Table 5.1 summarizes the objectives, strategies, and their linkages with watershed development programs in India.

Upgrading the rainfed production system involves integrated approaches to social and ecological management. There is a need for innovations in water management, which requires novel technologies and management practices, e.g., water harvesting and conservation agriculture (Rockström *et al.*, 2007). An integrated approach to rainwater management is necessary where the linkages are addressed between investments and risk reduction, between land, water, and crop, and between rainwater management and multiple livelihood strategies.

For improving rural livelihoods, the watershed approach is a logical unit for efficient management of natural resources, thereby sustaining rural livelihoods. There is a need for environment-friendly resource management practices to alleviate poverty through increased agricultural productivity (Wani *et al.*, 2008c). The current need for resource management in watershed development is use of high science tools and participatory approach. The current model of watershed management, as adopted by ICRISAT (International Crops Research Institute for the Semi-Arid Tropics) watershed consortium team, involves environment-friendly options and the use of new science tools, along with the concept of the consortium approach and emphasis on empowering farmers through capacity building (Wani *et al.*, 2002, 2006a, 2006b; Sreedevi *et al.*, 2004). In the policy front, there is need to strengthen the consortium approach to benefit full potential of participatory resource management. Although there are exceptions, most ongoing watershed development programs have concentrated on physical interventions such as contour bunding and check-dams that are intended to improve groundwater recharging and reduce land and soil degradation. These physical interventions are often not balanced against non-structural measures or measures to improve the production process or open up new livelihood opportunities. These measures include policy changes that bring about cropping pattern shifts and changes in livelihood strategies.

Equity is seen as a major policy issue, with past watershed programs often failing to reach the poorest households. Equity is also identified as critical for the success of collective action. The new common guidelines have tried to address the equity issue through institutionalizing the livelihoods dimension. However, as most of the proposed livelihood components are linked to irrigation water, the spread of livelihood benefits to marginal farmers will be limited, especially in areas that rely on groundwater. The equity safeguards provided in the guidelines are projected as effective in practice. The continued supply-side focus of the policies in the absence of demand management and clearly defined property rights in common resources are likely to perpetuate the inequities.

Furthermore, watershed development is not influenced by watershed policies alone. A range of other policies influence agriculture, water management, and land management. Power tariff pricing (which influences groundwater exploitation), the guaranteed purchase of rice and wheat, and other protection measures greatly influence

Table 5.1 Macro-level policies and their linkages with watershed management program

Policy	Objectives	Strategies	Priorities	Linkages with watershed management
Agricultural development policy	To propel a growth rate of more than 4% per annum – a growth that is efficient, equitable, demand-driven, and sustainable.	• Comprehensive national strategy for attaining lofty goals and targets. • Emphasized strengthening the watershed development program.	• Attempts to intensify integrated and holistic development of rainfed areas by conservation of rainwater. • All spatial components of a watershed will be treated as one geo-hydrological entity.	It reflects the observed commitment of the government to take up watershed development program more aggressively, including provision of the necessary financial and institutional support for its implementation.
Water policy	To address the newly emerging issues of water availability, quality, and inter-sectoral distribution.	Identifies water management as one of the most crucial elements in the development planning.	Rainwater harvesting, preventing soil erosion, providing sustainable irrigation and mitigating the problem of drinking water.	Fails to address watershed management.
Land policy	To protect the interest of the farming community and landless laborers.	Land reforms, land ceilings and restrictions to sell agricultural land.	• Consolidate the fragmented landholdings and distribute the donated and unutilized lands to landless laborers and small and marginal farmers. • Protect the interest of small and marginal farmers and discourage large farmers and thus bring social justice and equity in land distribution.	Supports watershed programs and these watershed programs provide opportunities to small and marginal farmers for collective action that allows a consistent treatment of adjoining pieces of land and reduces costs due to economies of scale.

(Continued)

Table 5.1 Continued

Policy	Objectives	Strategies	Priorities	Linkages with watershed management
Forest policy	• To maintain environment stability through preservation and, where necessary, restore ecological balance that has been adversely disturbed by serious depletion of the forests. • To prevent soil erosion and denudation in the catchment areas of rivers, lakes, and reservoirs in the interest of soil and water conservation for mitigating floods and drought and for reducing siltation of reservoirs. • To control further problem of sand dunes in the desert areas of Rajasthan, India and along the coastal tracts. • To expand the forest/tree cover in India through massive afforestation and social forestry programs. • To meet the growing demand of fuel-wood, fodder, minor forest produce, and small timber of the rural population. • To make the afforestation program a people's movement with the involvement of women.	• Focus on fuel-wood and fodder development on all degraded lands. • Community and village lands are given priority for afforestation and fodder development programs. • Control soil erosion and runoff, prevent desertification and improve micro-ecosystem.	• Encourage the participation of village community through panchayats and revenues were shared with communities to provide incentive to protect the forest resources. • Adequate grazing fees to discourage maintaining large herds of non-essential livestock.	• Forest policy objective and strategies are by and large consistent with those of the watershed development programs. • Rehabilitate, conserve, and manage degraded lands, and augment production of fuel and fodder through community participation.

the structure of incentives for watershed management in rainfed areas. While some policies (like water pricing) strive to improve the economic efficiency of water, agricultural price policies indirectly promote inefficient use of water. For example, subsidized power tariffs for agriculture are leading to widespread depletion and inequitable distribution of the groundwater resources (Shiferaw and Bantilan 2004; Reddy 2005).

5.4 WATERSHED DEVELOPMENT GUIDELINES

Several government departments and state governments undertook watershed development programs. Until 1997, watershed development projects have been carried out under different programs launched by the Government of India. Notably, the DPAP and DDP adopted the watershed approach in 1987. The Integrated Wasteland Development Project initiated by the National Wasteland Development Board in 1989 also aimed at developing wastelands based on the concept of watershed development. Since their inception, these programs were undertaken by the Ministry of Rural Development.

The other major program based on the watershed concept is NWDPRA under the Ministry of Agriculture. All these programs had their own guidelines, norms, funding patterns, and technical components based on their respective and specific aims (GOI 1994). In 1994, the Ministry of Rural Development issued a new comprehensive guideline for all its projects. It was realized that while the focus of these programs may have differed, the common objective of these programs has been land and water resource management for sustainable production. Therefore, common guidelines for all the programs under the Ministry of Rural Development were developed in 1994 and have been implemented since 1995. These guidelines were used by the central-sponsored schemes for the watershed development under the Ministry of Rural Development and the Ministry of Agriculture. Based on the common principles the Ministry of Agriculture developed a new guideline in 1997 for implementation of NWDPRA.

The 1994 guidelines of the Ministry of Rural Development were in operation for five years. This period has seen many successes as well as some failures in watershed development. Hence, greater flexibility of the guidelines was essential to enhance the robustness of the response to the regionally differentiated demands that characterize rural India. Since different ministries were involved in watershed development, it was decided to develop common guidelines. The 1994 guidelines were instrumental for developing the common guidelines. The Ministries of Agriculture and Rural Development jointly developed the 'Common Approach/Principles for Watershed Development' in 2000 (GOI 2000a). The two ministries and Ministry of Environment and Forest then adopted these guidelines as common principles for implementation of watershed development projects.

The Ministry of Agriculture brought out the new guidelines based on the 'Common Approach' in 2000 as 'WARSA – *Jan Sahbhagita*' (people participation), Guidelines for NWDPRA (GOI 2000b). A similar document of revised guidelines (Guidelines for Watershed Development) based on the common principles was also issued by the Ministry of Rural Development (GOI 2001a).

The new guidelines give more flexibility that was needed at village/watershed level. These guidelines, inter alia, envisage the convergence of different programs of the Ministry of Rural Development, Ministry of Agriculture, and other ministries and departments. Following the 73rd and 74th Amendments to the Constitution of India in early 1990s, the Panchayat Raj Institutions (PRIs) are mandated with an enlarged role in the implementation of developmental programs at the grassroot level, and accordingly their role has been more clearly brought out. The new guidelines also emphasize specific and focused project with destination, roadmap, and milestones. The 1994 guidelines were made more flexible, and workable with more participation of the community. The new guidelines provide more emphasis on local capacity building through various training activities and empowering community organizations.

5.4.1 The new common guidelines

Since 1994, several guidelines have been released focusing on different aspects of watershed development and implemented accordingly. As a result, the watershed development programs have had impacts such as increased water availability, reduced soil erosion, increased cropping intensity, more rural employment and increased crop productivity and incomes. However, these benefits have been largely confined to a few successful watershed programs. In fact, almost two-thirds of the watershed programs performed below average, as indicated by a meta analysis jointly undertaken by ICRISAT and Indian Council of Agricultural Research (ICAR). Therefore, at the Ministry level, there was apprehension about further investment to be made on watershed development programs in the country. Thus, ICRISAT in partnership with ICAR institutions, state agriculture universities, a number of state government departments, and NGOs, undertook the comprehensive assessment during 2006–08 and concluded that community watershed programs could serve as growth engines for the development of rainfed areas with prospects of doubling productivity[4]. The Comprehensive Assessment also highlighted the need for reform in institutional and policy front to ensure equity in benefit sharing among all sections of the community. It is in this context, that in coordination with the Planning Commission, an initiative has been taken to formulate Common Guidelines for watershed development projects in order to have unified perspective by all ministries. These guidelines are therefore applicable to all watershed projects in all departments/ministries concerned with watershed development projects.

The new common guidelines set the selection criteria and prioritization of watersheds based on a broad framework and states may incorporate any other relevant criteria within the prescribed framework:

- Extent of rainfed area
- Scarcity of drinking water
- Low productivity of crops
- Poverty index [people in the categories of below poverty line (BPL), scheduled castes (SC) or scheduled tribes (ST), etc.]
- Area owned by small and marginal farmers, SC/ST, BPL
- Contiguity to already treated/ongoing watersheds
- Extent of treatable common property resources
- Willingness of the villagers to participate, contribute, and support the program.

---

[4] The Ministry of Agriculture and Cooperation and the Ministry of Rural Development, jointly sponsored the Comprehensive Assessment.

The framework addresses most relevant issues and identifies broad indicators that must be followed by an implementing agency when selecting the project area. This framework can be complemented through participatory rural appraisal (PRA) exercise and demand-driven approaches. Another most important feature of the common guidelines is the development criteria for success of the watershed. Among others, the exit protocol for the PIAs is developed.

The larger question of livelihood security through natural resource management lies in the effective implementation of watershed programs. The appropriate information at all levels for suitable planning and execution is essential[5]. Therefore, the common guidelines reinstate the centrality of participatory process and community-based institutions for planning, implementation, and future management of the assets created by watershed projects (GOI 2008). It further extends the project duration from four years to seven years with a hike in the cost of ₹12,000 per ha in plain areas and ₹16000 per ha for hilly areas as per the Eleventh Five Year Plan. Emphasis has been laid on cluster approach of micro-watersheds with an average area of 1000–5000 ha as unit of implementation; multi-tier strategy based on ridge to valley approach[6] with the Forest Department and Joint Forest Management Committee playing an important role in the upper reaches mainly in hilly and forest areas.

The Common Guidelines emphasize creation of database both at national and state levels for scientific planning and monitoring which is essential to inform policy makers as well as planners about the current issues and debates of watershed programs. The common guidelines focus on livelihood security while ensuring resource conservation and regeneration and dedicated institutions at central, state, and district levels with professional experts and devolution of finances. The special feature of the common guidelines 2008 is the convergence with other schemes such as National Rural Employment Guarantee Scheme (NREGS), Bharat Nirman, and Backward Region Grant Fund (BRGF). Table 5.2 summarizes different guidelines.

5.5 INSTITUTIONAL ARRANGEMENTS FOR WATERSHED DEVELOPMENT

The institutional arrangements required for sustainable watershed management are equally varied and diverse. Watershed programs faced paradigm shift towards involving local village communities or institutions for implementing the projects. But village level institutions, in most cases, do not have relevant capacities to deal with complexities involved in natural resources management, which need necessary guidance initially to handle the responsibilities. Suitable institutional mechanisms should be placed to

---

[5] Inadequate access to the evaluation studies for the government supported watershed projects emerged as one of the important constraints while carrying out the comprehensive assessment of watershed projects, coordinated by ICRISAT.

[6] The approach is to identify an area, and first look at the forest and the hilly regions, in the upper water catchments wherever possible. The purpose of this approach is that all activities required to restore the health of the catchment area by reducing the volume and velocity of surface runoff, including regeneration of vegetative cover in forest and common land, afforestation, staggered trenching, contour and graded bunding, bench terracing, etc. (GOI 2008).

Table 5.2 Summary of watershed guidelines issued during 1994–2008 by the Ministry of Rural Development (MoRD), Government of India[a]

Item	MoRD guidelines (1994)	Hariyali guidelines (2004)	Neeranchal guidelines (2006)	The new common guidelines (2008)
Objectives	Economic development through regeneration of natural resources in drought-prone areas. Integrated treatment of both non-arable and arable lands on watershed basis.	Stressing the physical nature of watershed development and the central role of the Gram Panchayats in overall economic development.	Providing sustainable rural livelihoods through overall development centered around watershed development (harnessing, conserving, and developing natural resources, i.e., land, water, and biomass).	Sustainable rural livelihoods through natural resource managements.
Selection of watersheds	Where people's participation and voluntary contributions are forthcoming. The area should have an acute shortage of drinking water, preponderance of SC/ST population and wastelands. Only micro-watersheds of 500 ha each are selected. In case of more than one micro-watershed in a block, these need not be contiguous.	Same as 1994 Guidelines. Contiguity with existing watersheds is favored (and is one of the criteria used for watershed selection).	Same as 1994 guidelines. Positive history of women's agency and community action. Micro-watershed will be a part of the milli watershed (4–10 thousand ha) identified by the district watershed management team.	Same as 1994 Guidelines. Area of the project should not be covered under assured irrigation and productivity potential of the land need to be considered.
Institutional arrangement (a) National	MoRD	Same as 1994 Guidelines.	NASDORA (supported by apex stakeholders council/governing board).	NRAA
(b) State	Review committee under the chairmanship of the chief secretary.	Same as 1994 Guidelines.	State level governing board.	State Level Nodal Agency (SLNA)
(c) District	DRDA (District Rural Development Agency)/DPAP (Drought Prone Area Programme). A district watershed advisory committee which offers guidance on issues of implementation, including PIA selection.	Zilla Panchayat (ZP) or DRDA/DPAP	District watershed development agency; supported by district panchayat (i.e., ZP).	District Watershed Development Unit (DWDU)

(d) Watershed	Government or non-government organizations can be selected as PIAs at the watershed level. A multidisciplinary watershed development team to assist the PIA. Watershed association, which shall be a registered body, will be supported by a watershed committee (elected body) with representatives from SHGs, UGs, women, *Gram Panchayat*, and WDT	*Gram Panchayat* and NGO in the case of requirement. *Gram Sabha* will act as watershed association. Watershed committee is discontinued. A multidisciplinary watershed development team to assist the PIA directly.	The village watershed committee (VWC) (elected body) will implement the watershed, supported by milli watershed council, WDT, *Gram Sabha* and women's watershed councils. VWC would have maximum 20 members – 50% women and 33% SC/ST; and representation from SHGs, UGs, *Gram Panchayat*, and WDT. The VWC would function as a subcommittee of the *Gram Panchayat*.	Line departments, autonomous organizations under state/central government, institutes/research bodies, Intermediate Panchayats, and voluntary organizations (VOs) can be selected as PIAs. WDT, SHGs, UGs, and watershed committees to implement the project.
Funding pattern	₹4,000–4,500 per hectare at DRDA level Watershed works: 80% Social and human development: 10% Administrative overheads: 10%	₹6,000 per hectare (from 2001 onwards) at DRDA/ZP level Watershed works: 85% Social and human development: 5% Administrative overheads: 10%	₹12,000 per hectare at DRDA/ZP level Watershed works: 80% Social and human development: 8% Impact assessment: 2% Administrative overheads: 10%	₹12,000 per hectare Watershed works: 78% Social and human development: 10% Impact assessment: 2% Administrative overheads: 10%
Flow of funds	GoI, MoRD to DRDA-PIA	Same as 1994 Guidelines.	NASDORA – State Boards – DWDA–VWC	CLNA – SLNA – DWDU
Cost sharing	Compulsory, 5% for common property resources and 10% for private lands of general category, and 5% for SC/ST.	Same as 1994 Guidelines.	Same as 1994 Guidelines.	Same as 1994 Guidelines.
Time period	Four years	Five years (increased in 2001)	Eight years (2 + 4 + 2)	5–7 years (2 + 3 + 2)

(Continued)

Table 5.2 Continued

Item	MoRD guidelines (1994)	Hariyali guidelines (2004)	Neeranchal guidelines (2006)	The new common guidelines (2008)
Role of NGOs	Can be one of the implementing agencies for a group of 10 or 12 micro-watersheds.	Limited to group formation and social mobilization. PIA where Gram Panchayat and/or ZP capacity is not adequate.	Importance of NGOs is recognized and restored to pre-Hariyali level.	Can be one of the implementing agencies. However, not more than 25% of projects should be given to NGOs.
Watershed development fund (WDF)	Concerned WDT to take care with the help of the watershed committee. To support this activity, WDF is to be created and cost contributions will go to this fund. No operational rules were prepared.	Same as 1994 Guidelines.	Operation rules of the fund should be prepared by VWC and ratified by the Gram Sabha; 50% of the fund should go towards maintenance of common assets. Remaining should be used as revolving fund for giving loans to the villages who have contributed.	Same as 1994 Guidelines.
Gender	This item is addressed in the guidelines but the strategy has not been spelt out clearly.	Same as 1994 Guidelines.	Gender quota of 50% in VWC is introduced. Separate women watershed committee is introduced to support VWC.	Representation of 50% from women and vulnerable groups in watershed committee.
Equity	This item has been addressed by giving user rights for poor and SC/ST in common property resources. But no strategy is defined or suggested.	Same as 1994 Guidelines.	Livelihood component is added for the benefit of poor. But no specific strategy is defined for sharing the common resources like water.	Livelihood component was the thrust for the benefit of poor. Allotting usufruct rights.

Source: Modified from Reddy (2006).

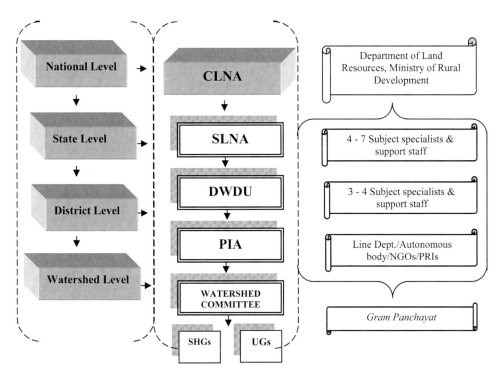

Figure 5.1   Institutional arrangements for integrated watershed management program
(Note: CLNA = Central Level Nodal Agency; SLNA = State Level Nodal Agency; DWDU =
District Watershed Development Unit; PIA = Project Implementing Agency; SHGs = Self-
help groups; UGs=User groups; NGOs=Non-government organizations; PRIs=Panchayat
Raj Institutions)

manage the dilemmas while implementing the project, which ultimately play an important role in determining efficiency and sustainability of the watershed development
programs.

Strong local level institutions can increase the viability and sustainability of watershed management programs by empowering the community to manage and maintain
the assets created under the project (Joshi *et al.*, 2004). However, strengthening and
empowering local institutions needs to be done through continuous process of capacity building which includes technical training and human resource development for
upgrading communication skills, building confidence and leadership, decision-making,
and conflict resolution. A number of institutions therefore are conceived and established at different levels (Figure 5.1). These institutions are created based on the
provisions of the common approach and principles for watershed management are
conceived and developed by the Ministries of Agriculture and Rural Development.

5.5.1   National Rainfed Area Authority

The National Rainfed Area Authority (NRAA) is a central level agency which supports
preparing strategic plans for watershed based development projects at state and district

level keeping in view specific agroclimatic and socioeconomic conditions. The NRAA is mandated to facilitate convergence of different schemes and projects of Government of India which are having similar objectives. The NRAA acts as an effective coordinating mechanism between all bodies/organizations/agencies/departments/ministries who are involved in watershed programs.

5.5.2   Central Level Nodal Agency

The Central Level Nodal Agency (CLNA) set up at the department or ministry level facilitates allocation of the budgetary outlay for the projects among the states keeping the specified criteria in the guidelines. The CLNA comprises professional multidisciplinary experts experienced in the fields of agriculture, water management, institution and capacity building along with representatives of different ministries. The CLNA should interact with state and district level agencies, facilitate, and ensure smooth flow of funds to the District Watershed Development Unit (DWDU) as per the fund flow norms as well as recommendations from the State Level Nodal Agency (SLNA).

5.5.3   State Level Nodal Agency

The state government is responsible to constitute the SLNA. The SLNA will sanction watershed projects for the state on the basis of approved state perspective and strategic plan as per procedure in vogue and oversee all watershed projects in the state within the parameters set out in the Common Guidelines. SLNA has a wide range of functions in the state.

5.5.4   District Watershed Development Unit

In districts, where the area under the watershed development projects is about 25,000 ha, the DWDU, a separate dedicated unit, is established at the district level, which will oversee the implementation of watershed program in each district and will have separate independent accounts for this purpose. The DWDU will identify potential PIAs in consultation with SLNA as per the empanelment process as decided by the respective state governments. The DWDU would facilitate coordination with relevant programs of agriculture, horticulture, rural development, animal husbandry, etc. with watershed development projects for enhancement of productivity and livelihoods.

5.5.5   Project implementing agency

The SLNA would evolve appropriate mechanisms for selecting and approving the PIAs, who would be responsible for implementation of watershed projects in different districts. These PIAs may include relevant line departments, autonomous organizations under State/Central Governments, government institutes/research bodies, intermediate panchayats, and voluntary organizations (VOs). However, the following criteria may be observed in the selection of these PIAs:

- They should preferably have prior experience in watershed-related aspects or management of watershed development projects.

- They should be prepared to constitute dedicated Watershed Development Teams (WDTs).

5.5.6 Watershed Committee

The Watershed Committee usually consists of 10–12 members; half of the members are representatives of SHGs and user groups (UGs), SC/ST community, women and landless persons in the village. The committee manages the project funds, and is responsible for coordination and liaising with the *Gram Panchayat*, PIA, WDT, and other agencies.

5.5.7 Self-help groups

Self-help groups are usually homogeneous groups consisting largely of landless individuals with common or similar sources of income such as animal husbandry, goat rearing, poultry, and agriculture labor. These are more often women's group having 15–20 members in each group. The primary activity of these groups is thrift and credit. Under the watershed guidelines, a revolving fund of an amount to be decided by the Nodal Ministry is allocated to each watershed project for supporting the SHG members to scale-up their activities or to invest in productive assets for increasing income.

5.5.8 User groups

User groups largely consist of those who are likely to derive direct benefits from a particular watershed work or activity such as different types of bunds, farm ponds, farm bunds, etc. The UGs usually formed around specific interventions. The UGs will be responsible for the operation and maintenance of all the assets created under the project in close collaboration with the *Gram Panchayat* and the *Gram Sabha*.

Institutional mechanisms installed in Sujala program seems to be effective in many activities due to the functional linkages between the elements involved in the project addressing post-project sustainability (Wani *et al.*, 2008a). This showed the importance of *Gram Panchayat* linkage and role in the watershed program for the success of the project. Among the watershed community-based organizations (CBOs), SHGs showed the potential to be sustainable in all the programs. The watershed implementing agency is more sustainable in Andhra Pradesh Rural Livelihoods Programme (APRLP) and Hariyali programs. Regarding participation of different sections of watershed community, Sujala program gets higher ranking as different sections of watershed community is involved in program management from the inception of the program. The Hariyali watersheds are ranked least while APRLP and IGWDP (Indo-German Watershed Development Project) watersheds fall between these two extremes with the latter ranked higher than the former (Wani *et al.*, 2008a). Hence, suitable institutional arrangements and linkages within the institutions are necessary to put in place, when the responsibility of managing natural resources is given to local communities to promote inclusiveness among the communities.

5.6 PROMOTING CLOSER INSTITUTIONAL LINKS

There have been significant changes in the options for local institutional development over the last decade. Although earlier guidelines highlighted the importance of PRIs, recent guidelines emphasized the role of local institutions including PRIs to enhance the benefits. The current policy environment is therefore more favorable to the development of local UGs with rights to plan, manage, and retain certain benefits. The new common guidelines for watershed development clearly illustrate this trend particularly well (GOI 2008). The common guidelines emphasize decentralization of powers to state, district, village, and community level. This is supported by financial allocations from Central Government to state /district level and then to village and community level organizations. Apart from institutional alignment for better outcomes in a watershed project, there is a need to account for principal types of macro-economic interventions influencing the farmers' decisions.

Over the years, watershed development has been threatened by the adoption of unsuitable technologies encouraged by subsidies (Kerr *et al.*, 1996). High subsidies for rural electricity encourage the use of electric pumps, leading to overexploitation of newly created groundwater resources in rehabilitated watersheds (Shiferaw and Bantilan 2004; Reddy 2005). In addition, irrigation subsidies cause farmers to shift cropping patterns to water-intensive crops, which results in further water scarcity; thus these crops should not be promoted in unfavorable regions. While subsidies could be justified under some conditions where market or institutional failures prevent socially desirable conservation, there is a need for careful appraisal of the equity and sustainability implications of policies that affect smallholder resource use and management decisions.

Subsidies are one form of incentive operating at the community level. They are intended to simultaneously support improved land management and generate employment opportunities and commonly take the form of contribution to the labor cost of constructing soil water conservation structures. Past experiences have indicated that it is difficult to operate without subsidies because communities act as though they are entitled to handouts, but not responsible for solving their own problems (Kerr *et al.*, 1996). Therefore, relevant questions in relation to the use of subsidies arise: What for? How much? How long? These questions need to be addressed keeping in mind that resource sustainability is at the center stage. One way of improving watershed benefits is through mobilizing social capitals at the community level. Since watershed development is a complex process involving a range of interest groups and distinct operations, promoting UGs and CBOs is the focus of the resource management debate.

Pricing policies for agricultural produce are a key factor influencing farmers' decisions. The most important fact of these policies is that the support prices for cereals such as wheat and rice encourage farmers to switch production away from traditional drought tolerant crops such as millet and sorghum to less water efficient crops such as rice. In many watersheds, gains arising from more efficient conservation of runoff are often offset by greater demands for irrigation water for water intensive crops (Sreedevi *et al.*, 2008). However, for sustainable water resource management, the strategy should include no incentives for growing water intensive crops in *rabi* (postrainy) and summer seasons.

One of the major constraints to the success of agricultural development and more specifically watershed development is high interest rate and lack of credit facilities for farmers. In general, small farmers turn to the informal lending market where interest rates of up to 60% are charged (Reddy *et al.*, 2008). Watershed projects make possible the introduction of new technologies; the use of which often requires a large initial investment but the lack of credit facilities and the high interest rates of the informal sector act as a disincentive to medium- and long-term investment. But there is concern that subsidizing credit schemes for particular projects may adversely affect their sustainability through directly altering economic viability (Turton *et al.*, 1998).

Another set of literature deals with issues of property rights in managing common pool resources. From an economic perspective, a rational farmer can only be expected to undertake resource-improving investments when the on-site discounted benefits that directly accrue to him/her from such investments are higher than discounted costs (Joshi *et al.*, 2004). When private resource-improving and conservation investments generate additional benefits off-site, to the community at large, the level of investment undertaken by the private farmer would be less than what would be socially desirable. This problem arises because of lack of excludability of undesirable effects which means that part of their decisions on resource use and production choices fall under the control of other farmers. Capturing such spillover social benefits requires special policies such as cost sharing, subsidies, and benefit transfer (Shah 2005). To address these problems, there is a need to strengthen institutional links at the grassroots level. As market prices and effective government regulation is missing to ensure sustainable management of these resources, households, and communities have to coordinate the supply and demand to avoid overexploitation.

5.7 DEALING WITH POLICY AND INSTITUTIONAL CONSTRAINTS

The Comprehensive Assessment on watershed programs indicated positive and significant effects for soil and water conservation and sustainable productivity growth in the rainfed regions (Wani *et al.*, 2008a). The study also noted that lack of appropriate institutional support is impeding in tapping potential benefits of the watershed programs. The isolated and piecemeal approach to watershed development has not been consistent with large-scale technology exchange and dissemination. It is indeed important to mention the role of people's participation in watershed development programs. Several experiences have already demonstrated that people's participation was recognized as important as the technical components of the watershed development programs (Wani *et al.*, 2003d, 2008a; Joshi *et al.*, 2005).

5.7.1 Collective action

The first generation watershed programs in the country were supply driven. In this approach, the implementing agency used to identify locations and decide various activities for implementation of the projects. This top-down approach did not match the needs of the stakeholders in the watershed. In the absence of people participation in the program, the potential benefits could not be realized and sustainability was a major concern wherever little benefits were achieved. Therefore, the involvement of the stakeholders in planning, development, and execution of the watershed activities is crucial for several reasons. The watershed is a community driven approach and hence it calls for community participation and collective action. The exclusion of an individual from using watershed services (e.g., drought control) is difficult, if not impossible. The *quasi-public good* feature implies that several individuals can use the services simultaneously without diminishing each other's use values. However, the distribution of investment costs and benefits and the presence of unintended spillover effects determine farmers' technology choice, land use pattern, and investment strategies in the watershed (Joshi *et al.*, 2004). These spillover effects impact on economic profit and utility of users of these services (e.g., soil conservation) will not necessarily enter the decision calculus of the supplier of the services. These services are typically characterized by economies of scale in production and consumption.

In current watershed development projects, collective action is more focused towards resource management and production and enhancement, while input and produce marketing get largely neglected. Extension of collective strategies to output marketing could lead to substantial benefits to smallholder and marginal farmers who now face high transaction cost in marketing their small marketable surplus. Providing institutional and infrastructure support to ease the information and marketing bottlenecks is critical for the success of watershed projects. There is a pressing need for innovative strategies that improve farm-gate prices. Such interventions have the potential to improve economic incentives for the poor and marginal groups to participate in collective action.

5.7.2   Bottom-up approach

Watershed programs involve activities which are able to cater to the specific needs of local people and certainly attract higher participation. These programs aim to contribute to the micro-environment and beneficiaries. Therefore, assessing the needs of the stakeholders together by the implementing agency and the stakeholders is necessary. Since watershed has diverse/heterogeneous communities or groups of beneficiaries, every group should appropriately be addressed in the watershed. Evidence shows that most of the watershed programs were not sensitive to the needs of women and landless vulnerable groups (Meinzen-Dick *et al.*, 2004; Sreedevi and Wani 2007). Therefore, there should be conscious efforts right from the beginning to ensure integration of small and marginal farmers, women, and landless laborers.

5.7.3   Capacity building

According to Wani *et al.* (2008c), training and capacity building is the weakest link in watershed programs. In fact, most stakeholders including policy makers do not have required knowledge about the watershed activities. Most stakeholders believe watershed programs as construction of rainwater harvesting structures and never go beyond to include productivity enhancement, income-generating activities, livestock-based activities, institutions, monitoring and evaluation mechanisms, wasteland development, market linkages, etc. Therefore, unlike the first generation watersheds, the social and human development component in the present watershed programs receives high attention and is instrumental in achieving intended goal.

5.7.4   Knowledge-based Entry Point Activity

Unlike cash-based subsidized entry point activities (EPAs), knowledge-based activities provide a sense of ownership on the assets created in the project. Subsidies in the form of wages for the construction of soil water conservation technologies can leave a legacy of dependency once support is withdrawn. Thus, knowledge-based EPAs can help achieving sustainability of the project.

It is now accepted that the problems related to water shortage in the rainfed systems are most appropriately addressed through the implementation of soil conservation and rainwater harvesting practices by adopting community watershed management strategy. To achieve this, community participation in program activities from planning, execution, and monitoring is critical for the success and sustainability of the interventions. However, mobilizing community participation is a challenging task and lack of community participation is identified as a major factor for lower or no impact of watershed programs (Farrington *et al.*, 1999; Kerr *et al.*, 2000; Wani *et al.*, 2003d; Joshi *et al.*, 2005).

Appropriately introduction of a watershed development program to the community has been recognized as an important activity and this is best done through EPA. An essential component of an EPA is building the rapport with the community, strengthening and sustaining it throughout the life of the program and beyond. To build a rapport between the PIA and the villagers before initiating the watershed programs, an EPA is envisaged. The EPA is identified through PRA. Realizing the importance of EPA, the Government of India watershed guidelines specifically allocate a financial budget of 4%, which works out as ₹0.4 million (US$8000) for a 1000-ha watershed (GOI 2008).

5.7.5   Empowering women and vulnerable groups

Since community participation plays an important role in determining the performance of watersheds, targeted activities should be economically beneficial to women and vulnerable groups (Sreedevi and Wani 2007). In order to restore active participation of marginal sections of the community, there should be more income-generating activities, and commercial scale activities which resulted in better participation as well as improved decision-making and better social status for women and landless families in the society (Joshi *et al.*, 2009). Sreedevi and Wani (2007) revealed that harnessing gender power by balancing activities for men and women, farmers, and landless people was found to be effective for enhancing the impact of community watershed programs.

To reduce drudgery for women, there should be specific interventions targeting drinking water supply, and efficient technologies for enhancing agricultural productivity through the operations undertaken by women. Targeted income-generating activities are must for women to get them additional cash in their hands which can enable them to improve their knowledge and social status and reduce workload accordingly.

5.8   SUSTAINABLE WATERSHED MANAGEMENT: ROLE OF COMMON GUIDELINES

Integrated watershed management approach is identified as a suitable approach to improve the rural livelihoods through increased productivity and efficient management of natural resources in the drylands (Wani *et al.*, 2003d, 2008a; Joshi *et al.*,

2009; Sreedevi and Wani 2009; Shieferaw *et al.*, 2009). However, lack of appropriate institutional support is impeding in tapping potential benefits of the watershed programs. The impacts have been identified in isolated cases due to lack of monitoring and evaluation process. Therefore, there is a need to concentrate more efficiently on market-led development holistic strategies than focusing on piecemeal approach.

The new common guidelines brought out by Government of India is the first set of guidelines that apply to watershed development projects across three ministries, viz., Ministry of Rural Development, Ministry of Agriculture, and Ministry of Environment and Forests. The common guidelines take into account significant lessons from the view point of policy formulation processes in the context of democratic setup within the country. The common guidelines reveal the difficulties in breaking out of the mindset of a fragmented view of schemes and programs and affecting a broad paradigm shift towards sustainable agriculture in general, and rainfed agriculture in particular, that involves simultaneous changes in a range of macroeconomic policies pertaining to technology, public expenditure in natural resource development, subsidies, pricing, etc. In the absence of such a shift, the basic agenda of sustainable agriculture could take diversion by the rapidly emerging policy prescriptions in favor of "privatization and corporatization", especially of small farm agriculture. In this context, the common guidelines offer a gradual expanding space for democratic intervention in the implementation and policy formulation processes.

5.8.1 Institutional responsibilities

The institutional arrangements suggested by the common guidelines strike a balance between different types of PIAs which may include department, VOs, NGOs, *Gram Sabhas, Gram Panchayats*, and CBOs created under watershed projects. It is imperative that the VOs/NGOs get their due share as PIAs, rather than getting relegated as agencies for community organization and awareness generation. The common guidelines also stipulate that not more than 25% of projects should be given to VOs/NGOs. This may be a good move on the part of common guidelines to identify and honor efficient and competent NGOs for effective implementation of the project.

5.8.2 Delegation of power to the states

The most critical feature of the common guidelines is the delegation of power to the states: the power of sanctioning and overseeing the implementation within the parameters of the common guidelines are to be vested with the state governments. This leaves substantial scope for calibration and fine-tuning of some of the concerns that may need an additional emphasis. A dedicated SLNA shall be constituted by the state government with an independent bank account for direct transfer of the financial assistance from the center. The SLNA will sign memorandum of understanding with the departments/nodal agencies that may be set up by the ministries in the Central Government. The common guidelines embody an unprecedented devolution of decentralization of powers to state, district, village, and community level. However, the issue of transparency and sharing of information or data and putting it in the public domain needs special attention.

5.8.3 Dedicated institutions

The most critical feature of the common guidelines is allotting dedicated institutions at various levels. These institutions have been assigned with specific functions. The SLNA and DWDU are the two major institutions at state and district level respectively which are key institutions in executing and monitoring watershed works. At the watershed level, institutional arrangements follow the 2001 revised guidelines. The Watershed Committee will receive and manage funds with guidance from *Gram Panchayat*. If the *Gram Panchayat* covers more than one village, subcommittees at village level are proposed. When a watershed consists of more than one *Gram Panchayat*, separate Watershed Committees will be organized for each *Gram Panchayat*. However, allocation and sharing of project funds between these Watershed Committees and *Gram Panchayats* may be a problem, since they will differ in area and requirements. Such aspects may have to be addressed in the course of preparation of the perspective plan and detailed project document or when the states draw up their own guidelines.

5.8.4 Convergence

The common guidelines make a special reference to convergence with other schemes such as NREGS, Bharat Nirman, and BRGF. They emphasize differential rates of cost-sharing privileging the resource-poor sections like SC and ST and clearly specify that the UGs in close collaboration with *Panchayats/Gram Sabha* should maintain structures and assets by using the Watershed Development Fund. The common guidelines suggest a compulsory amount of 5% for common property resources, 10% for private lands of general category, and 5% for SC/ST. Importantly, at least 50% of the Watershed Development Fund needs to be reserved for maintenance of assets created on community land or for common use under the project.

Convergence is becoming brand hallmark of any development project in recent years. In the context of watershed management approach, community watershed is used as an entry point to converge and to explicitly link watershed development with rural livelihoods and effective poverty reduction and in the process identify policy intervention at micro, meso, and macro levels (see Sreedevi and Wani 2009). For instance, APRLP has demonstrated the scope for issues related to suitable processes for change in micro practices, macro policies, convergence, and information and management systems. Convergence, therefore, can take place at different levels. For a successful convergence, socioeconomic institutional and policy needs are necessary to increase adoption of improved options by the rural people.

The process of convergence requires several components such as individual and community-based interventions, use of new science tools, empowerment of community and stakeholders, and consortium approach for technical backstopping. Therefore, convergence at the community level acts as a base flow to the bottom-up approach for promoting rural livelihoods. Convergence of crop-livestock based activities and other income-generating micro-enterprises in the watersheds by linking watershed development and research activities increases the effectiveness of holistic watershed programs through efficient use of conserved/harvested water and other natural resources for increasing production and income of the rural poor.

5.8.5 Consortium approach

There is a need for a multi-institutional consortium approach for technical backstopping to empower farmers and develop human and institutional resources through capacity building measures by integrating the activities of Krishi Vigyan Kendras (KVKs), farmers' training centers, NGOs, research organizations, and line departments of the state government for technical backstopping to undertake action research at watershed level. Consortium approach enables the addressing of equity, gender, sustainability, and improved livelihoods which are the pillars of inclusive and sustainable development (Wani *et al.*, 2002, 2009). For market-led development, the need for functional and effective linkages among watershed institutions and other institutions such as markets, banks, etc. is imperative for success of the program.

The common guidelines uphold the importance of consortium of resource organizations for capacity building support, which is a crucial component to achieve the desired results from watershed development projects. The common guidelines reinstate that the capacity building strategy and activities enumerated by NRAA, nodal agencies at the central level, and consortiums of resource organizations should be funded separately over and above the earmarked budget for institution and capacity building in the preparatory phase of the watershed development project. This not only strengthens the social and human resource development but also provides knowledge sharing opportunities for different actors in watershed management programs.

The common guidelines suggested key strategies for social and human development and that NRAA will collaborate with various resource organizations for developing national as well as state specific capacity building strategies. Emphasis has been laid on dedicated and decentralized institutional support and delivery mechanism and mechanism for effective monitoring and follow-up processes.

5.8.6 Addressing equity

Equity is an essential element in ensuring perennial benefits in the program. As indicated earlier, watershed program targeted overall development of local economy through natural resource development and productivity enhancement. Therefore, in order to distribute the benefits amongst all the beneficiaries, involvement of stakeholders is essential. Interestingly, landless and women have been inducted into the program through involvement in allied micro-enterprise activities. The SHGs have been promoted targeting women, assetless, and other socially and economically disadvantaged persons so as to minimize inequalities and social conflicts. They also set up micro-enterprises to provide supporting services to vulnerable sections. Further, 10% of the total budget was earmarked for livelihood activities for the assetless persons and 13% for production system and micro-enterprises. This is a positive move in equal distribution of watershed benefits.

Gender issue has been addressed by making provision to include women and SHG members in watershed committees. Various UGs have been suggested with due representation from women and vulnerable groups. SHGs have been dominant in project implementation at the grassroots level for planning, execution, and monitoring.

5.8.7   Project management

A positive step considering the watershed development program in three phases, viz., preparatory, works, and consolidation phase, can make a difference, if the progression from one phase to the next is made conditional on meeting the objectives, indicators, and targets of the previous phase, otherwise automatic progression would make little difference on the ground[7]. The duration of the project has been enlarged into minimum four to maximum seven years depending upon the activities and ministries/departments.

5.8.8   Post-project sustainability

Sustainable watershed management lies in the hands of communities. The common guidelines emphasize handing over responsibility to *Panchayats* and/or UGs as part of the withdrawal phase. This is important as there is a need to ensure actual performance or sustenance in the post-project phase. Continued long-term monitoring of the project impacts and the arrangements for future management is essential to address the sustainability issue. However, this needs to be ensured through proper capacity building and technical backstopping at the grassroots level. Social and human resource development is an essential component in ensuring the post-project sustainability.

5.9   OPERATIONALIZING POLICIES

The real challenge in achieving success in watershed development program lies with operationalizing policies. Special efforts are needed to enable these policies. However, the performance of a watershed depends on certain specific factors, for example, people's participation. While significant progress has been made in operationalizing a particular form of watershed management, much remains to be done for scaling-up the approach and seeing it translate into tangible benefits for communities. One of the key challenges lies in the formulation of appropriate institutional arrangements for more widespread application, given the isolation of different disciplines – and of research from development – within existing institutions. To move forward here, it is important to take a systematic look at the tasks and skill base required to operationalize watershed management program, and the degree to which existing institutions can be mobilized to fill the gap.

Another key challenge lies in forging stronger linkages between research and development, so that development is linked to and given at least equal status as research, and action research given equal weightage as more conventional empirical research. For this, university training, institutional mandates, incentive systems, and opportunities for social learning at local and institutional levels must be given close consideration if the integrated mandate embodied in integrated watershed management is to be enabled. In this direction, the new common guidelines have paved the way for new ideas through

---

[7] For example, fund releasing procedure is conditional on meeting the objective and proper certification and submission of documents after completion of each phase (Para 9 of Common Guidelines).

consortium approach in achieving the success of watershed development approach. For the first time in the history of watershed development, in CLNA, research organizations such as ICRISAT, Central Research Institute for Dryland Agriculture (CRIDA), Central Soil and Water Conservation Research and Training Institute (CSWCRTI) along with NGOs and departments and ministries are working together for proper operationalization of common guidelines. This is a remarkable institutional reform process which plays a major role in technology transfer as well as multidisciplinary integration for approaching watershed problems.

Most importantly, the mindset of different actors has to be tuned into the new policies. As the implementing departments and PIAs feel comfortable with earlier policies, there seems to be difficulty to adjust with new guidelines. The change should occur at various levels. However, it should begin with bottom level actors starting from farmers to PIAs, SLNA, and political representatives. The political commitment and the bureaucratic support should ensure the progress in new direction. The mindset of bottom level actors can be tapped with continuous consultation through capacity building and meetings. However, at the top level, it should be self-driven spontaneously.

To operationalize, people need more handholding. Local knowledge and skills need to be tapped through appropriate channels. People's involvement is a key in technology exchange as well as to spread knowledge among farmers' community. Therefore, people's participation has to be ensured in all phases of the project for communicating the goals of the project.

Climate change issues should be addressed. Climate-induced increase in surface temperatures can impact hydrological process of a watershed system and has potential implications on water quantity and quality at a regional scale. Further, increasing demand from population growth and economic development will lead to exacerbating water stress. Therefore, policy and institutional reform is necessary through: (a) clarifying rules governing roles of various stakeholders; (b) minimizing fragmentation and overlap of mandates of various agencies; (c) supporting decentralization and capacity building of local agencies; and (d) prioritizing watershed related research and technology development. In the new common guidelines the roles and responsibilities of CLNA, SLNA, DWDU, and PIAs have been specified. These responsibilities must be adhered to ensure transparency and sustainability of the project.

In addition, it is important to include a monitoring and evaluation system that seeks to ensure integration through periodic re-assessment. For the purposes of component integration, monitoring must assess the impacts of activities on diverse system components. Therefore, monitoring must address the impact of activities on diverse components (water, livestock, crop yield, and soil fertility). To operationalize this, it is important to consider all potential interactions between the activity conducted and different components, and to identify priority indicators from scientific and/or local perspectives that will be monitored for each. The recommendations of comprehensive assessment are worth noting here (Wani *et al.*, 2008a).

The recommendations include: (i) mid-term evaluation and impact assessment after program completion and post-project phase will enable PIAs to make mid-course corrections and government to adjust policies; (ii) a broad assessment to be made that takes into account total environmental and socioeconomic impacts rather than the current focus on income, productivity, water enhancement, and employment generation; (iii) baseline information and needs-assessment in uniform format must be undertaken before funds for works released. Further, only limited numbers of separate, tangible, and easily measurable indicators need to be tracked and current participatory monitoring, resource mapping, and social audit will enhance transparency and equity; and (iv) cost-effective and sustainable watershed development needs hydrological and environmental data from benchmark watersheds in each agroecoregion and district. This will also enable an assessment of impacts outside the watersheds. Such work needs adequate financial support.

In terms of multidisciplinary integration, it is important that interdisciplinary planning be done in detail, down to the level of activities, and the approach to be used to carry them out. In social terms it becomes critical as how to motivate and mobilize the community for balancing short- with long-term benefits, and farmer investments with project inputs. In economic terms, market opportunities should be identified prior to the selection of the agro-enterprises or crop varieties to be field-tested to counter the supply-driven emphasis of smallholder farming systems. Most importantly during the implementation phase, both intermediate planning and monitoring and evaluation should be done by multidisciplinary teams at project level and by multiple local stakeholders.

As an effort in operationalizing new common guidelines, model watersheds have been established in each state/district as sites of learning. As a first hand exercise, ICRISAT and ICAR institutes such as CSWCRTI are implementing these model watersheds in the country. ICRISAT is implementing 13 watersheds across nine states covering south, north, and western India while CSWCRTI covers north and northeastern states. The new components have been added in the implementation of model watersheds.

5.10   CONCLUSIONS

The issues discussed in this chapter are based on the information elicited from micro-level studies and macro-level changes with regard to policy and institutional structure in watershed management. It attempts to analyze the supporting role of existing policies and programs in augmentation of watershed benefits in the country. However, it is clear that watershed management has got good policy support from central and state governments in addressing the needs of poor sections of the society through watershed development approach. The discussion clearly reveals that common guidelines were helpful to evolve new approaches which accounts for varying needs of the community. However, the institutional support was not adequate during early phase of watershed management. This is due to underestimation of the role of UGs and other beneficiary groups in ensuring participation.

However, the new common guidelines were evolved taking into account lessons learnt from success and failure cases of earlier programs to guide watershed management program more effectively. They are focusing more on livelihood aspects taking into account all sections of the community in ensuring higher participation and gender equity. However, it is essential that benefits of all stakeholders should match their contributions and costs. Therefore, equity in benefit sharing contributes for greater collective action and participation which ensures sustainability of the program.

Besides all these, functional and effective linkages among watershed institutions and other institutions such as markets, banks, etc. are imperative for success of the program. Institutions at all levels need to further strengthen their capacities in order to successfully cope with contemporary challenges and to adopt innovative management styles. Capacity building is a multidimensional concept: it requires scientific as well as non-scientific competencies; it requires cooperation that enable knowledge sharing and mutual learning; and it requires institutionalized linkages between the producers of scientific knowledge and local knowledge. Capacity building measures should finally create conditions that are needed to make productive use of knowledge instead of solely creating that knowledge.

In the present system, inputs and produce markets are largely neglected. Extension of collective strategies to output marketing could lead to substantial benefits to smallholder and marginal farmers who now face high transaction cost in marketing their small marketable surplus. Therefore, future watershed policies need to reflect and influence the wider policy environment, especially policies related to agricultural development, agricultural input and output marketing, and other linked sectors like infrastructure. This can set the path of a sustainable and strong resilient rural economy.

REFERENCES

Falkenmark, M., and J. Rockström. 2004. Balancing water for humans and nature: the new approach in ecohydrology. London, UK: Earthscan.

FAOSTAT. 2005. http://faostat.fao.org/

Farrington, J., C. Turton, and A.J. James. (eds.) 1999. *Participatory watershed development in India: challenges for the 21st century.* New Delhi, India: Oxford University Press.

GOI. 1994. *Guidelines for watershed development.* New Delhi, India: Department of Land Resources, Ministry of Rural Development, Government of India.

GOI. 2000a. *Report of the inter-ministerial sub-committee on formulation of common approach/principles for watershed development.* New Delhi, India: Government of India.

GOI. 2000b. *Guidelines for National Watershed Development Project for Rainfed Areas (NWDPRA).* New Delhi, India: Ministry of Agriculture, Government of India.

GOI. 2001a. *Guidelines for watershed development.* (Revised 2001) New Delhi, India: Department of Land Resources, Ministry of Rural Development, Government of India.

GOI. 2001b. *Report of the working group on watershed development, rainfed farming and natural resource management for the Tenth Five Year Plan.* New Delhi, India: Planning Commission, Government of India.

GOI. 2007. *Report of the working group on natural resources management, Eleventh Five Year Plan (2007–2012), Volume I: Synthesis.* New Delhi, India: Planning Commission, Government of India.

GOI. 2008. *Common guidelines for watershed development projects.* New Delhi, India: Department of Land Resources, Ministry of Rural Development, Government of India.

Joshi, P.K., A.K. Jha, S.P. Wani, Joshi, Laxmi, and R.L. Shiyani. 2005. *Meta-analysis to assess impact of watershed program and people's participation.* Research Report 8. Comprehensive Assessment of Watershed Management in Agriculture. Patancheru, India: International Crops Research Institute for the Semi-Arid Tropics; and Manila, Philippines: Asian Development Bank.

Joshi, P.K., A.K. Jha, S.P. Wani, and T.K. Sreedevi. 2009. Scaling-out community watershed management for multiple benefits in rainfed areas. In *Rainfed agriculture: unlocking the potential*, ed. S.P. Wani, J. Rockström, and T. Oweis, 276–291. Comprehensive Assessment of Water Management in Agriculture Series. Wallingford, UK: CAB International.

Joshi, P.K., Vasudha Pangare, B. Shiferaw, S.P. Wani, J. Bouma, and C. Scott. 2004. Socioeconomic and policy research on watershed management in India–Synthesis of past experiences and needs for future research. *Global Theme on Agroecosystems, Report No. 7*. Patancheru, Andhra Pradesh, India: International Crops Research Institute for the Semi-Arid Tropics.

Kerr, J., G. Pangare, V.L. Pangare, and P.J. George. 2000. *An evaluation of dryland watershed development projects in India*. EPTD Discussion Paper No. 68. Washington, DC, USA: International Food Policy Research Institute.

Kerr, J., N.K. Sanghi, and G. Sriramappa. 1996. *Subsidies in watershed development projects in India: distortions and opportunities*. Gatekeeper Series 61. London, UK: International Institute for Environment and Development.

Meinzen-Dick, R., M. DiGregorio, and N. McCarthy. 2004. Methods for studying collective action in rural development. *Agricultural Systems* 82:197–214.

Molden, D. (ed.) 2007. *Water for food, water for life: a comprehensive assessment of water management in agriculture*. London, UK: Earthscan; and Colombo, Sri Lanka: International Water Management Institute.

Raju, K.V., Abdul Aziz, S.S. Meenakshi Sundaram, Madhushree Sekher, S.P. Wani, and T.K. Sreedevi. 2008. Guidelines for planning and implementation of watershed development program in India: a review. *Global Theme on Agroecosystems Report No. 48*. Patancheru, Andhra Pradesh, India; International Crops Research Institute for the Semi-Arid Tropics.

Reddy, V.R. 2005. Costs of resource depletion externalities: A study of groundwater overexploitation in Andhra Pradesh, India. *Environment and Development Economics* 10:533–556.

Reddy, V.R. 2006. Getting the implementation right. Can the proposed watershed guidelines help? *Economic and Political Weekly* 41(40):4292–4295.

Reddy, V.R., M. Gopinath Reddy, Y.V. Malla Reddy, and John Soussan. 2008. Sustaining rural livelihoods in fragile environments: resource endowments and policy interventions – a study in the context of participatory watershed development in Andhra Pradesh. *Indian Journal of Agricultural Economics* 63(2):169–187.

Rockström, J., N. Hatibu, T. Oweis, and S.P. Wani. 2007. Managing water in rainfed agriculture. In *Water for food, water for life: a comprehensive assessment of water management in agriculture*, ed. D. Molden, 315–348. London, UK: Earthscan; and Colombo, Sri Lanka: International Water Management Institute.

Rockström, J., L. Karlberg, S.P. Wani *et al.* 2010. Managing water in rainfed agriculture – the need for a paradigm shift. *Agricultural Water Management* 97(4):543–550.

Rosegrant, M., X. Cai, and S. Cline. 2002. *World water and food to 2025: Dealing with scarcity*. Washington DC, USA: International Food Policy Research Institute.

Rosegrant, M., W. Ringler, C. Benson *et al.* 2006. *Agriculture and achieving the Millennium Development Goals*. Washington, DC, USA: World Bank.

Seckler, D., and U. Amarasinghe. 2000. Water supply and demand, 1995 to 2025. In *IWMI Annual Report* 1999–2000, 9–17. Colombo, Sri Lanka: International Water Management Institute.

Seckler, D., U. Amarasinghe, D. Molden, R. de Silva, and R. Barker. 1998. *World water and demand and supply, 1990 to 2025: scenarios and issues*. Research Report 19. Colombo, Sri Lanka: International Water Management Institute.

SEI. 2005. Sustainable pathways to attain the millennium development goals – assessing the role of water, energy and sanitation. Document prepared for the UN World Summit, 14 September 2005, New York, USA. Stockholm, Sweden: Stockholm Environment Institute.

Shah, A. 2005. Economic rationale, subsidy and cost sharing in watershed projects. *Economic and Political Weekly* 40(26):2663–2671.

Sharma, R. 2002. Watershed development adaptation strategy for climate change. Presented in South Asia Expert Workshop on Adaptation to Climate Change for Agricultural Productivity. Organized by the Government of India, UNEP, and CGIAR in New Delhi, India.

Shiferaw, B., and C. Bantilan. 2004. Rural poverty and natural resource management in less-favoured areas: revisiting challenges and conceptual issues. *Journal of Food, Agriculture and Environment* 2(1):328–339.

Shiferaw, B., J. Okello, and V. Ratna Reddy. 2009. Challenges of adoption and adaptation of land and water management options in smallholder agriculture: Synthesis of lessons and experiences. In *Rainfed agriculture: unlocking the potential*, ed. S.P. Wani, J. Rockström, and T. Oweis, 258–275. Comprehensive Assessment of Water Management in Agriculture Series. Wallingford, UK: CAB International.

Shiklomanov, I. 2000. Appraisal and assessment of world water resources. *Water International* 25(1):11–32.

Sreedevi, T.K., B. Shiferaw, and S.P. Wani. 2004. *Adarsha watershed, Kothapally: understanding the drivers of higher impact*. Patancheru, Andhra Pradesh, India: International Crops Research Institute for the Semi-Arid Tropics.

Sreedevi, T.K., and S.P. Wani. 2007. Leveraging institutions for enhanced collective action in community watersheds through harnessing gender power for sustainable development. In *Empowering the poor in the era of knowledge economy*, ed. S. Mudrakartha, 27–39. Ahmedabad, Gujarat, India: VIKSAT.

Sreedevi, T.K., and S.P. Wani. 2009. Integrated farm management practices and upscaling the impact for increased productivity of rainfed systems. In *Rainfed agriculture: unlocking the potential*, ed. S.P. Wani, J. Rockström, and T. Oweis, 222–257. Comprehensive Assessment of Water Management in Agriculture Series. Wallingford, UK: CAB International.

Sreedevi, T.K., S.P. Wani, R. Sudi *et al.* 2008. Impact of watershed development in low rainfall region of Maharashtra: a case study of Shekta Watershed. *Global Theme on Agroecosystems Report No.* 49. Patancheru, Andhra Pradesh, India: International Crops Research Institute for the Semi-Arid Tropics.

Turton, C., Michael Warner, and Ben Groom. 1998. *Scaling up participatory watershed development in India: a review of the literature*. Network Paper No. 86. London, UK: Agricultural Research and Extension Network, Overseas Development Institute.

Wani, S.P., K.H. Anantha, T.K. Sreedevi, R. Sudi, S.N. Singh, and Marcella D'Souza. 2011. Assessing the environmental benefits of watershed development: evidence from the Indian semi-arid tropics. *Journal of Sustainable Watershed Science and Management* 1(1):10–20.

Wani, S.P., P.K. Joshi, K.V. Raju *et al.* 2008a. Community watershed as a growth engine for development of dryland areas. A comprehensive assessment of watershed programs in India. *Global Theme on Agroecosystems Report No.* 47. Patancheru, Andhra Pradesh, India: International Crops Research Institute for the Semi-Arid Tropics (ICRISAT).

Wani, S.P., P.K. Joshi, Y.S. Ramakrishna, T.K. Sreedevi, P. Singh, and P. Pathak. 2008b. A new paradigm in watershed management: a must for development of rain-fed areas for inclusive growth. In *Conservation farming: enhancing productivity and profitability of rainfed areas*, ed. Anand Swarup, Suraj Bhan, and J.S. Bali, 163–178. New Delhi, India: Soil Conservation Society of India.

Wani, S.P., A.R. Maglinao, A. Ramakrishna, and T.J. Rego. (ed.) 2003a. *Integrated watershed management for land and water conservation and sustainable agricultural production in Asia*. Proceedings of the ADB-ICRISAT-IWMI Annual Project Review and Planning Meeting, Hanoi, Vietnam, 10–14 December 2001. Patancheru, Andhra Pradesh, India: International Crops Research Institute for the Semi-Arid Tropics.

Wani, S.P., P. Pathak, L.S. Jangawad, H. Eswaran, and P. Singh. 2003b. Improved management of Vertisols in the semi-arid tropics for increased productivity and soil carbon sequestration. *Soil Use and Management* 19:217–222.

Wani, S.P., P. Pathak, T.K. Sreedevi, H.P. Singh, and P. Singh. 2003c. Efficient management of rainwater for increased crop productivity and groundwater recharge in Asia. In *Water productivity in agriculture: limits and opportunities for improvement*, ed. J.W. Kijne, R. Barker, and D. Molden, 199–215. Wallingford, UK: CAB International.

Wani, S.P., P. Pathak, H.M. Tam, A. Ramakrishna, P. Singh, and T.K. Sreedevi. 2002. Integrated watershed management for minimizing land degradation and sustaining productivity in Asia. In *Integrated land management in the dry areas*. Proceedings of a Joint UNU-CAS International Workshop, 8–13 September 2001, Beijing, China, ed. Z. Adeel. 207–230. Tokyo, Japan: United Nations University.

Wani, S.P., Y.S. Ramakrishna, T.K. Sreedevi *et al.* 2006a. Issues, concepts, approaches and practices in the integrated watershed management: experience and lessons from Asia. In *Integrated management of watershed for agricultural diversification and sustainable livelihoods in Eastern and Central Africa: lessons and experiences from semi-arid South Asia*. Proceedings of the International Workshop held at ICRISAT, Nairobi, Kenya, 6–7 December 2004, ed. B. Shiferaw, and K.P.C. Rao, 17–36. Patancheru, Andhra Pradesh, India: International Crops Research Institute for the Semi-Arid Tropics.

Wani, S.P., H.P. Singh, T.K. Sreedevi *et al.* 2003d. Farmer-participatory integrated watershed management: Adarsha watershed, Kothapally, India, an innovative and upscalable approach. A case study. In *Research towards integrated natural resources management: Examples of research problems, approaches and partnerships in action in the CGIAR*, ed. R.R. Harwood, and A.H. Kassam, 123–147. Washington, DC, USA: Interim Science Council, Consultative Group on International Agricultural Research.

Wani, S.P., P. Singh, K.V. Padmaja, R.S. Dwivedi, and T.K. Sreedevi. 2006b. Assessing impact of integrated natural resource management technologies in watersheds. In *Impact assessment of watershed development–issues, methods and experiences*, ed. K. Palanisami, and D. Suresh Kumar, 8–58.

Wani, S.P., T.K. Sreedevi, S. Dixit, K. Kareemulla, R. Singh, and K. Tirupataiah. 2009. Consortium approach for capacity building for watershed management in Andhra Pradesh: a case study. *Global Theme on Agroecosystems Report No. 51.* Patancheru, Andhra Pradesh, India: International Crops Research Institute for the Semi-Arid Tropics.

Wani, S.P., T.K. Sreedevi, T.S.V. Reddy, B. Venkateshvarlu, and C.S. Prasad. 2008c. Community watersheds for improved livelihoods through consortium approach in drought prone rainfed areas. *Journal of Hydrological Research and Development* 23:55–77.

Wani, S.P., T.K. Sreedevi, R. Sudi, P. Pathak, and Marcella D'Souza. 2010. Groundwater management an important driver for sustainable development and management of watersheds in dryland areas. In *2nd National Ground Water Congress,* 22 March 2010, New Delhi, 195–209. New Delhi, India: Government of India, Ministry of Water Resources.

Chapter 6

Application of new science tools in integrated watershed management for enhancing impacts

Suhas P. Wani, P.S. Roy, A.V.R. Kesava Rao, Jennie Barron, Kaushalya Ramachandran, and V. Balaji

6.1 INTRODUCTION

Insufficient scientific inputs in terms of research and development are responsible for low productivity of rainfed systems in the semi-arid tropics (SAT), in addition to biophysical and social constraints such as poor infrastructure, inherent low soil fertility, frequent occurrence of drought, severe degradation of natural resource base, and poor social and institutional networks (Wani *et al.*, 2003, 2009). Researchers and development workers apply high science tools mostly in well endowed areas as returns on the investments in terms of economic impact, successful experimentation, and adoption of new technologies are quick and assured. However, recent studies undertaken by the Asian Development Bank (ADB) have shown that the investments in rainfed areas are not as productive as in the well endowed areas but also more effective in reducing poverty in these hotspots of poverty. The International Crops Research Institute for the Semi-Arid Tropics (ICRISAT) in partnership with national agricultural research systems (NARSs) in Asia, for example, Central Research Institute for Dryland Agriculture (CRIDA), National Remote Sensing Centre (NRSC), State Agricultural Universities (SAUs) in India; Department of Agriculture (DoA) and Department of Land Development (DLD), Khon Kaen University (KKU), Thailand; Yunnan Academy of Agricultural Sciences (YAAS), The Guizhou Academy of Agricultural Sciences in China; Vietnam Academy of Agricultural Sciences (VAAS), Vietnam has applied new science tools such as simulation modeling, remote sensing, geographical information system (GIS), and information and communication technology (ICT) for enhancing the productivity of rainfed systems in the SAT through science-led development.

Watershed management is a process of formulating and carrying out a course of action involving manipulation of the natural system of a watershed to achieve objectives specific to the watershed such as control of soil erosion and land degradation, reclamation and rehabilitation of waste/degraded lands, land use changes consistent with land capability, or management of croplands, grasslands, and forests along with management of water resources. For a balanced participatory approach in the watershed development, all the stakeholders have to be involved at planning level itself for the smooth and efficient execution of the works in a timely manner. After understanding the requirements of the stakeholders, spatial technologies play a very crucial role in watershed planning. The tools along with spatial models help in visualizing the consequences of decisions taken before actually implementing them in the field. For example, computing runoff and sediment loads for locating a check-dam, efficiency of proposed soil conservation measures, location of industry and its non-point source pollution effects on various stakeholders, etc.

For executing any watershed program successfully without sacrificing the interests of stakeholders, the basic requirement is an account of natural resources, physiography, and socioeconomic data to assess the problems and prospects of the watershed. Role of modern technologies like remote sensing, GIS, internet, portable electronic devices, electronic sensors, and communication devices thus became vital in the gamut of total program planning, execution, monitoring, and evaluation.

In tropical rainfed areas, 80–85% farmers are small farm holders cultivating <2 ha each. To reach to the millions of small farm holders spread in the SAT across 3.65 million km^2 in Asia and 5 million km^2 in Africa to share knowledge and information about new technologies and products to improve productivity on their farms is indeed a gigantic task. Advances in space research enhanced the availability of spatial and temporal data. Processing of billions of data points to translate into knowledge and information to benefit policy makers, development investors, extension and development workers, and farmers has become feasible with the availability of advanced scientific tools, technologies, and combination of one or more of such tools (Diwakar and Jayaraman 2007; Wani et al., 2008; Kaushalya Ramachandran et al., 2009; Sreedevi et al., 2009).

In this chapter, the availability of geospatial technologies, simulation modeling, and ICT application and the impacts are assessed. Other new science tools in the areas of plant biotechnology, genetic transformation, crop management, social and institutional innovations are not covered. Each of the tools is described briefly and its applications with examples in integrated watershed management program (IWMP) are discussed in detail and this shows how efficiency of integrated watershed management could be enhanced.

6.2 NEW SCIENCE TOOLS FOR WATERSHED MANAGEMENT

6.2.1 Geographical information system (GIS)

Watershed level planning requires a host of inter-related information to be generated and studied in relation to each other. Remotely sensed data provides valuable and up-to-date spatial information on natural resources and physical terrain parameters. This is a very useful and essential tool in the planning and development of watersheds embracing all natural and socioeconomic facets. GIS is used in the development of digital database, in assessment of status and trends of the resources of an area/watershed, and to support and assess various resource management alternatives. Spectacular developments in the field of GIS to synthesize thematic information with collateral data have not only made this technology effective and economically viable but also an inevitable tool to arrive at sustainable development strategies for land and water resources management.

GIS is a tool that relates information to places. It stores spatial data in a topological framework defining the relationships between map elements (points, lines, polygons, and grid cells), facilitates convenient retrieval from the spatial database and supports analysis and modeling to be displayed as digital or hardcopy maps. By visualizing different types of data from different sources using digital maps, GIS cuts across communication boundaries and can become a medium for establishing a common language between otherwise contentious or disinterested groups. The ability of GIS to integrate and spatially analyze multiple layers of information is its core capability. During the initial phases of development, GIS was extensively used for data conversion/digitization of paper maps, storing and generating map prints with little focus on spatial analysis. Later, the scenario changed drastically wherein the spatial analysis took pivotal role in watershed planning. GIS also facilitates modelling to arrive at location specific solutions by integrating spatial and non-spatial data such as thematic layers and socio-economic data. With the simultaneous development of communication networks, the data storage boundaries were erased and new areas like collaborative mapping and web map services have been developed. The present GIS technology enables 'map anywhere and serve anywhere'. With recent developments, there is a leap in the development of spatial analysis tools and logical processing methods. This has enabled the development of numerous spatial algorithms, spatial modeling techniques, and better display and visualization of data. One such application that harnessed the benefits is watershed planning, wherein these techniques were efficiently used for land resources as well as water resources planning, watershed prioritization, and monitoring.

Multi-criteria spatial queries help us visualize the spatial patterns and spatial relationships to understand the phenomenon under study. With the progress in computing capabilities and availability of hardware, more functionality is added to the GIS and it became a very powerful tool for arranging and storing spatial and tabular data in a structured way. Spatial modeling is the application of analytical procedures with GIS. Models are coupled in different ways with GIS to produce spatial model outputs. Either spatial data is used as input to specific models or vice versa to understand spatial phenomenon. GIS has perhaps the best use in the field of agriculture, as it is the most widely prevalent activity on the earth. GIS is used to understand spatial dimensions of varied problems in agriculture especially when environmental variables such as climate, soil and water play a major role in production, constraints, and practices. Temporal data sets need to be analyzed, interpreted, and depicted suitably for better understanding of land related issues and this need can be fulfilled by GIS. Besides generation and spatial analysis of data, the important facet of this information is speedy real time public outreach. With the availability of RAID/BLADE servers that can serve data at faster rates, gigabit data transfer capabilities, and Mbps internet speeds, the outreach to outside world has improved tremendously. The spatial data with on-the-fly spatial analytical capabilities is now being served over internet. One such initiative by NRSC/ISRO is Bhoosampada, a portal wherein the natural resources information along with base details and very latest thematic information is being served through Web GIS. It has a provision to spatially analyze the data present, without the need for separate GIS package. Furthermore the output can be downloaded in a suitable output format also. The main aim of such attempts is to disseminate the information before the relevance is lost. Bhuvan is a Geoportal of ISRO showcasing the Indian imaging capabilities in a multi-sensor, multi-platform, and multi-temporal domain. It provides a gateway to explore and discover virtual earth in 3D space.

6.2.2 Remote sensing

Although remote sensing started during the 2nd World War, developments in satellite remote sensing started in early 1970s and have undergone significant improvements

Table 6.1 Suitability of various remote sensing sensors in watershed studies

Level of study	Suitable spatial resolutions	Sensors	Application potential
Basin level (1:250000 scale)	50 to 150 m	IRS-WiFS, AWiFS, LISS-I	Deriving overall base information on natural resources and land cover; large-scale monitoring of changes
Watershed level (1:50000 scale)	20 to 50 m	IRS-LISS-III, SPOT-MLA, TM, ETM	Deriving natural resources information for watershed prioritization, planning and monitoring
Sub-/Micro-watershed level (1:10000 or larger)	0.5 to 20 m	IRS-LISS-IV, SPOT-PLA, Cartosat-1/2, IKONOS, QuickBird, Worldview-2	Planning and execution, monitoring watershed developmental activities, detailed account of change occurrences, stereo data for DEM generation

in sensors and spatial, spectral, temporal, and radiometric resolutions. Satellite image resolutions increased from 80 m (coarse resolution: Landsat-MSS) in 1970s to 20–36 m (medium resolution: SPOT MLA/Landsat-TM/IRS-LISS II) in mid 1980s to 0.68–5.8 m during late 1990s (high resolution: QuickBird-PAN, MLA/IKONOS/IRS-PAN, LISS-IV, Cartosat-1/2). Corresponding with these developments, application of satellite data has extended from watershed level to sub-watershed and micro-watershed level (Table 6.1).

Simultaneously, stereo-satellite data was also made available from SPOT-PLA, IRS 1C/1D-PAN, IKONOS, and Cartosat-1/2. They enabled to develop digital elevation models (DEM) for the watersheds which is indispensable for topographic feature extraction, runoff analysis, slope stability analysis, landscape analysis, etc. DEM accuracy normally depends on base-height ratio and spatial resolution of the sensor. SPOT DEM accuracies generated from high resolution satellite imagery have absolute planimetric accuracy of 15 to 30 m and absolute elevation accuracy of 10 to 20 m (Anonymous 2004). In Cartosat-1, DEM of an accuracy of 3–4 m in height was achieved where spatial resolution is 2.5 m (Srivastava *et al.*, 2007). CartoDEM can be used as an input for planning developmental activities in watersheds. The geometric accuracy and information content of Ortho-images and DEM provided by the Cartosat-1 can be used for delineation of watershed boundaries at 1:25,000 and 1:50,000 scales, generation of contours at 10-m intervals, and generating thematic maps at 1:10,000 scale (Krishna Murthy *et al.*, 2008).

Besides, the latest developments in microwave interferometry from satellites like ERS-1/2 SAR, Radarsat and Envisat, and laser altimetry from aerial platforms enabled faster and precise generation of DEMs. Noteworthy developments in laser altimetry and its data processing capability enabled generation of DEM with centimeters accuracy under ideal condition. This sort of data is being used for canal, pipeline, road, and other fine spatial alignment planning works.

The high resolution (<6 m spatial resolution) satellite imagery (IRS-LISS IV, Cartosat-1/2, IKONOS, QuickBird) will be very useful for sub-watershed or micro-watershed level applications like mapping infrastructure (roads and drainage network), natural resources inventory (crops, soils, and groundwater potential), water resources (water bodies, natural springs, ponds), land use (single cropped areas, double cropped areas, wastelands, fallow lands, forest cover at level 4), etc. They can be employed at block or village level for management of disasters such as drought or flood damage, etc. and also for monitoring and impact assessment of developmental activities in the micro-watersheds. NRSA (2006) had demonstrated the utility of high resolution satellite data on the above mentioned activities in six micro-watersheds under crop production systems in different agroclimatic zones in India.

Advancements also took place in spectral resolutions, i.e., four spectral bands (Landsat MSS, IRS-1A/1B/1C/1D, SPOT) to seven bands (Landsat-TM) to 14 discrete spectral bands (ASTER). Simultaneous developments in ground-based observations helped to realize the importance of recording data in numerous narrow spectral bands and led to the development of satellite based hyperspectral remote sensing (Hyperion/HySI). Hyperspectral data provides unique capabilities to discern physical and chemical properties of natural resources which is not possible using broadband multispectral sensors. Some of the application areas in agriculture are crop stress (moisture, pest, nutrient) detection, yield prediction, soil quality, and agro-environmental health assessment.

6.3 CROP-GROWTH SIMULATION MODELING

Crop simulation models are mathematical, computer-based representations of crop growth and interaction with weather, soil, and other nutrients. They play an important role in scientific research and resource management, and have been used to understand, observe, and experiment with cropping systems. The strengths of models in general include the abilities to:

- Provide a framework for understanding a system
- Evaluate long-term impact of interventions
- Provide an analysis of the risks involved in adopting a strategy
- Provide answers quickly and more cheaply than is possible with traditional experimentation

The Decision Support System for Agrotechnology Transfer (DSSAT) is a software package integrating the effects of soil, crop phenotype, weather, and management options that allows users to ask "what if" questions and simulate results on a desktop computer. The DSSAT package incorporates models of 27 different crops with new tools to facilitate creation and management of experimental, soil, and weather data files. It also includes improved application programs for seasonal and sequence analyses that assess the economic risks and environmental impacts associated with irrigation, fertilizer and nutrient management, climate change, soil carbon sequestration, and precision management. Crop growth modeling software such as Agricultural Production Systems Simulator (APSIM) and InfoCrop is also widely used by various researchers.

Singh *et al.* (2009) have studied the yield gaps of important crops in various countries by simulating potential yields of sorghum, pearl millet, maize, soybean, groundnut and chickpea using DSSAT. They used InfoCrop software for rice and cotton and APSIM for pigeonpea potential yield estimations. They showed that the actual yields of food and other crops obtained by the farmers are much below the potential yields that can be obtained with improved management. Crop yields can be at least doubled from their current levels by the promotion and adoption of existing 'on-the-shelf' technologies available with the national and international research institutes. The governments need to provide more suitable policy environments and institutional support to promote greater adoption of new and improved technologies to benefit the poor farmers of rainfed areas and to meet the challenge of greater food needs of future.

Singh *et al.* (2009) have analyzed yield gaps for several crops in various countries including India, Thailand, Northern Vietnam, and West Asia and North Africa (WANA) region. Estimation of potential rainfed yields and yield gaps in Northern Vietnam was based on simulated yields, experimental station yields, and province yields – all obtained under rainfed situation. Potential yields of soybean, groundnut, and maize were simulated using DSSAT v3.5 crop models. The models were tested and validated using data of three experiments conducted at Than Ha watershed site in Hoa Binh province (Chuc *et al.*, 2005). Rainfed potential yields of crops were simulated using weather data of 28 years for the five locations (Vinh Phuc, Ha Nam, Ninh Binh, Ha Tay, and Phu Tho) and 10 years for the Hoa Binh. Long-term yield data of yield maximization trials were also available for each crop and benchmark site. These data were averaged over the time period and compared with mean simulated yields and province level mean yields for the benchmark sites to quantify the yield gaps for each crop.

As groundnut is more drought resistant during initial stages of its growth under rainfed conditions, the spring season for groundnut starts earlier as compared to soybean and maize. During spring season, simulated potential yields of groundnut across six provinces ranged from 3740 to 4700 kg ha^{-1} with an overall mean of 4170 kg ha^{-1} whereas the experimental potential yields ranged from 2550 to 3400 kg ha^{-1} with an overall mean of 3010 kg ha^{-1} (Figure 6.1). This indicates that even the experimental yields are below the simulated potential yields by about 1100 kg ha^{-1} in the provinces. The province yields of groundnut ranged from 1180 to 2200 kg ha^{-1} with an overall mean of 1520 kg ha^{-1}. The yield gap was 2650 kg ha^{-1} between the simulated and province yield and 1490 kg ha^{-1} between experimental and province yield.

During autumn-winter season, simulated potential and experimental potential yields were lower than those obtained during the spring season (Figure 6.1). Simulated potential yields ranged from 2910 to 3920 kg ha^{-1} with an overall mean of 3530 kg ha^{-1} whereas the experimental yields ranged from 2300 to 2800 kg ha^{-1} with an overall mean of 2620 kg ha^{-1}. Yield gap was 2010 kg ha^{-1} between the simulated and province yield and 1100 kg ha^{-1} between experimental and province yield. These results indicate that the groundnut yields in the six provinces during the spring and autumn-winter seasons can be more than doubled with improved management practices.

Crop simulation models are also used to understand the impacts of climate change on crop growth and productivity. Cooper *et al.* (2009) have looked at a *factorial combination* of climate change of five different temperature increases (1, 2, 3, 4, and 5°C) and three different percentage changes in seasonal rainfall (0%, +10%, and −10%) and compared the crop simulation outputs with a 'control' of the current climate. Their study indicated that predicted temperature increases have greater negative impacts on

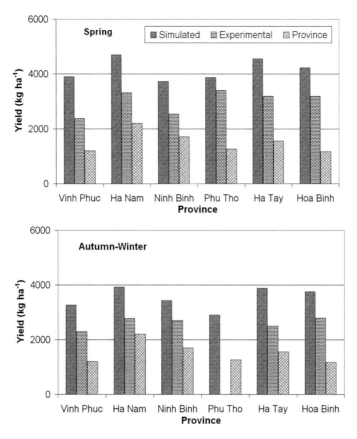

Figure 6.1 Simulated potential, experimental, and province mean pod yields and yield gap of rainfed groundnut in spring and autumn-winter seasons at selected sites in Northern Vietnam

crop production than relatively small changes in rainfall. They have shown that the ex ante analyses clearly illustrate both the challenges that climate risk poses as well as the opportunities that it can offer.

6.4 FIELD SENSORS AND DATA COMMUNICATION DEVICES

In watershed management, one important component is the collection and sharing of field data or ground information and integrating it into the processing and analysis of spatial data in real time, which helps in timely decision-making and taking up appropriate corrective measures. Field data collection typically consists of recording geographic location, photographs of the area at the sample points, notes on soils, crops, and land use and general details in a ground truth proforma. Collecting the data and putting it to use is normally done as a sequential process with a significant amount of time delay since the same scientists perform both the tasks and the entire ground truth data collection activity is normally allowed to be completed before starting the use of data.

Field data collection has undergone a number of changes from the days of hardcopy jottings on paper in the field to the use of laptops/palmtops in recent times. However, a combination of some of the recent technology trends promises to deliver significantly enhanced solutions in this area, which would benefit a wide range of users. The important technology areas impacting the field data collection process are described below.

6.4.1 Global Positioning System

Global Positioning System (GPS) is one of the important tools that brings location awareness to any application. While collecting and using any real time field data, the location from where it was collected is very important. GPS is a known electronic device to most of the tech-savvy people and has become an important tool for location awareness. Several location-based and location aware applications are being developed especially in emergency management, service and utility sectors. New developments and relaxations in security related matters have helped in improving the location accuracy to better than 15 m using ordinary code receivers. In differential mode, sub-meter accuracies are possible.

6.4.2 Automatic Weather Station

Collection of precise weather data at watershed level and transmission on real time basis is vital for resource management as well as for improving crop productivity. Automatic Weather Station (AWS) is an affordable way to get detailed weather information at the watershed areas. AWS records data on parameters such as rainfall, wind speed and direction, humidity, temperature, etc. Special sensors of particular interest can also be included in AWS, to measure soil temperature, leaf wetness, etc. AWS is a very compact, modular, rugged, powerful, and low-cost system. The AWS system consists of a compact datalogger, data transmitter, antenna, GPS, solar panel, and sensors. Power requirements are minimum and hence do not pose any operational problems. Sensors on AWS collect data at specified time interval and store the data in its memory. Logged weather data is transmitted at prescribed time slots through geostationary communication satellite systems. Datalogger, power supply, and battery are housed in a weather proof enclosure.

The AWS data finds extensive applications in agricultural monitoring (drought/crop condition assessment), crop management, disaster management (flood forecasting), and in other fields like transport. Near-real time information on weather and crops allows the calculation of water requirements of crops and hence invaluable for drought monitoring and management. Integration of relevant spatial and non-spatial information of natural resources and socioeconomic aspects related to agricultural drought is required for generation of a spatial decision support system and AWS data would be a value-addition for drought management.

6.4.3 Mobile devices

Mobile devices, which are of interest to field data collection process are Personal Digital Assistants (PDAs) and cell phones. PDAs are basically palm-size devices which originally started as high-end organizers, but quickly added a number of features like bigger LCD screens, color, keyboard, stylus, handwriting recognition, higher speed wired and wireless data connectivity to desktop systems, etc. With time, as processor power grew, their operating systems evolved and now compact Windows operating systems are adapted to these devices. Thus, desktop applications (word processing, spreadsheets, email clients, web browser, etc.) are made available on PDAs also. This forms a handy device to record and store field level information in an organized way.

Cellular phones, on the other hand, evolved from being primarily wireless voice communication devices to encompass various features like organizer, messaging, camera, music player, and Bluetooth connectivity. Over time these mobile phones became powerful tools with many other features like larger screen, deployment of custom applications, and web browser. Integrated mobile devices are also commonly equipped with a digital camera, which can be used to capture necessary field photographs for storing as well as sharing by email. Thus it forms an important component for communicating data wirelessly to any part of the world. Public wireless networks serving the common man like the cellular networks based on GSM and CDMA technologies have become widespread and ubiquitous in recent times.

PDA phone with GPS is the resultant of convergence of the PDA, cellular phone, and GPS technologies with a built-in camera. These PDA devices are becoming increasingly powerful with the deployment of powerful processors and larger memory. They also have bigger color touch screens and full QWERTY keypads for better inputting of data. With these powerful configurations, it is now possible to deploy rich Graphical User Interface (GUI) applications, which were considered to be difficult just a few years ago.

6.5 DATA STORAGE AND DISSEMINATION

Latest development in server technology has enabled availability of blade servers with RAID capabilities at a very cost effective price. These servers act as storehouses for storing the data in a safe and efficient way and can serve clients via network sharing and World Wide Web, in near-real time.

Internet is all pervasive and cost-effective technology where a number of applications are specifically designed to use the Internet and the related IP-based protocols to communicate and exchange data with one another, thereby optimizing the costs as well as ensuring widespread geographical reach. Internet connectivity on current PDA devices is easily ensured with an appropriate subscription to GPRS/EDGE feature from the wireless network service provider. Almost all present-day organizations have an Ethernet local area network in place for data communication among the various computer systems including servers, workstations, and desktop PCs. The same network is also invariably used to implement a number of intranet applications in addition to the traditional client-server based applications and databases on servers.

6.6 SPATIAL TECHNOLOGIES IN RAINFED AGRICULTURE AND WATERSHED MANAGEMENT

6.6.1 Characterization of production systems in India

Production systems (PSs) based approach to agricultural research was found to be more relevant at ICRISAT during the 1990s and the SAT was divided into 29 PSs.

A GIS database of PS maps consisting of soils, climate, crops and other socioeconomic variables was used. It was further proposed to refine these PSs using GIS to be able to compare with the national agroecological zones (AEZs) so that these PSs are useful for up-scaling and down-scaling of technologies (Johansen 1998).

Out of the 12 PSs in Asia, India has 10 types of PSs. Further, 12 were delineated in Latin America and 5 in Africa. A PS is defined by the environmental resources, geography, and important issues, or constraints to, and opportunities for improving productivity and sustainable agriculture (ICRISAT 1994).

Preliminary definition of these PSs required that they assist in the prioritization essential for development of ICRISAT Medium Term Plans. It also allowed for better focusing of projects to particular PSs and of activities within projects. To identify the target regions and priority areas and allocate resources in PS research, the ability of GIS, which can analyze multiple layers of information and provide answers spatially, became evident.

Soil being the basis of life on earth and for agriculture, information on soil attributes was the most important input variable for any PS assessment. Production system-wise soil attributes were mapped and described to help researchers identify target locations for research and technology transfers. The NBSS&LUP map based on soil taxonomy was used in a GIS to provide soil information along with PS boundaries and district boundaries and area was estimated for each suborder in all the PSs.

Out of the 11 soil orders of soil taxonomy, seven occur in the 10 PSs in India. The Entisols are the most pervasive of all soils and occur in all the PSs. Alfisols (suborder Ustalfs) and Vertisols (suborder Usterts) are found in 8 of the 10 PSs, but Alfisols occupy a total area of $615016 \, \mathrm{km}^2$ and Vertisols $470148 \, \mathrm{km}^2$ in all the PSs with maximum area (Figure 6.2). This helped in understanding the soil types and their attributes in all the PSs of India to appropriately devise technologies and provide more options to farmers of the SAT.

6.6.2 Land use mapping for assessing fallows and cropping intensity

To delineate rainy season fallows in the state, data obtained from the Indian remote sensing satellite were analyzed. A deductive approach including delineation of agricultural land and forests from temporal satellite data was employed to identify (rainy season) fallow. Three sets of satellite data corresponding to three periods, namely mid-, late, and (postrainy) *season* were used. While mid-season satellite data provide information on agricultural lands, which were lying unutilized along with those agricultural lands that have been supporting crops, the satellite data of *season*, on the other hand, exhibited spatial distribution pattern of the land supporting crops. These lands include the areas, which were lying fallow during *season* in addition to the lands that were cultivated during *season*, and are now supporting crops. In contrast, satellite data acquired during late season showed agricultural lands that were laid fallow during *season* and the areas where crops were planted (Figure 6.3). Madhya Pradesh is covered by two WiFS (Wide Field Sensor) images. Owing to the presence of persistent cloud cover during *season*, the availability of cloud-free space borne multispectral data has been the major problem. However, very short repetivity and tandem operation of the IRS-1C and IRS-1D satellites, along with the IRS-P3 satellite, enabled acquiring

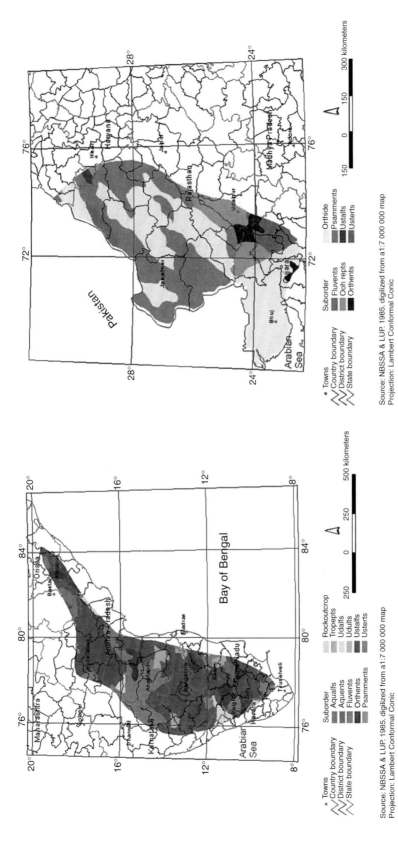

Figure 6.2 Distribution of different soil orders in the production systems in India (See color plate section)

Source: NBSSA & LUP. 1985. digilized from a1:7 000 000 map
Projection: Lambert Conformal Conic

Source: NBSSA & LUP. 1985. digilized from a1:7 000 000 map
Projection: Lambert Conformal Conic

Mid-rainy season image

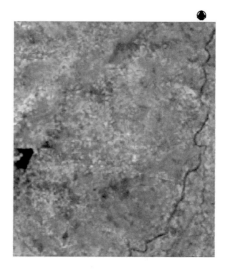

Late-rainy season image

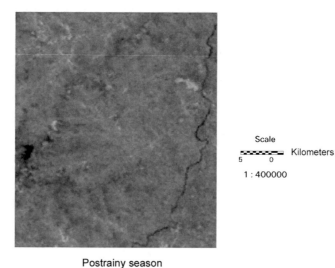
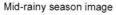

Scale

5   0   Kilometers

1 : 400000

Postrainy season

Figure 6.3 A close view of WiFS images of part of Vidisha district, Madhya Pradesh during mid-rainy, late-rainy, and postrainy seasons (See color plate section)

virtually cloud-free WiFS data of September from IRS-1D and IRS-P3 satellites. The situation remains more or less same even during post-monsoon period also. Consequently, cloud-free WiFS data were not available and out of two images covering the former state of Madhya Pradesh, one image for October was used. Satellite data acquired during peak growing period of crops help identification of land where crops have been taken.

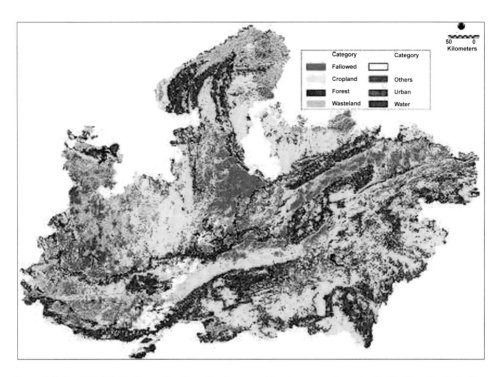

Category		Category	
Fallowed		Others	
Cropland		Urban	
Forest		Water	
Wasteland			

50 0
Kilometers

Figure 6.4 Spatial distribution of various land use and land cover categories in Madhya Pradesh (See color plate section)

Digital multispectral data from WiFS aboard IRS-1D/-P3 over the area acquired during the *season* of 1999–2000 and *season* of 2000–01 was utilized for deriving information on fallow lands. In addition, Survey of India topographic maps at 1:250,000 scales were also used (Figure 6.4). The approach essentially involved preparation of the mosaic of WiFS digital data covering entire state, preliminary digital analysis, ground truth collection, map finalization, and generation of area statistics.

Basically, a deductive approach was employed for delineation of fallow lands. Based on past experience, initially areas akin to fallow lands were identified after displaying the digital multispectral data onto color monitor of Silicon Graphics work station. Besides, topographic maps were used for exclusion of the areas with rock, outcrops, scrubs, hills, etc. Furthermore, other categories like forestland, crop land, wasteland, water, and settlements were also broadly delineated. Doubtful areas were located in the topographic maps of 1:250,000 scale for further verification in the field.

The second generation of Indian remote sensing satellites (IRS-1C and IRS-1D) have better resolution and wide applicability. The WiFS sensor provides reflectance data in red and near infrared bands at 188 m spatial resolution and at 5 days revisit, covering a swath of about 812 km, and is useful in deriving regional level crop information. Frequent availability of the WiFS data due to shorter revisit period also facilitates the monitoring of crops (Kasturirangan *et al.*, 1996). WiFS data was found to be suitable for deriving regional information on the spatial distribution of rice (*Oryza sativa*) crop grown in the Godavari delta of East and West Godavari districts and pulse crops cultivated in the rice-fallow fields of the Krishna delta of Krishna and Guntur districts of Andhra Pradesh, India (Navalgund *et al.*, 1996). In the present study, WiFS data of 1999 and 1999/2000 seasons were used to derive the regional level information on spatial distribution of rice and rice-fallow lands in the South Asian countries of Bangladesh, India, Nepal, and Pakistan.

Reflectance spectra of plant canopies are a combination of the reflectance spectra of plants and of the underlying soil (Guyot 1990). When a plant canopy grows, soil contribution progressively decreases. Thus, during the active vegetative growth phase, visible and middle infrared reflectance decreases and near infrared reflectance increases. During senescence, opposite phenomenon occurs. Maximum reflectance from vegetation is sensed when crop canopy fully covers the ground, which coincides mostly with the beginning of reproductive phase. Hence, satellite data corresponding to this stage were selected to discriminate rice crop during the season.

6.6.3   Spatial distribution of rainy season fallows in Madhya Pradesh

As pointed out earlier, a deductive approach including delineation of agricultural land and forests from temporal satellite data was employed to identify fallow in Madhya Pradesh. Three sets of satellite data corresponding to three periods, namely mid-season, late-*kharif* (rainy season), and *rabi* (postrainy season) were used. While mid-season satellite data provides the information on agricultural lands, which were lying unutilized along with those agricultural lands that have been supporting crops, the satellite data of *rabi*, on the other hand, exhibits the spatial distribution pattern of the land supporting crops. These lands include the areas, which were lying fallow during the season, and are now supporting crops. Contrastingly, the satellite data acquired during late season show the agricultural lands that were lying fallow during the season and the areas where crops were planted.

It was estimated that 2.02 million ha accounting for 6.57% of the total area of the state were under fallow (Figure 6.5). Madhya Pradesh is endowed with well distributed rains ranging from 700 to 1200 mm. Vertisols with good moisture holding capacity can be used to grow short-duration soybean by adopting sound land management practices (Dwivedi *et al.*, 2003). ICRISAT-led consortium through funding from Sir Dorabji Tata Trust (SDTT) and Sir Ratan Tata Trust (SRTT) in selected districts in Madhya Pradesh have initiated concerted farmer participatory research and development (PR&D) trials using broad-bed and furrow (BBF) system to alleviate waterlogging short-duration soybean and maize cultivars during rainy season and minimum tillage for *rabi* chickpea to minimize rainy season fallows.

6.6.4   Spatial distribution and quantification of rice-fallows in South Asia: Potential for legumes

Rice, the most extensively grown crop in South Asia, is cultivated on approximately 50 million ha. Despite growing demands for food production because of an increasing

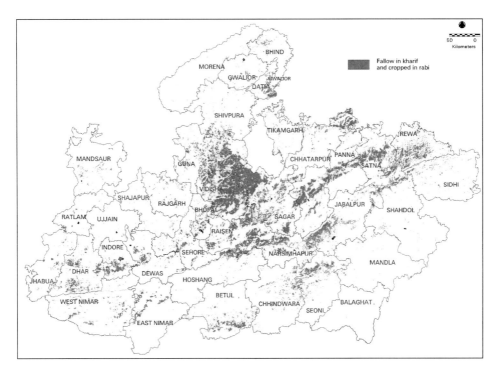

Figure 6.5 Spatial distribution of rainy season fallows in districts of Madhya Pradesh (See color plate section)

population in South Asia, there is little scope for expansion of cropping into new areas and therefore an increase in cropping intensity, along with improvement of yields, needs to take place on existing agricultural lands. Rice-fallows present considerable scope for crop intensification and diversification if the appropriate technology is applied. But there has been limited information on the area of rice-fallows available and on the potential technologies that could be implemented.

This study describes the use of satellite remote sensing and GIS technology to develop an accurate and updated quantification and spatial distribution of rice-fallow lands and a corresponding classification of their potential and constraints for post-rice legumes cultivation in South Asia (Bangladesh, India, Nepal, and Pakistan). These rice-fallows represent diverse soil types and climatic conditions and most of these areas appear suitable for growing either cool season or warm season legumes.

Introducing appropriate legumes in rice-fallows is likely to have significant impact on the national economies through increased food security, improved quality of nutrition to humans and animals, poverty alleviation, employment generation, and contribution to the sustainability of these cereal-based PSs in South Asia. This would also provide guidance to policy makers and funding agencies to identify critical research areas and to remove various bottlenecks associated with effective and sustainable utilization of rice-fallows in South Asia.

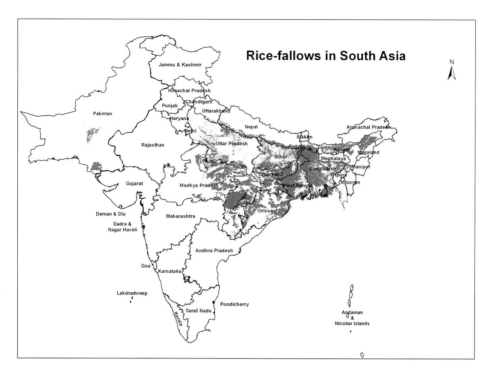

Figure 6.6   Spatial distribution of rice-fallows in the Indo-Gangetic Plains of South Asia (See color plate section)

Satellite image analysis estimated that rice area during 1999 season was about 50.4 million ha. Rice-fallows during 1999/2000 season were estimated at 14.29 million ha in Bangladesh, India, Nepal, and Pakistan. This amounts to nearly 30% of the rice-growing area (Figure 6.6). These rice-fallows offer a huge potential niche for legumes production in this region. Nearly 82% of the rice-fallows are located in the Indian states of Bihar, Madhya Pradesh, West Bengal, Orissa, and Assam. The GIS analysis of these fallow lands has indicated that they represent diverse soil types and climatic conditions; thus a variety of both warm season legumes [such as soybean (*Glycine max*), mung bean (*Vigna radiata*; green gram), black gram (*Vigna mungo*), pigeonpea (*Cajanus cajan*), and groundnut (*Arachis hypogaea*)] and cool season legumes [such as chickpea (*Cicer arietinum*), lentil (*Lens culinaris*), khesari (*Lathyrus sativus*; grass pea), faba bean (*Vicia faba*), and pea (*Pisum sativum*)] can be grown in this region (Subbarao *et al.*, 2001).

An economic analysis has shown that growing legumes in rice-fallows is profitable for the farmers with a benefit-cost ratio exceeding 3.0 for many legumes. Also, utilizing rice-fallows for legume production could result in the generation of 584 million person-days employment for South Asia. Technological components of rainfed cropping, especially for chickpea crop, have been identified. These include the use of short-duration chickpea varieties, block planting so as to protect the crop from grazing animals, sowing using rapid minimum tillage as soon as possible after harvesting rice, seed priming for 4 to 6 hours with the addition of sodium molybdate to the priming water at $0.5\,g\,L^{-1}\,kg^{-1}$ seed and *Rhizobium* inoculum at $5\,g\,L^{-1}\,kg^{-1}$ seed, and application of manure and single superphosphate. Yield of chickpea following rice ranged from $0.4\,t\,ha^{-1}$ to $3.0\,t\,ha^{-1}$ across various rice-fallow areas in eastern India. More than six thousand farmers who have been exposed to this technology are now convinced that a second crop can be grown without irrigation in rice-fallows. Similar results have been obtained for the Barind region in Bangladesh. Seed priming has been shown to substantially improve the plant stand for chickpea in rice-fallows in the Barind regions of Bangladesh (Harris *et al.*, 1999). Rainfed cropping in rice-fallow areas increased incomes and improved food security and human nutrition (Subbarao *et al.*, 2001). In a number of villages in Chhattisgarh, Jharkhand, and Madhya Pradesh in India, the on-farm farmers' participatory action research trials sponsored by the Ministry of Water Resources, Government of India showed significantly enhanced rainwater use efficiency through cultivation of rice-fallows with a total production of ₹5600 to $8500\,kg\,ha^{-1}$ for the two crops (rice + chickpea) benefiting farmers with increased average net income of ₹51000 to $84000\,ha^{-1}$ (US$1130 to $1870\,ha^{-1}$) (Singh *et al.*, 2010).

6.6.5 GIS mapping of spatial variability of soil micronutrients at district level

Spatial variability of secondary nutrient sulfur and micronutrients boron and zinc in selected rainfed districts of Karnataka in South India was studied using GIS. Stratified random sampling methodology described by Sahrawat *et al.* (2008) was used for collecting soil samples from each watershed. About 30,000 soil samples were collected and analyzed for soil nutrients including boron, sulfur, and zinc content. Village-level geographical coordinates were obtained using a GPS. The IDW method in the ArcGIS 9.0 software for interpolation was standardized in this study. Nutrient availability maps for 15 districts were generated for all nutrients including boron, sulfur and zinc (Figure 6.7). All maps of predicted surfaces are classified into two classes viz., deficient and sufficient. Boundary limits of nutrient availability for the critically low, low, and normal classes were obtained from standard results (Sahrawat *et al.*, 2007).

Through the standardized GIS-based interpolation method, agricultural extension personnel and farmers in watersheds can be provided with reliable and cost-efficient soil analysis results of selected districts for developing balanced nutrient management strategies at *taluk* level. However, due to limitations in the IDW method, the generated maps are to be used only at district or *taluk* level and not for predicting the nutrient availability at single field level.

6.6.6 Assessment of seasonal rainfall forecasting and climate risk management options for peninsular India

Uncertainty of the climate and weather in the SAT has adverse effects on crop production and farmer income. Farmers are traditionally risk averse and conservative in adopting high input improved technologies because of the uncertainties in production associated with variable climate. Seasonal climate prediction before onset of the season could help them in taking appropriate decisions to minimize losses in low rainfall years and harness the potential in the normal or high rainfall years. With

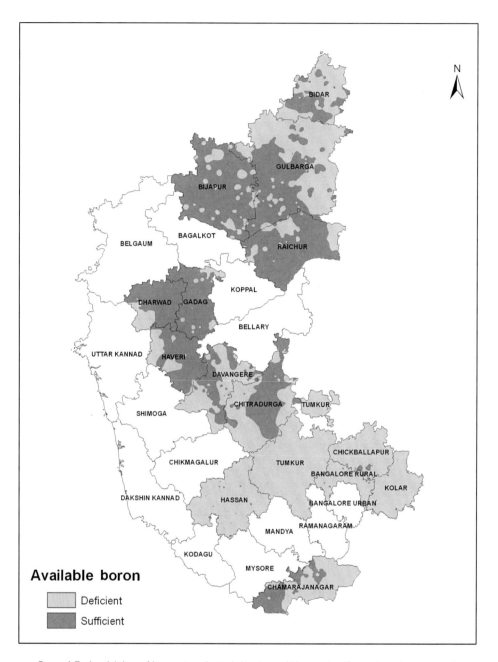

Figure 6.7 Availability of boron in selected districts of Karnataka (See color plate section)

the technical input from the International Research Institute on Climate Prediction, a pilot project was carried out in Nandyal and Anantapur in Andhra Pradesh to assess the value and benefit of seasonal climate prediction at district scale to the farmers (Rao *et al.*, 2007). Using Global Circulation Model (GCM) predictor-based model output statistical (MOS) technique, the probabilistic seasonal rainfall prediction for 2003 was communicated to the farmers at a lead time of more than a month to take up appropriate cropping decisions for the two districts. Seasonal climate prediction for Nandyal proved accurate and the farmers derived significant benefit by adopting double cropping in the region as compared to the single crop. Farmers in Anantapur had mixed experience as the rains started late in the district. The farmers who adopted groundnut/short-duration pigeonpea intercrop were benefited and those who followed groundnut/medium-duration pigeonpea intercrop incurred losses as compared to the sole groundnut system.

6.6.7 Baseline studies to delineate watershed

Accurate delineation of a watershed plays an extremely important role in the management of the watershed. The delineated boundaries form the nucleus around which the management efforts such as land use, land cover change, soil types, geology, and river flows are analyzed and appropriate conclusions drawn. Digital elevation models provide good terrain representation from which watershed boundary can be delineated automatically using GIS technology. There are various data sources for generation of DEM. Usually, the height contours mentioned in topographical maps are digitized and are used for generation of DEM. Besides, photogrammetric techniques using stereo data from aerial or satellite platforms can also be used for DEM generation. In this context, data acquired across the path from satellites like IRS-1D and SPOT has shown temporal variation in terrain radiometry leading to poor DEM accuracies. To improve cross image correlation between stereo pair imagery, Cartosat-1 launched with two cameras beaming along the path with which DEM of 3–4 m height accuracy (Srivastava *et al.*, 2007) was achieved. Further, processing techniques like stereo strip triangulation has greatly improved throughput of DEM generation with limited ground control points and short time. Besides, the latest developments in interferometry and laser altimetry enable faster and precise generation of DEMs, especially the recent developments in laser altimetry and its data processing enable generation of DEM with centimeters accuracy under ideal condition.

The techniques for automated watershed delineation have been available since mid 1980s and have been implemented in various GIS systems and custom applications (Garbrecht and Martz 1999). Figure 6.8 portrays the Cartosat-1 data and the DEM derived therefrom along with LISS-IV multispectral data. In Figure 6.9, the perspective view generated from the DEM draped with LISS-IV multispectral data is presented. The field of view was 50, pitch was −5, and the azimuth was 328. This sort of analysis helps in understanding the watershed terrain in a perspective way. Besides, extraction of watershed boundaries from Cartosat-1 DEM in an automated way is also possible. For such extraction of watershed boundaries, identification of pour points (watershed outlet) is a prerequisite. Further, the DEM needs to be filled for sinks so that the runoff is accumulated as a concentric flow and passes through one of the outlets.

Watershed characterization involves inventorying and assessment of natural resources which are essential prerequisites of any watershed management activity. For example, watershed managers need timely and reliable information on soils, crops, groundwater potential, and land use. Similarly, an assessment of the properties of

High 187 m

Low 0 m

Drainage layer draped over DEM generated from Cartosat data

FCC of IRS P6 LISS IV MSS + Cartosat merged data

Cartosat data

Figure 6.8 Satellite data and DEM of watershed in part of Nalgonda district, Andhra Pradesh (See color plate section)

Figure 6.9 False Color Composite (FCC) draped over DEM for perspective view of watershed in part
of Nalgonda district, Andhra Pradesh (See color plate section)

the soils and their response to management is required in agriculture and forestry, for decision-making in planning, and for many other engineering works. It has been proved beyond doubt that remotely sensed data can be effectively used to prepare maps on various themes such as land use/land cover, soil distribution, geomorphology, etc., which in turn form the basic tools for designing a proper management strategy. High resolution remotely sensed data when used in conjunction with conventional data can provide valuable inputs such as watershed area, size and shape, topography, drainage pattern, and landforms for watershed characterization and analysis (Obi Reddy *et al.*, 2001).

Prioritization of watersheds helps in focusing the implementation activities on a few watersheds that urgently need attention. Watershed prioritization is simply ranking of different sub-watersheds of a watershed according to the order in which they have to be taken up for treatment and soil and water conservation measures or to improve crop productivity. This also helps to avoid spreading too thin, the limited financial resources available for implementation over the entire area. Remote sensing derived inputs were considered for prioritizing the watershed when it is based on natural resources limitations or potentials in a watershed (Sharma 1997; Rao *et al.*, 1998; Saxena *et al.*, 2000; Khare *et al.*, 2001; Sekhar and Rao 2002; NRSA 2006).

The prioritization of watersheds in India is on the basis of natural resources status, and socioeconomic, biophysical, and other criteria. During initial stages, soil erosion control was the prime concern for watershed prioritization. Various methods were developed in this regard for watershed prioritization like sediment yield modeling (Sharma 1997) or erosion-proneness of land units (Sekhar and Rao 2002). Subsequently, land productivity was also considered through identification of critical areas (NRSA 2006). In latest guidelines for prioritization of watersheds the combination of natural resources, problem areas, and socioeconomic conditions (agricultural laborers, schedule caste and schedule tribe population, distribution of below poverty line families) were considered for prioritization.

Geospatial data and multi-criteria based prioritization of watersheds help in making unbiased choice of target areas for development. The multi-layer geospatial analysis results in the generation of composite mapping units which could further be processed through multi-criteria analysis to arrive at the end result. GIS and IT tools at watershed level have been successfully used to establish a strong baseline information system and prioritization (Khan *et al.*, 2001; Thakkar and Dhiman 2007; Diwakar and Jayaraman 2007). Success of conservation measures whether it is vegetative or structural, depends upon the selection of suitable sites. Various factors such as physiography, soil characteristics, and topographic features of the terrain have to be considered to arrive at a decision regarding sites for conservation measures. Computer-based database management systems for terrain and elevation modeling and GIS have enhanced the potential of remotely sensed data in identifying suitable locations for conservation measures.

6.6.8   Regional-scale water budgeting for SAT India

A soil water balance model (WATBAL) (Keig and McAlpine 1974) was used to estimate the available soil water spatially (2.5 arc minutes 4.5 km approximately) and temporally (monthly) using the above pedo-climatic datasets to run WATBAL. Input data for the WATBAL model are the precipitation and potential evapotranspiration (PET) as gridded interpolated surfaces from point data. The interpolated climatic surfaces are available at monthly temporal resolution. Maximum soil water-holding capacity is extracted from the Digital Soil Map of the World and its derived soil properties (FAO 1996).

For prioritization and selection of target regions for watershed development, first-order water budgeting using GIS-linked water balance model was used for the selected states in central and peninsular India. Such a simulation model used with monthly rainfall and soils data generated outputs that can be effectively used to prioritize the regions and strategies for improved management of rainwater (Figure 6.10). Once the target region is selected, then the selection of appropriate benchmark sites using second-order water budgeting with more detailed simulation models can be applied. The GIS map produced using this methodology shows the potential of various regions in central and peninsular India for the amount of water surplus available for water harvesting and groundwater recharging.

6.6.9   Spatial water balance modeling of watersheds

In partnership with Michigan State University, USA, we have attempted to integrate topographic features of watersheds with hydrological models. Automation of terrain analysis and use of DEMs have made it possible to quantify topographic attributes of the landscape for hydrological models. These topographic models, commonly called digital terrain models, partition the landscape into series of interconnected elements, based on topographic characteristics of landscape and are usually coupled to a mechanistic soil water balance model. Partitioning between vertical and lateral movement at a field-scale level helps to predict complete soil water balance and consequently available water for plants over space and time.

Data generated in the black soil watershed (BW 7) on-station experiment at ICRISAT, Patancheru, India was used for validating the model developed at Michigan

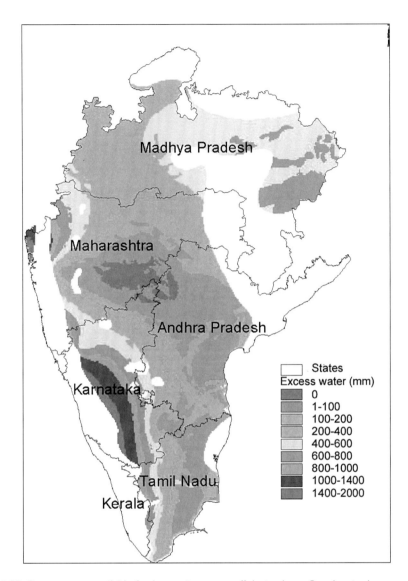

Figure 6.10 Excess water available for harvesting as runoff during June–October in the states of SAT India (See color plate section)

State University. This partnership research led to the development of SALUS-TERRAE, a digital terrain model for predicting the spatial and temporal variability of soil water balance. A regular grid DEM provided the elevation data for SALUS-TERRAE. We have successfully applied the SALUS-TERRAE, which was a functional spatial soil water balance model, at a field scale to simulate the spatial soil water balance and identified how the terrain effects the water routing across the landscape. The model provided excellent results as compared with the field-measured soil water content (Figure 6.11).

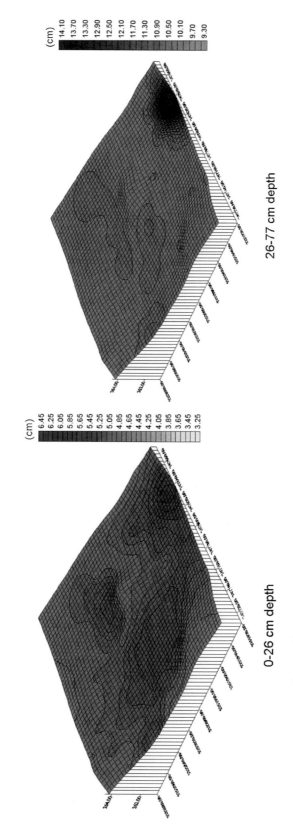

Figure 6.11 Soil water content on day-2 for scenario I (uniform soil type, high rainfall, no restricting soil layer)

6.7   INTEGRATED WATERSHED MANAGEMENT FOR LAND AND WATER CONSERVATION AND SUSTAINABLE AGRICULTURAL PRODUCTION IN ASIA

6.7.1   Assessment of agroclimatic potential

Maximizing agricultural production from rainfed areas in a sustainable manner is the need of the day to feed the ever-increasing population. Knowledge on agroclimatology is a valuable tool in assessing the suitability of a watershed for rainwater harvesting and crop planning. Role of climate assumes greater importance in the semi-arid rainfed regions where moisture regime during the cropping season is strongly dependent on the quantum and distribution of rainfall vis-à-vis the soil water-holding capacity and water release characteristics. In spite of cultivation of high-yielding varieties, improved cultural practices and plant protection measures and favorable weather are essential for good harvests (Rao *et al.*, 1999). A thorough understanding of the climatic conditions helps in devising suitable management practices for taking advantage of the favorable weather conditions and avoiding or minimizing risks due to adverse weather conditions. Agroclimatic analysis and characterization of watersheds need to be carried out using databases having long-period weather data and agroclimatic datasets need to be developed at individual watershed level. Agroclimatic analysis of the watersheds is based on the concepts of rainfall probability, dry and wet spells, water balance, length of growing period (LGP), droughts, crop-weather modeling and climate variability and change. Enhancing climate awareness among the rural stakeholders using new IT tools is the need of the hour.

6.7.2   Climatic water balance

Availability of water in right quantity and at the right time and its management with suitable agronomic practices are essential for good crop growth and yield. To assess water availability to crops, soil moisture should be taken into account and the net water available through soil moisture can be estimated using water balance technique. Simple single-layer water balance model of Thornthwaite and Mather (1955) outputs various water balance elements like actual evapotranspiration (AET), water surplus, and water deficit based on rainfall, PET, and soil water-holding and release properties. PET (i.e., amount of water that is lost into the atmosphere through evaporation and transpiration from a short green crop, completely shading the ground, of uniform height and with adequate water status in the soil profile) can be estimated using the modified FAO-Penman-Monteith method (Allen *et al.*, 1998). Water balance though simple, is a powerful tool to quantify water deficit, water surplus, and runoff potential, to delineate the rainfed LGP, dry and wet spells during the crop growth period, and to monitor moisture stress leading to drought in watersheds.

6.7.3   Climatic water balance of watersheds in China, Thailand, Vietnam, and India

Weekly water balances of selected watersheds in China, Thailand, and Vietnam were completed based on long-term agrometeorological data and soil type. The water balance components included PET, AET, water surplus, and water deficit. PET varied

Table 6.2 Annual water balance characters (all values in mm)[a]

Country	Location	Rainfall	PET	AET	WS	WD
	Xiaoxingcun	641	1464	641	Nil	815
China	Lucheba	1284	891	831	384	60
	Wang Chai	1171	1315	1031	138	284
Thailand	Tad Fa	1220	1511	1081	147	430
Vietnam	Vinh Phuc	1585	1138	1076	508	62
	Bundi	755	1641	570	186	1071
	Guna	1091	1643	681	396	962
India	Junagadh	868	1764	524	354	1240
	Nemmikal	816	1740	735	89	1001
	Tirunelveli	568	1890	542	Nil	1347

[a]PET = Potential evapotranspiration; AET = Actual evapotranspiration; WS = Water surplus; WD = Water deficit.

from about 890 mm at Lucheba in China to 1890 mm at Tirunelveli in South India (Table 6.2). AET values are relatively lower in the watersheds in China and India compared to those in Thailand and Vietnam. Varying levels of water surplus and water deficit occur in the watersheds. Among all the locations, Tirunelveli in India has the largest water deficit (1347 mm) level and no water surplus. Chine in Vietnam has the largest water surplus level of 907 mm. These analyses defined the dependability for moisture availability for crop production and opportunities for water harvesting and groundwater recharge.

6.7.4 Rainfed length of growing period

Knowledge on the date of onset of rains will help plan better the agricultural operations, particularly, land preparation and sowing. Length of rainy season is the duration between onset and end of agriculturally significant rains. Rainfed LGP is defined as length of the rainy season plus the period for which the soil moisture storage at the end of rainy season and the postrainy season and winter rainfall can meet the crop water needs. Therefore, LGP depends on not only the rainfall distribution but also the soil type, soil depth, and water retention and release characteristics of the soil. This assumes greater importance from a watershed perspective where soil depth in a toposequence can also alter the LGP across the watershed being highest in the low-lying regions and lowest in the upper reaches of the watersheds.

Agroclimatic characterization of selected watersheds in Nalgonda, Mahabubnagar, and Kurnool districts of Andhra Pradesh based on water balance and rainfed LGP (Kesava Rao *et al.*, 2007) indicated that the beginning and end of the crop-growing season varied across the years in the watersheds. In all the watersheds, the end was more variable compared to the beginning; however, there was no definite relationship between the beginning and length of growing season. Nemmikal (medium-deep Vertisol) and Nandavaram (deep Vertisol) watersheds provide greater opportunity for double cropping. Appayapally, Thirumalapuram, and parts of Nemmikal watersheds having medium-deep Alfisols, provide opportunity for double cropping with relatively short-duration crops, but are more suitable for intercropping with medium-duration crops like pigeonpea and castor (*Ricinus communis*). Kacharam, Mentapally,

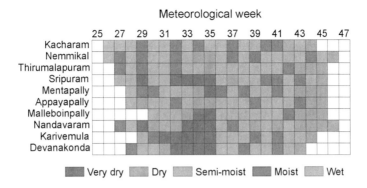

Figure 6.12 Drought monitoring at benchmark watersheds in Andhra Pradesh during 2004 (See color plate section)

Sripuram, Malleboinpally, and Karivemula have medium-deep Alfisols and provide greater potential for sole cropping during rainy season with crops of 120 to 130 days duration and intercropping with short- to medium-duration crops to make better use of soil water availability. Early season drought occurs at Karivemula and Thirumalapuram and early and mid-season droughts occur at Nandavaram. These sites would require crop/varieties tolerant to early or mid-season droughts depending upon the location. Mentapally, Malleboinpally, Nemmikal, and Appayapally have greater potential for water harvesting.

6.7.5 Drought monitoring at watersheds

Based on the weather data generated by the AWS and using the simple water balance model, weekly moisture stress conditions were monitored at selected benchmark watersheds during 2004 in Andhra Pradesh (Figure 6.12). The analysis indicated that among the 10 watersheds, the longest crop growing period of about 21 weeks was observed at Nemmikal while Karivemula and Devanakonda had only 16 weeks. Kacharam, Nemmikal, Thirumalapuram and Appayapally experienced good moisture conditions. Sripuram, Nandavaram, and Devanakonda experienced severe drought conditions before flowering period. Karivemula experienced a disastrous drought of 5-week duration. At most locations, growing period ended by 1st week of November, two-weeks early compared to the normal. Near-real time monitoring of moisture conditions at watershed level offers great scope in drought management for stabilizing crop yields.

6.7.6 Weather forecasting for agriculture

Day-to-day agricultural operations are weather sensitive; hence farmers show keen interest to know the weather in advance. Weather forecasts provide guidelines for seasonal planning and selection of crops and day-to-day management practices. Weather forecasts for agricultural operations are required in terms of rainfall and its intensity, air temperature, wind speed and direction, humidity, and sunshine/radiation. All three types of weather forecasts, viz., short, medium, and long range, are being issued by the India Meteorological Department (IMD).

One of the major functions of weather forecasts is to provide need-based information to enable the farmers to decide on taking a positive action, evasive action, or no action at all. Weather-based advisories can help farmers in minimizing the loss of inputs mainly seed, diesel, fertilizer, pesticide, labor, and time. Recommendations of land preparation for nursery and sowing will be of great help to farmers. IMD in collaboration with several organizations is implementing Agromet Advisory Services on an experimental basis at about 125 locations in India. Improvement in the accuracy of forecasts and providing appropriate advisory will result in increased economic returns. A state-of-the-art Integrated Forecasting and Communication System was implemented during September 2010 at IMD, New Delhi that is expected to provide more accurate weather data. Weather alerts by E-mail are being planned.

An understanding of the distribution and magnitude of biophysical resources of watersheds is required to develop technology intervention plans for the management of natural resources and to increase agricultural productivity in an area. Characterization of agroclimatic and other biophysical resources such as soils and vegetation resources of the watersheds helps in planning and in quantifying the impacts made during the project period as well as at the termination of the project.

6.7.7 Watershed monitoring

Repetitive nature of satellite data enables monitoring change and assists in understanding the effect of management activity undertaken. Projects like Integrated Mission for Sustainable Development (IMSD), National Agricultural Technology Project (NATP), and Sujala watershed project demonstrated the operationalization of remote sensing in the sphere of watershed management, ranging from resource appraisal to implementation and monitoring (NRSA 1995, 2002; Kaushalya Ramachandran et al., 2010; Rao et al., 2010). Cyclic revisit of space-borne sensors enables to repetitively cover the same watershed at regular time intervals to detect, monitor, and evaluate the changes occurring in the treated watersheds. Satellite images of watersheds acquired during pre- and post-treatment periods offer a rich source of information about the process of implementation of the program and its impact. Changes like increase in crop land, cropping intensity, clearing of natural vegetation, change in surface water spread/levels, afforestation, etc. could be monitored using multi-date satellite images.

6.7.8 Satellite images for impact assessment

Remotely sensed data has the advantages of providing synoptic view and large area coverage, which helps in obtaining a bird's-eye view of the ground features. Satellites, which orbit around the earth, provide a vantage point to find, measure, map, and monitor the earth's natural resources. Remotely sensed data potentially offer a rich source of information about conditions on the earth surface that change over time. Measuring and evaluating changes in a landscape over time is an important application of remote sensing. With the launch of Indian Remote Sensing (IRS) satellites, data availability both in the multispectral and panchromatic domains with varieties of spatial resolution is assured for the user community. The repetitive coverage of the same area over a period of time provides a good opportunity to monitor the land resources and evaluate land cover changes through a comparison of multi-temporal images acquired for the same area at different points of time. Changes like increased area under cultivation, conversion of annual crop land to horticulture, change in surface water body, afforestation, soil reclamation, etc. could be monitored through satellite remote sensing. Due to large area coverage at different points of time, the technology facilitates evaluating the ground realities at any given point of time.

The satellite images from different space platforms have various sensors in the visible and infrared region and are good for assessing the dynamics of watershed development, type of vegetation, crop vigor, crop growth, green biomass, and soil and water characteristics of a watershed. However, these sensors have a constraint of not being able to sense the earth's surface during cloud cover conditions. This is particularly a constraint while imaging in the optical region of the electromagnetic spectrum during the *kharif* season.

6.7.9 Monitoring and evaluation of NWDPRA watersheds using remote sensing

During the first phase of the project, 60 watersheds were identified for impact evaluation in Madhya Pradesh, Maharashtra, Orissa, Rajasthan, Tamil Nadu, and Uttar Pradesh. Similarly, evaluation of 62 NWDPRA (National Watershed Development Project for Rainfed Areas) watersheds treated during the 9th Five Year Plan period was taken up during the second phase in Andhra Pradesh, Gujarat, Haryana, Karnataka, Madhya Pradesh, Maharashtra, Orissa, Rajasthan, Tamil Nadu, Uttaranchal (Uttarakhand), Uttar Pradesh, and West Bengal. Evaluation of identified watersheds was carried out using remote sensing technique by considering the parameters such as *cropped area*: change in area extent of agricultural crops, cropping pattern, extent of wetland and irrigated crops; *plantations*: increase in agricultural and forest plantations; *wastelands*: change in aerial extent; *alternate land use*: switching over from marginal crop land to agro-horticulture and agroforestry; *water body*: change in number and aerial spread; and *biomass*: overall changes in biomass or canopy cover or productivity.

Satellite remote sensing data of identified watersheds pertaining to pre- and post-treatment periods were analyzed. The analysis involved geometric corrections, digitization, and extraction of the study area from the satellite imagery, preparation of land use/land cover maps of two periods data, preparation of normalized difference vegetation index (NDVI) images for both data sets, and quantification of improvements in the arable and non-arable lands using time-series analysis of both data sets. Digital analysis of satellite data was carried out at the Regional Remote Sensing Service Centre (RRSSC), ISRO. The analysis involved geometric correction of image data with respect to reference map to start with, digitization of watershed boundary, land use/land cover mapping, and NDVI generation and image comparisons (Figure 6.13). Geometric correction of IRS-LISS-III sensor data covering the study area was done through acquisition of ground control points (GCPs) from 1:50,000 reference map with respect to corresponding satellite images followed by computation of polynomial transformation model with two-way relationship, followed by output image generation through resampling techniques to obtain rectified final image. Image-to-image

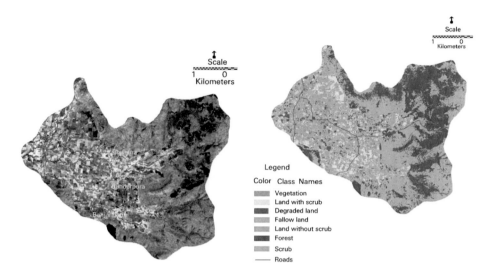

Figure 6.13 NDVI Image of Guna watershed in Madhya Pradesh (See color plate section)

registration of two-date satellite data was done by identifying accurate common GCPs on both images for computing yet another transformation model followed by re-sampling, resulting in co-registered images for comparative analysis.

Change detection is a process of determining and evaluating difference in a variety of surface phenomena over time while using geospatial data sets of multiple dates. Changes can be determined by comparing spectral responses at the same spatial location amongst a set of two or more multispectral data acquired at different points of time. There are many change detection algorithms using digital techniques such as image differencing, image rationing, principal component analysis, and comparison of classified images.

6.7.10 Monitoring and impact assessment of Adarsha watershed

Adarsha watershed in Kothapally is bound by geo-coordinates 17°21′ to 17°24′ N and 78°5′ to 78°8′ E and forms part of Shankarpally mandal (an administrative unit) of Ranga Reddy district, Andhra Pradesh, India. Vertisols and associated Vertic soils occupy 90% of the watershed area. However, Alfisols do occur to an extent of 10% of the watershed area. The main (rainy season) crops grown are sorghum (*Sorghum bicolor*), maize (*Zea mays*), cotton (*Gossypium* sp.), sunflower (*Helianthus annuus*), mung bean (green gram), and pigeonpea. During (postrainy season) wheat (*Triticum aestivum*), rice, sorghum, sunflower, vegetables, and chickpea are grown (Figure 6.14). The mean annual rainfall is about 800 mm, which is received mainly during June to October.

A number of watershed case studies using satellite data are available in addition to the centrally sponsored initiatives (Wani *et al.*, 2003; Sreedevi *et al.*, 2009; Kaushalya Ramachandran *et al.*, 2009, 2010; Roy *et al.*, 2010). For Adarsha watershed,

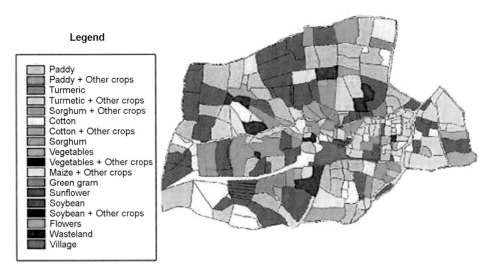

Legend

- Paddy
- Paddy + Other crops
- Turmeric
- Turmetic + Other crops
- Sorghum + Other crops
- Cotton
- Cotton + Other crops
- Sorghum
- Vegetables
- Vegetables + Other crops
- Maize + Other crops
- Green gram
- Sunflower
- Soybean
- Soybean + Other crops
- Flowers
- Wasteland
- Village

Figure 6.14 Land use and cropping pattern of Adarsha watershed, Kothapally, Andhra Pradesh (See color plate section)

Kothapally in Andhra Pradesh, thematic maps were prepared by enhancing the low resolution multispectral data with high resolution panchromatic data by a process of merging to obtain information on hydrogeomorphological conditions, soil resources, and present land use/land cover. The maps have been generated through a systematic visual interpretation of IRS-1B/-1C/-1D LISS-II and -III data in conjunction with the collateral information in the form of published maps, reports, wisdom of the local people, etc. supported by ground truth. The information derived on the lithology of the area and geomorphic and structural features in conjunction with recharge condition and precipitation was used to infer groundwater potential of each lithological unit.

In addition, derivative maps, namely, land capability and land irrigability maps were generated based on information on soils and terrain conditions according to criteria from the All India Soil and Land Use Survey Organization (All India Soil and Land Use Survey 1970). Land use/land cover maps have been prepared using monsoon and winter crop growing seasons and summer period satellite data for delineating single-cropped and double-cropped areas apart from other land use and land cover categories. Furthermore, micro-watersheds and water bodies have been delineated and the drainage networks have also been mapped (Figure 6.15). Slope maps showing various slope categories have been prepared based on contour information available at 1:50,000 scale topographical sheets. Rainfall data were analyzed to study the rainfall distribution pattern in time and space. Demographic and socioeconomic data were analyzed to generate information on population density, literacy status, economic backwardness, and the availability of basic amenities.

Since the watershed very often experiences drought, apart from alternate land use based on potential and limitations of natural resources, various drought proofing measures such as vegetative barriers, contour bunding, stone check-dams, irrigation water management, horticulture, groundwater development with conservation

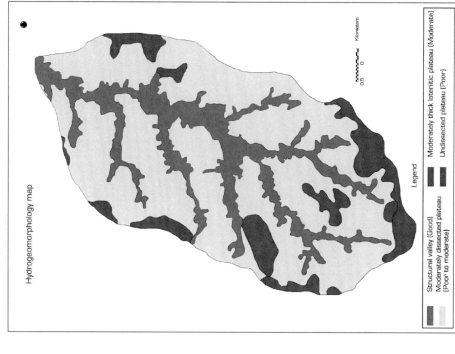

Soil map

Legend

Kakulur series
Parwada series
Alur series
Ghanpur series
Rangpur series
Kommeta series
Burgapalli series
Narayanpur series

Kilometers
0.5   0

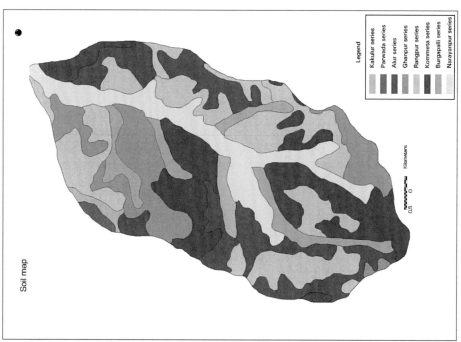

Hydrogeomorphology map

Legend

Structural valley (Good)
Moderately dissected plateau (Poor to moderate)
Undissected plateau (Poor)
Moderately thick lateritic plateau (Moderate)

Kilometers
0.5   0

Figure 6.15 Thematic maps depicting soils, land use pattern, and proposed drought proofing measures in Adarsha watershed, Kothapally, Andhra Pradesh

Suggested drought proof works

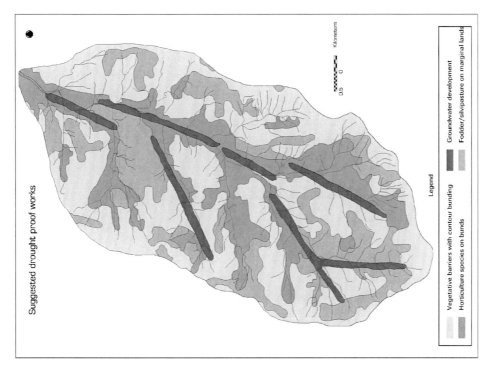

Legend

Vegetative barriers with contour bunding	Groundwater development
Horticulture species on bunds	Fodder/silvipasture on marginal lands

Kilometers
0.5 0

Optimal land use plan

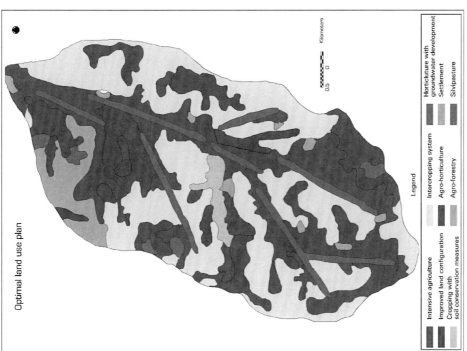

Legend

Intensive agriculture	Intercropping system	Horticulture with groundwater development
Improved land configuration	Agro-horticulture	Settlement
Cropping with soil conservation measures	Agro-forestry	Silvipasture

Kilometers
0.5 0

Figure 6.15 Continued

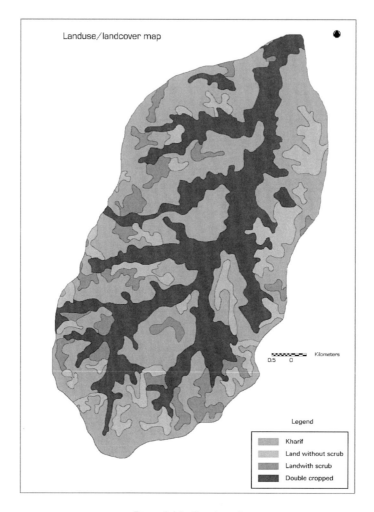

Figure 6.15 Continued

measures, and silvipasture in marginal lands have been undertaken. The suggested optimal land use practices are intensive agriculture, intercropping system, improved land configuration, agro-horticulture, horticulture with groundwater development, and silvipasture. Soon after implementation of the suggested action plan, the watershed underwent transformation, which was monitored regularly. Such an exercise not only helps in studying the impact of the program, but also enables resorting to midcourse corrections, if required. Parameters included under monitoring activities are land use/land cover, extent of irrigated area, vegetation density and condition, fluctuation of groundwater level, well density and yield, cropping pattern and crop yield, occurrence of hazards, and socioeconomic conditions. Land use/land cover parameters

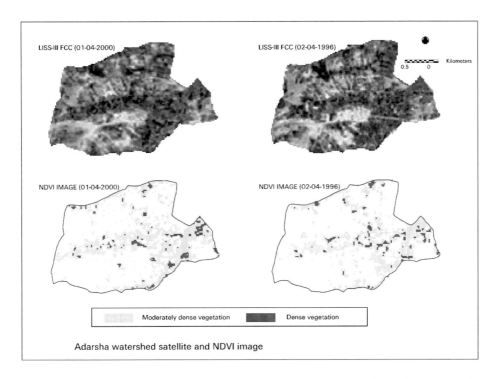

Figure 6.16 FCC and NDVI image of Adarsha watershed, Kothapally, Andhra Pradesh (See color plate section)

include changes in the number and aerial extent of surface water bodies, spatial extent of forest and other plantations, wastelands, and cropped area.

NDVI has been used to monitor the impact of implementation of action plan. NDVI images of 1996 and 2000 reveal an increase in the vegetation cover, which is reflected in improvement in the vegetation cover. The spatial extent of moderately dense vegetation cover, which was 129 ha in 1996, has risen to 152 ha in 2000. Though the satellite data used in the study depicts the terrain conditions during 1996, implementation activities started only in 1998. It is, therefore, obvious that it will take considerable time for detectable changes in terrain and vegetation conditions (Dwivedi *et al.*, 2003) (Figure 6.16).

In another study from CRIDA, use of GIS and remote sensing capabilities to evaluate watershed projects in Rangareddy and Nalgonda districts of Andhra Pradesh has been showcased. A suite of thirty-nine sustainability indicators was constructed to assess the sustainability of watershed development program in four villages at three spatial levels: household, field, and watershed. The multidisciplinary and transdisciplinary approach has helped to identify critical indicators for evaluation of watershed projects (Kaushalya Ramachandran *et al.*, 2010) (Figure 6.17).

Using GIS and survey data, the watersheds in India, Thailand, and Vietnam were characterized for the distribution of natural resources like soils, climate, water resources, and land use systems at the initiation of the watershed projects. In India,

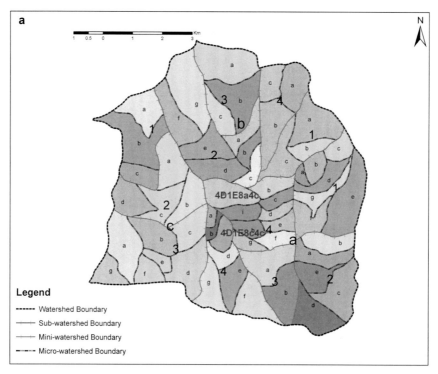

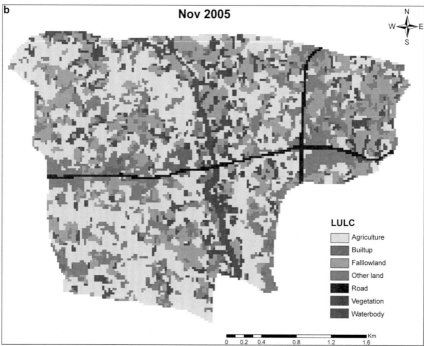

Figure 6.17 Use of GIS to delineate micro-watershed and map two sustainability indicators – land use/land cover (LULC) and NDVI and Cob-web diagram showing impact of watershed development program on agricultural productivity in Pamana micro-watershed, Rangareddy district, Andhra Pradesh (See color plate section)

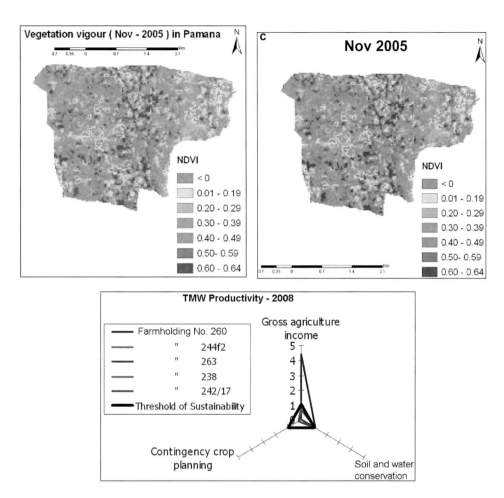

Figure 6.17 Continued

the watersheds in Andhra Pradesh (Kothapally, Malleboinpally, Appayapally, Thirumalapuram, Nemmikal, and Kacharam) and Madhya Pradesh (Lateri and Rignodia) were characterized; also Tad Fa watershed in Thailand and Thanh Ha watershed in Vietnam were characterized. Using remote sensing and GIS technology it was observed that significant improvements in the vegetation cover in Kothapally watershed in Andhra Pradesh and Lateri watershed in Madhya Pradesh with the introduction and adoption of improved resource management and crop production technologies over the period of five years occurred.

6.8 TECHNOLOGY INTEGRATION

A vast amount of technology encompassing different domains exists. As long as they are individual tools for collection of data or processing of data, their use is limited.

Hence they need to be integrated into a total solution system that can take care of most of the operational requirements as well as decisions to a great extent. In the following sub-sections, a few concepts about integration of various technologies have been discussed. Finally, the concept for achieving total solution has been presented.

6.8.1 Field data transmission

Even though the field data collection methods make use of IT products as and when they become mainstream (for example, laptop and handheld computers, GPS receivers, etc.), an integrated and comprehensive process formulation driven by a 'total solution' approach is emerging. As a result of such revolutions in ICT, access to the Internet via mobile-device based web browsers and other IP protocol based applications became possible. This provides a huge opportunity to develop customized applications on integrated PDA devices for specific end uses like those for field data collection and communicate to base server in real time. In this direction a system was developed at NRSC keeping in view 'total solution' approach to realize the mobile device based field data collection application (Figure 6.18).

It consists of configuration of mobile device prior to field visit, mobile device application, wireless network services, automated data receiving server program, central data storage (repository), and LAN based application to utilize the stored field data. The solution highlights the importance of prior planning and preparation of reference data to be carried on the mobile device, which includes the discipline, parameters to be collected, project information, team members information, etc. This also ensures that the mobile application is flexible and configurable enough to support field data collection activity for a variety of disciplines/end uses. It also uses a central data repository

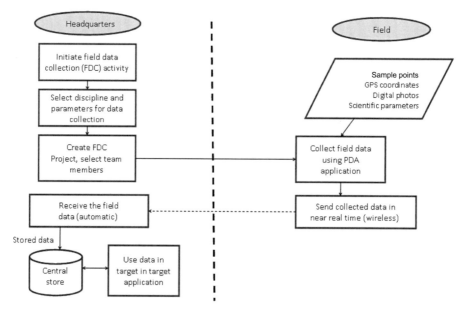

Figure 6.18 Total solution to field data collection system

to store all the reference and sample data that continuously accumulates with each field data collection activity.

The central data repository is a critical component for ensuring systematic data organization and management for the process. The deployment of this solution enables the near-real time transmission of collected data directly from the field to the base headquarters for initiating immediate further action. The scope of this solution can be enhanced with the implementation of additional functionalities like visualization, historical studies, data mining, data extraction, and GIS export.

6.8.2   Sensor Web

It is a physical platform for a sensor which is aerial or terrestrial, fixed or mobile, and data collected by the sensor could be accessible in real time via wireless networks and Internet. At times the term "Sensor Web" is used to refer to sensors connected to the Internet for a real time application. The purpose of a Sensor Web system is to extract knowledge from the data it collects and use this information to intelligently react and adapt to its surroundings. With the vast development in computer and telecommunication markets, the price of state-of-the-art electronic chips became very affordable and ushered the development of vertical applications. Even the multi-directional sensors-to-sensor communication is possible with the recent developments. The various sensors can be integrated together using a protocol similar to TCP/IP (used for networking various computers) and make them to share information among themselves and act as a single system. In essence, the Sensor Web is a macro-instrument comprising a number of sensor platforms.

The major advantage of the Sensor Web is that the sensors can be placed in very remote and harsh environments, where it is very difficult to collect data under direct human supervision. Further, the data can be collected continuously and delivered to the needy in near-real time basis. This has immense potential in applications related to agriculture, medical, life safety, emergency management, and so on. The Sensor Web is now focusing much on applications of this technology. This Sensor Web approach allows for various complex behaviors and operations such as on-the-fly identification of outlying sensor, mapping of vector fields from measured scalar values and interpreting them locally, and detection of critical events. As the Sensor Web infrastructure becomes more common in various user communities, there will be a demand for associated sensors to populate these systems. As a result, the combined exponential growth of both computer and telecommunication technologies will contribute to a similar explosive growth of sensor technology.

6.8.3   Spatial simulation modeling

The action plan for watershed essentially aims at reducing soil loss, improving ground or surface water harvesting, and improving crop productivity. Spatial modeling and integration of point models in spatial domain have greater significance in watershed studies to achieve the above-mentioned goals. They can enhance the impacts of agricultural research in watershed development. Simulation modeling using the surface and groundwater balance models and crop growth model enables to optimize the use of water resources in the watershed and to minimize the gap between the achievable yield and potential yield. Assessing long-term impacts of various management options on carbon sequestration, environmental balance, land degradation, etc. could be assessed using simulation modeling approaches, which would otherwise, not be possible using conventional approaches on a routine basis (Sreedevi *et al.*, 2009).

Temporal acquisition of satellite data during crop growing season enables to monitor the crop growth with the help of biophysical parameters such as leaf area index, soil/crop moisture, NDVI, etc. and when coupled with spatio-dynamic modeling facilities in GIS, scenario generation is quite possible for crop intensification analysis besides the sustainability assessment of the systems. There is a need to incorporate these dynamic parameters in refining prioritized watersheds for effective utilization of resources.

Baseline data generated using above tools forms the basic input to characterize the watershed spatially and also provides necessary inputs for spatial models after proper translation. While preparing any action plan aiming at overall development of watershed it is essential to visualize the impact of interference done with the existing environment. Better Assessment Science Integrating point and Nonpoint Sources (BASINS) and Soil and Water Assessment Tool (SWAT) are some of the comprehensive models available in GIS environment that help in modeling the watershed environment and visualizing the future scenarios. To run the above continuous simulation models, updation of information on climate (rainfall, PET, radiation, temperature, wind velocity, LGP), soils (organic carbon, nutrients, bulk density, pH, etc.), crops (cropping intensity, crops and their growth attributes, phenology, yield and yield attributes, pattern, cultivars, inputs applied), major plant nutrient uptake data, socioeconomic data (income sources, labor sources, input, output/income, infrastructure, etc), runoff and soil loss measurements and groundwater level (Wani 2002) is essential. For this, the Sensor Web, GPS, and communication networks are useful.

6.8.4 Use of ICT in watershed management

It is increasingly realized that facilitation of knowledge flow is a key in fostering new rural livelihood opportunities using modern ICTs. The concept adapted is one of intelligent intermediation for facilitation of flows of information and knowledge. The community center managed by the PIAs (project implementation agencies) functions as a Rural Information Hub connecting participating villages (or groups of villages, as the case may be) and also with other Internet connected websites (Figure 6.19). It is operated or managed by a rural group [women or youth self-help groups (SHGs)] identified by the village watershed council through a consultative process. The activities in this module are planned to adopt a hub-and-spokes model for information dissemination among the participants and stakeholders. The electronic network across select nuclear watersheds enables sharing of experience and best practices.

6.8.5 Intelligent watershed information system

The previous sub-sections discuss about the application of technologies in watershed related activities. There is a need to integrate these components into an intelligent information system for efficient management of watersheds. The spatial data of

Figure 6.19 Information and communication technology services enabled at Addakal, Mahabubnagar district, Andhra Pradesh

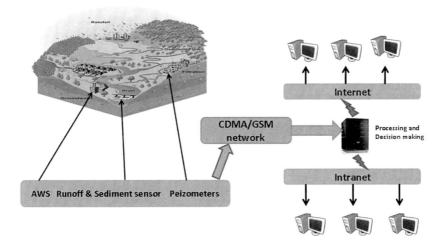

Figure 6.20 Technology integration for watershed management

watersheds (slope, soils, crops, land cover, wastelands, etc.) along with field data collected with field Sensors Web (runoff, sediment loss, nutrient loss, etc.) and AWSs can be directly communicated to the central server using mobile communication (CDMA/GSM) or WAP enables networks. These inputs could be translated into the input format required to run point or spatial models. Further, by suitably processing the above input data with simulation models, various scenarios can be generated and validated with field data.

To achieve this, initially a semantic network could be generated keeping the goal and objectives of the watershed program, which could be translated to automated decision-making system using adoptive algorithms like Artificial Neural Networks (ANN) and Decision Trees (DT) that could help in dynamically prioritizing watersheds. A schema towards achieving an Intelligent Watershed Information System (IWIS) for effective watershed management is depicted in Figure 6.20. The decisions generated from the above system have to be communicated to the stakeholder (farmers/extension staff) in a reasonable time frame.

6.9   SUMMARY AND CONCLUSIONS

Application of new science tools in rainfed agriculture opens up new vistas for development through IWMPs. These tools can help in improving the rural livelihoods and contributing substantially to meet the millennium development goals of halving the number of hungry people by 2015 and achieving food security through enhanced use efficiency of scarce natural resources such as land and water in the tropical countries. Till now rainfed areas of the SAT did not get much benefit of new science tools but the recent research using these tools such as simulation modeling, remote sensing, GIS as well as satellite-based monitoring of the natural resources in the SAT has shown that not only the effectiveness of the research is enhanced substantially but also the cost efficiency and impact are enhanced. The remarkable developments in space technology currently offer satellites which provide better spatial and spectral resolutions, more frequent revisits, stereo viewing, and on-board recording capabilities. Thus, the high spatial and temporal resolution satellite data could be effectively used for watershed management and monitoring activities at land ownership level. By using crop simulation modeling approach, yield gap analyses for the major crops in Asia, Africa, and WANA regions revealed that the yields could be doubled with the existing technologies if the improved crop land, nutrient, and water management options are scaled-out.

Similarly, technology application domains could be easily identified for better success and greater adoption of the particular technologies considering the biophysical as well as socioeconomic situations. GIS helped in speedy analysis of voluminous data and more rationale decision in less time to target the investments as well as to monitor the large number of interventions in the SAT. The satellite-based techniques along with GIS helped in identifying the vast fallow areas (2 million ha) in Madhya Pradesh during the rainy season. Similarly, 14 million ha rice-fallows in the Indo-Gangetic Plain offer excellent potential to grow second crop on residual soil moisture by using short-duration chickpea cultivars and simple seed priming technology. These techniques are also successfully used for preparing detailed thematic maps, watershed development plans, and continuous monitoring of the natural resources in the country in rainfed areas. Further, such data could be of immense help in tracking the implementation, applying midcourse corrections, and for assessing long-term effectiveness of the program implemented. The synergy of GIS and Web Technology allows access to dynamic geospatial watershed information without burdening the users with complicated and expensive software. Further, these web-based technologies help the field data collection and analysis in a collaborative way. However the availability of suitable software for watershed studies and their management in open GIS platform is very limited. Hence, there is a requirement to strengthen this area through collaborative efforts between various line organizations.

Use of ICT in IWMP can bridge the existing gap to reach millions of small farm holders who have no access to new technologies for enhancing agricultural productivity on their farms. Use of smart sensor network along with GIS, remote sensing, simulation modeling and ICT opens up new opportunities for developing intelligent watershed management information systems. However, it calls for a new partnership involving corporates, development agencies, researchers from various disciplines and most importantly to reach millions of small farm holders in rainfed areas of the world. Application of new science tools in IWMP have helped to substantially enhance productivity as well as income from rainfed agriculture and improved livelihoods of the rural people.

REFERENCES

Allen, R.G., L.S. Pereria, Dirk Raes, and Martin Smith. 1998. Crop evapotranspiration–Guidelines for computing crop water requirements. *FAO Irrigation and Drainage Paper 56.* Rome, Italy: FAO.

All India Soil and Land Use Survey. 1970. Soil survey manual. New Delhi, India: All India Soil and Land Use Survey, Ministry of Agriculture, Government of India.

Anonymous. 2004. www.satimagingcorp.com/SPOT_DEM_Product_Description_v1-1.pdf (Accessed on 18 Jan 2010.)

Chuc, N.T., Piara Singh, K. Srinivas *et al.* 2005. Yield gap analysis of major rainfed crops of northern Vietnam using simulation modeling. *Global Theme on Agroecosystems Report No. 26.* Patancheru, Andhra Pradesh, India: International Crops Research Institute for the Semi-Arid Tropics.

Cooper P., K.P.C. Rao, P. Singh *et al.* 2009. Farming with current and future climate risk: Advancing a "Hypothesis of Hope" for rainfed agriculture in the semi-arid tropics. *Journal of SAT Agricultural Research,* Volume 7. (ejournal.icrisat.org)

Diwakar, P.G., and V. Jayaraman. 2007. Space enabled ICT applications for rural upliftment – experience of participatory watershed development. IETE Technical Review 24(4):313–321.

Dwivedi, R.S., K.V. Ramana, S.P. Wani, and P. Pathak. 2003. Use of satellite data for watershed management and impact assessment. In *Integrated watershed management for land and water conservation and sustainable agricultural production in Asia:* proceedings of the ADB-ICRISAT-IWMI Project Review and Planning Meeting, 10–14 December 2001, Hanoi, Vietnam, ed. S.P. Wani, A.R. Maglinao, A. Ramakrishna, and T.J. Rego, 149–157. Patancheru, Andhra Pradesh, India: International Crops Research Institute for the Semi-Arid Tropics.

FAO. 1996. Digital soil map of the world and derived soil properties. CDROM Series. Rome, Italy: FAO.

Garbrecht, J., and L.W. Martz. 1999. Digital elevation model issues in water resources modeling. In *Proceedings of the 19th ESRI Users Conference*, San Diego, California, USA. (http://proceedings.esri.com/library/userconf/proc99/proceed/papers/pap866/p866.htm.)

Guyot, G. 1990. Optical properties of vegetation canopies. In *Applications of remote sensing in agriculture,* ed. M.D. Steven, and J.A. Clark, 19–44. Cambridge, UK: University Press.

Harris, D., A. Joshi, P.A. Khan, P. Gothkar, and P.S. Sodhi. 1999. On-farm seed priming in semi-arid agriculture: development and evaluation in maize, rice and chickpea in India using participatory methods. *Experimental Agriculture* 35:15–29.

ICRISAT. 1994. *ICRISAT now, sowing for future.* Patancheru, Andhra Pradesh, India: International Crops Research Institute for the Semi-Arid Tropics.

Johansen, C. 1998. Production system concepts. In *GIS analysis of cropping systems:* proceedings of an international workshop on harmonization of databases for GIS analysis of cropping systems in Asia region, 18–19 August 1997, ICRISAT, Patancheru, India, ed. S. Pande, C. Johansen, J. Lauren, and F.T. Bantilan Jr., 67–74. Patancheru, Andhra Pradesh,

India: International Crops Research Institute for the Semi-Arid Tropics; and Ithaca, New York, USA: Cornell University.

Kasturirangan, K., R. Aravamudam, B.L. Deekshatulu, George Joseph, and M.G. Chandrasekhar. 1996. Indian remote sensing satellite (IRS)-1C – The beginning of a new era. *Current Science* 70:495–500.

Kaushalya Ramachandran, U.K. Mandal, K.L. Sharma *et al.* 2009. Methodology for evaluating livelihood security of farm households in treated watersheds. *Indian Journal of Soil Conservation* 37(2):151–163.

Kaushalya Ramachandran, U.K. Mandal, K.L. Sharma, and B. Venkateswarlu. 2010. Evaluation methodology for post-facto assessment of watershed development project–a multi-disciplinary, multi-scale and multi-aspect approach. *Natural Resource Management & Livelihood Series*. Hyderabad, India: CRIDA.

Keig, G., and J.R. McAlpine. 1974. Watbal: A computer system for the estimation and analysis of soil moisture regimes from simple climatic data. *Technical Memorandum 74/4*. Australia: Division of Land Resources, Commonwealth Scientific and Industrial Research Organization.

Kesava Rao, A.V.R., S.P. Wani, Piara Singh, M. Irshad Ahmed, and K. Srinivas. 2007. Agroclimatic characterization of APRLP-ICRISAT nucleus watersheds in Nalgonda, Mahabubnagar and Kurnool districts. *Journal of SAT Agricultural Research*, Volume 3, Issue 1. (ejournal. icrisat.org)

Khan, M.A., V.P. Gupta, and P.C. Moharana. 2001. Watershed prioritization using remote sensing and geographical information system: a case study from Guhiya, India. *Journal of Arid Environments* 49:465–475.

Khare, Y.D., N.T. Srivastara, A.S. Deshpande, R.M. Tamhane, and A.K. Sinha. 2001. Remote sensing and GIS based integrated model for watershed prioritization – An approach based on physical and socio-economic indices. Presented at the National Symposium on Advances in Remote Sensing Technology with special emphasis on high resolution imagery, SAC, Ahmedabad, Dec 11–13, 2001.

Krishna Murthy, Y.V.N., S. Srinivasa Rao, D.S. Prakasa Rao, and V. Jayaraman. 2008. Analysis of dem generated using Cartosat-1 stereo data over Mausanne Les Alpiles – cartosat scientific appraisal programme (Csap Ts – 5). *The International Archives of the Photogrammetry, Remote Sensing and Spatial Information Sciences*, Vol. XXXVII, Part B1, pp. 1343–1348.

NRSA. 1995. *Integrated mission for sustainable development*. Technical guidelines. Hyderabad, India: NRSA.

NRSA. 2002. *Integrated mission for sustainable development*. Project report. Hyderabad, India: NRSA.

NRSA. 2006. *Atlas – National agricultural technology project*. Development of regional scale watershed plans and methodologies for identification of critical areas for prioritized land treatment in the watersheds of rainfed rice, oilseeds, pulses, cotton and NCRL production systems. Hyderabad, India: NRSA.

Navalgund, R.R., J.S. Parihar, L. Venkataratnam *et al.* 1996. Early results from crop studies using IRS-1C data. *Current Science* 70:568–574.

Obi Reddy, G.P., A.K. Maji, C.V. Srinivas, and K.S. Gajbhiye. 2001. Terrain characterization and evaluation for land form – soil resources analysis using remote sensing and GIS applications: a case study. Presented at the National Symposium on Advances in Remote Sensing Technology with special emphasis on high resolution imagery, SAC, Ahmedabad, Dec 11–13, 2001.

Rao, B.R.M., K. Sreenivas, M.A. Fyzee, and T. Ravi Sankar. 1998. Evaluation of IRS-1C PAN data for mapping soil resources. *NNRMS Bulletin* 22:68–71.

Rao, G.G.S.N., A.V.R. Kesava Rao, Y.S. Ramakrishna, and U.S. Victor. 1999. Resource characterisation of drylands: climate. In *Fifty years of dryland agricultural research in India* ed. H.P. Singh, Y.S. Ramakrishna, K.L. Sharma, and B. Venkateswarlu, 24–40. Hyderabad, India: CRIDA.

Rao, K.V., B. Venkateswarlu, K.L. Sahrawat *et al.* (ed.) 2010. *Proceedings of national workshop-cum-brain storming on rainwater harvesting and reuse through farm ponds: Experiences, issues and strategies,* 21–22 April 2009, CRIDA, Hyderabad, India. Hyderabad, India: CRIDA.

Rao, V.N., P. Singh, J. Hansen, T. Giridhara Krishna, and S.K. Krishna Murthy. 2007. Use of ENSO-based seasonal rainfall forecasting for informed cropping decisions by farmers in the SAT India. In *Climate prediction and agriculture: advances and challenges.* ed. M.V.K. Siva Kumar, and James Hansen, 165–179. New York: Springer Berlin Heidelberg; and Geneva: WMO.

Roy, P.S., T. Ravisankar, and K. Sreenivas. 2010. Advances in geospatial technologies in integrated watershed management. Presented at the National Symposium on Use of High Science Tools in Integrated Watershed Management, 1–2 February 2010, New Delhi, India.

Sahrawat K.L., T.J. Rego, S.P. Wani, and G. Pardhasaradhi. 2008. Stretching soil sampling to watershed: evaluation of soil-test parameters in a semi-arid tropical watershed. *Communications in Soil Science and Plant Analysis* 39:2950–2960.

Sahrawat K.L., S.P. Wani, T.J. Rego, G. Pardhasaradhi, and K.V.S. Murthy. 2007. Widespread deficiencies of sulphur, boron and zinc in dryland soils of the Indian semi-arid tropics. *Current Science,* 93(10):1–6.

Saxena, R.K., K.S. Verma, G.R. Chary, R. Srivastava, and A.K. Barthwal. 2000. IRS-1C data application in watershed characterization and management. *International Journal of Remote Sensing* 21(17):3197–3208.

Sekhar, K.R., and B.V. Rao. 2002. Evaluation or sediment yield by using remote sensing and GIS: A case study from Phulang vagu watershed, Nizamabad district, AP, India. *International Journal of Remote Sensing* (20):4499–4509.

Sharma, T. 1997. An integrated approach to sustainable development of a watershed in India. Using remote sensing and GIS space applications for sustainable development. In *Proceedings of the High-level Seminar on Integrated Space Technology Applications for Poverty Alleviation and Rural Development,* 21–24 October 1996, Bangkok, Thailand, 35–42. UNESCAP Reference No.: ST/ESCAP/1817. Division: Information and Communications Technology and Disaster Risk Reduction.

Singh, P., P.K. Aggarwal, V.S. Bhatia *et al.* 2009. Yield gap analysis: modelling and achievable yields at farm level. In R*ainfed agriculture: unlocking the potential,* ed. S.P. Wani, J. Rockström, and T. Oweis, 81–123. Comprehensive Assessment of Water Management in Agriculture Series. Wallingford, UK: CAB International.

Singh, P., P. Pathak, S.P. Wani, and K.L. Sahrawat. 2010. Integrated watershed management for increasing productivity and water use efficiency in semi-arid tropical India. In *Water and agricultural sustainability strategies,* ed. Manjit S. Kang, 181–205. The Netherlands: CRC Press.

Sreedevi, T.K., S.P. Wani, A.V.R. Kesava Rao, P. Singh, and I. Ahmed. 2009. New science tools for managing community watersheds for enhancing impact. *Journal of SAT Agricultural Research,* Volume 7. (ejournal.icrisat.org)

Srivastava, P.K., T.P. Srinivasan, A. Gupta *et al.* 2007. Recent advances in Cartosat-1 data processing. Presented at ISPRS Hannover Workshop 2007. (www.commission1.isprs.org/hannover07/paper/Srivastava_etal.pdf; accessed on 11 Jan 2010.)

Subbarao, G.V., J.V.D.K. Kumar Rao, J. Kumar *et al.* 2001. *Spatial distribution and quantification of rice-fallows in South Asia – potential for legumes.* Patancheru, Andhra Pradesh, India: International Crops Research Institute for the Semi-Arid Tropics.

Thakkar, A.K., and S.D. Dhiman. 2007. Morphometric analysis and prioritization of mini-watersheds in Mohr watershed, Gujarat using remote sensing and GIS techniques. Journal of the Indian Society of Remote Sensing 35(4):313–321.

Thornthwaite, C.W., and J.R. Mather. 1955. The water balance. Publications in climatology. Vol. VIII, No. 1. New Jersey, USA: Drexel Institute of Technology, Laboratory of Climatology.

Wani, S.P. 2002. data needs, data collection, and data analysis for on-farm trials. In *Improving management of natural resources for sustainable rainfed agriculture: proceedings of the Training Workshop on On-farm Participatory Research Methodology*, 26–31 July 2001, Khon Kaen, Bangkok, Thailand, ed. S.P. Wani, T.J. Rego, and P. Pathak, 43–45. Patancheru, Andhra Pradesh, India: International Crops Research Institute for the Semi-Arid Tropics.

Wani, S.P., P.K. Joshi, K.V. Raju *et al.* 2008. Community watershed as a growth engine for development of dryland areas. A Comprehensive Assessment of Watershed Programs in India. *Global Theme on Agroecosystems Report No. 47*. Patancheru, Andhra Pradesh, India: International Crops Research Institute for the Semi-Arid Tropics.

Wani, S.P., J. Rockström, and T. Oweis. (ed.) 2009. *Rainfed agriculture: unlocking the potential*. Comprehensive Assessment of Water Management in Agriculture Series. Wallingford, UK: CAB International.

Wani, S.P., H.P. Singh, T.K. Sreedevi *et al.* 2003. Farmer-participatory integrated watershed management: Adarsha watershed, Kothapally India, an innovative and upscalable approach. A Case Study. In, *Research towards integrated natural resources management: Examples of research problems, approaches and partnerships in action in the CGIAR*, ed. R.R. Harwood, and A.H. Kassam, 123–147. Washington, DC, USA: Interim Science Council, Consultative Group on International Agricultural Research.

Chapter 7

Soil and water conservation for optimizing productivity and improving livelihoods in rainfed areas

Prabhakar Pathak, P.K. Mishra, Suhas P. Wani, and R. Sudi

7.1 INTRODUCTION

Soil and water are the most valuable natural resources to meet the basic needs of food, feed, and fiber for human beings. However, conserving soil and water resources is a growing challenge as they are under increasing stress to produce more food for the ever growing population. The loss of soil surface layer, which contains most nutrients and organic matter, reduces fertility. In addition, high runoff water causes moisture stress in the later part of the season, leading to low and variable crop productivity especially under rainfed conditions. Globally, total area affected by moderate to serious soil erosion is estimated around 1028 million ha, of which 748 million ha is due to water erosion and the rest by wind erosion. In Asia and Africa, 673 million ha area is impacted by erosion (Oldeman *et al.*, 1991). It is estimated that 186 million ha area is affected by chemical and physical degradation, which reduce vegetative cover and exacerbate soil erosion (Oldeman *et al.*, 1991). In Asia, South America, and Africa soil erosion rates are the highest with estimated average of $30–40\,t\,ha^{-1}\,yr^{-1}$, while in Europe and North America average rates are somewhat lower at about $17\,t\,ha^{-1}\,yr^{-1}$. A sustainable rate of soil loss (rate of soil loss is equal to rate of soil formation) is thought to be about $1\,t\,ha^{-1}\,yr^{-1}$ (Pimental *et al.*, 1995).

The high erosion hazards have serious on-site and off-site impacts on productivity, ecosystem services, and environmental quality. In rainfed regions where population growth and poverty level are high and external inputs for farming are meager due to economic reasons, the erosion impacts on agricultural productivity are generally very high both in the short- and the long-term. El-Swaify (1993) reported the results from the long-term experiments on crop responses to different levels of soil erosion (Table 7.1). Clearly, the erosion-induced changes in soil quality and the resulting unfavorable root proliferation combine to reduce water and nutrient use efficiency by crops on the eroded soils. These unfavorable effects have implications especially for the rainfed farming systems, since inefficient use of stored soil water further exacerbates the already costly water loss by uncontrolled runoff. The net results could be frequent crop failures in systems that are mainly dependent on seasonal rainfall.

In recent years, off-site (or downstream) erosion impacts have received increasing attention. This is partly due to mounting concerns with sediment-based nonpoint source pollution and its detrimental effects on water quality. In the tropics, increased encroachment of human populations and activities in the upper reaches of river basins and watersheds have significantly accelerated sediment production and delivery to

Table 7.1 The changes in water and fertilizer use efficiency by maize as a result of erosion and restorative fertilization on an Oxisol[a]

Erosion level (cm)	Fertility level	Water use efficiency (kg stover dm^{-3} of water)	Fertilizer/amendment use efficiency (kg stover kg^{-1} of elemental added amendment)
0	Low	0.42	40.0
	Intermediate	0.48	8.6
	Optimum	0.60	2.6
	Average	0.50	17.0
10	Low	0.25	24.0
	Intermediate	0.36	5.6
	Optimum	0.51	2.3
	Average	0.37	11.0
35	Low	0.07	3.8
	Intermediate	0.17	4.0
	Optimum	0.37	1.9
	Average	0.20	3.2

[a]Source: El-Swaify (1993).

low-lying lands. Runoff and eroded sediments cause siltation of waterways, dams, and reservoirs thus reducing the efficiency of hydroelectric power generating plants. Runoff also causes burial and flooding of low-lying lands, property, life, and shoreline fisheries and reefs and destruction of roads, terraces, and other structures. This upsets the balance involving sediment removal and deposition in beds and banks of rivers or streams serving as water resources and transport network (El-Swaify *et al.*, 1982).

Water scarcity is undoubtedly the most critical issue in rainfed agriculture. The demand for fresh water is increasing globally at an accelerated rate especially for agriculture and other sectors including domestic, energy, and industrial uses. It is estimated that approximately $7100\,km^3\,yr^{-1}$ water is consumed globally to produce food, of which $5500\,km^3\,yr^{-1}$ is used in rainfed agriculture and $1600\,km^3\,yr^{-1}$ in irrigated agriculture (De Fraiture *et al.*, 2007; Molden *et al.*, 2007). The analysis also predicts large increases in the amount of water needed to produce food by 2050, ranging from 8500 to $11,000\,km^3\,yr^{-1}$, depending on the assumptions regarding the improvements in rainfed and irrigated agricultural systems. However, the rainfall in rainfed regions generally occurs in short torrential downpours. Large portion of this water is lost as runoff. The current rainfall use efficiency for crop production is low ranging from 30 to 55%; thus annually large percentage of seasonal rainfall goes unproductive, lost either as surface runoff, evaporation, or deep drainage. Groundwater levels are depleting in the rainfed regions and most rural rainfed areas are facing general water scarcity and drinking water shortages. Though the problem of water shortages and land degradation have been in existence in the past also, the pace of natural resource degradation has greatly increased in recent times due to the burgeoning population and the increased exploitation of natural resources.

In rainfed agriculture, accelerated water demand can be met through efficient rainwater conservation and management. For this both in-situ and ex-situ rainwater management play crucial roles in increasing and sustaining the crop productivity.

The Comprehensive Assessment of water management in agriculture describes a large untapped potential for upgrading rainfed agriculture and calls for increased water investments in the sector (Molden *et al.*, 2007; Rockström *et al.*, 2007). Yield gap analyses carried out by Comprehensive Assessment for major rainfed crops in semi-arid regions in Asia and Africa reveal large yield gaps with farmers' yields being a factor of two to four times lower than achievable yields for major rainfed crops (Rockström *et al.*, 2007).

To achieve the vast potential, rainfed agriculture needs to be upgraded. Soil and water conservation should be used as entry point activity for upgrading rain-fed agriculture through a more holistic approach based on converging all the aspects of natural resource conservation, their efficient use, production functions and income-enhancement avenues through value-chain and enabling policies (Wani *et al.*, 2003; Rockström *et al.*, 2007, 2010). Thus, soil and water conservation play a critical role in increasing and sustaining agricultural productivity in rainfed areas in the fragile agroecosystems.

This chapter reviews in brief in-situ and ex-situ soil and water conservation practices, which have been found promising for improving productivity and controlling land degradation in different rainfed regions of Asia and Africa. An "integrated watershed management approach" for enhancing the impacts of soil and water conservation is highlighted. The key factors, which would facilitate the greater adoption of soil and water conservation practices by the farmers are also discussed.

7.2   SOIL AND WATER CONSERVATION PRACTICES

In the past, most soil conservation programs were based on the introduction of practices and measures aimed mainly at conserving soil by slowing down and safely disposing of runoff. All of these are technically sound and there will be a place for them in the future. However, these measures take up valuable space and can be costly and time consuming to maintain. Farmers therefore, are usually reluctant to adopt such measures and they frequently fail for the lack of their maintenance. Strategies should therefore aim at retaining and using rainwater where it falls. If this is done, the chances of healthy plant growth and better yields are increased, while effects of drought and crop failure are decreased. This strategy is also expected to greatly reduce runoff and thereby soil erosion.

Based on experiences from the various rainfed regions of Asia and Africa, the soil and water conservation practices for the different rainfed regions are given in Table 7.2. It clearly shows that for different regions the problems of soil and water conservation are quite different. This information could be useful in determining the appropriate soil and water conservation practices for various regions. This classification and related information also assists in utilizing the research and field experience of one place to other places of identical soil, climatic, and topographic conditions.

7.2.1   In-situ soil and water conservation

In-situ soil and water conservation measures are important for effective conservation of soil and water at the field level. The main aim of these practices is to reduce or prevent either water erosion or wind erosion, while achieving the desired moisture

Table 7.2 Soil and water conservation problems and recommended technologies for different rainfed regions of Asia and Africa

Annual rainfall (mm)	Problems	Recommended technology	General remarks
≤500	Extreme moisture stress and drought, overgrazing, improper land management, shifting sand dunes, wind erosion	Contour cultivation with conservation furrows, ridging sowing across slopes, off-season tillage, minimum tillage, inter-row water harvesting system, small water harvesting structures, vegetative barriers, contour bunds, field bunds, mulching, scoops, tied ridges indigenous methods such as *khadin*	Major focus needs to be given on in-situ soil and water conversation with low-cost technologies. Vegetative barrier along with appropriate land use systems should be used to control wind erosion.
>500–≤750	Sheet erosion, ravine lands, shortage of moisture, recurring droughts, moderate to high runoff overgrazing, siltation of reservoirs and tanks, lack of adequate groundwater recharge	Contour cultivation with conservation furrows, compartment bunding, ridging, sowing across slopes, minimum tillage, zingg terrace, off-season tillage, broad-bed and furrow (BBF), contour/graded border strips, scoops, tied ridges, mulching, inter-row water harvesting system, small basins, stone bunds, field bunds, graded bunds, contour bunds, vegetative bunds, small gully control structures, runoff water harvesting structures	On Vertisols and associated soils major emphasis needs to be given on in-situ soil and water conversation. On Alfisols and associated soils emphasis needs to be given on both in-situ and ex-situ soil and water conversation.
>750–≤1200	High sheet and gully erosion, ravine lands, high runoff, waterlogging, poor workability of soils, moisture stress particularly during postrainy and summer seasons, siltation of reservoirs and tanks, downstream flooding	BBF (for Vertisols and associated soils), flat-on-grade cultivation, conservation furrows, sowing across slopes, field and main drains, conservation tillage, contour/graded border strips, small basins, stone bunds, field bunds, vegetative bunds, graded bunds, modified contour bunds, *Nadi* Zingg terrace, gully control structures, runoff water harvesting and groundwater recharging structures	Emphasis needs to be given on both in-situ and ex-situ soil and water conservation practices. On Vertisols and other heavy soils, graded type soil and water conservation practices which provide balance between moisture conservation and waterlogging need to be adopted. Good potential for harvesting runoff and groundwater recharging.

Table 7.2 Continued

Annual rainfall (mm)	Problems	Recommended technology	General remarks
>1200	High soil erosion, gully formation, waterlogging, poor workability of soils, shortage of water during postrainy and summer seasons, siltation of reservoirs and tanks, downstream flooding	BBF (Vertisols), field bunds, stone bunds, vegetative bunds, flat-on-grade cultivation, field and main drains, conversation tillage, contour/graded border strips, modified contour bunds, gully control structures, runoff water harvesting and groundwater recharging structures, graded bunds	Emphasis on controlling soil erosion and safe disposal of excess runoff water. Excellent potential of harvesting runoff and groundwater recharge.

for sustainable production. The suitability of any in-situ soil and water management practice depends greatly upon soil, topography, climate, cropping system, and farmers resources. Some of the promising in-situ soil and water conservation practices from the different rainfed regions are discussed below.

7.2.1.1 Contour cultivation and conservation furrows

In several rainfed regions, the up and down cultivation is still a common practice. This results in poor rainfall infiltration and accelerated soil erosion. Contour cultivation or cultivation across the slope is a simple method of cultivation, which can effectively increase rainfall infiltration and reduce runoff and soil loss on gently sloping lands. The contour cultivation involves performing cultural practices such as plowing, planting, and cultivating on the contours (Figure 7.1). It creates a series of miniature barriers to runoff water when it flows along the slope. Mishra and Patil (2008) reported that this system in farmers' fields on Alfisols of Kabbalanala watershed near Bengaluru, India increased soil moisture during the cropping season from 35th to 43rd weeks over farmers' practice of up and down cultivation (Figure 7.2). Contour cultivation conserved the rainwater and reduced the runoff and soil loss, and increased the yields of sesame, finger millet, and groundnut in the Alfisols at Bengaluru.

The effectiveness of this practice was greater when the crops were fertilized with nitrogen (N) and phosphorus (P) nutrients and other improved practices were implemented (Krishnappa *et al.*, 1999). This practice resulted in 35% and 22% increase in sorghum and *Setaria* yield, respectively on Vertic inceptisols and 66% increase in sorghum yield on Alfisols over the up and down method of cultivation.

In most situations the effectiveness of contour cultivation can be greatly enhanced by adding conservation furrows into the system. In this system in addition to contour cultivation, a series of furrows are opened on contour or across the slope at 3.0–7.5 m apart (Figure 7.3). The spacing between the furrows and its size can be chosen based on the rainfall, soils, crops and topography (Pathak *et al.*, 2009a). The furrows can be made either during planting time or during interculture operations using traditional plow. Generally, two passes in the same furrow may be needed to obtain the required

Figure 7.1   Contour cultivation at Kurnool watershed in Andhra Pradesh, India (See color plate section)

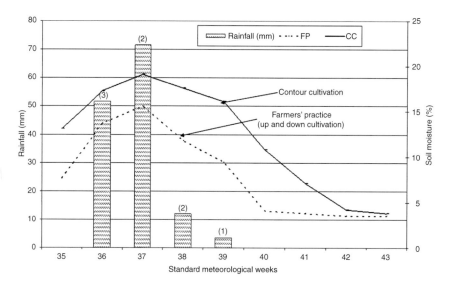

Figure 7.2   Soil moisture as influenced by farmers' practice (FP) and contour cultivation (CC) (Note: Number of rainy days is given in parenthesis) (Source: Mishra and Patil 2008)

furrow size. These furrows harvest the local runoff water and improve soil moisture in the adjoining crop rows, particularly during the period of moisture stress. One of the major advantages of this system is that it provides stability to contour cultivation particularly during moderate and big runoff events. Using the farmer participatory approach, Pathak *et al.* (2009a) reported on the performance of the practices followed

Figure 7.3 Conservation furrow system at Hedigonda watershed, Haveri, Karnataka, India; (right) conservation furrows prepared with local implements; and (left) groundnut crop with conservation furrows (See color plate section)

Table 7.3 Crop yields in different land and water management systems at Sujala watersheds in different districts of Karnataka, India[a]

District	Crop	Yield with farmers' practice (t ha^{-1})	Yield with contour cultivation with conservation furrows (t ha^{-1})	Increase in yield (%)
Haveri	Maize	3.35	3.89	16
Dharwad	Soybean	1.47	1.80	23
Kolar	Groundnut	1.23	1.43	16
Tumkur	Groundnut	1.25	1.50	21
	Finger millet	1.28	159	24

[a]Source: ICRISAT (2008).

by farmers (flat cultivation) as compared with contour cultivation along with conservation furrows based on the results of 121 trials conducted in farmers' fields in four districts of Karnataka during 2006–08 (Table 7.3). Contour cultivation along with conservation furrows was found promising both in terms of increasing crop yields and better adaptation by farmers. This land and water management system increased the crop yields of maize, soybean, and groundnut by 16–21% over the farmers' practice.

Contour cultivation along with conservation furrows was also found economically profitable to farmers (Table 7.4). Due to this system the benefit-cost ratio increased by 12 to 23% compared to farmers' practice of flat cultivation. The average benefit-cost ratio in contour cultivation with conservation furrow system was 1.94 and in farmers' practice it was 1.66 with overall average increase of 17%. One major advantage of this system is its very low cost. The average additional expenditure incurred for implementing contour cultivation along with conservation furrow system was only ₹400 ha^{-1}. Results from these large number of trials suggest that there is good possibility of getting good returns on the investments made on this simple land management system.

Table 7.4 Benefit-cost ratio for different crops and land management systems in Sujala watersheds, Karnataka, India[a]

District	Watershed	Crop	Benefit-cost ratio		
			Farmers' practice	Contour cultivation with conservation furrows	Increase (%)
Haveri	Aremallapur	Maize	2.00	2.32	16
Dharwad	Anchatageri	Soybean	1.84	2.26	23
Kolar		Finger millet	1.18	1.32	12
Tumkur	Belaganahalli	Groundnut	1.23	1.43	16
Mean			1.56	1.83	16.75

[a]Source: ICRISAT (2008).

7.2.1.2 Tied ridges

Tied ridges or furrow diking is a proven soil and water conservation method under both mechanized and labor-intensive systems, and is used in many rainfed areas of the world. Tied ridging results in the formation of small earthen dikes or dam across the furrow of a ridge furrow system. It captures and holds runoff water in place until it infiltrates into the soil. Tied ridges are most effective when constructed on the contour. Under mechanized systems, the furrow dykes are usually destroyed by tillage operations and need to be reconstructed each season. They also obstruct cultivation and other field operations.

Morin and Benyamini (1988) determined the optimum requirements for the implementation of furrow dykes. They used rainfall simulator to determine infiltration characteristics, storm intensity distribution, runoff amount and rate and by combining infiltration function and rainfall intensity pattern predicted the long-term runoff using rainfall probability distribution. Simulation models to determine the effects of tied ridging on runoff were developed by Krishna and Arkin (1988) and William *et al.* (1988) for various cropping systems. When combined with crop modeling, the potential effect of tied ridges on crop yield can be determined. Krishna and Gerik (1988) reported that tied ridges are most effective in the annual rainfall range of 500–800 mm. These models can be effectively used to determine the optimum size and other details of tied ridges and their possible impacts on crop yield, runoff, and soil loss under different rainfall and topographic conditions.

The tied ridge system reduced runoff and soil loss and also increased crop yield. For example, Vogel (1992) reported that tied ridging reduced the soil loss $<0.5 \, t \, ha^{-1} \, yr^{-1}$ whereas the soil loss under conventional tillage system was up to $9.5 \, t \, ha^{-1} \, yr^{-1}$ on a sandy soil in Zimbabwe. Njihia (1979) reported from Katumani in Kenya that tied ridging resulted in producing maize in low rainfall years, whereas the flat planted crops gave no grain yield.

El-Swaify *et al.* (1985) summarized the experiences of tied ridging in Africa and reported that under certain circumstances, the system has been beneficial not only in reducing runoff and soil loss but also for increasing crop yields. However, during the high rainfall years or in years with long wet periods, significantly lower yields were

Figure 7.4 Scoops with sorghum crop on an Alfisol at ICRISAT, Patancheru, India (Source: Pathak and Laryea 1995a)

reported from tied ridges system than from graded systems, which reduced the ponding of water on the soil surface (Dagg and Macartney 1968). Under such conditions, tied ridging enhanced waterlogging, resulting in the development of anaerobic conditions in the rooting zone and excessive leaching of N fertilizer (Kowal 1970). Jones and Stewart (1990) expressed serious concerns about overtopping of tied ridges and emphasized that this system should be so designed that the tied are lower than ridges, which themselves should be graded so that excessive runoff is drained along the furrows and not down the slope. Further, a support system of conventional contour bunds/furrows must be installed to manage the runoff from big storms.

7.2.1.3 Scoops (or pitting)

Scoops have been extensively used in the Asian, Australian, and African semi-arid tropics (SAT) as an in-situ soil and water conservation system. Scoops on agricultural land involve the formation of small basin depression at closely spaced interval to retain runoff water and eroded sediments from rainstorm (Figure 7.4). Scoops can be made manually or by machine. The commonly used machine for making scoops is a tractor-drawn chain diker equipment, which is extensively used in Australia, USA, and Africa. In India, at Hagari in Bellary district, Karnataka intercultivation by hoes was practiced successfully for scooping purpose in a cost-effective manner. Scoops helped in reducing the runoff by 50% and soil loss by 65%. In Bijapur, Karnataka, implementation of pitting increased sorghum yield by 12% (Mishra and Patil 2008). In Australia, scoops are used to promote vegetation on grazing land. In the SAT areas of Africa, farmers have shown interest in using these techniques for range improvement in Baringo, Kenya (Smith and Chitchley 1983). The implementation of chain diking treatment reduced runoff by 46% compared to the non-diked treatment on a fine sandy loam soil. Scoops did not appear to hinder subsequent tillage operations.

Pathak and Laryea (1995a) conducted studies to arrive at an optimum design (shape, size, and spacing) of scoops for greater stability and increased soil and water

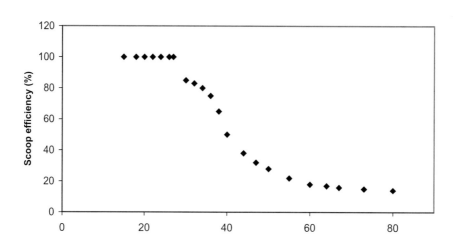

Figure 7.5 Effect of rainfall amounts on the performance of scoops and flat land on an Alfisol (Source: Pathak and Laryea 1995a)

conservation. Experiments were conducted both under simulated (using rotating disc type rainfall simulator) and under natural rainfall conditions to study the effects of various parameters, viz., shape and size of scoops, rainfall amount and intensities, slope, soil texture, soil type, surface cover and others, on the performance of scoops. The relative performances of scoops with other land management systems were also studied. The effect of rainfall amount on the performance of scoops and flat land in terms of runoff is shown in Figure 7.5 where scoop efficiency (P_e) is defined as follows:

$$P_e = (R_f - R_p/R_f) * 100 \qquad\qquad (1)$$

where, R_f is the runoff (mm) from flat land, and R_p is the runoff (mm) from the scooped land.

Scoop efficiency is a measure of the effectiveness of scoops in controlling runoff compared to the flat land treatment. Rainfall amount greatly influences scoop efficiency. The results showed that the scoops were efficient in controlling runoff only from small- and medium-size storms (20–40 mm h^{-1}) and their effectiveness for big storms (i.e., rainfall >50 mm) was relatively low.

Overall, runoff and soil loss under the scoop treatment were significantly lower than from the flat land surface. However, the comparative advantage of scoops over the flat land treatment for reducing runoff and soil loss varied considerably under various rainfall and soil cover conditions (Table 7.5).

The salient results from the several experiments conducted under simulated and natural rainfall were:

- Scoops and tied ridges significantly reduced runoff and soil loss compared with flat seedbed cultivation. Using runoff and soil loss from the flat land as a basis for

Table 7.5 Runoff and soil loss under scoop and flat land treatments from the application of 46 mm rainfall on an Alfisol at ICRISAT, Patancheru, India[a]

Treatment	Runoff (mm)			Soil loss (kg ha^{-1})		
	Scoops	Flat	SE	Scoops	Flat	SE
Bare surface						
Rainfall intensity 28 mm h^{-1}	16	27	±3.1	1781	2906	±210
Rainfall intensity 65 mm h^{-1}	26	34	±4.2	4969	8344	±479
Surface with mulch (60% cover)						
Rainfall intensity 28 mm h^{-1}	6	14	±2.1	750	1875	±138
Rainfall intensity 65 mm h^{-1}	15	24	±1.9	2063	2813	±291

[a]Source: Pathak and Laryea (1995a).

comparison, scoops reduced seasonal runoff by 69% and soil loss by 53%, while runoff in the tied ridge system was reduced by 39% and soil loss by 28%.

- Scoops are relatively more stable than tied ridges, particularly during high-intensity rainfall and runoff conditions.
- On Alfisols, scoops reduced runoff and soil loss significantly over flat cultivation during the early part of the crop-growing season.
- The stability of scoops can be greatly enhanced by providing a graded outlet system in the field. Scoops are recommended only for low and medium rainfall areas (annual rainfall ≤800 mm) for increasing crop yields over flat cultivation.
- Scoops are effective in conserving runoff and soil loss only up to 5–6% land slopes. On higher slopes the chances of breaching of scoops increases substantially.
- The effectiveness of scoops is greatly influenced by the texture of surface soil layer. On very sandy soils (sand >93%), the effectiveness of scoops is extremely low.

7.2.1.4   Broad-bed and furrow and related systems

On Vertisols and associated soils, the problem of waterlogging and water scarcity occurring during the same cropping season is quite common. For such a situation, there is a need for an in-situ soil and water conservation and proper drainage technology that can protect the soil from erosion throughout the season and provide control at the place where the rain falls. A raised land configuration broad-bed and furrow (BBF) system, has been found to satisfactorily attain these goals (Figure 7.6). The BBF system consists of a relatively raised flat bed or ridge approximately 95 cm wide and shallow furrow about 55 cm wide and 15 cm deep (Figure 7.7). The system is laid out on a grade of 0.4 to 0.8% for optimum performance. This BBF system is most effectively implemented in several operations or passes. After the direction of cultivation has been set out based on the topographic survey (Figure 7.7), furrows are made by an implement attached to two ridgers with a chain tied to the ridgers or a multipurpose tool carrier called "tropicultor" to which two ridgers are attached (Figure 7.6). A bed former is used to further shape the broad-beds. If there are showers before the beginning of the rainy season, another cultivation is done after showers to control weeds and improve

Figure 7.6 The broad-bed and furrow (BBF) system at ICRISAT, Patancheru, India: (top) BBF formation with tropicultor; and (bottom) groundnut crop on BBF (See color plate section)

the shape of the BBFs. Thus at the beginning of the growing season, the seedbed is receptive to rainfall and, importantly, moisture from early rains is stored in the surface layer without disappearing into deep cracks of the Vertisols. The BBFs formed during the first year can be maintained by reshaping for the long-term (more than 30 years). This will save considerable cost as well as preferentially improve the health of the soil on the bed.

Different land and water management systems, viz., BBF at 0.6% slope, BBF at 0.4% slope, flat on grade at 0.6% slope, and traditional flat system with monsoon fallow system at the International Crops Research Institute for the Semi-Arid Tropics (ICRISAT), Patancheru, Andhra Pradesh, India showed that runoff, soil loss, and peak runoff rates were significantly reduced in BBF treatments (Table 7.6). The BBF system

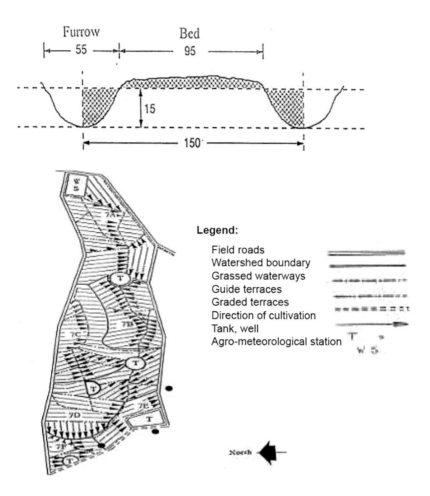

Figure 7.7   Broad-bed and furrow (BBF) system dimension (in cm) and field layout based on topographic map (Source: Pathak *et al.*, 2009a)

was found more efficient in controlling soil and water losses as compared to the flat on grade system. But the BBF system at 0.6% slope within existing farmers' field bunds was found most efficient in reducing runoff and soil loss. On an average, this system reduced annual runoff to one-third, soil loss to one-eleventh, and peak runoff rate to half when compared with the traditional system.

After perfecting the BBF system at ICRISAT, Patancheru, this technology was taken up for large-scale adoption by farmers in Madhya Pradesh. This state has a large area of deep black soils (Vertisols) and they are kept fallow during the rainy season and the crops are sown during the postrainy season. The rainy season fallow area covers about 1.83 million ha of the 13.2 million ha total crop area in Madhya Pradesh. Five districts of Madhya Pradesh, viz., Vidisha, Sagar, Guna, Sehore, and Raisen have large percentage of area under rainy season fallow. Farmer participatory

Table 7.6 Effect of alternative land management systems on annual runoff, soil loss, and peak runoff rate on Vertisols at ICRISAT, Patancheru, India (average annual values from 1975–80)[a]

| Treatment | Rainfall (mm) | Runoff | | Peak runoff rate (cum sec^{-1} ha^{-1}) | Soil loss (t ha^{-1}) |
		mm	% of rainfall		
Broad-bed and furrow at 0.6% slope	810	116	17.3	0.11	1.12
Broad-bed and furrow at 0.6% slope with farmers field bunds	808	76	9.4	0.07	0.58
Broad-bed and furrow at 0.4% slope	853	91	10.7	0.07	0.86
Flat-on-grade at 0.6% slope	812	173	17.6	–[b]	1.35
Traditional flat, monsoon fallow	806	220	27.3	0.16	6.67

[a]Source: Pathak *et al.* (1985).
[b]Data not available; problem with recorder.

Table 7.7 Mean soybean yield in improved and traditional management system in Madhya Pradesh during 2007–09[a]

| District | Grain yield (t ha^{-1}) | | Increase in yield over farmers' practice (%) |
	Improved system	Farmers' practice	
Guna	1.7	1.46	16
Raisen	2.28	1.56	45
Vidisha	2.23	1.72	30
Indore	2.90	2.51	15
Sehore	2.50	2.09	19
Mean	2.32	1.87	24

[a]Source: ICRISAT (2008).

research-cum-demonstrations were taken up to enhance crop yields in these five districts of Madhya Pradesh.

In total, 140 farmer participatory action research-cum-demonstrations were conducted in 17 villages on enhancing water use efficiency (WUE) through increased crop yields during 2007-09. With BBF, improved varieties and application of micro and secondary nutrients (50 kg ha^{-1} zinc sulfate for zinc and 2.5 kg ha^{-1} agribor for boron) significantly increased crop yields by 16 to 45% with an average increase of 24% due to improved technology over farmers' practice (Table 7.7).

Farmers requested a simpler BBF maker for greater adoption of this technology. A customized user-friendly tractor-drawn modular inclined plate planter-cum-BBF maker was designed and developed for use by farmers for increased adoption of the BBF system (Figure 7.8). This equipment was designed for easy and efficient planting with BBF making simultaneously, which saves the additional cost of operation for forming BBF. The adoption of BBF system by the farmers increased substantially with increased availability of the new BBF maker and seed drill units.

The BBF system has been found quite promising for Ethiopian highlands with Vertisols covering an area of 7.6 million ha. These soils are usually cultivated with

Figure 7.8 Development of modular type seed drill-cum-BBF maker at CIAE, Bhopal, India: (top) for intercrop sowing; and (bottom) sowing with new implement

low-yielding food crops that are normally planted during the later part of the rainy season to avoid waterlogging damage to crops. The practice leaves a great proportion of the bare land resulting in high runoff and soil loss (60 to $100\,t\,ha^{-1}\,yr^{-1}$). During 1991 the BBF system was introduced by ICRISAT along with its three partners, viz., International Livestock Research Institute (ILRI), Alemaya University, and Ethiopian Agricultural Research Organization. Experiments on BBF system were conducted at the research stations in Ethiopian highlands. The BBF system increased wheat yield significantly compared to flat cultivation (Table 7.8). Following on-station trials the

Table 7.8 Effect of different BBF systems on grain yield of wheat (cv ET-13) at two locations in Ethiopia, 1991[a]

Location	Treatment	Bed height (cm)	Yield (kg ha^{-1})	Increase in yield over control (%)
Ginchi	Flat cultivation (control)	–	835 ($\pm$75)*	–
	Normal BBF	13	979 ($\pm$45)	17
	Raised BBF	26	1221 ($\pm$45)	46
Akaki	Flat cultivation (control)	–	960 ($\pm$62)*	–
	Normal BBF	13	1286 ($\pm$73)	34
	Raised BBF	26	1481 ($\pm$73)	54

[a]Source: Srivastava *et al.* (1993).

BBF system was tested at several on-farm sites in Ethiopian highlands in collaboration with farmers selected from peasant associations. For making the BBFs in farmers' fields, a simple broadbed maker was developed. This implement greatly facilitated the adoption of the BBF technology by the Ethiopian farmers. Since 1998, the Government of Ethiopia has introduced market liberalization policies and strategies to achieve food self-sufficiency. In response to the policy change the diffusion and adoption process was also strengthened. The Ministry of Agriculture and several non-government organizatons (NGOs) including Sasakawa Global 2000 took part in diffusing the BBF system along with other improved technologies. In order to reach their food production targets, the Ethiopian government in 2004 initiated a new program for promoting the adoption and use of the BBF and other improved systems through price subsidies, increased access to credit, and increased training. The uptake of BBF system in Ethiopia was recently assessed by a multidisciplinary team of ICRISAT scientists along with officials from Ministry of Agriculture, Ethiopia. In 2010, the BBF system was adopted by more than 120,000 farmers in highlands of Ethiopia. The Government of Ethiopia has prepared a 5-year plan to increase this number enormously in the next five years (2011–15). The appropriate technology, simple implement for making BBF, and sustained substantial support by the Ministry of Agriculture (through appropriate policies) seems to be the key drivers of the rapid uptake of BBF system in recent years in Ethiopia.

Channappa (1994) reported that the graded furrowing at the time of sowing the crop at 1.5 to 3 m intervals was found to increase and stabilize yield levels over years by 8 to 10%, apart from better rainwater management during low as well as high intensity rains. Modified technique known as paired row in pigeonpea–finger millet intercrop with a furrow in between the pigeonpea rows and 8 to 10 rows of finger millet was found to be the best intercrop as well as inter-terrace management practice for the Alfisol regions of Karnataka state in India. The relative performance of different bed systems, i.e., flat bed, BBF, narrow bed and furrow, and raised-sunken bed was studied on the black soils at Indore. The results indicated that maize yield was maximum (2.01 t ha^{-1} and WUE of 8.81 kg ha^{-1} mm^{-1}) in the BBF system, followed by raised-sunken bed and flat bed systems. In the Vertisols of Bellary, Karnataka, the BBF system proved effective in conserving the rainwater, increasing the soil water in the profile and thus the winter sorghum grain yield increased by 23.7% and safflower yield by 7.7% as compared to flat bed sowing.

Variation of BBF system has been used in North America (Phillips 1963) and in Central Africa. A variation known as camber-bed system was used in Kenya. An extension of BBF system developed in India (Pathak *et al.*, 1985) has been made on similar soils in Ethiopia. Another variation using small ridges was developed at Agricultural Research Center for Semi-Arid Tropics (CPATSA) of the Brazilian Enterprise for Agricultural Research (EMBRAPA) located at Petrolina, Brazil (Lal 1985, 1986). Some of the major benefits of the semi-permanent BBF system are given below:

- The raised bed portion acts as an in-situ 'bund' to conserve more moisture and ensures soil stability. The shallow furrows provide good surface drainage to promote aeration in the seedbed and root zone, and also prevent waterlogging of crops on the bed.
- The BBF design is quite flexible for accommodating crops and cropping systems with widely differing row spacing requirements.
- Precision operations such as seed and fertilizer placement and mechanical weeding are facilitated by the defined traffic zone (furrows), which saves energy, time, and cost of operation and inputs.
- The system can be maintained for the long-term (25–30 years).
- It reduces runoff and soil loss and improves soil properties over the years.
- It facilitates double cropping and increases crop yields.

7.2.2 Bunding

Bunding is one of the most commonly used methods for the conservation of soil and water on agricultural lands. A bund is a mechanical measure where an embankment or ridge of earth is constructed across a slope to control runoff and minimize soil erosion. The experiences with some of the most commonly used bunding systems are discussed below.

7.2.2.1 Contour bunding

Contour bunding is one of the extensively used soil and water conservation technique in several rainfed areas of Asia and Africa. In India during 1947–79 contour bunds have been constructed on about 21 million ha of agricultural lands costing about US$ 30 ha^{-1}; this figure constitutes about 90% of the total expenditure on soil conservation on agricultural lands in India. Contour bunding involves the construction of small bunds across the slope of the land along a contour so that the long slope is reduced to a series of small ones. Each contour bund is provided with an elevated spillway at the lower end of the field for the safe disposal of excess water. The contour bund acts as a barrier to the flow of water down a hillside and thus increases the time so that water concentrates in an area, thereby allowing more water to be absorbed into the soil profile.

Extensive studies conducted on the alluvial soils of Gujarat showed that 1.3 m^2 cross section bunds spaced at 1.83 m vertical interval are suitable for lands having slope between 6 and 12%. For slopes less than 6%, contour bunds with cross section of 0.9 to 1.3 m^2, spaced at 0.9 to 1.2 m vertical interval were found to be effective (Bhumbla *et al.*, 1971). In the Alfisols of Hyderabad, contour bunding recorded increase in crop yields of sorghum, pigeonpea, and pearl millet and reduced runoff and soil loss.

Figure 7.9 Conventional contour bund system with water stagnation at ICRISAT, Patancheru, Andhra Pradesh, India

Contour bunding in agricultural watersheds of many regions were found to reduce runoff and soil erosion considerably. In Dehradun region, runoff observations from a 55-ha agricultural watershed treated with contour bunds have shown that runoff volume and peak runoff rate were reduced to 62 and 40% respectively (Ram Babu *et al.*, 1980). Research conducted in the Doon valley, India with an annual rainfall of 1680 mm, indicted 88% reduction in soil loss from the area with contour bunding treatment as compared to the cultivated fallow (Gurmel Singh *et al.*, 1990). The studies undertaken on lateritic hills with 8 to 10% slope in the heavy rainfall area with finger millet as a test crop indicated that contour bunding is more effective in reducing soil loss. Gund and Durgude (1995) reported that lowest runoff was observed in contour bunding supported by live bunding of *subabul* (*Leucaena leucocephala*) at Bengaluru. However, in India the use of contour bunds to retain runoff has not always been found to be effective. When used mainly as a soil and water conservation measure in areas with 750–1250 mm rainfall on deep Vertisols, it was found that disadvantage of waterlogging in the vicinity of the bund both uphill and downhill exceeded the advantage of increased cropping intensity from the stored moisture in the dry season (Pathak *et al.*, 1987).

7.2.2.2 *Modified contour bunds*

Well-designed and maintained conventional contour bunds on Alfisols and other light-textured soils undoubtedly conserve soil and water, and for this purpose contour bunds are perhaps efficient. However, the associated disadvantages – mainly water stagnation (particularly during the rainy season) (Figure 7.9) causing reduction in crop yields – outweigh any advantage from the view point of soil and water conservation.

Figure 7.10 Gated-outlet contour bund system at ICRISAT, Patancheru, India; (inset) gated-outlet

The modified contour bunds with gated-outlets (Pathak *et al.*, 1989a) have shown promise because of the better control on ponded runoff water (Figure 7.10). This system involves constructing embankments on contours with gated-outlet at the lower end of the field. The gated-outlet system allows the runoff to be stored in the field for a desired period, which is then released at a predetermined rate through the spillway, thus reducing the time of water stagnation behind the bund that would have no adverse effects on crop growth and yield and also facilitates the water infiltration into soil to its optimum capacity. The results on the comparison of gated-outlet contour bunds with the other alternative land management systems are shown in Table 7.9. The conventional contour bunds and gated-outlet contour bunds were found to be most effective in controlling runoff and soil loss. However, only contour bunds with gated outlets were found to be more effective in increasing yield and this system produced highest crop yields and provided adequate control of runoff and soil loss. The benefits of this system are given below:

- The problem of prolonged water stagnation around the contour bund and bund breaching are reduced in the gated outlet contour bund system. This results in better crop growth and higher crop yields.
- More timely tillage and other cultural operations are possible in the gated-outlet contour bund system because of better control on ponded runoff water.
- Gated-outlet contour bund system involves low cost for modification and is simple to adopt.

7.2.2.3 Graded bunding

Graded bunding on grades varying from 0.2 to 0.4% is generally practiced in areas with more than 600 mm annual rainfall to drain the excess runoff water into the grassed

Table 7.9 Grain yield, runoff, and soil loss from different land management systems in Alfisol watersheds at ICRISAT, Patancheru, India[a]

Land management systems	Crop	Grain yield (kg ha^{-1})	Runoff (mm)	Soil loss (t ha^{-1})
Conventional contour bund	Sorghum/	2520	75	0.97
	Pigeonpea	710		
	Pearl millet/	2230		
	Pigeonpea	730		
Modified contour bund gated-outlet	Sorghum/	3020	160	0.92
	Pigeonpea	970		
	Pearl millet/	2730		
	Pigeonpea	1010		
Broad-bed and furrow	Sorghum/	2740	289	3.61
	Pigeonpea	880		
	Pearl millet/	2400		
	Pigeonpea	920		
Contour cultivation with field bunds	Sorghum/	2810	215	3.35
	Pigeonpea	910		
	Pearl millet/	2510		
	Pigeonpea	920		

[a]Source: Pathak *et al.* (1987).

waterway. Gurmel Singh *et al.* (1990) reported that graded bund can reduce the soil loss by 86% compared to the cultivated fallow. Similar results were also reported by Kale *et al.* (1993), wherein soil loss in the graded bund plot was 9.71 t ha^{-1} compared to 18.92 t ha^{-1} in the control treatment. In the deep black soils of Bellary, India the increase in yield of sorghum, cotton, and safflower from graded bunds was 14, 25, and 12%, respectively. Chittaranjan *et al.* (1997) conducted a study on the semi-arid Vertisols of South India. The results revealed that graded bunds with farm pond at the tail end are the most suitable soil and water management measures compared to contour bund and conservation ditch.

7.2.2.4   Field bunding

Field bunding is traditionally practiced by a large number of farmers. Stabilizing and strengthening of the existing field bunds will not allow the fragmentation of fields of small farmers. This is acceptable to one and all. Singh *et al.* (1973) evaluated various practices for conserving soil moisture, viz., field bunding, field bunding + land shaping, basin listing, deep furrow and control (no bunding and no land shaping) with pearl millet as the test crop. The results of three-year study showed that field bunding plus land shaping practice gave the highest pearl millet grain yield.

7.2.2.5   Compartmental bunding

Compartmental bunding is extensively practiced in several rainfed areas of Asia and Africa. This is done by dividing fields into small land parcels of square or rectangle shapes, by providing small bunds. In deep black soils depending upon the land slopes, the entire field is laid out into small bunded compartments varying in size from 6 m × 6 m to 10 m × 10 m. Rains received during the rainy season are collected in these bunded areas; these are slowly harrowed and land is prepared into a good seed bed for raising postrainy season crops (Mishra et al., 2002a).

Selveraju and Ramaswami (1997) recorded significantly higher sorghum and pigeonpea grain yields in the intercropping system with compartmental bunding as compared to flat cultivation. A study on the best cultivation practices for effective rainwater management, revealed that the seed and stalk yields of castor were significantly influenced by different land treatments (control, compartmental bunding, opening of ditches across the slope, interculturing and forming ridges at last interculturing) (Anonymous 1998). The highest seed yield of castor (1219 kg ha^{-1}) was recorded under the compartmental bunding treatment, which was at par with treatment of opening of ditches across the slope.

More et al. (1996) reported that there was beneficial effect of compartmental bunding on in-situ moisture conservation. This was reflected on the better performance of the winter sorghum. The overall percent increase in grain and fodder yield by compartments was 38 and 50%, respectively over the control.

7.2.2.6 Vegetative barriers

Vegetative barriers or vegetative hedges or live bunds have drawn greater attention in recent years because of their long life, low cost, and low maintenance needs. In several situations, the vegetative barriers are more effective and economical than the mechanical measures, viz., contour and graded bunds.

Vegetative barriers can be established either on contour or on moderate slope of 0.4 to 0.8%. In this system, the vegetative hedge acts as a barrier to runoff flow, which slows down the runoff velocity, resulting in the deposition of eroded sediments and increased rainwater infiltration. It is advisable to establish the vegetative hedges on small bunds. This increases the effectiveness particularly during the first few years when the vegetative hedges are not so well established. The key aspect of design of vegetative hedge is the horizontal distance between the hedge rows which mainly depends on rainfall, soil type, and land slope. Species of vegetative barrier to be grown, number of hedge rows, plant to plant spacing, and method of planting are very important and should be decided based on the main purpose of the vegetative barrier. If the main purpose of the vegetative barrier is to act as a filter to trap the eroded sediments and reduce the velocity of runoff, then the grass species such as vetiver, sewan (Lasiurus sindicus), sania (Crotalaria burhia), and kair (Capparis aphylla) could be used. But if the purpose of the vegetative hedges is to stabilize the bunds, then plants such as Gliricidia could be effectively used (Figure 7.11). The Gliricidia plants grown on bunds not only strengthen the bunds while preventing soil erosion, but also provide N-rich green biomass, fodder, and fuel. The cross section of earthen bund can also be reduced. A study conducted at ICRISAT, Patancheru indicated that by adding the N-rich green biomass from the Gliricidia plants planted on the bund at a spacing of 0.5 m apart for a length of 700 m could provide about 30–45 kg N ha^{-1} yr^{-1} (Wani and Kumar 2002).

In the shallow Alfisols of Anantapur (mean annual rainfall 570 mm), vetiver alone increased groundnut yield by 11% and with contour cultivation the yield increased up to 39% with greater conservation of rainwater (Mishra and Patil 2008). However,

Figure 7.11 *Gliricidia* plants on bunds and aerial view of a watershed with *Gliricidia* on graded bunds at ICRISAT, Patancheru, India (See color plate section)

in Alfisols at Bengaluru (mean annual rainfall 890 mm), combination of graded bund and vetiver conserved more soil and water and was better than other treatments. In the shallow Alfisols of Hyderabad (mean annual rainfall 750 mm), *Cenchrus* or vetiver barriers along with a small section bund recorded higher yields over conventional mechanical measures. In the Vertisols of Deccan Pleateau at Bellary, the vegetative barrier proved effective in conserving soil and rainwater and increasing the soil water availability in the profile. The increased water availability has resulted in better plant growth with increased grain yield of winter sorghum by 35% over control (Table 7.10).

Table 7.10 Effect of vegetative barrier on resource conservation and sorghum grain yield during 1988–89 to 1996–97 in Vertisols at Bellary, India[a]

Treatment	Slope (%)			Average
	0.5	*1.0*	*1.5*	
Runoff (mm)				
Up and down cultivation (control)	49.65	54.81	59.14	55.53 –
Vegetative barrier	22.69	39.86	44.10	35.55 (36%)
Soil loss (kg ha^{-1})				
Up and down cultivation (control)	1053	2167	1712	1644 –
Vegetative barrier	500	1372	1027	966 (41%)
Grain yield (kg ha^{-1})				
Up and down cultivation (control)	911	685	475	690 –
Vegetative barrier	1149	848	787	928 (35%)

[a]Source: Rama Mohan Rao *et al.* (2000).
Data is average of eight years for 100 mm rainfall.

The vegetative barrier reduced the runoff by 36% and soil loss by 41% over control. The vegetative barrier was more effective (Rama Mohan Rao *et al.*, 2000) at higher slope (1.5%) and increased winter sorghum grain yield by 66% at 1.5% slope, 25% at 1.0% slope, and 26% at 0.5% slope. In Bellary with 500 mm mean annual rainfall, the exotic *vetiver* was less effective than the native grass (*Cymbopogan martinii*). Vetiver requires higher rainfall (>650 mm) and can perform better in well drained red soils with neutral pH as compared to areas with low rainfall and soils with pH in the alkaline range (>8.5) such as at the Bellary site. The native grass (*C. martinii*) is also not grazed by animals and can be used for thatch making, in addition to its medicinal use.

In areas with long dry periods, vegetative hedges may not survive or perform well. The establishment of vegetative barriers in very low rainfall areas, and the maintenance in high rainfall areas, could be the main problems. Proper care is required to control pests, rodents, and diseases for optimum growth and survival of both vegetative hedges and main crops (Rama Mohan Rao *et al.*, 2000).

7.2.3 Tillage

Most of the soils in the rainfed regions are fragile and structurally unstable when wet. A major consequence of the lack of stability of their aggregates is the tendency of many soils, to exhibit rapid surface sealing during rainfall, crusting and in some cases hardening of a considerable depth of soil profile during subsequent drying cycles. Tillage on such poor soils helps to increase pore space and also keeps the soil loose so as to maintain higher level of infiltration. Laryea *et al.* (1991) found that cultivation of the surface greatly enhanced water intake of soil particularly in the beginning of rainy season. In the absence of cultivation, the highly crusting Alfisols produce as much or even more runoff than the low permeable Vertisols under similar rainfall situations. Larson (1962) stated that pulling a tillage implement through soil results in the total porosity and thickness of the tilled area being greatly increased temporarily. Surface

Table 7.11 Effect of subsoiling on root density 89 days after emergence of maize (Deccan Hybrid 103) on an Alfisol, ICRISAT, Patancheru, India during rainy season 1984[a]

Soil depth (cm)	Root density (cm cm^{-3})		
	Subsoiling	Normal tillage	SE±
0–10	0.55	0.42	0.072
10–20	0.29	0.21	0.022
20–30	0.20	0.09	0.034
30–40	0.15	0.10	0.028
40–50	0.12	0.06	0.016
50–60	0.14	0.05	0.039

[a]Source: Pathak and Laryea (1995a).

Table 7.12 Effect of normal and deep primary tillage on sorghum yield, runoff, and soil loss on Alfisols at ICRISAT, Patancheru, India[a]

Tillage practices	Sorghum yield (kg ha^{-1})	Runoff (mm)	Soil loss (t ha^{-1})
Normal tillage (mold board plowing 12 cm deep)	2160	285	3.27
Deep tillage (cross chiseling 25 cm deep)	2720	195	2.86
LSD (P = 0.05)	386	44.0	0.702

[a]Source: ICRISAT (1985).

roughness and micro depressions thus created play greater role in higher retention of water (Unger and Stewart 1983).

On many soils in the semi-arid tropics (SAT), intensive primary tillage has been found necessary for creating favorable root proliferation and enhancing rainfall infiltration. Deep tillage with plow, followed by chiseling (Channappa 1994) opens the hard layers and increase the infiltration and water storage capacity and this results in better crop growth with higher yields on Alfisols at Bengaluru, Karnataka, India. Similarly, on Alfisols in farmers' fields in Coimbatore, Tamil Nadu, India, deep plowing with chisel plow and disc plow plus cultivator increased the soil water stored in the profile at different stages of sorghum growth as compared to soil cultivation with cultivator once or twice, i.e., reduced tillage operations (Manian *et al.*, 1999). Primary tillage carried out in the Alfisols at ICRISAT, Patancheru (Pathak and Laryea 1995b) improved the soil physical properties with better root development (Table 7.11). Also deep tillage reduced runoff and soil loss, and increased the soil water; sorghum yield was increased by 26% over normal tillage (Table 7.12). The positive effects of deep tillage on rainwater conservation, better root development, and increased crop yields were observed for 2 to 5 years after deep tillage, depending on the soil texture and rainfall.

On Alfisols, the problems of crusting, sealing, and hardening are more encountered during the early part of the crop growing season when the crop canopy is not yet fully developed. Pathak *et al.* (1987) studied the effectiveness of shallow tillage imposed as

Table 7.13 Effects of inter-row cultivation (shallow tillage) in addition to normal tillage on runoff, soil loss, and grain yield in 1981–83 on an Alfisol, ICRISAT, Patancheru, India[a]

Year	Rainfall (mm)	Tillage treatment	Runoff (mm)	Soil loss (t ha^{-1})	Grain yield (kg ha^{-1})			
					Sole sorghum	Intercrop		Sole pearl millet
						Sorghum	Pigeonpea	
1981	1092	Normal[b]	246	5.0	2350			
		Additional[c]	223	4.9	2360			
		SE	±10.6	±0.34	±50			
1982	780	Normal	159	3.1		2260	925	
		Additional	120	2.6		2620	920	
		SE	±8.0	±0.33		±25	±41	
1983	990	Normal	231	4.2				2620
		Additional	196	4.0				2970
		SE	±12.3	±0.24				±32

[a] Source: Pathak *et al.* (1987).
[b] Two inter-row cultivations.
[c] Two additional shallow inter-row cultivations.

secondary inter-row cultivation in breaking up the crust and improving infiltration and soil moisture conservation. Results showed that additional shallow tillage effectively reduced runoff and soil loss in all years (Table 7.13). In some years, it was also effective in reducing moisture loss through evaporation by acting as dust mulch. However, a significant increase in crop yield was obtained only in low and normal rainfall years.

On Alfisols, the off-season tillage serves several useful purposes and should be done whenever feasible. At ICRISAT, Patancheru, the off-season tillage has been found to be helpful in increasing rainwater infiltration and in decreasing weed problems. In most years, off-season tillage alone increased crop yields by 7–9% over the control. Also, it significantly reduced the early season runoff and soil loss. Furthermore, the off-season tillage has been found to minimize the loss by evaporation of stored water by "mulching" effect and thus allowing the acceleration of planting operations and extension of the growing season (Pathak *et al.*, 1987). Similar results were also reported by Mishra and Patil (2008).

7.2.3.1 Zero tillage or minimum tillage or conservation tillage

Sub-optimal subsoil conditions have significant impact on soil water and nutrient regime, base exchange between soil and atmosphere, and crop growth. Adverse effects on crop growth have agronomic and economic implications, while those on soil water and aeration regime lead to ecological and environmental problems of regional and global importance. Soil surface management in general and tillage practices in particular play major role in magnitude and seasonal trends in gaseous emission from soils. Minimum tillage is an ecological approach to resource conservation and sustainable production.

However, the tillage in SAT soils is critically dependent upon available draft power and soil moisture. Timeliness of tillage operations is important, as the rainfall is erratic and the limited water-holding capacity of some of the soils may make them either too

Table 7.14 Effect of different tillage practices and amendments on grain yield (kg ha^{-1}) of rainy season maize and postrainy chickpea on Vertisols at ICRISAT, Patancheru, India[a]

Tillage Practices	1983–84		1984–85	
	Maize	Chickpea	Maize	Chickpea
Flat configuration				
Zero tillage (including chemical weed control)	3500	330	2320	340
15 cm deep primary tillage (normal tillage)	4030	990	2970	970
30 cm deep primary tillage	4390	1160	3140	1060
BBF configuration				
15 cm deep primary tillage (normal tillage)	4380	1150	3320	1090
15 cm deep primary tillage, cross plowing, and reformation of beds every year	4290	1160	3110	1030
30 cm deep primary tillage	4240	1050	3300	1170
30 cm deep primary tillage (without blade hoeing before sowing second crop)	4210	830	3280	1060
30 cm deep primary tillage + application of phosphogypsum at 10 t ha^{-1}	4710	1280	3270	1060
Crop residue incorporation at 5 t ha^{-1} with 30 cm deep primary tillage[b]	5010	1240	3240	1250
SE	±133	±49	±105	±56

[a]Source: ICRISAT (1987).
[b]Chopped dry rice straw incorporated in 1983–84; chopped dry maize stalk incorporated in 1984–85.

wet or too dry to cultivate. Conservation tillage techniques that lower energy inputs and prevent the structural breakdown of soil aggregates have been used particularly in USA, Australia, and in experimental station trials of developing countries of the SAT. In conservation tillage, it is still necessary to follow the accepted and recognized cultural practices of fertilization, pest control, and correct planting time and also use improved varieties, it reduces production costs, greatly reduces energy needs, ensures better soil water retention, reduces runoff, and wind erosion, ensures little or no damage from machinery, and saves labor (Young 1982).

The success of mechanized conservation tillage depends largely on herbicides (which may be expensive and hazardous in nature for use by the resource-poor farmers of the SAT). Crop residues left on the soil surface protect it against the impact of torrential rains, and no-till planting equipment allows precision sowing through trash. Unfortunately, most of the farmers in the SAT use crop residues to feed their animals and to construct fences and buildings. In most parts of semi-arid India, animals are allowed to roam freely on the field after crops have been harvested. Consequently, most of the residue left over is consumed by these animals (Laryea *et al.*, 1991).

Notwithstanding, a comparison between different tillage practices (Table 7.14) on a Vertisol at ICRISAT, Patancheru showed on flat land, the highest yield of maize-chickpea from 30 cm deep primary tillage treatment while zero-tilled plots gave the lowest yield. On BBF landform, incorporation of 5 t ha^{-1} crop residue with deep primary tillage (30 cm) gave on average the highest yield of maize and chickpea. There were no significant differences among the other treatments for maize or chickpea yields.

On Alfisols at ICRISAT, Yule *et al.* (1990) while comparing the effects of tillage (i.e., no-till, 10 cm deep till, 20 cm deep till), amendments (i.e., bare soil, rice straw mulch applied at 5 t ha^{-1}, farmyard manure applied at 15 t ha^{-1}), and the use of perennial species (e.g., perennial pigeonpea, *Cenchrus ciliaris*, and *Stylosanthes hamata* alone or in combination) on runoff and infiltration found that straw mulch consistently reduced runoff compared with bare plots. Tillage produced variable responses in their study. Runoff was reduced for about 20 days after tillage, but the tilled plots had more runoff than no-tilled treatments during the remainder of the cropping season, suggesting some structural breakdown of the soil aggregates in the tilled plots. On average, straw mulch and tillage increased annual infiltration by 127 and 26 mm, respectively. These results of Yule *et al.* (1990) indicate that mulching or keeping the soil covered (as in the case of *Stylosanthes*) should be an important component in the cropping systems of the SAT.

Studies conducted in the semi-arid regions of Africa also indicate that some of the conservation tillage systems, particularly no-till techniques give lower yield than conventional tillage methods. For example, Huxley's (1979) no-till experiments at Morogoro in Tanzania showed that no-tilled maize yielded two-thirds to three-quarters the amount of that in cultivated soil. Furthermore, Nicou and Chopart (1979) conclude in their studies in Senegal, West Africa that in order to be effective, straw mulch in conservation tillage systems needs to be applied in sufficient quantity to cover the surface of the soil completely so that it can fully protect the soil against evaporation and runoff. Straw tends to be used for animal feed in most parts of the SAT, particularly in India, Senegal, and Mali. Therefore while mulches appear to be useful theoretically, from a practical point of view it is difficult to see how they can be used in the present conditions of SAT agriculture. It is even debatable if production of more biomass through breeding will induce farmers in the region to apply residue to their soils or induce them to sell their extra residues in view of the attractive prices offered for fodder during the dry season.

7.2.4   Ex-situ soil and water conservation (runoff harvesting and supplemental irrigation)

The mean annual rainfall in most rainfed regions is sufficient for raising one or in some cases, two good crops in a year. However, the onset of rainfall and its distribution are erratic and prolonged droughts are frequent. A large part of rain occurs as high intensity storms, resulting in sizeable runoff volumes. In most rainfed regions harvesting of excess runoff and storage into appropriate structure as well as recharging groundwater is very much feasible and a successful option for increasing and sustaining the productivity of rainfed agriculture through timely and efficient use of supplemental irrigation. In the areas with annual rainfall >500 mm, this approach could be widely adopted to enhance the cropping intensity, diversify the system into high value crops, increase the productivity and incomes from rainfed agriculture and at the same time, create assets in the villages. Different types of runoff harvesting and groundwater recharging structures are currently used in various regions. Some of the most commonly used runoff harvesting and groundwater recharging structures are earthen check-dam, masonry check-dam, stop check-dam, farm ponds, tank, sunken pits, recharge pits, loose boulder, gully checks, drop structure, and percolation pond (Figure 7.12).

Figure 7.12 Commonly used water harvesting and groundwater recharging structures (Source: Pathak *et al.*, 2009a) (See color plate section)

Designing these structures requires estimates of runoff volume, peak runoff rate, and other hydrological parameters, which are generally not available in most of the rainfed regions. Due to non-availability of the data many times these structures are not properly designed and constructed resulting in higher costs and often failure of the structures. Studies conducted by ICRISAT scientists have shown that the cost of water harvesting and groundwater recharging structures varies considerably with type of structures (Figure 7.13) and selection of appropriate location. Selection of appropriate location for structures also can play very important role in reducing the cost of structures (Figure 7.14).

Figure 7.12 Continued

Pathak *et al.* (2009b) reported that considerable information on various aspects of runoff water harvesting and supplemental irrigation could be obtained by using various models (Pathak *et al.*, 1989b; Ajay Kumar 1991), viz., runoff model, water harvesting model (Sireesha 2003), and model for optimizing the tank size (Sharma and Helweg 1982; Arnold and Stockle 1991). These models can assess the prospects of runoff water harvesting and possible benefits from irrigation. The models also can be used to estimate the optimum tank size, which is very important for the success of the water harvesting system. The information generated can also help in developing

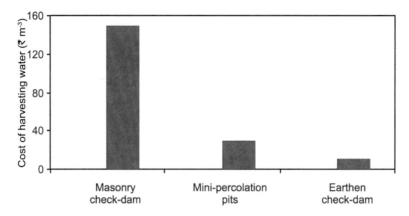

Figure 7.13 Cost of harvesting water in different structures at Kothapally watershed, Andhra Pradesh, India (Source: Pathak *et al.*, 2009a)

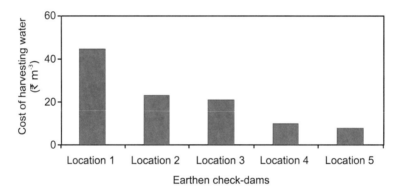

Figure 7.14 Cost of water harvesting at different locations in Lalatora watershed, Madhya Pradesh, India (Source: Pathak *et al.*, 2009a)

strategies for scheduling supplemental irrigation particularly in cases where more than one drought occurs during the cropping season.

Rainfed agriculture has traditionally been managed at the field scale. Supplemental irrigation systems, with storage capacities generally in the range of 20–100 mm of irrigation water, even though small in comparison to irrigation storage, require planning and management at the catchment scale, as capturing local runoff may impact other water users and ecosystems. Legal frameworks and water rights pertaining to the collection of local surface runoff are required, as are human capacities for planning, constructing, and maintaining storage systems for supplemental irrigation and moreover, farmers must be able to take responsibility for the operation and management of the systems. Supplemental irrigation systems also can be used in small vegetable gardens during the dry seasons to produce fully irrigated cash crops. Supplemental irrigation is a key strategy, still underused, for unlocking the rainfed productivity potential and water productivity.

Table 7.15 Effect of supplemental irrigation and fertilizer on sorghum grain yield (kg ha^{-1}), Sahel, 1998–2000[a]

Treatment	1998 Mean[b]	SD	1999 Mean[b]	SD	2000 Mean[b]	SD	1998–2000 Mean[b]	SD
C	666[a]	154	238[a]	25	460[a]	222	455[a]	232
I	961[a]	237	388[b]	182	787[b]	230	712[b]	320
F	1470[b]	254	647[c]	55	807[b]	176	975[c]	404
IF	1747[b]	215	972[d]	87	1489[c]	123	1403[d]	367

[a]Source: Fox and Rockström (2003).
C = Control; I = Irrigation application; F = Fertilizer application; IF = Supplemental irrigation and fertilizer application.
SD = Standard deviation.
[b]Test of treatment effect. Mean values in a column followed by different letters are significantly different at the 5% level using the Student-Newman-Keul's test.

7.2.4.1 Crop responses to supplemental irrigation

Good response to supplemental irrigation had been reported from several parts of the SAT of Africa (Carter and Miller 1991; Jenson *et al.*, 2003; Oweis and Hachum 2003; Barron 2004; Rockström *et al.*, 2007). On-farm research in the semi-arid locations in Kenya (Machakos district) and Burkina Faso (Ouagouya) indicates a significant scope of improving water productivity in rainfed farming through supplemental irrigation, especially if the practice is combined with soil fertility management (Oduor 2003). From the experiments conducted in the Sahel region, Fox and Rockström (2003) reported that supplemental irrigation alone resulted in sorghum grain yield of 712 kg ha^{-1}, while supplemental irrigation combined with fertilizer application resulted in grain yield of 1403 kg ha^{-1}, which was higher than the farmer's normal practice by a factor of 3 (Table 7.15).

Barron (2004) reported from the studies made in Kenya that the water productivity for maize was 1796 m^3 t^{-1} of grain with supplemental irrigation and 2254 m^3 t^{-1} of grain without supplemental irrigation, i.e., decrease in water productivity by 25%. The study concluded that the water harvesting system for supplemental irrigation of maize was both biophysically and economically viable. However, the viability of increased water harvesting implementation at the catchment scale needs to be assessed so that other downstream uses of water remain uncompromised.

Excellent responses to supplemental irrigation have been reported from several locations in India (Singh and Khan 1999; Gunnell and Krishnamurthy 2003). Vijayalakshmi (1987) reported that the effect of supplemental irrigation was largest in rainy season sorghum and pearl millet and yields increased by 560 and 337% respectively and for pigeonpea the yield increased by 560%, but a comparatively lesser response in case of groundnut where the yield increased by only 32% (Table 7.16). For postrainy season crops, an increase by 123% for wheat, 113% for barley, 345% for safflower, and 116% for rapeseed were reported for crops grown at several research stations in India. Havanagi (1982) reported similar crop yield responses to supplemental irrigation in Bengaluru.

Table 7.16 Effect of supplemental irrigation on crop yields at different locations in India[a]

Crop	Irrigation (cm)	Yield (t ha^{-1})	Yield increase due to irrigation (%)	Location
Short-duration rainy season crops				
Sorghum	1.6	2.51	560	Hyderabad
Maize	1	2.66	15	Jhansi
	2	4.43	40	
Finger millet	5	2.32	43	Bengaluru
Soybean	8	2.05	14	Indore
Long-duration rainy season crops				
Castor	5	1.32	31	Hyderabad
Pigeonpea	3	0.17	240	Jhansi
(sole crop)	5	0.33	560	
Tobacco	4	1.30	58	Dantiwada
Postrainy season crops				
Wheat	2	1.58	35	Dehradun
	4	2.06	78	
	6	2.60	123	
Rape seed	1	0.35	40	Ranchi
	3	0.46	84	
	5	0.54	116	

[a] Source: Vijayalakshmi (1987).

Singh and Khan (1999) also summarized the yield responses of crops to supplemental irrigation at different locations in India; the data indicated that one supplemental irrigation at critical stage of crop growth considerably increased the crop yields. Introduction of high value crops such as hybrid cotton under protective irrigation further helps in enhancing the income of dryland farmers. Due to better moisture availability through supplemental irrigation, crops respond to the application of higher rates of nutrients. In an experiment carried out on medium deep black soils at Bijapur, Karnataka, India, the responses of horticultural crops, viz., *ber* (jujube), guava, and fig to supplemental irrigation was studied. The highest (122.6%) response to supplemental irrigation was recorded in guava and the lowest (41.7%) in fig (Radder *et al.*, 1995).

On SAT Alfisols, excellent benefits have been reported from supplemental irrigation at ICRISAT, Patancheru (Pathak and Laryea 1990). As shown in Table 7.17, good yield responses to supplemental irrigation were obtained on Alfisols in both rainy and postrainy seasons. The average water application efficiency (WAE) for sorghum ($14.8\,\text{kg}\,\text{mm}^{-1}\,\text{ha}^{-1}$) was more than that for pearl millet (8.7 to $10.1\,\text{kg}\,\text{mm}^{-1}\,\text{ha}^{-1}$). Tomatoes responded very well to water application with an average WAE of $186.3\,\text{kg}\,\text{mm}^{-1}\,\text{ha}^{-1}$.

On SAT Vertisols, Srivastava *et al.* (1985) found that the average WAE was largest for chickpea ($5.5\,\text{kg}\,\text{mm}^{-1}\,\text{ha}^{-1}$), followed by chili ($4.0\,\text{kg}\,\text{mm}^{-1}\,\text{ha}^{-1}$) and safflower ($2.0\,\text{kg}\,\text{mm}^{-1}\,\text{ha}^{-1}$). They concluded from their experiments that irrigation was profitable for sequential crops of chickpea and chili on Vertisols.

Table 7.17 Mean grain yield response of cropping systems to supplemental irrigation on an Alfisol watershed, ICRISAT, Patancheru, India, 1981–84[a]

One irrigation of 40 mm	Increase due to irrigation (kg ha^{-1})	WAE (kg mm^{-1} ha^{-1})	Two irrigations (40 mm each)	Increase due to irrigation (kg ha^{-1})	WAE (kg mm^{-1} ha^{-1})	Combined WAE (kg mm^{-1} ha^{-1})
Intercropping system						
Pearl millet			Pigeonpea			
2353	403	10.0	1197	423	5.3	6.8
Sorghum			Pigeonpea			
3155	595	14.9	1220	535	6.7	9.4
Sequential cropping system						
Pearl millet			Cowpea			
2577	407	10.2	735	425	5.3	6.9
Pearl millet			Tomato			
2215	350	8.8	26250	14900	186.3	127.1

[a]Source: Pathak and Laryea (1990).

Pathak _et al._ (2009b) critically analyzed the crop response to supplemental irrigation from different regions. The following key points emerge from the analysis:

- To get the maximum benefit from supplemental irrigation, factors that limit crop productivity must be removed; responsive cultivars, fertilizers, and other recommended package of practices should be followed.
- The best responses to supplemental irrigation were obtained when irrigation water was applied at the critical stages of crop growth.
- On Alfisols and other sandy soils, the best results from limited supplemental irrigation were obtained during the rainy season. On these soils, the additional benefits from one or two supplemental irrigations during postrainy season were found to be limited.
- On Vertisols in medium to high rainfall areas, pre-sowing irrigation for postrainy season crops was found to be most beneficial.
- Crop responses to supplemental irrigation on lighter soils were found better than on heavier soils in the low and medium rainfall areas. However, this was not true for high rainfall areas (<850 mm).
- To get the maximum benefit from the available water, growing high value crops, viz., vegetables and horticultural crops are getting popular even with poor farmers.

7.2.5 Indigenous soil and rainwater conservation practices

Indigenous knowledge is the local wisdom that people have gained through inheritance from their ancestors. It is a people derived science and represents people's creativity, innovations, and skills. Indigenous technological knowledge pertains to various cultural norms, social roles, or physical conditions. Such knowledge is not a static body of wisdom, but instead consists of dynamic insights and techniques, which are changed

Table 7.18 Some documented indigenous soil and water conservation measures in semi-arid India[a]

Categories	Indigenous soil and water conservation measures
Agronomic, tillage practices	Cultivation and sowing across the slope, wider row spacing and deep interculturing, mixed cropping, Cover cropping, application of organic manure, strip cropping, green manuring, conservation furrows with traditional plow, deep plowing, summer plowing, and repeated tillage during monsoon season
Bunding and terracing (mechanical and vegetative barrier)	Vegetative barrier, stone bunding, compartmental bunding, peripheral bunding/field bunding, conservation bench terrace, strengthening bunds by growing grasses, bund farming of pulse crops in *kharif* under rainfed situation, earthen bunds, stone-cum-earthen bunding and live bunding by raising cactus
Soil amendment/mulching	Application of tank silt, sand mulching, gravel sand mulching, and retention of pebbles on the soil surface
Erosion control and runoff diversion structures	Sand bags as gully check, loose boulder checks, stone waste weir, waste weir, brushwood structure across the bund, grassed waterways, and *nala* plugging
Water harvesting, seepage control, and groundwater recharge	Seepage control by lining farm ponds with white soil, harvesting of seepage water, wells as runoff storage structures, farm pond percolation pond/tank, groundwater recharging through ditches and percolation pits, dug wells, *haveli/Bharel* system, *bandh* system of cultivation, earthen check-dams, field water harvesting, *Nadi* farming system, and rainwater harvesting in *Kund/Tanka*

[a]Source: Mishra *et al.* (2002b).

over time through experimentation and adoption to environmental and socioeconomic changes. This knowledge is based on hundreds and sometimes thousands of years of adoption, while bearing odds and evens of the time.

Traditional knowledge and practices have their own importance as they have stood the test of time and have proved to be efficacious to the local people. Many indigenous soil and water conservation practices are practiced in different countries. They need to be scientifically evaluated to qualify as modern technological knowledge for wider adoption by addressing the researchable issues. A detailed study of indigenous technical knowledge on soil and water conservation in India was taken up by Mishra *et al.* (2002b). Some documented indigenous practices from different rainfed regions of India are presented in Table 7.18.

In Africa also, cultivators apply a wide range of techniques, both mechanical and agronomic practices, such as crop rotation, crop mixtures, application of manure, protection of N-fixing trees, terrace building, pitting systems, drainage ditches, and small dams in valley floors, to conserve soil and water and to prevent soil degradation. Reij (1991) has attempted to assess current knowledge on indigenous soil and water conservation in Africa. Several examples of indigenous soil conserving practices in the tropical region of Africa presented in Table 7.19. These indigenous techniques are not an exception and they are applied over large parts of the continent. Several reports create an impression that African indigenous soil and water conservation practices are at peril and have no future because these techniques are increasingly abandoned due to several reasons such as political instability, population density, and efficiency of market

Table 7.19 Indigenous soil conserving practices in the tropical region of Africa

Country	Rainfall (mm)	Indigenous SWC techniques	Major crops	Reference
Burkina Faso (Central)	400–700	Stone lines, stone terraces, planting pits (Zey)	Sorghum, millet	Savonnet (1958); Reij (1991)
Burkina Faso (South)	700–800	Stone lines	Sorghum, millet	
Burkina Faso (Southwest)	1000	Contour stone bunds on slopes, drainage channels	Sorghum, millet	Pradeau (1975)
Mali (Djenne-Safara)	400	Pitting systems	Sorghum, millet	Ayers (1989)
Sudan (Djebel Marra)	600–1000	Bench terraces	Millet/sorghum	Miehe (1986)
Tanzania (Uluguru mountains)	1500	Ladder terraces		Temple (1972)
Tchad (Ouddai)	250–650	Various earth-bunding systems with upslope wing walls, in drier regions with catchment area (water harvesting)		Sommerhalter (1987)

forces. However, there are many locations where indigenous techniques continue to be maintained and even expanded. In some instances, indigenous techniques, abandoned some decades ago have been revived recently.

7.3 ENHANCING THE IMPACTS OF SOIL AND WATER CONSERVATION AND WATER HARVESTING INTERVENTIONS THROUGH INTEGRATED WATERSHED APPROACH

To maximize the benefits from soil and water conservation interventions, a more integrated approach is needed. In rainfed agriculture, where water is a highly variable production factor, risk reduction through integrated soil and water management is a key to unlocking the potential of managing crops, soil fertility, and pests and allowing for diversification. For rainfed agriculture, watershed provides a logical hydrological scale for effectively managing soil erosion, rainfall, runoff, and groundwater. Results from the several integrated watershed programs clearly indicated excellent opportunities of implementing soil and water conservation, water harvesting, groundwater recharging, and supplemental irrigation at the watershed scale. The key advantage of this approach is that these interventions can be implemented both at farmers' field level as well as community level. Also, the watershed-based community organizations and institutions assist in sustainable management of soil conservation and water harvesting structures.

Although the integrated watershed program includes multi-faceted activities, soil and water conservation, water harvesting, groundwater recharging and its efficient utilization have been the key components of most watershed programs in India and other Asian countries. Results from some key watershed programs with reference to these aspects are discussed.

In Asia, ICRISAT in partnership with the national agricultural research systems (NARS) has developed an innovative and up-scalable consortium model for managing watersheds holistically (Wani et al., 2003). The approach uses rainwater management as an entry point activity starting with in-situ conservation of soil and rainwater, harvesting the excess runoff, and groundwater recharging and converging the benefits of stored rainwater into increased productivity by using improved cultivars and suitable nutrient, pest, and land and water management practices. The consortium strategy brings together institutions from the scientific, non-government, government, and farmers' groups for knowledge management. Convergence allows integration and negotiation of ideas among actors. Cooperation enjoins all stakeholders to harness the power of collective actions. Capacity building engages in empowerment for sustainability. This approach of integrated and participatory watershed development and management has emerged as the cornerstone of rural development in the SAT. It ties together the biophysical notion of a watershed as a hydrological unit with the social aspect of community and its institutions for sustainable management of land, water, and other resources. At ICRISAT benchmark watersheds in India, Thailand, Vietnam, and China, community- and farmer-based soil and water conservation interventions control soil loss and improve the surface and groundwater availability. Findings in most of the watershed sites reveal that open wells located near water harvesting structures have significantly higher water levels compared to those away from the structures. Improved water availability in the watershed not only resulted in increased crop productivity but significant shift in area under cultivation took place towards high-value cereals, cash crops, vegetables, flowers, and fruits.

At Kokriguda watershed, Koraput district, Orissa, India various soil and water conservation measures were implemented to improve the water availability and control soil erosion. Water Users' Association was constituted to maintain the various structures. Open wells registered water table rise by 0.32 m and crop yields increased by 15% in finger millet to 38% in upland paddy. Due to these interventions, area under remunerative crops like vegetables increased from 2 to 35 ha, conveyance efficiency from 23 to 95%, and overall irrigation efficiency from 20 to 43% (Patnaik et al., 2004). In Rajiv Gandhi Watershed program in Madhya Pradesh, India, over 0.7 million water harvesting structures were constructed. The program ran on a mission mode and had over 19% peoples' contribution in monetary terms. There has been 59% increase in irrigated area and 34% decrease in wasteland area where the mission has worked. Agricultural production in the project villages increased by 37% during rainy season and by 30% during postrainy season. Over 3000 villages have reported accretion in groundwater. At Fakot in Tehri Garhwal district, India, a 370-ha watershed was treated with various water harvesting and soil conservation measures. Consequently, paddy and wheat yields increased by $1.65\,t\,ha^{-1}$ and $1.93\,t\,ha^{-1}$ respectively. These measures considerably reduced runoff and soil loss from 42.0 to 0.7% and 11.0 to $2.7\,t\,ha^{-1}$, respectively. The benefit-cost ratio considering 25 years project life has been worked out as 2.71 at 12% discount rate (Sharda and Juyal 2007).

7.4 STRATEGIES FOR IMPROVING ADOPTION OF SOIL AND WATER CONSERVATION PRACTICES BY FARMERS

Despite being effective in increasing crop yield and having positive effects on soil quality, the adoption of most improved soil and water conservation practices is limited.

Farmers do not operate as independent decision-makers, but rather are subjected to and influenced by a variety of factors. Thus many a times, the decision to adopt and use soil and water conservation practices is made not only in the context of personal and family circumstances, but also in response to government policies, institutional arrangements, community attitudes, and customs.

In seeking workable conservation prescriptions, research institutions, governments, and aid agencies should cooperate closely and fully with local farmers, extension personnel, and community leaders. Such an approach permits information exchange about what already works well, what might work well, and what would be required to make proposed new soil and water conservation techniques feasible and acceptable. Some of the key points, which can facilitate the greater adoption of soil and water conservation techniques, are:

- *Participatory research and demonstration:* Participatory research and demonstrations are very useful to show potential adopters that soil and water conservation technologies and techniques are appropriate for farming systems employed in their community. Field demonstrations are also useful to show potential adopters the type of technical skills they must possess to effectively implement recommended soil and water conservation programs on their farm. Before farmers adopt the technology, it must be adequately demonstrated in terms of its benefits as well as its limitations. Farmers must have enough time to assess the improved technology and compare this with what they have become familiar and have been practicing for a long time.
- *Increased emphasis to rainwater management:* In addition to soil conservation emphasis should be given to rainwater management. This will enhance the adoption of soil and water conservation practices by farmers, as this will provide both short- and long-term benefits to the farmers.
- *Short-term and visible benefits to farmers:* Profitability is assessed in the context of financial return to investment, savings in time and labor, modifications needed in the management of farm activities to integrate innovations, increased risk of failure associated with adoption, and many other factors. Unless the economic return associated with adoption is high enough to compensate adopters for all of these costs, farmers will not adopt any recommended technologies. They evaluate all soil and water conservation technologies and techniques in the context of short- and long-term return to investment. Conservation practices that produce short-term benefits will be more readily adopted than those that produce long-term benefits. The recommended technologies or practices should be able to provide farmers with sufficient benefits, especially cash benefits. This should also be adequately addressed and explained to farmers.
- *Selecting the right technologies with full technical and other assistance:* The right soil and water conservation technology which gives farmers both short- and long-term benefits should be identified. Also, all the assistance and other help should be provided in effectively implementing the technology. For example, if BBF system is recommended, the appropriate implement for BBF making and planting should be provided.
- *Encourage more farmer-to-farmer transfer:* This can facilitate the adoption of new soil and water conservation technologies.

- *Government policy to promote adoption:* Government policy has a very important role to play in the adoption of new practices. For example, in China, terracing has been promoted by the government as the main soil and water conservation practice for which subsidies are provided. In India, contour bunding was promoted by the government during 1974–89.
- *Increased farmers' perception of environmental problems and their effects:* Perception of soil erosion does not mean that farmers are motivated to reduce it. Farmers, without assistance, cannot be expected to know that the erosion of fine, nutrient-rich particles of soil reduces soil fertility. Farmers' awareness of environmental problems has been one of the most important factors to affect adoption and continued use of soil and water conservation technologies and techniques at the farm level. Efforts should be made to increase the awareness of soil erosion, efficient utilization of water and its effects, both on-site and off-site.
- *Improved farmers' perception of the recommended technology:* Lack of access to information about problems and possible solutions can prevent adoption of soil and water conservation technologies and techniques because potential adopters are not informed of alternatives to the existing production systems.
- *Technology which reduces risk:* Small farmers tend to avoid adopting technologies and techniques that increase the level of risk. Efforts should be made to recommend the technology, which reduces risk.
- *Integrated watershed approach:* Implementation of soil and water conservation practices in integrated watershed mode for greater impact and increased adoption.
- *Training and capacity building:* This is important for effective implementation of technology in the fields.

7.5 CONCLUSIONS

Fast deterioration of natural resources is one of the key issues, threatening sustainable development of rainfed agriculture as most rainfed regions are facing multifaceted problems of land degradation, water shortage, acute poverty, and escalating population pressure. Improved and appropriate soil and water management practices are most important for sustainable and improved livelihoods in the rainfed areas. This is because other technological interventions such as improved varieties, fertilizers, etc. are generally not so effective where soil is degraded and water is severely limited. For in-situ soil and land water conservation practices such as contour cultivation, conservation furrows, tied ridges, scoops, BBF system, contour bunding, graded bunding, field bunding, compartmental bunding, vegetative barriers, and tillage systems, considerable body of research knowledge and experiences exist. The real challenge is to identify appropriate technologies, implement and execute strategies for different rainfed regions with different socioeconomic, soil, crop, rainfall, and topographic conditions.

Physical erosion-control measures have been effectively used in the past and the need for them will continue in the future too. However, if emphasis is first placed on rainwater management, the need for physical conservation works can be greatly reduced and many of the problems faced in the past could be overcome. Evidence shows that this approach improves the adoption of soil and water conservation practices by farmers as it provides both short- and long-term benefits.

Conservation tillage or zero tillage is probably one of the most effective systems for soil and water conservation. However, the performance of this practice in many rainfed regions has been poor. In addition to its poor performance, there are also several constraints (demands of crop residues for animal feed, high cost of new tools and equipment, and high level of management) to adoption of no-till farming, clearly indicating that there is need for more research on how these tillage systems will perform in short- and long-term on different soil types. So far results indicate that for rainfed regions minimum tillage appears better compared to no-till farming.

Studies have indicated that water harvesting and supplemental irrigation systems make a lot of difference through enhanced water use efficiency and these systems are affordable even for small-scale farmers. However, policy frameworks, institutional structures, and human capacities similar to those for full irrigation infrastructure are required to be successfully applied for water harvesting and supplemental irrigation systems in rainfed agriculture. Due to the high initial cost, favorable government policies and the availability of credit may be essential for popularization of efficient irrigation system. Impressive benefits have been reported from supplemental irrigation both in terms of increasing and stabilizing crop productivity from many rainfed regions of Asia and Africa. The best response to supplemental irrigation was obtained when water was applied at the critical stage of crop growth. Even small amounts of water applied at critical growth stage were highly beneficial. To get the maximum benefits from supplemental irrigation, other improved inputs such as responsive cultivars and fertilizers must be used. Majority of the soil and water conservation projects have in the past had narrow focus and now a more holistic approach of integrated watershed management is required to ensure sustainability and overall improvement in livelihoods. Integrated watershed management approach enables to have "win-win" situations for sustaining productivity, controlling land degradation, and improving livelihoods of the community. Some successful watershed development models, e.g., "consortium model for managing watersheds holistically" have high potential in conserving soil and water and bringing favorable changes in rainfed areas for sustainably improving livelihoods.

The adoption of soil and water conservation practices is still a major problem in most rainfed regions. Clearly these technologies require greater and sustained support from the implementing agencies than generally required for other improved agricultural technologies, viz., crop varieties, fertilizers, etc. Finally, farmers, scientists, policy makers, and government must work together to enhance the adoption of soil and water conservation technologies for producing adequate amounts of food, feed, and fiber, and to meet the challenge of sustaining the natural resource base.

REFERENCES

Ajay Kumar. 1991. *Modeling runoff storage for small watersheds*. MTech Thesis, Asian Institute of Technology, Bangkok, Thailand.

Anonymous. 1998. Study on best cultivation practices for effective rain water management. *Annual Research Report*. SK Nagar, Gujarat, India: All India Co-ordinated Research Project for Dry Land Agriculture, Gujarat Agricultural University.

Arnold, J.G., and C. Stockle. 1991. Simulation of supplemental irrigation from on-farm ponds. *Journal of Irrigation and Drainage Engineering* 117(3):408–424.

Ayers, A. 1989. *Indigenous soil and water conservation in Sub-Saharan Africa: the circle of Djenne, Central Mali*. MSc Thesis, University of Reading, UK.

Barron, J. 2004. *Dry spell mitigation to upgrade semi-arid rainfed agriculture: water harvesting and soil nutrient management*. PhD Thesis, Natural Resources Management, Department of Systems Ecology, Stockholm University, Stockholm, Sweden.

Bhumbla, D.R., S.C. Mandal, K.G. Tejwani *et al*. 1971. Soil technology. In *Review of soil research in India*, ed. J.S. Kanwar, and S.P. Rayachowdhury, 137–168. New Delhi, India. Indian Society of Soil Sciences.

Carter, D.C., and S. Miller. 1991. Three years with an on-farm macro-catchment water harvesting system in Botswana. *Agricultural Water Management* 19:191–203.

Channappa, T.C. 1994. *In situ* moisture conservation in arid and semi-arid tropics. *Indian Journal of Soil Conservation* 22(1&2):26–41.

Chittaranjan, S., W.S. Patnaik, R.N. Adhikari *et al*. 1997. Soil and water conservation measures in Vertisols of semi-arid regions of South India. *Journal of Agricultural Engineering, ISAE* 34(1):32–42.

Dagg, M., and J.C. Macartney. 1968. The agronomic efficiency of NIAE mechanized tied ridge system of cultivation. *Experimental Agronomy* 4:279–294.

De Fraiture, C., D. Wichelns, J. Rockström *et al*. 2007. Looking ahead to 2050: Scenarios of alternative investment approaches. In *Water for food, water for life: A comprehensive assessment of water management in agriculture*, ed. D. Molden, 91–145. London, UK: Earthscan; and Colombo, Sri Lanka: International Water Management Institute

El-Swaify, S.A. 1993. Soil erosion and conservation in the humid tropics. In *World soil erosion and conservation*, ed. D. Pimental, 233–235. UK: Cambridge University Press.

El-Swaify, S.A., E.W. Dangler, and C.L. Amstrong. 1982. *Soil erosion by water in the tropics*. Research and Extension Series 024, Hawaii Institute of Tropical Agriculture. Honolulu, Hawaii: University of Hawaii Press.

El-Swaify, S.A., P. Pathak, T.J. Rego *et al*. 1985. Soil management for optimized productivity under rainfed conditions in the semi-arid tropics. *Advances in Soil Science* 1:1–64.

Fox, P., and J. Rockström. 2003. Supplemental irrigation for dry-spell mitigation of rainfed agriculture in the Sahel. *Agricultural Water Management* 61(1):29–50.

Gund, M., and A.G. Durgude. 1995. Influence of contour and subabul live bunds on runoff, soil loss and yield of pearl millet + pigeonpea intercropping. *Indian Journal of Soil Conservation* 23(2):171–173.

Gunnell, Y., and A. Krishnamurthy. 2003. Past and present status of runoff harvesting systems in dryland peninsular India: A critical review. *Ambio* 32(4):320–323.

Gurmel Singh, G. Shastry, and Bhardwaj. 1990. Watershed responses to conservation measures under different agroclimatic regions of India. *Indian Journal of Soil Conservation* 18(3):16–22.

Havanagi, G.V. 1982. Water harvesting and life saving irrigation for increasing crop production areas of Karnataka. In *Proceedings of the Symposium on Rainwater and Dryland Agriculture*, 159–170. New Delhi, India: Indian National Science Academy (INSA).

Huxley, P.A. 1979. Zero-tillage at Morogoro, Tanzania. In *Soil tillage and crop production*, ed. R. Lal, 259–265. Proceedings No. 2. Ibadan, Nigeria: International Institute of Tropical Agriculture.

ICRISAT. 1985. *Annual Report 1984*. Patancheru, Andhra Pradesh, India: International Crops Research Institute for the Semi-Arid Tropics.

ICRISAT. 1987. *Annual Report 1986*. Patancheru, Andhra Pradesh, India: International Crops Research Institute for the Semi-Arid Tropics.

ICRISAT. 2008. *Watershed Project, Terminal Report* 2008. Establishing participatory research-cum-demonstrations for enhancing productivity with sustainable use of natural resources in Sujala watersheds of Karnataka, 2005–2008. Global Theme on Agroecosystems. Patancheru, Andhra Pradesh, India: International Crops Research Institute for the Semi-Arid Tropics.

Jenson, J.R., R.H. Bernhard, S. Hansen *et al.* 2003. Productivity in maize based cropping systems under various soil-water-nutrient management strategies in a semi-arid Alfisol environment in East Africa. *Agricultural Water Management* 59:217–237.

Jones, O.R., and B.A. Stewart. 1990. Basin tillage. *Soil Tillage Research* 18:249–265.

Kale, S.R., V.G. Salvi, and P.A. Varade. 1993. Runoff and soil loss as affected by different soil conservation measures under *Eleusine coracana* (ragi) in lateritic soils, West Coast Konkan Region (Maharashtra), India. *Indian Journal of Soil Conservation* 21(3):11–15.

Kowal, J. 1970. The hydrology of small catchment basin at Samuru, Nigeria. IV: Assessment of soil erosion. *Niger Agriculture* 7:134–147.

Krishna, J.H., and G.F. Arkin. 1988. *Furrow-diking technology for agricultural water conservation and its impact on crop yield in Texas.* Texas Water Resource Research Institute Pub. TR-140. College Station, USA: Texas A&M University.

Krishna, J.H., and Gerik, T.J. 1988. Furrow-diking technology for dryland agriculture. In *Challenges in dryland agriculture – a global perspective. Proceedings of International Conference on Dryland Farming*, ed. P.W. Unger, T.U. Sheed, W.R. Jordan, and R. Jensen, 258–260. College Station, USA: Texas Agricultural Experiment Station.

Krishnappa, A.M., Y.S. Arun Kumar, Munikappa *et al.* 1999. Improved *in situ* moisture conservation practices for stabilized crop yields in drylands. In *Fifty years of dryland agriculture research in India*, ed. H.P. Singh, Y.S. Ramakrishna, K.L. Sharma, and B. Venkateswarlu, 291–300. Hyderabad, India: CRIDA.

Lal, H. 1985. Animal drawn wheeled carrier: An appropriate mechanization for improved farming system. *Agricultural Mechanization in Asia, Africa and Latin America* 16(1):38–44.

Lal, H. 1986. Development of appropriate mechanization for the "W" form soil management system. *Soil and Tillage Research* 8, Special issue, 145–160.

Larson, W.E. 1962. Tillage requirements for corn. *Journal of Soil and Water Conservation* 17:3–7.

Laryea, K.B., P. Pathak, and M.C. Klaij. 1991. Tillage systems and soils in the semi-arid tropics. *Soil and Tillage Research* 20:201–218.

Manian, R., G.J. Baby Meenakshi, K. Rangaswamy *et al.* 1999. Effect of tillage operations in improving moisture conservation practices for stable sorghum yield. *Indian Journal of Dryland Agricultural Research and Development* 14 (2):64–67.

Miehe, S. 1986. *Acacia albida* and other multipurpose trees on the Fur farmlands in the Jebel Marra Highlands, Westen Darfur, Sudan. *Agroforestry Systems* 4:89–119.

Mishra, P.K., and S.L. Patil. 2008. *In situ* rainwater harvesting and related soil and water conservation technologies at the farm level. Presented at the International Symposium on Water Harvesting in June 2008 at TNAU, Coimbatore, India.

Mishra, P.K., G. Sastry, M. Osman *et al.* 2002a. *Dividends from soil and water conservation practices: A brief review of work done in rainfed eco-regions.* Hyderabad, India: CRIDA.

Mishra, P.K., G. Sastry, M. Osman *et al.* 2002b. *Indigenous technical knowledge on soil and water conservation in semi-arid India.* Hyderabad, India: CRIDA.

Molden, D., Karen Frenken, Randolph Barker *et al.* 2007. Trends in water and agricultural development. In *Water for food, water for life: A comprehensive assessment of water management in agriculture*, ed. D. Molden, 56–89. London, UK: Earthscan; and Colombo, Sri Lanka: International Water Management Institute.

More, S.M., S.P. Malik, S.S. Despande *et al.* 1996. *In-situ* moisture conservation for increased productivity of winter sorghum. *Journal of Maharashtra Agricultural Universities* 20(1):129–130.

Morin, J., and Y. Benyamini. 1988. Tillage method selection based on runoff modeling. In *Challenges in dryland agriculture – a global perspective. Proceedings of International Conference on Dryland Farming*, ed. P.W. Unger, T.U. Sheed, W.R. Jordan, and R. Jensen, 251–254. College Station, USA: Texas Agricultural Experiment Station.

Nicou, R., and J.L. Chopart. 1979. Water management methods in sandy soil of Senegal. In *Soil tillage and crop production*, ed. R. Lal, 248–257. Proceedings No. 2. Ibadan, Nigeria: International Institute of Tropical Agriculture.

Njihia, C.M. 1979. The effect of tied ridges, stover mulch and farm yard manure on water conservation in a medium potential area, Katumanu, Kenya. In *Soil tillage and crop production*, ed. R. Lal, 295–302. Ibadan, Nigeria: International Institute of Tropical Agriculture.

Oduor, A. 2003. *Hydrological assessment of runoff catchment schemes: A case of water harvesting using underground innovative tanks for supplementary irrigation in semi-arid Kenya.* MSc Thesis, UNESCO-IHE, Delft, The Netherlands.

Oldeman, L.R., R.T.A. Hakkling, and W.G. Sombreck. 1991. *World map of the status of human-induced soil degradation. An exploratory note.* Wageningen: International Soil Reference and Information Centre.

Oweis, T., and A. Hachum. 2003. Improving water productivity in dry areas of West Asia and North Africa. In *Water productivity in agriculture: limits and opportunities for improvements*, ed. J.W. Kijne, R. Barker, and D. Molden, 179–198. Wallingford, UK: CAB International.

Pathak, P., and K.B. Laryea. 1990. Prospects of water harvesting and its utilization for agriculture in the semi-arid tropics. In *Proceedings of Symposium of the SADCC (Southern African Development Coordination Conference)*, 266–278. Gaborone, Botswana: Land and Water Management Research Program.

Pathak, P., and K.B. Laryea. 1995a. Soil and water conservation in the Indian SAT: principles and improved practices. In *Sustainable development of dryland agriculture in India*, ed. R.P. Singh, 83–94. Jodhpur, India: Scientific Publishers.

Pathak, P., and K.B. Laryea. 1995b. Soil and water management options to increase infiltration and productivity on SAT soils. In *Eighth EARSAM Regional Workshop on Sorghum and Millets*, ed. S.Z. Mukuru, and S.B. King, 103–111. Patancheru, Andhra Pradesh, India: International Crops Research Institute for the Semi-Arid Tropics.

Pathak, P., K.B. Laryea, and S. Singh. 1989a. A modified contour bunding system for Alfisols of the semi-arid tropics. *Agricultural Water Management* 16:187–199.

Pathak, P., K.B. Laryea, and R. Sudi. 1989b. A runoff model for small watersheds in the semi-arid tropics. *Transactions of American Society of Agricultural Engineers* 32:1619–1624.

Pathak, P., S.M. Miranda, and S.A. El-Swaify. 1985. Improved rainfed farming for the semi-arid tropics: implications for soil and water conservation. In *Soil erosion and conservation*, ed. S.A. El-Swaify, W.C. Moldenhauer, and L. Andrew, 338–354. Ankeny, Iowa, USA: Soil Conservation Society of America.

Pathak, P., P.K. Mishra, K.V. Rao *et al.* 2009a. Best bet options on soil and water conservation. In *Best bet options for integrated watershed management*, Proceedings of the Comprehensive Assessment of Watershed Programs in India, 25–27 July 2007, ICRISAT, Pantancheru, India ed. S.P. Wani, B. Venkateswarlu, K.L. Sahrawat, K.V. Rao, and Y.S. Ramakrishna, 75–94. Patancheru, Andhra Pradesh, India: International Crops Research Institute for the Semi-Arid Tropics.

Pathak, P., K.L. Sahrawat, S.P. Wani, R.C. Sachan, and R. Sudi. 2009b. Opportunities for water harvesting and supplemental irrigation for improving rainfed agriculture in semi-arid areas. In *Rainfed agriculture: unlocking the potential*, ed. S.P. Wani, J. Rockström, and T. Oweis, 197–221. Comprehensive Assessment of Water Management in Agriculture Series. Wallingford, UK: CAB International.

Pathak, P., S. Singh, and R. Sudi. 1987. Soil and water management alternatives for increased productivity on SAT Alfisols. In *Soil conservation and productivity, Proceedings of the*

IV International Conference on Soil Conservation, 533–550. Maracay, Venezuela: Soil Conservation Society of Venezuela.

Patnaik, U.S., P.R. Choudary, S. Sudhishri *et al.* 2004. *Participatory watershed management for sustainable development in Kokriguda watershed, Koraput, Orissa.* Sunabeda, Orissa: CSWCRTI, Research Center.

Phillips, R.L. 1963. Surface drainage system for farm lands (eastern United States and Canada). *Transactions of American Society of Agricultural Engineers* 6:313–317, 319.

Pimental, D., C. Harvey, P. Resosudarmo *et al.* 1995. Environmental and economic cost of soil erosion and conservation benefits. *Science* 267:1117–1122.

Pradeau, C. 1975. Adaptabilite d'une agriculture topicals traditionnelle. *Etudes Rurales* 58:7–28.

Radder, G.D., C.J. Itnal, and M.I. Belgaumi. 1995. Protective irrigation (life saving) – Principles and practices. In *Sustainable development of dryland agriculture in India*, ed. R.P. Singh, 207–215. Jodhpur, India: Scientific Publishers.

Ram Babu, M.M. Srivastava, G. Sastry *et al.* 1980. *Studies on hydrological behavior of small watersheds under different land uses.* Dehradun, India: CSWCRTI.

Rama Mohan Rao, M.S., S.L. Patil, S.K.N. Math *et al.* 2000. Effect of different agronomic and mechanical measures in reducing soil and water losses in the Vertisols of semi-arid tropics of South India. In *International Conference on Managing Natural Resources for Sustainable Agricultural Production in the 21st Century*, Extended summary, Vol. 3, 1227–1228. New Delhi, India: Resource Management.

Reij, C. 1991. Indigenous soil and water conservation in Africa. *Gatekeeper Series No. 27.* Sustainable Agriculture Program. London, UK: International Institute for Environment and Development.

Rockström, J., N. Hatibu, T. Oweis *et al.* 2007. Managing water in rainfed agriculture. In *Water for food, water for life: a comprehensive assessment of water management in agriculture*, ed. D. Molden, 315–348. London, UK: Earthscan; and Colombo, Sri Lanka: International Water Management Institute.

Rockström, J., Louise Karlberg, S.P. Wani *et al.* 2010. Managing water in rainfed agriculture – The need for a paradigm shift. *Agricultural Water Management* 97:543–550.

Savonnet, G. 1958. Methods employees par certaines populations de la Haute Volta poor letter contre l'erosion. *Notes Africaines* 78:38–40.

Selveraju, R., and C. Ramaswami. 1997. Influence of fallowing and seasonal land configuration on growth and yield of sorghum (*Sorghum bicolor*) + pigeonpea (*Cajanus cajan*) intercropping in Vertisols under varying seasonal precipitation. *Indian Journal of Soil Agronomy* 42(3):396–400.

Sharda, V.N., and G.P. Juyal. 2007. Rainwater harvesting, groundwater recharge and efficient use in high rainfall areas. In *Ensuring water and environment for prosperity and posterity, Souvenir*, 59–70. 10th Inter-Regional Conference on Water and Environment (ENVIROWAT 2007), organized by Indian Society of Water Management in collaboration with Indian Society of Agricultural Engineers and International Commission on Agricultural Engineering.

Sharma, P.N., and O.J. Helweg. 1982. Optimal design of a small reservoir system. *ASCE Journal of Irrigation and Drainage* IR4:250–264.

Singh, K.B., R.S. Chauhan, V.D. Sharma *et al.* 1973. To study the effect of different mulches on barley. In *Progress Report (Rabi 72–73)*, 7–9. Bichpuri, Agra, India: AICRPDA, RBS College.

Singh, R.P., and M.A. Khan. 1999. Rainwater management: water harvesting and its efficient utilization. In *Fifty years of dryland agricultural research In India*, ed. H.P. Singh, Y.S. Ramakrishna, and B. Venkateswarlu, 301–313. Hyderabad, India: Central Research Institute for Dryland Agriculture (CRIDA).

Sireesha, P. 2003. *Prospects of water harvesting in three districts of Andhra Pradesh*. MTech Thesis, Center for Water Resources, Jawaharlal Nehru Technological University (JNTU), Hyderabad, India.

Smith, P.D., and W.R.S. Chitchley. 1983. The potential of runoff harvesting for crop production and range rehabilitation in semi-arid Baringo. In *Soil and water conservation in Kenya*. Proceedings of Second National Workshop, Nairobi, Kenya, March 10–13, 1982, ed. D.B. Thomas, and W.M. Senga, 324–331. Nairobi, Kenya: Development Studies, University of Nairobi.

Sommerhalter, T. 1987. Boden-und Wasserkonservierende Masznamen im Quaddai Geographique des Tschad. Mission Report.

Srivastava, K.L., Abiye Astatke, Tekalign Mam *et al.* 1993. Land, soil and water management. In *Improved management of Vertisols for sustainable crop–livestock production in the Ethiopian highlands: Synthesis report 1986–92*, ed. Tekalign Mamo, Abiye Astatke, K.L. Srivastava, and Asgelil Dibabe, 75–84. Addis Ababa, Ethiopia: Technical Committee of the Joint Vertisol Project.

Srivastava, K.L., P. Pathak, J.S. Kanwar *et al.* 1985. Watershed-based soil and rain water management with special reference to Vertisols and Alfisols. Presented at the National Seminar on Soil Conservation and Watershed Management, 5–7 Sep 1985, New Delhi, India.

Temple, P.H. 1972. Soil and water conservation policies in Uluguru mountains, Tanzania. *Geografiska Annaler* 54A(3–4):110–123.

Unger, P.W., and B.A. Stewart. 1983. Soil management for efficient water use: An overview. In *Limitations of efficient water use on crop production*, ed. H.M. Taylor, W.R. Jordan, and T.R. Sinclair, 419–460. Madison, Wisconsin, USA: American Society of Agronomy.

Vijayalakshmi, K. 1987. Soil management for increasing productivity in semi-arid red soils: Physical aspects. In *Proceedings of the Consultants' Workshop on the State of the Art and Management Alternatives for Optimizing the Productivity of SAT Alfisols and Related Soils*, 1–3 December 1983, ed. P. Pathak, S.A. El-Swaify, and Sardar Singh, 109–115. Patancheru, Andhra Pradesh, India: International Crops Research Institute for the Semi-Arid Tropics.

Vogel, H. 1992. Effects of conservation tillage on sheet erosion from sandy soil at two experimental sites in Zimbabwe. *Applied Geography (UK)* 12(3):229–242.

Wani, S.P., and M.S. Kumar. 2002. On-farm generation of N-rich organic material. In *A training manual on integrated management of watersheds*, ed. S.P. Wani, P. Pathak, and T.J. Rego, 29–30. Patancheru, Andhra Pradesh, India: International Crops Research Institute for the Semi-Arid Tropics.

Wani, S.P., H.P. Singh, T.K. Sreedevi *et al.* 2003. Farmer-participatory integrated watershed management: Adarsha watershed, Kothapally, India – An innovative and upscalable approach. In *Research towards integrated natural resources management: Examples of research problems, approaches and partnerships in action in the CGIAR*, ed. R.R. Harwood, and A.H. Kassam, 123–147. Washington, DC, USA: Interim Science Council, Consultative Group on International Agricultural Research.

William, J.R., G.L. Wistrand, V.W. Benson *et al.* 1988. A model for simulating farm disk management and use. In *Challenges in dryland farming – A global perspective. Proceedings of International Conference on Dryland Farming*, ed. P.W. Unger, T.U. Sheed, W.R. Jordan, and R. Jensen, 244–257. College Station, USA: Texas Agricultural Experiment Station.

Young, H.M. 1982. *No tillage farming*. Wisconsin, USA: No-till farmer Inc.

Yule, D.F., G.D. Smith, and P.J. George. 1990. Soil management options to increase infiltration in Alfisols. In *Proceedings of the International Symposium on Water Erosion, Sedimentation and Resource Conservation*, 180–187. Dehradun, Uttar Pradesh, India: Central Soil and Water Conservation Research and Training Institute.

Chapter 8

Rainwater harvesting improves returns on investment in smallholder agriculture in Sub-Saharan Africa

B.M. Mati, W.M. Mulinge, E.T. Adgo, G.J. Kajiru, J.M. Nkuba, and T.F. Akalu

8.1 INTRODUCTION

Rainwater harvesting (RWH) has great potential to contribute to poverty reduction efforts by improving agricultural productivity and profitability in rainfed areas in Africa and Asia (Rockström *et al.*, 2010). **Rainwater harvesting** *is about collecting, conserving, storing, and utilizing rainwater for various purposes (domestic use, livestock watering, or agriculture).* Rainwater is best collected where it falls, through in-situ conservation and/or channeling excess runoff water in a guided manner from catchment areas into storage reservoirs for various uses, especially for agriculture. But efforts to exploit the rainwater potential is a real challenge in the rural areas of sub-Saharan Africa (SSA) where over 70% of the population live, most of them engaged in smallholder agriculture. Three core assets for these people include land, water, and human capital, albeit they are beset with various constraints. However, smallholder agriculture relies on traditional agronomic and/or agro-pastoral practices poorly matched to declining land space and technologies. At the continental level, Africa has about 874 million ha of land considered suitable for agricultural production, but 83% of it faces water and/or soil-related constraints to productivity (NEPAD 2003; World Bank 2006).

The development of agricultural water management (AWM) in SSA is fairly invisible in comparison to its potential, and compared with other regions of the world. For instance, only about 3% of the total farmed area has been developed for AWM in SSA (World Bank 2006). The World Development Report (2008) states that only 4% of Africa's annual renewable water resources have been developed for agriculture, water supply, and hydropower use, compared to 70–90% in developed countries. Moreover, RWH has great potential to improve water use efficiency by storing and utilizing in supplemental irrigation at critical stages of the crop. However, the average available water storage capacity in Africa is about $200\,m^3$ per person per year, compared to North America, where it is $5,961\,m^3$ per person per year (Global Risk Network 2008). Thus, the much hyped statement that 70% of water in the region is used for agriculture (specifically irrigation) is inaccurate, as it considers only the amount of water mobilized from developed infrastructure. The real problem is that minimal developmental work has been done, especially to tap and utilize rainwater. Currently, water withdrawals for agriculture account for 3.6% of total renewable water resources in Africa and if all land suitable for AWM were to be developed, it would consume only

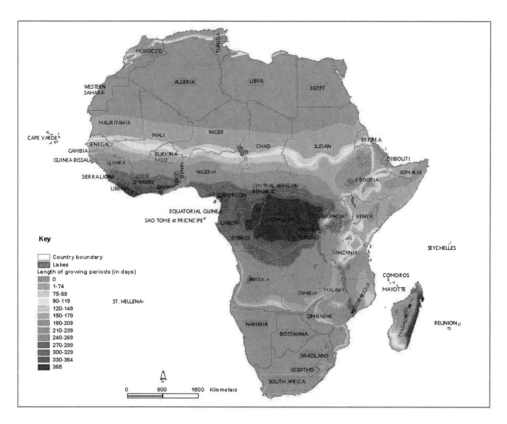

Figure 8.1 Major agroclimatic zones and length of the growing period in Africa (Data source: FAO 2002) (See color plate section)

12% of the "available water" (FAO 2003; Darghouth 2007). Thus, SSA experiences severe economic water scarcity, caused by lack of human and infrastructure capacity to satisfy water demand (Molden *et al.*, 2007) including the failure to adequately harness rainwater. This can be attributed as part of the reason why SSA missed the first agricultural green revolution.

Meanwhile, about 70% of the total land area of SSA comprises arid, semi-arid, and dry subhumid zones (Figure 8.1). Crop and livestock production in these zones are threatened by a multiplicity of natural and human-induced constraints, including low and erratic rainfall, which ranges from less than 100 mm in desert zones to about 800 mm in the Savannah, while the soils are generally highly weathered, of low fertility and prone to erosion. Moreover, agriculture relies on traditional agronomic and/or agro-pastoral practices poorly matched to declining land space and generally, drylands are inhabited by the poorest people. Just a little push, as with exposure to good ideas or technical support, and mobilization with or without funded projects, is sometimes all that is needed for farmers to improve the productivity of smallholder agriculture, and RWH offers great scope.

Focusing on SSA, this chapter presents quantified evidence that RWH earns returns on investment for smallholder agriculture, with cascade benefits on poverty reduction, creation of employment, and mitigation of the impacts of climate change. The main purpose is to inform, create awareness, and advocate for RWH as a necessary intervention for the transformation of smallholder agriculture and rural livelihoods in SSA.

8.2   THE NEED TO RESPOND TO THE THREAT OF CLIMATE CHANGE

Climate change is expected to adversely impact the rural poor reliant on rainfed agriculture in SSA. Africa as a whole is warmer by 0.5°C than it was 100 years ago, putting extra strain on land and water resources. The six warmest years recorded in Africa all occurred since 1987 (Global Risk Network 2008). Over the last three decades, farmers have been facing weather-related events such as drought, prolonged dry spells, erratic rainfall, and floods. But this vulnerability persists even where the risk of adverse weather is well known or is a regular occurrence. The average incidence of serious drought has been on the increase, with seven serious droughts being experienced in Africa during 1980–1990 and 10 droughts during 1991–2003, while drought-induced crop failures are common in the region (FAO 2005). The threat of climate change and that caused by poorly distributed water resources is expected to increase the vulnerability of land users to cope with these shocks. There is broad agreement that one of the biggest climate change impacts will be on rainfall, making it more variable and less reliable (Lenton and Muller 2009). Thus, RWH interventions are needed to improve the availability of water through harvesting, storage, and its efficient use with improved water management practices to adequately respond to the climate change impact. The potential for RWH in SSA is enormous (Figure 8.2). It is estimated that the gross volume of harvestable runoff is about 5,195 km^3 (Malesu *et al.*, 2006a). Moreover, there is a wide selection of RWH technologies and practices to suit nearly every situation, and these are well-known, validated, and documented.

8.3   POLICIES AND INSTITUTIONAL FRAMEWORKS

The policy and institutional framework influencing the development of AWM and thus RWH in SSA are well espoused by the New Partnership for Africa's Development (NEPAD), Comprehensive Africa Agriculture Development Programme (CAADP), which recommended among others, *"extending the area under sustainable land management and reliable water control systems, especially small-scale water control, building up soil fertility and moisture holding capacity of agricultural soils and expansion of irrigation"* as one of three "Pillars" (NEPAD 2003). There are also regional policies such as the Southern Africa Development Community (SADC) Water Policy and Land and Water Management for the Nile Basin Initiative. In most countries, there are national policies that support water development for agriculture, albeit most countries are in the process of reviewing their policies. For instance, in a study of policies (Mati *et al.*, 2007) covering Eritrea, Kenya, Madagascar, Malawi, Mauritius, Rwanda, Sudan, Tanzania, and Zimbabwe, it was found that most countries have developed national and sector policies to address poverty reduction, achievement of

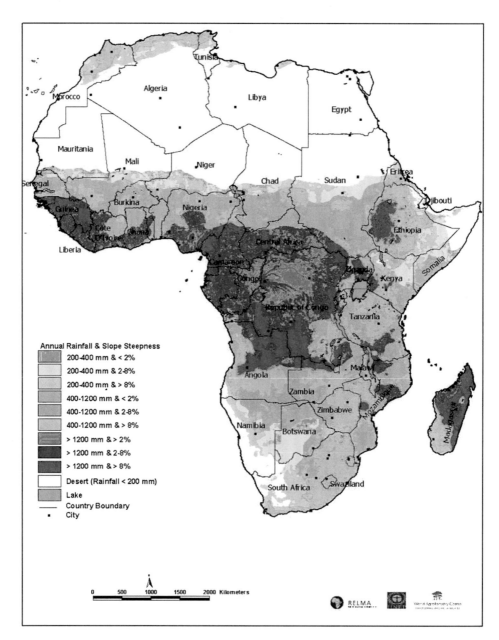

Figure 8.2 Relative potential of rainwater harvesting in Africa (Source: Malesu *et al.*, 2006) (See color plate section)

economic growth, attainment of increased agricultural productivity and food security and securing environmental sustainability. However, the policies on water and land management place greater emphasis on drinking water and sanitation, paying less attention to water for agricultural purposes (Mati 2006). A critical review of the various policies showed that there is no specific policy document that addresses RWH in its broad sense in any of the nine countries. Instead, existing policies had statements on AWM scattered across different ministries or sectors. The scattering of AWM issues across several sectors had resulted in unavoidable overlapping of policies, duplication of efforts, and inefficient use of resources as well as the lack of clear ownership of AWM issues. However, there is need to support various African countries in their efforts to fast-track their policy reforms and improve their infrastructural and institutional frameworks so as to make them responsive to RWH for smallholder agriculture and the changing demands for modernization of the sector.

8.4 WHY FOCUS ON RAINWATER HARVESTING?

Africa's water crisis can be greatly reduced if the vast potential of RWH is harnessed. RWH offers great potential for improving the productivity of rainfed agriculture, which accounts for the livelihoods of more than 80% of smallholders in SSA. However, the productivity of rainfed agriculture is low. For instance, the average maize yield in SSA is about $1\,t\,ha^{-1}$ during the last 30 years, while during the same period the yields in Asia have risen to $4\,t\,ha^{-1}$ and in USA the average yield is around $9\,t\,ha^{-1}$ (World Bank 2006). Yield gap analysis undertaken from different crops in Asia and Africa have showed that farmer crop yields are lower by two- to four-fold as compared to achievable potential yield (Wani et al., 2003, 2009; Rockström et al., 2007; Singh et al., 2009). Crop productivity is affected more by management aspects of water, soil, and agronomy than by the natural resource base (Place et al., 2005). Nutrient depletion is common in Africa and represents a significant loss of natural capital, valued at about US$1 to 3 billion annually. Moreover, the combination of low and erratic rainfall, high evaporation losses compounded by rudimentary agronomic management results in most of the rainfall getting lost as surface runoff and evaporation. There are many causes for this very low performance but the most critical is the variability in the availability of soil moisture for crop use. It is well documented that in typical farms, especially when only the hand hoe is used for tillage, only a small fraction of the rain falling on the field reaches and remains in the soil long enough to be useful to crops while up to 70% can be lost as unproductive evapotranspiration and runoff. So, the poor crop yields and crop failures are not so much the result of low rainfall but due to wastage of valuable rainwater. The effects of this poor "rainfall capture" is linked to the recurrent floods coupled with long periods of water shortage and droughts experienced over much of the continent. Consequently, only 19.7% of the rainfall in SSA becomes available water for crops (ECA 1995) and thus, the experienced water scarcity is not absolute but rather "infrastructural scarcity".

In the context of agriculture, RWH is designed to make more of the rainwater available for use by crops. RWH minimizes the risk of crop failure during droughts, intra-seasonal droughts, and floods. RWH for agriculture is therefore, not an alternative to irrigation, but rather a means of making more water available from the same rainfall for different uses (including irrigation) in agriculture. Also, RWH and storage if well managed can increase food production and farm incomes (Reij et al., 1996; Mati et al., 2005). Barron and Rockström (2003) observed that maize yield can be tripled with RWH through conservation agriculture. Similarly, Mulinge et al. (2010)

noted that supplemental irrigation with harvested rainfall could bridge the prolonged dry spells in Lare, Kenya. They obtained that supplemental irrigation using the harvested water increased yields of onions from $1.6\,t\,ha^{-1}$ to $11.9\,t\,ha^{-1}$ and for kales, from $6.4\,t\,ha^{-1}$ to $15.8\,t\,ha^{-1}$. Therefore, better control and management of rainwater is a win-win solution which helps to increase water available for different uses while reducing the negative consequences caused by surface runoff.

Attempts by smallholder farmers to fully engage RWH best practices are constrained by the disconnection between available knowledge, its applicability and adoption, as well as lack of capital and support from value chains (Chisenga and Teeluck 2006). The private sector, government, non-government organizations (NGOs), and other development agencies have provided sporadic support, which at times is inadequate. Consequently, water management projects over the past three decades have performed poorly in many countries. In particular, several communal irrigation schemes have performed below their potential (Kauffman *et al.*, 2003; Inocencio *et al.*, 2007). In addressing these constraints, farmers apply a vast array of low-input, ecologically friendly agricultural technologies. Recent reviews have cataloged numerous technologies and approaches on AWM (Reij *et al.*, 1996; Hatibu *et al.*, 2006; Mati 2007). Most assessments of these technologies register substantial increase in farmer yields and incomes, often exceeding double those achieved by conventional methods (Reij and Waters-Bayer 2001; Penning de Vries *et al.*, 2005). However, in SSA, most of the reporting on agricultural water issues has been negative, yet so much has changed over the last few decades. Positive data that combines both the technical and socioeconomic components is relatively scarce.

There is need for data to quantify AWM interventions which have worked at the project scale, in order to guide policy and investment decisions in the region. Some recent work on this aspect has yielded important pointers to the fact that investments in RWH are achieving substantial gains in crop productivity and incomes to smallholder farmers. Mati (2010) identified 12 broad categories of AWM practices which improved the productivity and profitability of smallholder agriculture and four of them were RWH-based. In nearly all the cases, water management with technology increased crop yield levels by factors ranging from 20% to over 500%, while net returns on investment increased up to ten-fold. Also, it was observed that these gains were linked to poverty reduction, employment creation, and environmental conservation.

8.5 OPTIONS FOR RAINWATER HARVESTING

It is obvious that RWH constitutes a key ingredient to the success of the agricultural productivity in SSA. Unlike the Asian green revolution which found in place a relatively well developed water management infrastructure, the African agricultural sector is predominantly rainfed, even in ecological zones which by necessity should be fully or partially irrigated. Again, unlike fertilizers and improved seeds, which can be imported, purchased, and distributed in small packages to farmers, water has to be sourced locally, is niche relevant, and requires higher technological input. RWH services such as surveying, design, layout, and construction of water supply and application equipment require some level of specialized expertise and cost. This does not imply that the farmer is completely helpless. There are many RWH technologies for which the farmer can survey, layout, and construct using own labor at the farm level with minimum training and facilitation. Mati (2007) listed 100 technologies/practices but even that list is not exhaustive. Some common interventions which include individual technologies/practices, are: (i) RWH from surface runoff and storage in ponds, pans, and underground tanks (blue water); (ii) Rooftop RWH (blue water); (iii) Small earth dams and weirs (blue water); (iv) Flood flow storages and sand/subsurface dams (blue water); (v) Runoff harvesting and storage in soil profile (green water); (vi) Spateflow diversions for supplemental irrigation (blue water); (vii) In-situ RWH (soil and water conservation); (viii) Conservation tillage systems for rainwater conservation; (ix) RWH for livestock in dry areas (drinking water, fodder); (x) Improving efficiencies, e.g., RWH with micro-irrigation; and (xi) Socioeconomic issues impacting on RWH.

8.5.1 RWH from surface runoff and storage in ponds, pans, and tanks (blue water)

RWH is used to provide drinking water for people and livestock, as well as for supplemental irrigation to drought-proof crop production in the dry areas. In SSA, some examples include: (i) rooftop RWH and storage in either surface or underground tanks (Gould and Nissen-Peterssen 1999); (ii) runoff harvesting from open surfaces and paths, roads, rocks, and storage in structures such as ponds or underground tanks (Guleid 2002; Nega and Kimeu 2002; Mati 2005); (iii) flood flow harvesting from valleys, gullies, ephemeral streams and its storage in ponds, weirs, and small dams; and (iv) flood flow harvesting from ephemeral water courses and its storage within sand formations as subsurface or sand dams (Nissen-Peterssen 2000, Mati 2007). Pans and ponds are particularly popular in community scale projects (Figure 8.3). They are excavated where the catchment is appropriate and have been used for RWH in many parts of SSA especially for livestock watering. The cost of construction can be relatively low since these structures utilize local knowledge in site selection, and community labor through such programs such as food for work thereby reducing the costs even further (Natea 2002; Mati and Penning de Vries 2005). The main difference between ponds and pans is that ponds receive some groundwater contribution, while pans rely solely on surface runoff. Thus, pans which range in size from about 5,000 to 50,000 m^3 are constructed almost anywhere as long as physical and soil properties permit. When properly designed and with good sedimentation basins, the water collected can be used for livestock watering or to supplemental irrigation of crops.

One case study (Mulinge *et al.*, 2010) is drawn from Lare Division of Nakuru District of Kenya. Lare is a dry area with seasonal effective rainfall for the two crop growing seasons; the long rains receiving 145 mm and the short rains about 100 mm respectively, amounts which are inadequate for crop production. The area has no perennial rivers. Thus, before RWH was introduced, water shortages were common, while irrigation was not feasible. Water harvesting was introduced in 1998 with community training, using modules designed to empower farmers to be able to do their own site selection, calculate water storage capacities, construct and maintain water pans, and how to use the water for supplemental irrigation of crops suited to respective conditions. Starting with about 409 household pans in 1998, adoption rates were high, reaching over 2,000 households by 2004. Generally, the system involves harvesting runoff from roads, footpaths, home compounds, and other surfaces and its

Figure 8.3 An excavated pan for holding runoff in Koibatek, Kenya

storage in manually excavated earthen pans of capacities ranging from about $9\,m^3$ to $8,000\,m^3$. Most of the pans were not lined as the soils are naturally impermeable, hence little seepage losses are experienced. The pans get filled within a few rainfall days and retain water throughout the year. The water is used for supplemental irrigation mainly of kales, cabbages, tomatoes, onions, carrots, and leeks as well as for livestock watering and domestic use. Due to the limited volumes of water available, the average irrigated area under supplemental irrigation was $703\,m^2$. However, the average landholding is 1.9 ha per household most of which is still used to grow maize, beans, and potato under rainfed conditions. For water application from the pans, most farmers use bucket while selected use treadle pumps. Gross margins calculated from the costs and returns from crops grown in the farms were higher under supplementary irrigation with water harvesting than under rainfed (Figure 8.4). Negative gross margin ($-US\$116\,ha^{-1}$) was obtained from the returns of kales under rainfed conditions. The increment in gross margin with supplemental irrigation was US\$110, 625, 1,428, and $4,603\,ha^{-1}$ for cabbage, kales, tomatoes, and onions, respectively. This increase was attributed to additional dry season cultivation, increased use of purchased inputs, and higher quality produce capable of fetching a better price at market.

Another success case is that of Minjar Shenkora, Ethiopia (Akalu and Adgo 2010) in which water harvesting with storage in individually owned plastic-lined ponds achieved high adoption rates with resultant improvements in agricultural productivity, incomes, and livelihoods. The area faces prolonged dry spells, food insecurity, lack of potable drinking water within realistic distance from homesteads, and heavy burden on women who fetch water. The runoff harvesting and storage in ground-lined tanks was introduced in the area in 2004 starting with 308 households and by 2008, over 7,618 households had adopted the technology. The ponds are of standard size and design,

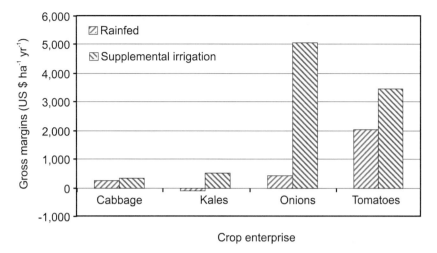

Figure 8.4 Gross margins for vegetables comparing conventional and supplemental irrigation from water harvesting at Lare, Kenya (Source: Mulinge *et al.*, 2010)

Figure 8.5 Plastic-lined pond for rainwater harvesting in Minjar Shekora, Ethiopia

usually trapezoidal in shape, measuring 8 m × 8 m at the top, 5 m × 5 m at the bottom and 3 m in depth, giving a total storage volume of about 102 m³ (Figure 8.5). The plastic lining or geo-membranes are factory-manufactured with standard shape and size to fit these dimensions. The cost of the geo-membranes was subsidized up to 85% by the government, thus making them affordable to the poor. Participatory approaches were used to identify beneficiary farmers, site selection and layout of ponds, with the assistance of technical staff. However, the excavation was done by the farmers and checked by technical staff for compliance. The farmers would then purchase the geo-membrane, install it, and ensure safe use and management of the pond. This water

Figure 8.6   Roof catchment with semi-underground tank at Mwembe, Same District, Tanzania

is used for supplemental irrigation of seedlings and vegetables, as well as for domestic use and livestock watering. Growing of fruit trees has been made possible due to availability of the seedlings. The harvested water is used to irrigate onion seedlings, making them available for planting at the onset of the rains. The area has become a major source of marketable onions.

8.5.2   Rooftop rainwater harvesting

Rainwater harvesting from impervious roofs (clay tiles and galvanized iron roofs) is one of the easiest ways of providing drinking water at the household level (Figure 8.6). It is a popular method adopted to secure water for domestic use, because it provides water at home, is affordable, easy to practice regardless of the physical or climatic conditions, and can be designed to suit different conditions (available finances, roof area, family size). It reduces women's burden of collecting water for domestic use, allowing them time for other productive activities (RELMA 2007). Since the structure is family owned, maintenance is usually very good and no water conflicts occur. Surface tanks may vary in size from 1 m^3 to more than 40 m^3 for households and up to 100 m^3 or more for schools and hospitals. The tank size is dependent on the rainfall regime and the demand. Areas with seasonal rainfall will require larger tanks (25 m^3 to 35 m^3) and a roof probably exceeding 100 m^2 would be required if total household water demand is to be met throughout the dry period. Another benefit of surface tanks (compared to subsurface tanks) is that water can be extracted easily through a tap just above the base of the tank. If placing it on a stand or base elevates the tank, water can be piped by gravity to where it is required. In addition, construction of such water tanks makes use of locally available materials and local artisans, thus creating employment (Gould and Nissen-Peterssen 1999).

Generally, an area receiving just 200 mm annual rainfall has as much potential (and higher priorities) as one receiving 2,000 mm for rural households in Africa. For example, simple arithmetic assuming per capita rural water consumption at 20 L day^{-1} shows an annual water demand of 7.3 m^3 per person per year, which could be supplied by a roof catchment of 36.5 m^2, if only 200 mm of rainfall per annum were available. Therefore, all that is required is the presence of roofs to provide the necessary catchments. In countries where settlements have been mapped such as Kenya, it is possible to show where rooftop RWH can be targeted. High rainfall areas do not necessarily preclude the need for rooftop RWH because of poor levels of development, and neither do low population areas, where scattered settlements may mean centralized piped systems are uneconomically viable. However, the main reason this technology has not been widely adopted is the relatively high costs involved (by local standards). On an average, the per capita daily water requirement in rural areas is 20 L. Using Kenyan estimates (Mati 2005), the cost of tank construction per capita is about US$150 (equivalent to about US$0.07 L^{-1}); as a tank can last up to 30 years or more, the investment is considered cost-effective. Another problem has been structural failure, especially of concrete built tanks due to various reasons such as use of low cement, aggregate mixes, poor quality sand, bad workmanship, poor curing process during construction, and generally poor management (e.g., some families drain the tank completely dry). However, well constructed and maintained surface tanks provide a durable and long-lasting source of clean water for households, schools, and communities.

8.5.3   Small earth dams and weirs

A lot of rainwater gets lost from streams and rivers during the rainy season, especially in dry areas prone to flush floods. When larger quantities of water are desired, earthen dams are preferred. An earthen dam is constructed either on-stream or off-stream, where there is a source of large quantities of channel flow. The dam wall is normally at least 2 m to 5 m high and has a clay core and stone aprons and spillways to discharge excess runoff (Figure 8.7). Volume of water ranges from hundreds to tens of thousands of cubic meters. Reservoirs with a water volume less than 5,000 m^3 are usually called ponds. Due to the high costs of construction, earthen dams are usually constructed through donor-funded projects. However, there have been cases of smallholder farmers digging earthen dams manually in Mwingi District in Kenya (Mburu 2000). Earthen dams can provide adequate water for irrigation projects as well as for livestock watering. Low earthen dams, called "*malambo*", are common in the Dodoma, Shinyanga, and Pwani regions of Tanzania (Hatibu *et al.*, 2000). It involves dam construction to collect water from less than 20 km^2 for a steep catchment to 70 km^2 for flat catchment. Some of these are medium-scale reservoirs used for urban or irrigation water supply. Sediment traps and delivery wells may help to improve water quality but, as with water from earthen dams, it is usually not suitable for drinking without being subject to treatment.

8.5.4   Sand and subsurface dams

The semi-arid zones of SSA are criss-crossed by seasonal rivers, which carry a lot of sand and hence the term "sand river". Sand rivers (*lugga, wadi,* and *khor*) comprise

Figure 8.7 A small weir across a dry valley conserves rainwater in Machakos, Kenya

ephemeral water courses, which tend to be dry most of the year (Nissen-Peterssen 2000). However, they are subject to flooding during the rainy season, providing an opportunity for RWH. The construction of an embankment across the sand river can store the water in the voids within the sand. The most convenient way to harvest water in a sand river is by either sand or subsurface dams. Local materials for construction are usually available and the only extra cost is that of cement and labor. Local people are usually trained on how to identify a suitable site and to construct the dams. It is also a cost-effective method for providing water for drinking and also for irrigation. For instance, subsurface dams in Machakos District of Kenya cost the community about US$0.20–0.30 per m^3 of water (Nissen-Peterssen 2000), but these costs are easily recoverable in the long run. Because the water is stored under the sand, it is protected from significant evaporation losses and is also less liable to contamination. Another advantage of sand river storage is that it normally represents an upgrading of a traditional and, hence, socially acceptable water source. Nissen-Peterssen (2000) distinguished between three types of subsurface dams: (i) sand dam built of masonry, (ii) subsurface dams built of stone masonry, and (iii) subsurface dams built of clay. The construction of river intakes and hand-dug wells with hand pumps in the river bank can further help to improve the quality of water.

8.5.5   Runoff harvesting and storage in soil profile

Runoff from land, roads, and paved areas is harvested and channeled into specially treated farmlands for storage within the soil profile for crop production. The cropped area may be prepared as planting pits, basins, ditches, bunded basins (*majaluba*), semi-circular basins (demi-lunes), or simply plowed land (Hai 1998; Ngigi 2003; Mati

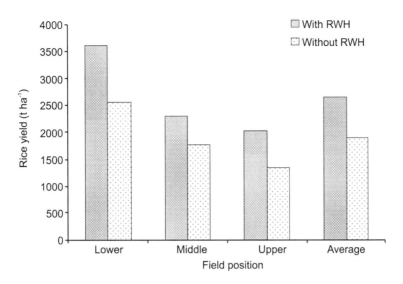

Figure 8.8 Paddy rice yields from bunded basins with and without runoff or rainwater harvesting (RWH) in Shinyanga, Tanzania (Source: Kajiru and Nkuba 2010)

2005). Storing rainwater in the soil profile for crop production is sometimes referred to as "green water" and forms a very important component for agricultural production. The design of a run-on facility (e.g., semicircular bund, *negarim*, zai pit) depends on many factors including catchment area, volume of runoff expected, type of crop, soil depth, and availability of labor (Hatibu and Mahoo 2000). The source of water could be small areas or "micro-catchments" or larger areas such as external catchments. The latter involves runoff diversion from larger external catchments such as roads, gullies, open fields into micro-basins for crops, ditches, or fields (with storage in soil profile) including paddy production where the profile can hold water relatively well.

In Tanzania, farmers make excavated bunded basins, locally known as "*majaluba*" which hold rainwater for the crop season. This system is practiced in the semi-arid areas where rainfall amounts range from 400 to 800 mm per year. About 35% of the rice in the country is produced this way under smallholder individual farming in Shinyanga, Dodoma, Tabora, and the Lake Regions. In many cases, *majaluba* utilize direct rainfall, but sometimes, farmers combine the system with runoff harvesting from external catchments. Generally, rice yields are higher, attaining 3.43 t ha^{-1} with the use of harvested water for irrigation as compared to 2.17 t ha^{-1} obtained without supplemental irrigation (Figure 8.8). Thus water harvesting and recycling can enhance food security, reduce poverty, and improve labor productivity (Kajiru and Nkuba 2010). Examples for water harvesting include trapezoidal bunds, semicircular and contour bunds, planting pits, *negarims*, T-basins and various types of channeling and conservation of runoff (Critchley *et al.*, 1999; Duveskog 2001; Mati 2005). A popular method that combines soil fertility management with water harvesting is called "*tumbukiza*" pit, whereby crop-water is conserved in small basins. *Tumbukiza* pits are constructed by digging huge pits, which measure at least 0.6–0.9 m in diameter and with similar dimensions

Figure 8.9 Excavated bunded basin for rainfed rice at Mwasonge in Mwanza, Tanzania

in depth. A fruit tree or fodder crop, especially banana or napier grass, is usually grown in the pit (Mati 2007).

Water harvesting with supplemental irrigation has multiple benefits for farmers in areas with droughts and prolonged dry spells (Hatibu *et al.*, 2000). In Shinyanga region of Tanzania, farmers make excavated bunded basins locally known as *majaluba* (Figure 8.9). These are made to hold water harvested from surface runoff for crop production. About 35% of the rice in Tanzania smallholder individual farming is produced this way. The study found that runoff diversions with *majaluba* had increased paddy rice yields significantly from 2.17 t ha^{-1} under rainfed conditions to 3.43 t ha^{-1} with bunded basins *(majaluba)*, which amounted to 58% increase in yield. Meanwhile, in Shinyanga region of Tanzania where farmers grow rice with water harvesting using *majaluba*, household incomes increased by 67% from US$430 ha^{-1} without runoff harvesting to US$720 ha^{-1} with the technology. The study found that runoff harvesting should be encouraged and accompanied by use of inputs such as fertilizers and manures, improved seeds, good agronomic practices, value addition, and access to markets. The main constraint was that with or without runoff harvesting, the *majaluba* system is predominantly rainfed with water storage in the soil profile (green water). Consequently, climatic uncertainties and prolonged dry spells adversely affect the system. It was noted that improvements could be made to allow water harvesting with storage infrastructure so as to drought proof the system.

In Ethiopia, water harvesting and storage in small ponds for supplemental irrigation of vegetables and seedlings at Minjar Shenkora obtained average net incomes of US$155 per 100 m^2 plot from onion seedlings, while incomes from bulb onions grown in the field provided equivalent of US$1,848 ha^{-1}, adding up to US$2,003 ha^{-1}, from

Figure 8.10 Sunken beds for in-situ water conservation for vegetable production in Mbeere, Kenya

onion crop alone. Comparatively, net incomes from teff and wheat were US$523 ha^{-1} and US$525 ha^{-1}, respectively. Thus, water harvesting with small storage ponds could make major contributions to household incomes and rural poverty reduction. These results are consistent with the findings of Gezahegn Ayele *et al.* (2006) and Nega and Kimeu (2002) who assessed small-scale water harvesting technologies in Ethiopia and found that returns on investment were high.

8.5.6 In-situ water harvesting and conservation

The SSA region has large areas of agricultural land, where productivity is constrained by slope steepness and high runoff rates, causing soil erosion, loss of nutrients and water. Examples include the highlands of Ethiopia, Madagascar, Rwanda, Burundi, Kenya, and Tanzania. Conservation of water through various types of structures and agronomic management can improve water productivity by 50–100% (Mati 2010). Techniques like terracing, ditches, earth and stone bunds, sunken beds (Figure 8.10), and vegetative barriers are normally defined as soil and water conservation structures but are sometimes referred to as "in-situ RWH". They are primarily promoted to reduce soil erosion and are also used to improve rainfall infiltration and conservation in the soil profile. In-situ water harvesting means rainwater is conserved on the same area where it falls, whereas water harvesting systems involve a deliberate effort to transfer runoff water from a "catchment" to the desired area or storage structure. The actual technologies may include stone bunds, ditches, earth bunds, *fanya juu* terraces, vegetative strips, trash lines, vegetative barriers, bench terraces, hedges, and all types of terracing land.

In a case study of Anjenie watershed of Ethiopia, long-term terracing increased yields of teff, barley, and maize (Table 8.1). In contrast, cultivation on the steep

Table 8.1   Net profits from teff, barley, and maize crop with and without terraces in Anjenie watershed, Ethiopia

Crop/Treatment	Gross revenue (US$ ha⁻¹)	Total expenses[a] (US$ ha⁻¹)	Net profit (US$ ha⁻¹)
Teff			
Terraced	292.6	271.7	20.9
Un-terraced	144.1	256.3	−112.2
Barley			
Terraced	382.3	197.1	185.2
Un-terraced	98.5	139.6	−41.1
Maize			
Terraced	245.7	280.2	−34.5
Un-terraced	102.2	203.0	−100.8

[a]Total expenses excluding family labor.

Figure 8.11   Radical (steep back-slope bench) terraces in Gichumbi District, Rwanda

un-terraced hillsides had negative gross margins. This explains part of the reason small-holder farmers have remained poor because they base their decisions on crop yields even when very low. These results are in line with the findings of Vancampenhout *et al.* (2006) on the positive effects of soil conservation on the yields of field crops in the highlands of Ethiopia.

In another example, terracing in Buberuka region of Rwanda has been very effective in controlling soil erosion on very steep slopes (Figure 8.11). Reverse-slope bench terraces were constructed in a method locally known as "radical terracing". On the more gentle slopes, the hedgerow system with trees and grasses was planted along the contour in a method referred to as "progressive terracing". The project, which was

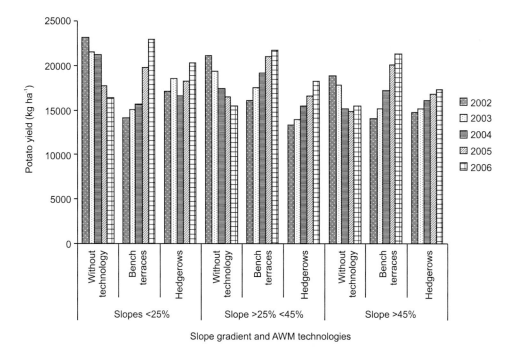

Figure 8.12 Potato yields during five years on bench terraces and hedgerows at different slope gradients in Buberuka region, Rwanda (Source: Kagabo and Nsabimana 2010)

completed in 2004 brought 953 ha under radical terracing and 1,350 km of progressive contour hedges. In addition, 3,000 people were trained in the construction of bench terraces.

The project had 18,600 households considered as direct beneficiaries. From 2002, the year bench terracing was established, potato yields increased from $14\,t\,ha^{-1}$ in 2002 to $23\,t\,ha^{-1}$ by 2006. Inversely, on sloping lands without conservation measures, potato production decreased steadily from year to year regardless of the steepness of the land or field due to soil erosion (Figure 8.12). It was noted that beyond 45% of slope steepness, there was little variation in potato production with time on farms with hedgerows as conservation measures. The initial investment for constructing radical terraces was quite high, costing about US$725 ha^{-1} in 2002, hence the need for project support to poor farmers. However, the returns on investment have been high, achieving an average income of US$1,715 ha^{-1} from potato crop alone (Kagabo and Nsabimana 2010).

8.5.7 Spateflow diversion and utilization

Spate irrigation or floodwater diversions involves techniques in which flood water is used for supplemental irrigation of crops grown in low-lying lands, sometimes far way from the source of runoff. Spate irrigation has a long history in the Horn of Africa, and still forms the livelihood base for rural communities in the arid parts of Eritrea,

Figure 8.13 Farmers divert spateflow for supplemental irrigation of crops at Makanya, Tanzania (See color plate section)

Ethiopia, Kenya, Somalia, and Sudan (Critchley *et al.*, 1992; Negassi *et al.*, 2000; SIWI 2001). It is also practiced in other dry areas, for instance, in Tanzania, spate irrigation increased rice yield from 1 to $4\,t\,ha^{-1}$ using RWH systems (*majaluba*) (Gallet *et al.*, 1996). Although spateflow irrigation has high maintenance requirement, its applicability is valid for large areas of the Sahel and the Horn of Africa, where other conventional irrigation methods may not be feasible. One example is drawn from Tanzania's Makanya village (Figure 8.13) in Same District where annual rainfall is less than 500 mm, and inadequate for crop production (Mati *et al.*, 2008). Here, spateflow emanates from the Pare mountains which lie about 16 km above the dry lowlands, with the runoff finding its way to Makanya through gullies and ephemeral streams. For years the local people at Makanya suffered from food insecurity and were reliant on sisal estates for casual labor. The diversion and utilization of spateflow from the Makanya gully was started in 1958 and steadily by 2008 about 2,228 farmers had put about 680 ha under supplemental irrigation using the spateflow, to grow crops like maize, beans, lablab, vegetables, and cotton. In another example from North Kordofan in the Sudan (Mati *et al.*, 2008) in an area at the edge of the Sahara desert, water scarcity is a major problem as rainfall is low, ranging from 150 to $250\,mm\,yr^{-1}$. Spateflow diversions are used to grow crops and improved productivity has been observed especially of sorghum crop.

8.5.8　Conservation agriculture and RWH

Conservation tillage, an essential component of conservation agriculture (CA), constitutes land cultivation techniques which try to reduce labor, promote soil fertility,

Figure 8.14 A farmer prepares tied ridges enriched with manure in Kitui, Kenya

and enhance soil moisture conservation. Conservation agriculture is now recognized as the missing link between sustainable soil management and reduced cost of labor, especially during land preparation, and holds the potential to increase crop production, and reduce soil erosion. It can be implemented through: (i) minimum or zero soil turning, (ii) permanent soil cover, (iii) stubble mulch tillage, and (iv) crop selection and rotations. It may also include pot-holing, infiltration pits, strip tillage, or tied ridging (Figure 8.14). It has been gaining acceptance in countries such as Tanzania, Madagascar, Zambia, and Zimbabwe in Africa (Biamah *et al.*, 2000; Nyagumbo 2000).

One example of application of CA is drawn from Tanzania's Bukoba and Missenyi districts of Kagera region (Kajiru and Nkuba 2010) where mulching practiced by smallholder farmers increased the average maize yields from 2.50 t ha^{-1} to 3.40 t ha^{-1} (Figure 8.15). However, the increase in yield varied with zone, and was the highest in the low rainfall zone (170%) and the lowest in medium rainfall zone (13%). Tanzania has been fostering the adoption of CA because of its potential to address three areas of crucial importance to smallholder farmers, i.e., demand on household labor, food security through increased and sustainable crop yields and household income (Mariki 2004; Lofstrand 2005). For best results, CA practices such as mulching must be accompanied by requisite agronomic practices such as use of fertilizers, manures, pesticides, and high quality seed, as well as proper water application and management.

Another example is from the Alaotra Lake Region of eastern-central Madagascar (Rakotondralambo *et al.*, 2010). The study area includes the districts of Ambatondrazaka, Amparafaravola and Andilamena, and it covers vast cascading plateaux with altitudes that vary from 750 to 1260 m asl. The climate of Alaotra is temperate-tropical with rainy season from November to April, and monthly rainfall ranging from 100

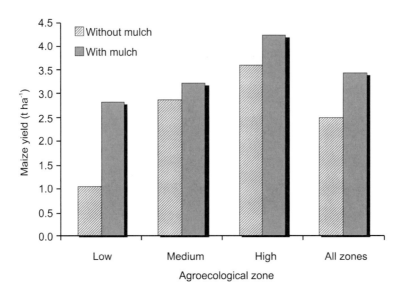

Figure 8.15 Maize yields with and without mulch in Missenyi district of Tanzania (Source: Kajiru and Nkuba 2010)

to 350 mm. However, the rainfall distribution is erratic and average maximum temperature is 26°C; and the main climatic factor limiting agricultural production is the long dry season. Alaotra basin is one of the biggest rice-producing zones in Madagascar. Annual production can be up to 200,000 t, although traditional land management techniques prevail including plowing, stubble mulch planting, and zero tillage. In the lowlands, the main activity is rice production, which is rotated with dry cropping of cassava and sweet potato and extensive cattle breeding under zero-grazing. Maize and beans occupy 5,905 ha and 7,430 ha respectively. The main innovation in Alaotra Lake Region is cropping systems based on direct seeding with permanent soil cover (Figure 8.16). This has helped improve soil condition especially in the rice fields with poor water control, by enhancing water conservation. For instance during the 2006–07 cropping season, 1,401 ha of land was under permanent CA. Meanwhile, the research continued on CA by working with farmers especially for monitoring the impacts of the introduction of new rice varieties in paddy fields with poor water control and of cassava and *Brachiaria* sp. intercropping system on degraded hillsides.

8.5.9 Soil fertility management in supporting RWH efforts

Throughout the SSA region, there are many projects that have been implemented with soil fertility at the helm. A well-known success is that of the fertilizer subsidies in Malawi. In the recent past, Malawi has made headlines for the increased production of staple maize, which turned the country from food deficit to a net food exporter in less than five years. The farmers were trained to use both fertilizers and manures for achieving high crop yields. Combined with organic manures and improved seeds,

Figure 8.16 Permanent soil cover (left); and relay intercropping (right) in Alaotra, Madagascar

the yields of maize and other crops have increased, thereby improving the availability of food (Mati *et al.*, 2008). Nearly all the other agronomic practices discussed above require a soil fertility management component, but the problem of declining soil fertility persists, in part due to the high costs of fertilizers. Organic materials, such as residues and manures though commonly used, are limited especially in agro-pastoral systems, due to competition with livestock feed. One of the most urgent requirements for improving dryland agriculture in SSA is to make fertilizers affordable, especially to the poor. This has partly been the secret to Malawi's food success, and was also credited with the success of the green revolution in Asia. The same should be extended to other countries of SSA (Tabo *et al.*, 2006).

8.5.10 Water for livestock

Water for livestock sometimes gets ignored in developmental projects targeting crop-based systems. It is quite common to see livestock walk very long distances in search of water; livestock water need within an area should be part of and benefit from soil and water conservation initiatives. By providing livestock with water close to home, time and resources wasted in searching for water can be saved, and livestock productivity enhanced. Water shortage for livestock is particularly acute in the Horn of Africa countries covering Eritrea, Ethiopia, Kenya, Sudan, Somalia, as well as in many Sahelian countries, and this is associated with frequent drought occurrence. For instance, in a study covering 160 pastoralists in different locations of northern Kenya, drought was perceived by 94% of the sampled population to be the principal livelihood challenge (Nyamwaro *et al.*, 2006). In contrast, the lack of pasture was listed as a challenge by 63% of the respondents. Thus, to enhance drought resilience

Figure 8.17   Livestock water from small earthen dam at Rukoki valley, Kayonza District, Rwanda

for livestock, the provision of water for both drinking and fodder production tops the list of interventions required.

In the dry lowlands of Makanya village in Same District of Tanzania, excavated pans or ponds locally known as "*Charco*" dams are constructed on selected sites located on relatively flat topography for use for livestock watering. They are small rectangular pans constructed by hand or by machinery with depth of up to 3 m. The design is simple and can be implemented at village level with minimal engineering requirements. *Charco* dams receive their runoff mostly from outlying areas of a rangeland; thus contour bunds are constructed to divert runoff into the dam. *Charco* dams are also commonly found in Shinyanga, Dodoma, Arusha, Tabora, Singida, and Mwanza regions of Tanzania (Hatibu *et al.*, 2000). A similar technology is used in Rwanda for livestock watering in the rangelands. Since around 2001, the Umutara Community Development Programme excavated over 40 small earth dams locally known as "valley dams" to conserve runoff water in dry water courses for livestock watering (Figure 8.17). The system is common in the relatively dry Kayonza District of Rwanda. The survival and health of livestock greatly improved, and the water storage has been especially useful for drought mitigation.

8.5.11   Socioeconomic issues in RWH

Another aspect of these interventions is their applicability for smallholder agriculture and micro-irrigation projects. The success of these interventions depends to a large extent on the operational framework against which they are implemented, especially the inclusion of the farmers in planning, implementation, and management of the systems. Other supportive aspects include implementation of AWM interventions as part of a more inclusive integrated watershed management, and thus the institutionalization of management structures such as water users associations (WUAs). Whenever

Table 8.2 Typical costs for selected rainwater harvesting technologies[a]

Technology	Typical example	Cost	Unit
Underground tanks	Concrete dome shaped tank	7	US$/m^3
	Brick dome shaped tank	9 to 14	US$/m^3
	Bottle shaped tank	4	US$/m^3
	Ferrocement tank	12 to 15	US$/m^3
	Ball shaped plastic tank	160	US$/m^3
Aboveground tanks	Brick tank	93	US$/m^3
	Ferrocement tank	30 to 70	US$/m^3
	Plastic tank	130	US$/m^3
Runoff open reservoirs	Plastic lined	3	US$/m^3
	Cement lined	5	US$/m^3
	Unlined	1.5	US$/m^3
	Lined oval tank	8	US$/m^3
Runoff closed reservoirs	Concrete dome shaped underground tank	7	US$/m^3
	Brick dome shaped underground tank	9 to 14	US$/m^3
	Bottle shaped underground tank	4	US$/m^3
	Ferrocement underground tank	13	US$/m^3
	Hemi spherical underground tank	23	US$/m^3
	Sausage shaped tank with cement lining	16	US$/m^3
In-situ	Human land preparation	113	US$/ha
	Draft animal power land preparation	53	US$/ha
Sand or subsurface dams	Sand dam	0.8	US$/m^3
	Subsurface dam	0.7	US$/m^3
Rock catchments	Open rock dam with stone gutters	71	US$/m^3
	Closed rock dam with stone gutters	89	US$/m^3
	Open rock dam with tank	110	US$/m^3
	Rock catchment tank with stone gutters	46	US$/m^3
	Stone gutters	2	US$/m^3

[a]Source: Nissen-Peterssen (2000); Desta et al. (2005).
[b]Local material and labor can be provided by the community.

possible, AWM interventions should target to provide water for multiple purposes and enhance cost-effectiveness. Capacity building for all cadres of stakeholders and local ownership are necessary for success. Generally, the costs of construction of various water harvesting structures can be relatively high (Table 8.2). This is partly the reason that most RWH projects are implemented at community scale to take account of economies of scale. Although operating costs of RWH systems can be relatively low once constructed, maintenance of community-based systems faces many challenges especially if the water requires pumping. Generally, it is best to implement RWH systems that utilize gravity flow for water distribution, or smaller, household-based systems, where individual farmers are responsible for the operations and maintenance (Akalu and Adgo 2010; Mulinge et al., 2010).

The major threats to enhancing AWM in SSA include negative perceptions about the returns to investment from AWM including from irrigation. It is established that smallholder water management, especially where the farmer has some level of individual autonomy in decision-making, is highly profitable as well as sustainable. Another constraint is the high initial investment required, as sometimes, supporting infrastructure such as roads, stores, and processing facilities, may have to be constructed first. Moreover, the poorest and most vulnerable communities tend to be located in the driest and remotest (far from roads, towns) part of the country where transaction costs of any activity tend to be high. This therefore, poses a challenge as to where to allocate resources, especially given the slim chances of payback from such vulnerable groups. The trial and error tendency of farmers exposed to irrigation and/or water harvesting for the first time can lead to mistakes, which could discourage both the farmers and investors. However, even with these limitations, the benefits of managing water for agriculture optimally far outweigh the threats, especially as there is increased food security, wealth creation, poverty reduction, and improved livelihoods for beneficiaries.

8.6 CONCLUSIONS

The development of RWH technologies and practices from the technical point of view is an aspect of AWM that has remained poorly funded over a long period. Also missing has been serious assessment of economic returns of the various approaches under different conditions, so that choice of interventions can be more evidence-based and strategic. Because the best returns will come from improved capture and use of rainwater where it falls, one under-funded technical opportunity is the integration of soils mapping, climate data, and agroecological mapping to assess potentials for enhancing soil water available for plant growth (the 'green' water approach). This together with the economic assessment should be followed-up for the development of robust strategies for integrated management of soil water and soil health in response to market demands. Another technical quick-win is the integration of RWH with rural infrastructure development. Potential exists for combining RWH, supplemental irrigation, improved drainage and strategic storage of water, designed to avoid failure of rural and irrigation infrastructure while increasing water availability for domestic, livestock, and irrigation needs.

However, critical attention is required in policy framework to promote and support private sector investments. For example, CA requires the availability of advanced implements for planting with no-tillage as well as chemicals for weed control. Availability of these inputs requires integrated approaches to the manufacturing and/or importation, supply, and operation by the private sector, which must be underpinned by public policy. To be effective, scalable, and sustainable, interventions designed to promote RWH must deal with policy, economic, and technical issues in an integrated manner and should use the evaluation of past investments as a starting point. It should be understood that factors that limit or facilitate the scaling-up and effective utilization of RWH go beyond the limited realm of agricultural water experts and institutions. Opportunities for RWH could be enhanced through leveraging public–private sector partnerships and inclusion of infrastructural components for RWH systems.

8.7 WAY FORWARD

It is evident that RWH is sustainable but the limited adoption currently evidenced in SSA calls for urgent re-think of the investment strategies of farmers, governments, and development organizations. IMAWESA (2007) has, through extensive stakeholder consultations, identified five entry points to promote AWM. They include: (i) exploiting the potential of rainfall, (ii) providing secure rights to access land and water, (iii) adoption of innovative financing for smallholder farmers, (iv) developing human capacity, and (v) harmonizing policies and institutions dealing with AWM. This chapter elaborates briefly on these entry points to highlight quick-win investment opportunities.

8.7.1 Support rainwater harvesting

There are as many RWH technologies and practices to choose from as there are different agroecological and socioeconomic needs (Hatibu and Mahoo 2000; Negassi *et al.*, 2000; Oweis *et al.*, 2001; SIWI 2001; Reij and Waters-Bayer 2001; Mati 2005, 2007, 2010; IWMI 2006). From these options, some five broad systems can be identified to form the basis for extensive project-scale adaptation to meet different needs and conditions.

8.7.1.1 Optimizing rainwater harvesting (Integrated watershed management)

Agricultural productivity in SSA is constrained by loss of water as surface runoff, soil erosion, loss of nutrients, and poor investment in water management. Productivity can thus be increased through integrated watershed management that takes into account the full water budget for an area, as well as its use, output, and cost-benefit ratio. RWH is an important component of the system as it balances the losses while enhancing the gains. Thus, support should be directed to watershed management options which:

- Optimize infiltration – the main purpose being to reduce non-productive depletion of the rainwater through evaporation and runoff, while reducing erosion and increasing recharge of groundwater;
- Increase the water-holding capacity of soil within the root zone – to make most of the captured water available to plants;
- Ensure an efficient water uptake (i.e., high ratio of transpiration/evapotranspiration) by beneficial plants – achieved through appropriate agronomic and crop husbandry practices; and
- Optimize the productivity of water used by plants, in terms of value of products – through the choice of crops with sufficient demand in accessible markets.

8.7.1.2 Runoff harvesting, diversions, and storage in soil profile

This system is suitable in areas where crops are grown on soils with large storage capacities but the direct rainfall is not large enough to fill this capacity. Extra water is added from areas external to the field with crop using a "Tail-to-Mouth Approach to RWH", which integrates interventions that range from conservation of rainwater to supplementary irrigation. This practice ensures that investments in RWH and supplementary irrigation start by managing the field for optimum capture and utilization of rainwater that falls directly on the field (treating the "tail"), before applying additional water. The irrigation portion of the system is normally called Spate Irrigation

(Critchley *et al.*, 1992; Negassi *et al.*, 2000; SIWI 2001). In Tanzania, the use of this RWH system has led to an increase in the yield of rice from 1 to $4\,t\,ha^{-1}$.

8.7.1.3 Small individual water storages in ponds, pans, and tanks

There are situations where the nature of the soil profile and rainfall distribution would make using the soil as a storage medium, inadequate in meeting crop water requirements. Under these situations, RWH would be beneficial if the design includes some storage structures. These structures may include ponds, pans, and lined tanks. The main constraints for smallholders includes initial capital investment especially where there is need for pumping the stored water.

8.7.1.4 Medium-scale storage

This involves the impounding of the rainfall runoff through construction of weirs, sand dams, and other forms of medium size storage. It is interesting that many irrigation schemes in SSA often suffer from shortage of water because they are based on direct of-the-river diversion without storage. Thus, the storage of peak stream flows will reduce this problem which is a major contributor to the inefficiencies of the few irrigation works that are currently installed in SSA.

8.7.1.5 Rainwater harvesting for underground storages

This involves channeling rainfall runoff into recharge basins of underground water systems so that installed wells can yield longer into the dry season.

8.7.2 Provide secure rights to access land and water

From a smallholder's point of view, water cannot be treated separately from land – as linkages between the two are self-evident – land without water is of little use in a semi-arid climate as is access to water without land. Land provides the pathway to water for example by motivating smallholder farmers to invest with confidence in management practices and technologies that enable them to improve the capture and conservation of rainwater. Therefore, to promote and support adoption and investments in RWH means fully recognizing the significance of land and its influence on water governance so the real issue is one of 'land and water governance' and not just 'water governance'. In this case actions are required to improve security of access and use rights to land and water resources.

8.7.3 Adoption of innovative financing for smallholder farmers

Rainwater harvesting of any type requires investment that ranges from implements for CA to medium size storage reservoirs for supplementary irrigation. However, financing of these basic investments is usually limited and high levels of performance are demanded. Public sector funding is still the main player in financing of capital investment for AWM schemes. However, what is required from the public sector is to focus more on the promotion and development of rural financial markets that encourage innovative ways for smallholders to self-finance investments in RWH. To stimulate rural financial markets, strategic interventions are needed from the public sector to facilitate private investment. Action is required in the following aspects:

- The credit worthiness of farmers' needs to be raised through the evaluation, recognition, registration, and protection of land and other fixed assets.
- Facilitate investment in innovative low-cost credit schemes. For example, strengthening the legal framework that would help to improve the performance of savings and credit schemes, micro-financing as well as commercial lending.
- Make credit cheaper by reducing risks currently faced by lending institutions. One way is to increase repayment by linking production with markets such that investments in water management form an integral part of a robust market chain. Without such linkages the investments are unlikely to succeed.
- Ensure economic efficiency of investments by introducing and paying attention to economic efficiency in the selection of RWH systems for investment support.

8.7.4   Interactive capacity strengthening for RWH

RWH interventions have to be accompanied by targeted capacity building for farmers and other support agents, e.g., artisans, material suppliers, extension workers, and even researchers. Capacity building for RWH will target the following groups of actors: (i) Policy makers through awareness creation campaigns, learning exchanges across countries and implementation support; (ii) Managers and implementers of projects through hands-on training on aspects of RWH relevant to their activities, process documentation, and implementation support across projects; (iii) Researchers and other scientists through thorough academic programs that range from short-term training, MSc and PhD programs; (iv) Extension workers and other local leaders through hands-on training; (v) Farmers through actual practical training on farmers' fields; and (vi) Local-level institutions, e.g., WUAs so as to build strong institutions that provide sustainable management of watershed-scale interventions.

8.7.5   Enhance policy support

It is evident from the issues requiring attention in all the other four entry points that everything evolves around having the correct policy support. The current policy framework is dominated by a focus on irrigated agriculture based on direct extraction of water from rivers and underground systems with little attention to rainwater capture and storage. A study conducted by IMAWESA (2007) covering nine countries examined 78 policies that were deemed to have implications on AWM in the target countries. The study found that most of the policies that touch on water management pay only passing attention to water for agricultural purposes. More specifically, no country had a policy or strategy dedicated to RWH. Therefore, work is required to produce evidence-based support to policy-making processes at national, regional, and international levels. The focus should be on integration of policies, institutional frameworks, and investments to achieve synergies in RWH.

8.7.6   Recommendations

Many suggestions have been made regarding how to upscale RWH to bring about economic and environmental benefits to SSA. Technologies as well as financing are needed, but these alone are not sufficient to facilitate acceptability and buy-in across the stakeholders spectrum. More specifically, there is need to improve the following:

- In order to maximize the benefits of RWH in agriculture it should be accompanied by proper use of farm inputs such as fertilizers, improved varieties, and other good agronomic practices.
- Promote policy and institutional support/reforms for RWH (improve visibility of RWH in legislation, strategic plans, and activities).
- Invest in building human resource capacity across the value chain.
- Support research and development with strong participation of farmers.
- Mobilize financial/investment support to RWH (from local and international sources, private sector).
- Provide implementation support to existing RWH initiatives and grassroots organizations (e.g., WUAs).
- Create awareness across the stakeholder base to improve the acceptability and political clout of RWH.
- Support learning alliances to mobilize knowledge capture, sharing, and utilization within and across countries.
- Improve the business orientation to RWH innovative incentives (e.g., through public–private partnership).
- Address governance issues (security of tenure and access to water and land).
- Identify and implement viable and sustainable RWH projects with a clearly defined business plan particularly with respect to increasing farmers' incomes that are socially acceptable and environmentally sustainable.
- Investigate and support methods of storing runoff water (blue water) in sub-catchments in upland areas to mitigate the impact of droughts and prolonged dry spells.
- The accrued benefits from the use of runoff harvesting should be tapped to increase the farmers' income by facilitating access to markets and postharvest technologies.
- There is need for further research on hydrogeology of many parts of Eastern and Southern Africa to facilitate other methods of RWH, e.g., to recharge shallow wells.
- In some situations, it is difficult to implement runoff harvesting by individual farmers, thus organizing communities in groups and assisting them could make it possible to achieve better economies of scale.

REFERENCES

Akalu, T.F., and E.T. Adgo. 2010. Water harvesting with geo-membrane lined ponds: Impacts on household incomes and rural livelihoods in Minjar Shenkora district of Ethiopia. In *Agricultural water management interventions delivers returns on investment in Africa. A compendium of 18 case studies from six countries in Eastern and Southern Africa*, ed. B.M. Mati, 42–52. VDM Verlag.

Baron, J., and J. Rockström. 2003. *Water harvesting to upgrade smallholder farming: Experiences from on-farm research in Kenya and Burkina Faso*. Nairobi, Kenya: RELMA.

Biamah, E.K., J. Rockström, and J. Okwach. 2000. *Conservation tillage for dryland farming. Technological options and experiences in Eastern and Southern Africa*. Nairobi, Kenya: RELMA.

Chisenga, J.K., and J.P. Teeluck. 2006. Initial scoping of status of knowledge management in Kenya, Madagascar, Malawi, Mauritius and Tanzania. Mission Report. *SWMnet Working Paper 10*. Nairobi, Kenya: IMAWESA. (www.imawesa.net/publications/Working papers)

Critchley, W., R. Cooke, T. Jallow, J. Njoroge, V. Nyagah, and Saint-Firmin, E. 1999. *Promoting farmer innovation: Harnessing local environmental knowledge in East Africa*. Workshop Report No. 2. Nairobi, Kenya: UNDP-Office to Combat Desertification and Drought and RELMA.

Critchley, W., C. Reij, and A. Seznec. 1992. *Water harvesting for plant production. Volume II: Case studies and conclusions for sub-Saharan Africa*. World Bank Technical Paper Number 157. Africa Technical Department Series.

Darghouth, S. 2007. Reengaging in agricultural water management in Sub-Saharan Africa. Edition GSDM/CIRAD.

Desta, L., V. Carucci, A. Wendem-Agenehu, and A. Abebe. 2005. Community based participatory watershed development: a guideline, part 1. Addis Ababa, Ethiopia: Ministry of Agriculture and Rural Development.

Duveskog, D. 2001. *Water harvesting and soil moisture retention. A study guide for Farmer Field Schools*. Sida, Nairobi: Ministry of Agriculture and Farmesa.

ECA. 1995. *Water resources: Policy assessment, Report of UNECA/WMO*. Addis Ababa, Ethiopia: Economic Commission for Africa.

FAO. 2002. TERRASTAT. Global land resources GIS models and databases for poverty and food insecurity mapping (double CD-ROM). FAO Land and Water Digital Media Series.

FAO. 2003. AQUASTAT. Global information system of water and agriculture. Rome, Italy: FAO.

FAO. 2005. Food security and agricultural development in sub-Saharan Africa: Building a case for more public support. Main Report. Rome, Italy: FAO and NEPAD.

Gallet, L.A.G., N.K. Rajabu, and M.J. Magila. 1996. Rural poverty alleviation: Experience of SDPMA. Presented at a workshop on Approaches to Rural Poverty Alleviation in SADC Countries, Cape Town, South Africa, January 31 to February 3, 1996.

Gezahegn Ayele, Gemechu Ayana, Kiflu Gedefe, Mekonnen Bekele, Tilahun Hordofa, and Kdane Georgis. 2006. Water harvesting practices and impacts on livelihood outcomes in Ethiopia. Addis Ababa, Ethiopia: EDRI.

Global Risk Network. 2008. Africa@Risk. Geneva, Switzerland: World Economic Forum.

Gould, J., and E. Nissen-Peterssen. 1999. *Rainwater catchment systems for domestic supply: design, construction and implementation*. UK: Intermediate Technology Publications.

Guleid, A.A. 2002. Water-harvesting in the Somali National Regional State of Ethiopia. In *Workshop on the Experiences of Water Harvesting in Drylands of Ethiopia: Principles and Practices*, ed. M. Haile, and S.N. Merga, 45-49. DCG Report No. 19. Dryland Coordination Group.

Hai, M.T. 1998. *Water harvesting: an illustrative manual for development of microcatchment techniques for crop production in dry areas*. Technical Handbook No. 16. Nairobi, Kenya: RELMA.

Hatibu, N., and H.F. Mahoo. 2000. *Rainwater harvesting for natural resources management. A planning guide for Tanzania*. Technical Handbook No. 22. Nairobi, Kenya: RELMA.

Hatibu, N., K. Mutabazi, E.M. Senkondo, and A.S.K. Msangi. 2006. Economics of rainwater harvesting for crop enterprises in semi-arid areas of East Africa. *Agricultural Water Management* 80(1–3):74–86.

Hatibu, N., H.F. Mahoo, and G.J. Kajiru. 2000. The role of RWH in agriculture and natural resources management: From mitigating droughts to preventing floods. In *Rainwater harvesting for natural resources management. A planning guide for Tanzania*, eds. N. Hatibu and H.F. Mahoo, 58–83. Technical Handbook No. 22. Nairobi, Kenya: RELMA.

IMAWESA. 2007. *Agricultural water management, a critical factor in the reduction of poverty and hunger: Principles and recommendations for action to guide policy in Eastern and Southern Africa*. Regional synthesis report of the Policy Study on AWM in ESA. (www.asareca.org/swmnet/imawesa)

Inocencio, A., D. Merrey, M. Tonosaki, A. Maruyama, I. de Jong, and M. Kikuchi. 2007. *Costs and performance of irrigation projects: a comparison of Sub-Saharan Africa and other developing regions.* IWMI Research Report 109. Colombo, Sri Lanka: IWMI.

IWMI. 2006. *Agricultural water management technologies for small scale farmers in Southern Africa: An inventory and assessment of experiences, good practices and costs.* Final report produced by the International Water Management Institute (IWMI), Southern Africa Regional Office, Pretoria, South Africa. For Office of Foreign Disaster Assistance, Southern Africa Regional Office, United States Agency for International Development.

Kagabo, D.M., and J.D. Nsabimana. 2010. Radical terracing and hedgerow practices for improved potato production in Buberuka Region of Rwanda. In *Agricultural water management interventions bearing returns on investment in Eastern and Southern Africa.* Nairobi: Kenya: IMAWESA.

Kajiru, G.J., and J.M. Nkuba. 2010. Assessment of runoff harvesting with *'majaluba'* system for improved productivity of smallholder rice in Shinyanga, Tanzania. *Agricultural water management interventions delivers returns on investment in Africa. A compendium of 18 case studies from six countries in Eastern and Southern Africa,* ed. B.M. Mati, 65–75. VDM Verlag.

Kauffman, J.H., S. Mantel, J. Ringersma, J.A. Dijkshoom, G.W.J. van Lynden, and D.L. Dent. 2003. Making better use of green water under rain-fed agriculture in sub-Saharan Africa. In *Proceedings of the symposium and workshop on water conservation technologies for sustainable dryland agriculture in Sub-Saharan Africa* held at Bloemfontein, South Africa, 8–11 April 2003, 103–108.

Lenton, R., and M. Muller. 2009. *Integrated water resources management in practice: better water management for development.* Global Water Partnership. London, UK: Earthscan.

Lofstrand, F. 2005. *Conservation agriculture in Babati district, Tanzania. Impacts of conservation agriculture for small-scale farmers and methods for increasing soil fertility.* Master of Science Thesis, Swedish University of Agricultural Science, Department of Soil Science, Uppsala, Sweden.

Malesu, M., E. Khaka, B. Mati *et al.* 2006. *Mapping the potentials for rainwater harvesting technologies in Africa. A GIS overview of development domains for the continent and nine selected countries.* Technical Manual No. 7. Nairobi, Kenya: World Agroforestry Centre (ICRAF), Netherlands Ministry of Foreign Affairs.

Mariki, W.L. 2004. The impact of conservation tillage and cover crops on soil fertility and crop production in Karatu and Hanang Districts in Northern Tanzania. *TFSC/GTZ Technical Report 1999–2003.* Arusha, Tanzania: TFSC/GTZ.

Mati, B.M. 2005. *Overview of water and soil nutrient management under smallholder rain-fed agriculture in East Africa.* Working Paper 105. Colombo, Sri Lanka: International Water Management Institute. (www.iwmi.cgiar.org/pubs/working/WOR105.pdf)

Mati, B.M. 2006. Preliminary assessment of policies and institutional frameworks with bearing on agricultural water management in Eastern and Southern Africa. Report of a baseline study. *SWMnet Working Paper 11.* Nairobi, Kenya: IMAWESA.

Mati, B.M. 2007. *100 ways to manage water for smallholder agriculture in Eastern and Southern Africa.* SWMnet Proceedings 13. Nairobi, Kenya: IMAWESA. (www.asareca.org/swmnet/imawesa)

Mati, B.M., N. Hatibu, I.M.G. Phiri, and J.N. Nyanoti. 2007. Policies and institutional frameworks impacting on agricultural water management in Eastern and Southern Africa (ESA). *Synthesis report of a rapid appraisal covering nine countries in the ESA.* Nairobi, Kenya: IMAWESA.

Mati, B.M., and F.W.T. Penning de Vries. 2005. Bright spots on technology-driven change in smallholder irrigation. Case studies from Kenya. In *Bright spots demonstrate community successes in African agriculture. Working Paper 102*, 27–47. Colombo, Sri Lanka: International Water Management Institute.

Mati, B.M., D. Siame, and W.M. Mulinge. 2008. *Agricultural water management on the ground: Lessons from projects and programmes in Eastern and Southern Africa.* Experiences shared by 15 IFAD-funded programmes/projects in Eastern and Southern Africa. IMAWESA Working Paper 16. Nairobi, Kenya: IMAWESA.

Mati, B.M. 2010. Agricultural water management delivers returns on investment in Eastern and Southern Africa: A regional synthesis. In *Agricultural water management interventions delivers returns on investment in Africa. A compendium of 18 case studies from six countries in Eastern and Southern Africa*, ed. B.M. Mati, 1–29. VDM Verlag.

Mburu, C.N. 2000. Farmer innovators in Kenya. In *Farmer innovation in land husbandry.* Proceedings of Anglophone Regional Workshop, Feb 6–11, 2000, Mekelle, Tigray 68–70. Ethiopia, ed. M. Haile, A. Waters-Bayer, M. Lemma, 68–70.

Molden, D., K. Frenken, R. Barker *et al.* 2007. Trends in water and agricultural development. In *Water for food, water for life: a comprehensive assessment of water management in agriculture*, 57–89. London, UK: Earthscan; and Colombo, Sri Lanka: International Water Management Institute.

Mulinge, W.M., J.N. Thome, and F.M. Murithi. 2010. Water harvesting with small storage ponds: impacts on crop productivity and rural livelihoods in Lare, Kenya. In *Agricultural water management interventions delivers returns on investment in Africa. A compendium of 18 case studies from six countries in Eastern and Southern Africa*, ed. B.M. Mati, 128–139. VDM Verlag.

Natea, S. 2002. The experience of CARE Ethiopia in rainwater harvesting systems for domestic consumption. In W*orkshop on The Experiences of Water Harvesting in Drylands of Ethiopia: Principles and Practices.* ed. M. Haile, and S.N. Merga, 19–28. *DCG Report No. 19.* Dryland Coordination Group.

Nega, H., and P.M. Kimeu. 2002. Low-cost methods of rainwater storage. *Results from field trials in Ethiopia and Kenya.* Technical Report No. 28. Nairobi, Kenya: RELMA.

Negassi, A., B. Tengnas, E. Bein, and K. Gebru. 2000. *Soil conservation in Eritrea. Some case studies.* Technical Report No. 23. Nairobi, Kenya: RELMA.

NEPAD. 2003. *Comprehensive Africa Agriculture Development Program (CAADP).* Midrand, South Africa: NEPAD.

Ngigi, S.N. 2003. *Rainwater harvesting for improved food security. Promising technologies in the Greater Horn of Africa.* Nairobi, Kenya: Kenya Rainwater Association.

Nissen-Peterssen, E. 2000. *Water from sand rivers. A manual on site survey, design, construction and maintenance of seven types of water structures in riverbeds.* Technical Handbook No. 23. Nairobi, Kenya: RELMA.

Nyagumbo, I. 2000. Conservation technologies for smallholder farmers in Zimbabwe. In *Conservation tillage for dryland farming. Technological options and experiences in Eastern and Southern Africa*, ed. E.K. Biamah, J. Rockström, and J. Okwach, 70–86. Nairobi, Kenya: RELMA.

Nyamwaro, S.O., D.J. Watson, B. Mati *et al.* 2006. *Assessment of the impacts of the drought response program in the provision of emergency livestock and water interventions in preserving pastoral livelihoods in northern Kenya.* Report of an ILRI multidisciplinary scientific team of consultants assessing the emergency drought response project in northern Kenya. Nairobi, Kenya: ILRI.

Oweis, T., P. Prinz, and A. Hachum. 2001. *Water harvesting. Indigenous knowledge for the future of the drier environments.* Aleppo, Syria: International Centre for Agricultural Research in the Dry Areas.

Penning de Vries, F.W.T., B. Mati, G. Khisa, S. Omar, and M. Yonis. 2005. Lessons learned from community successes: A case for optimism. In *Bright spots demonstrate community successes in African agriculture. Working Paper 102*, 1–6. Colombo, Sri Lanka: International Water Management Institute.

Place, F., F.M. Muriithi, and C.B. Barret. 2005. A comparison and analysis of rural poverty between western and central Kenyan highlands. In *Rural markets, natural capital and dynamic poverty traps in East Africa*. Policy Brief No. 4.

Rakotondralambo, A., A.F.V. Ravelombonjy, and B.M. Mati. 2010. Performance analysis of cropping systems based on permanent soil cover in Alaotra Lake region, Madagascar. In *Agricultural water management interventions bearing returns on investment for smallholders in Eastern and Southern Africa*. Regional study. Nairobi, Kenya: IMAWESA.

Reij, C., I. Scoones, and C. Toulmin. (ed.) 1996. *Sustaining the soil. Indigenous soil and water conservation in Africa*. London, UK: Earthscan.

Reij, C., and A. Waters-Bayer. 2001. *Farmer innovation in Africa. A source of inspiration for agricultural development*. London, UK: Earthscan.

RELMA. 2007. *Good to the last drop: Capturing Africa's potential for rainwater harvesting*. RELMA Review Series Rainwater Harvesting in Africa, Issue 2. Nairobi, Kenya: RELMA.

Rockström, J., Nuhu Hatibu, Theib Oweis, and S.P. Wani. 2007. Managing water in rainfed agriculture. In *Water for food, water for life: a comprehensive assessment of water management in agriculture,* ed. D. Molden, 315–348. London, UK: Earthscan; and Colombo, Sri Lanka: International Water Management Institute.

Singh, P., P. Pathak, S.P. Wani, and K.L. Sahrawat. 2009. Integrated watershed management for increasing productivity and water-use efficiency in semi-arid tropical India. *Journal of Crop Improvement* 23(4):402–429.

SIWI. 2001. *Water harvesting for upgrading of rain-fed agriculture. Problem analysis and research needs*. Report II. Stockholm, Sweden: Stockholm International Water Institute.

Tabo, R., A. Bationo, K. Diallo Maimouna, O. Hassane, and S. Koala. 2006. Fertilizer microdosing for the prosperity of small-scale farmers in the Sahel: Final report. *Global Theme on Agroecosystems Report No. 23*. Niamey, Niger: International Crops Research Institute for the Semi-Arid Tropics.

Vancampenhout, K., J. Nyssen, D. Gebremichael *et al.* 2006. Stone bunds for soil conservation in the northern Ethiopian highlands: Impacts on soil fertility and crop yield. *Soil and Tillage Research* 90(1–2):1–15.

Wani, S.P., T.K. Sreedevi, P. Pathak *et al.* 2003. Minimizing land degradation and sustaining productivity by integrated water management: Adarsha Watershed, Kothapally, India. In *Integrated watershed management for land and water conservation and sustainable agricultural production in Asia,* Proceedings of the ADB-ICRISAT-IWMI Project Review and Planning Meeting, Hanoi, Vietnam, 10–14 Dec 2001, ed. S.P. Wani, A.R. Maglinao, A. Ramakrishna, and T.J. Rego, 79–96. Patancheru, Andhra Pradesh, India: International Crops Research Institute for the Semi-Arid Tropics.

Wani, S.P., J. Rockström, and T. Oweis. (ed.) 2009. *Rainfed agriculture: unlocking the potential*. Comprehensive Assessment of Water Management in Agriculture Series. Wallingford, UK: CAB International.

World Bank. 2006. *Reengaging in agricultural water management: challenges, opportunities and trade-offs*. Water for Food Team, Agriculture and Rural Development Department (ARD). Washington, DC, USA: World Bank.

Chapter 9

Management of emerging multinutrient deficiencies: A prerequisite for sustainable enhancement of rainfed agricultural productivity

K.L. Sahrawat, Suhas P. Wani, A. Subba Rao, and G. Pardhasaradhi

9.1 INTRODUCTION

Soil, water, vegetation, and production systems constitute the most important natural resources in an agroecosystem. In the rainfed production systems, the importance of water shortage and associated stress cannot be overemphasized especially in the semi-arid tropical (SAT) regions (Pathak *et al.*, 2009; Passioura and Angus 2010; Rockström *et al.*, 2010; Sahrawat *et al.*, 2010a; Sharma *et al.*, 2010). However, apart from water shortage, soil infertility is also the issue for crop production and productivity enhancement in much of the SAT regions of the world (El-Swaify *et al.*, 1985; Black 1993; Zougmore *et al.*, 2003; Sahrawat *et al.*, 2007, 2010a; Bationo *et al.*, 2008; Singh 2008; Twomlow *et al.*, 2008a; Bekunda *et al.*, 2010).

Apart from deficiencies of the major nutrients nitrogen (N) and phosphorus (P), the deficiencies of secondary nutrients especially of sulfur (S) and micronutrients have been reported with increasing frequencies from the intensified irrigated production systems (Kanwar 1972; Pasricha and Fox 1993; Takkar 1996; Scherer 2001, 2009; Fageria *et al.*, 2002; Singh 2008). While in the irrigated systems the deficiencies of various plant nutrients have been diagnosed through soil and plant testing and managed through the fertilization of crops, little attention has been paid to diagnosing the deficiencies of secondary nutrients such as S and micronutrients in dryland rainfed production systems (Sahrawat *et al.*, 2010a). In general, very little attention has been devoted to determine the fertility status of farmers' fields and hence to diagnose the nutrient problems in the rainfed production systems. Although, the information on the soil fertility status not only can help in enhancing crop productivity through balanced nutrient management, but also can promote judicious use of external inputs of nutrients (Wani 2008).

This apparent paradox of lack of application of adequate amounts of nutrients from external inputs (Katyal 2003; Bationo *et al.*, 2008) despite the common knowledge that the soil resource base in the rainfed systems of the SAT regions is relatively fragile and marginal compared to that under the irrigated production systems (El-Swaify *et al.*, 1985; Black 1993; Rego *et al.*, 2003; Sahrawat *et al.*, 2007, 2010a; Sharma *et al.*, 2009a, 2009b) is inexplicable. In the rainfed systems, water shortage has been the primary focus of research and developmental activities in these areas and soil infertility has largely been ignored (El-Swaify *et al.*, 1985; Wani *et al.*, 2003; Sahrawat *et al.*, 2010a, 2010b) or has not been addressed in an integrated manner (Wani *et al.*, 2002, 2009; Rockström *et al.*, 2007, 2010).

However, even in water-limiting environments there is potential to enhance agricultural productivity through efficient management of soil, water and nutrients in an integrated manner (Twomlow *et al.*, 2008a; Wani *et al.*, 2009; Sahrawat *et al.*, 2010a). To achieve the potential of productivity in water-limited environments, a concept of water-limited potential yield seems very appropriate as this forms the basis to reach the attainable yield in these environments by management of constraints other than just water shortage (Passioura 2006; Singh *et al.*, 2009). For example, in Australia, farmers have adopted the notion of water-limited potential yield as a benchmark for yield and if farmers find that their crops are performing below the benchmark, they look for the reasons and attempt to improve their management accordingly (Passioura and Angus 2010). We emphasize that in the concept of water-limited potential yield in the rainfed systems, natural resource management (NRM) in general and soil fertility management in particular need to be paid due attention alongside water stress management in view of the fragile nature of the soil resource base (Wani *et al.*, 2009; Sahrawat *et al.*, 2010a, 2010b).

Moreover, it is commonly believed that at relatively low yields of crops in the rainfed systems, the deficiencies of major nutrients, especially N and P are important for the SAT soils (El-Swaify *et al.*, 1985; Rego *et al.*, 2003; Sharma *et al.*, 2009a) and little attention has been devoted to diagnose the extent of deficiencies of the secondary nutrients such as S and micronutrients in various crop production systems (Sahrawat *et al.*, 2007, 2010a) on millions of small and marginal farmers' fields.

It is duly recognized and emphasized that the productivity of SAT soils is low due to water shortages. Although low fertility is also an issue, in practice the deficiencies of major nutrients (N and P) are considered important. Moreover, the input of major nutrients to dryland production systems is meager compared to that in the irrigated systems (Burford *et al.*, 1989; Rego *et al.*, 2005; Wani *et al.*, 2009). Also, due to low productivity of the rainfed crops, it is generally assumed that the mining of micronutrient reserves in soils is much less than in irrigated production systems (Rego *et al.*, 2003).

For sustained increase in dryland productivity, soil and water conservation measures need to be integrated with plant nutrition, and choice of crops and their management (Burford *et al.*, 1989; Wani *et al.*, 2003; Passioura 2006; Passioura and Angus 2010; Sahrawat *et al.*, 2010b). The on-going farmer participatory integrated watershed management program at ICRISAT (International Crops Research Institute for the Semi-Arid Tropics) provided the opportunity to implement nutrient management strategy with soil and water conservation practices in farmers' fields in the Indian semi-arid tropics. For efficient and judicious use of nutrients through fertilizer inputs, assessing the soil's inherent nutrient status is a prerequisite (Sahrawat 2006).

Therefore, in this chapter the literature on the general fertility status of soils in the rainfed systems is reviewed and analyzed with emphasis on the diagnosis and management of the deficiencies of secondary and micronutrients in the rainfed systems of the SAT regions. Preference has been given to the results reported from the on-farm research in the SAT regions. First, the results on the fertility status of SAT soils are dealt, followed by the response of various food crops to balanced nutrient management considering the various nutrient deficiencies under the on-farm conditions. The role of soil testing in the diagnosis of nutrient deficiencies is demonstrated and the importance of integrated approach in which both water shortage and multi-nutrient deficiencies are simultaneously addressed is emphasized for sustainable enhancement of crop production and productivity in the rainfed systems.

9.2   SOIL DEGRADATION – ORGANIC MATTER AND NUTRIENT STATUS OF SAT SOILS

For the purpose of this chapter, we define soil degradation as the decline or loss of soil functions to produce goods of value to humans; and undoubtedly, soil degradation is at the heart of stagnant productivity, perpetuation of hunger, and malnutrition, and environmental security loss (Lal 1997, 2007; Sanchez 2002; Bationo *et al.*, 2008; Stringer 2009; Bekunda *et al.*, 2010). Soil degradation entails loss of soil (including organic matter and nutrients therein) as well as deterioration in its physical, chemical, and biological properties, and is a major threat to the sustainability of the agricultural systems (Bationo *et al.*, 2008; Sahrawat *et al.*, 2010b).

Soil organic matter is critical to soil fertility and water cycle management in the agroecosystems and its importance cannot be overemphasized in the SAT regions where soils are marginal and water shortage is the major stress to production systems (Bossio *et al.*, 2007). The maintenance of soil organic matter at a threshold level, depending on the soil type and climatic factors, is critical for the physical, chemical, and biological integrity of the soil and for the soil to perform its agricultural productivity and environmental functions on a sustainable basis (Pathak *et al.*, 2005; Bationo *et al.*, 2008; Sahrawat *et al.*, 2010b). To maintain soil organic matter status, there is need to add organic materials including manures, organic and crop residues, on a regular basis (Bationo and Mokwunye 1991; Edmeades 2003; Harris 2002; Bationo *et al.*, 2008; Ghosh *et al.*, 2009; Materechera 2010).

Agricultural production related activities as a part of the NRM practice impact soil quality. The negative effects on soil quality that lead to soil degradation can be classified in two broad categories: (i) caused by soil loss due to water and wind erosion (Lal 1995; Pimentel *et al.*, 1995; den Biggelaar *et al.*, 2004a, 2004b; Montgomery 2007; Sahrawat *et al.*, 2010b), and (ii) as a result of deterioration in physical, chemical, and biological properties of the soil (Pathak *et al.*, 2005; Poch and Martinez-Casanovas 2006; Sahrawat *et al.*, 2010b). The effects of soil loss on crop productivity vary widely depending on soil and NRM practices, and crop. Among the soil characteristics, soil organic matter status, clay, soil depth, etc. are important. The causes of physical, biological, and chemical degradation of soil include loss of organic matter, salinization and alkalization, waterlogging, and the contamination of water resources. Both types of soil degradation result in the loss of organic matter and nutrients and are major constraints to maintenance of soil quality, fertility, and agricultural productivity (van Asten 2003; Bellamy *et al.*, 2005; Pathak *et al.*, 2005; Singh 2008; Wani *et al.*, 2009; Bekunda *et al.*, 2010; Materechera 2010; Sahrawat *et al.*, 2010b; Verhulst *et al.*, 2010).

Bationo *et al.* (2008) and Bekunda *et al.* (2010) extensively reviewed the various causes that hamper agricultural production and productivity and overall agricultural development in sub-Saharan Africa (SSA). The most important constraints included low soil fertility, fragile ecosystems, rainfall dependence, insufficient research, inadequate extension services, postharvest crop losses, insufficient market, and lack of consistent provisions for agricultural policies and land tenure. Overdependence on rainfall and associated water shortage related problems along with soil infertility constitute the major constraints to sustainable increase in agricultural productivity.

The fundamental biophysical cause for the declining per capita food production in smallholder farms in SSA during the past 3–5 decades was solely ascribed to soil fertility depletion including the loss of soil organic matter, and major plant nutrients (N, P, and K). The application of major nutrients from external sources remains dismally low (Sanchez *et al.*, 1997; Rego *et al.*, 2005; Bationo *et al.*, 2008; Bekunda *et al.*, 2010). The main factors contributing to soil fertility depletion were identified as erosion by water and wind, especially in the semi-arid and arid regions. For example, Sterk *et al.* (1996) reported a total loss of $45.9\,t\,ha^{-1}$ soil by wind erosion in the arid region of Niger. The loss of soil organic matter and major nutrients by erosion varies widely, but remains a major threat to soil fertility and environmental quality (for review see Bationo *et al.*, 2008).

Moreover, nutrients are removed by crops and unless their pool is replenished by addition there is depletion in nutrient reserves, eventually leading to nutrient deficiencies. To put it simply, for sustained productivity at a high level, the maintenance of soil fertility on a long-term basis is a prerequisite. And for sustained fertility, it is essential that organic matter and nutrients removed in harvest or produce plus those lost through various physical, biological, and chemical processes are replenished through external addition on a regular basis such that soil organic matter status is maintained and nutrient balances are not negative in the longer term (Rego *et al.*, 2003; Wani *et al.*, 2007; Sahrawat *et al.*, 2010b). The intensification of production systems without adequate investment to sustain the system, results in the loss of fertility (Katyal 2003; Morris *et al.*, 2007; Sahrawat *et al.*, 2010a, 2010b). The effects of loss of soil fertility (organic matter and nutrients) are in the longer term manifested as reduced crop yields and quality due to reduced soil quality (Lal 1997; Carpenter 2002; den Biggelaar *et al.*, 2004a, 2004b; Pathak *et al.*, 2005; Sahrawat *et al.*, 2008a; Sharma *et al.*, 2009a, 2009b).

Soil organic matter and major plant nutrient (N, P, and K) depletion remains a major constraint to long-term agricultural sustainability in much of the rainfed agricultural systems in the SAT regions of Asia and SSA. Negative nutrient balances (nutrient added minus nutrient harvested in crop) relative to mostly major plant nutrients have been reported as the nutrient removal exceeds input over a long period of time with concomitant decline in soil organic matter status. Organic matter depletion is particularly acute in the rainfed systems where the external inputs of organic matter and nutrients is far lower than the loss or removal (Burford *et al.*, 1989; Sahrawat *et al.*, 1991; Black 1993; Bationo *et al.*, 1998, 2008; Stoorvogel and Smaling 1998; Rego *et al.*, 2003; Bijay-Singh *et al.*, 2004; Bekunda *et al.*, 2010).

Since 1999, ICRISAT and its partners have been conducting systematic and detailed studies on the diagnosis and management of nutrient deficiencies in the semi-arid regions of Asia with emphasis on the semi-arid regions of India under the integrated watershed management program (Wani *et al.*, 2009). Under this program, first a soil sampling methodology was developed to take representative soil samples in a watershed. The methodology is based on stratified random sampling of the watershed considering the soil types including topography, major crops, and farmers' landholding size (for details see Sahrawat *et al.*, 2008b). During these studies, soil samples were collected from farmers' fields in a farmer participatory manner, processed and analyzed

Table 9.1 Critical limits in the soil of plant nutrient elements to separate deficient samples from non-deficient samples[a]

Plant nutrient	Critical limit (mg kg^{-1})
Sodium bicarbonate-extractable P	5
Ammonium acetate-extractable K	50
Calcium chloride-extractable S	8–10
Hot water-extractable B	0.58
DTPA-extractable Zn	0.75

[a]The data gleaned from various literature sources (for details see Rego *et al.*, 2007; Sahrawat *et al.*, 2007).

for soil chemical fertility parameters in the ICRISAT central analytical laboratory. The soil test results were shared with farmers and recommendations were developed for balanced nutrient management (BN) using the critical limits in the soil for various plant nutrients (Sahrawat 2006; Rego *et al.*, 2007; Sahrawat *et al.*, 2007) (Table 9.1) for the follow-up on-farm crop response studies. However, it must be stated that the critical limits of major, secondary, and micronutrient elements in the soil as well as in plant tissue vary with crop, soil type (especially clay and organic matter status), and agroclimatic conditions especially availability of irrigation water and status of other nutrients (other nutrients than the nutrient studied) in the soil (Mills and Jones 1996; Takkar 1996; Reuter and Robinson 1997; Fageria *et al.*, 2002; Sahrawat 2006; Rattan *et al.*, 2009; Scherer 2009; Tandon 2009).

The soil test results for pH, organic carbon (C), and extractable P, potassium (K), S, boron (B), and zinc (Zn) of a large number of soil samples collected from farmers' fields in the SAT regions of Indian states of Andhra Pradesh (3650), Karnataka (22867), Madhya Pradesh (341), and Rajasthan (421) showed that the results varied with district in a state and had a wide range in soil chemical fertility parameters (Table 9.2). The soil analysis was carried out following methods described in Sahrawat *et al.* (2010a).

These first results on the fertility status of farmers' fields at a large scale showed that the samples were generally low in organic C (used as a proxy for N supplying capacity of a soil), low to medium in Olsen extractable P, medium to high in exchangeable K, and generally low in calcium chloride extractable S, hot water extractable B, and DTPA extractable Zn (Table 9.2). The results clearly demonstrate that soils are not only low in organic C and Olsen-P but also low in secondary nutrients such as S and micronutrients such as B and Zn. The number of farmers' fields sampled from 14 districts of Karnataka was fairly large and based on these some plausible conclusions can be drawn for the prevalence of plant nutrient problems in the state, which is the second largest state in the country with rainfed agriculture after Rajasthan. The mean organic C content in the soil samples was 0.45%; Olsen-P was deficient in 47% of the 22867 farmers' fields sampled, exchangeable K was deficient only in 16% fields, extractable S in 83% fields, hot water extractable B in 66% fields, and DTPA extractable Zn was deficient in 61% of the sampled farmers' fields.

In Andhra Pradesh, B deficiency was most prevalent (in 85% of 3650 farmers' fields sampled), followed by S, which was deficient in 79% of the farmers' fields and Zn was deficient in 69% of the farmers' fields; Olsen-P was deficient in 38% of the fields and K only in 12% of the fields (Table 9.2). In Madhya Pradesh, B deficiency

Table 9.2 Chemical characteristics of soil samples collected from farmers' fields in the SAT regions of India[a]

District (No.of fields)	Parameter	pH	Organic C (%)	Olsen-P (mg kg⁻¹)	Exch. K (mg kg⁻¹)	Extractable nutrient elements (mg kg⁻¹)		
						S	B	Zn
Andhra Pradesh								
Adilabad (63)	Range	6.4–8.9	0.27–1.33	0.2–48.8	46–549	2.0–142.2	0.10–0.74	0.22–2.90
	Mean	8.2	0.62	6.9	204	12.2	0.34	0.62
	% Deficient			60	2	76	92	75
Anantapur (593)	Range	5.4–9.6	0.11–1.45	0.6–42.4	14–352	0.2–117.3	0.02–1.40	0.14–5.00
	Mean	7.5	0.30	7.7	73	4.5	0.21	0.59
	% Deficient			33	31	94	98	83
Kadapa (114)	Range	5.3–8.8	0.11–0.79	0.2–25.4	17–387	1.7–41.9	0.04–3.02	0.24–5.20
	Mean	7.4	0.27	3.9	80	6.6	0.39	0.76
	% Deficient			75	43	85	81	67
Khammam (102)	Range	5.1–8.8	0.32–1.50	0.2–57.8	31–856	3.6–71.9	0.12–1.22	0.28–6.80
	Mean	6.8	0.70	8.5	180	10.6	0.39	1.09
	% Deficient			60	2	67	87	45
Kurnool (331)	Range	5.6–9.7	0.09–1.06	0.4–36.4	33–509	1.4–53.8	0.04–2.04	0.08–4.92
	Mean	7.9	0.34	7.6	144	6.3	0.37	0.45
	% Deficient			42	5	85	79	91
Mahabubnagar (1035)	Range	5.3–10.2	0.08–2.18	0.2–247.7	16–1263	1.2–801.0	0.02–4.58	0.12–35.60
	Mean	7.4	0.42	12.6	119	16.2	0.30	1.11
	% Deficient			25	10	60	88	59
Medak (258)	Range	5.0–9.1	0.09–3.00	0.5–75.1	11–978	1.7–431.0	0.08–1.84	0.24–3.26
	Mean	7.7	0.49	8.0	161	12.4	0.57	0.78
	% Deficient			45	11	78	59	57
Nalgonda (441)	Range	5.0–9.2	0.12–1.36	0.2–50.4	21–379	1.4–140.3	0.02–1.48	0.08–16.00
	Mean	7.6	0.42	8.9	120	10.2	0.30	0.82
	% Deficient			31	7	78	90	66
Prakasam (492)	Range	6.4–9.3	0.12–1.30	0.2–41.7	28–697	0.6–19.2	0.02–1.86	0.20–10.8
	Mean	8.4	0.43	5.7	205	4.1	0.45	0.53
	% Deficient			56	1	94	71	88

District								
Ranga Reddy (121)	Range	5.1–8.2	0.15–1.56	0.2–60.0	24–405	1.1–81.6	0.06–1.24	0.30–5.72
	Mean	6.7	0.50	8.9	92	3.7	0.26	1.16
	% Deficient			39	17	98	98	35
Warangal (100)	Range	6.1–9.4	0.08–0.84	0.2–53.4	21–280	1.8–48.9	0.10–1.42	0.26–3.88
	Mean	7.8	0.41	16.0	118	9.4	0.38	0.96
	% Deficient			14	5	77	84	50
State total (3650)	Range	5.0–10.2	0.08–3.00	0.2–247.7	11–1263	0.2–801	0.02–4.58	0.08–35.6
	Mean	7.6	0.41	9.1	129	9.6	0.34	0.81
	% Deficient			38	12	79	85	69
Karnataka								
Bengaluru Rural (2223)	Range	5.0–9.5	0.01–1.31	0.3–220.8	9–847	0.9–94.5	0.10–5.12	0.14–235
	Mean	6.4	0.41	18.9	93	5.4	0.39	1.47
	% Deficient			16	30	94	68	34
Bidar (1189)	Range	5.6–8.7	0.19–1.98	0.6–118.6	18–2297	1.0–181.3	0.12–2.96	0.16–18
	Mean	7.6	0.63	8.5	221	7.2	0.56	0.94
	% Deficient			49	1	84	65	55
Bijapur (1395)	Range	6.7–9.2	0.00–1.21	0.1–91.9	24–2613	0.9–4647.4	0.02–18.22	0.15–10.4
	Mean	8.2	0.44	3.9	225	38.5	0.93	0.58
	% Deficient			80	3	77	46	85
Chamaraja Nagara (818)	Range	5.1–9.7	0.05–1.85	0.2–77.5	25–738	0.4–119.4	0.08–3.80	0.14–6.4
	Mean	7.8	0.43	9.6	188	5.6	0.63	0.77
	% Deficient			40	3	90	57	62
Chikkaballapur (2257)	Range	5.0–9.9	0.07–1.42	0.2–430.8	4–1650	0.5–470.0	0.06–1.98	0.06–21.5
	Mean	6.9	0.39	18.0	95	9.1	0.38	1.15
	% Deficient			37	34	80	80	52

(Continued)

Table 9.2 Continued

District (No. of fields)	Parameter	pH	Organic C (%)	Olsen-P (mg kg⁻¹)	Exch. K (mg kg⁻¹)	Extractable nutrient elements (mg kg⁻¹)		
						S	B	Zn
Chitradurga (1489)	Range	5.1–10.1	0.03–1.36	0.2–480.0	12–1953	0.8–291.8	0.04–6.94	0.08–40.5
	Mean	7.8	0.40	7.0	137	7.3	0.63	0.64
	% Deficient			54	15	86	64	80
Davangere (1500)	Range	5.0–9.0	0.04–1.38	0.0–138.8	11–510	0.9–945.0	0.06–6.30	0.04–11.2
	Mean	7.0	0.51	13.1	109	12.7	0.54	0.74
	% Deficient			34	13	77	66	74
Dharwad (1129)	Range	5.1–9.3	0.17–1.99	0.2–207.0	36–2344	1.4–715.0	0.10–12.48	0.24–24.3
	Mean	7.4	0.65	9.3	220	9.7	0.82	0.98
	% Deficient		31	53	1	79	39	44
Gadag (655)	Range	5.0–9.2	0.04–1.41	0.0–65.6	27–526	1.0–223.3	0.08–9.62	0.06–4.9
	Mean	8.1	0.44	5.3	178	7.4	0.88	0.44
	% Deficient			65	2	85	36	90
Gulbarga (2811)	Range	5.1–10.0	0.01–2.50	0.0–97.3	14–1722	0.4–12647	0.02–24.90	0.10–14.8
	Mean	8.0	0.46	7.1	244	27.6	0.64	0.52
	% Deficient		65	58	2	79	66	87
Haveri (1532)	Range	5.1–10.5	0.08–3.60	0.1–143.0	25–3750	0.3–120.3	0.08–8.44	0.20–34.1
	Mean	7.7	0.51	12.4	133	7.0	0.71	0.81
	% Deficient		65	42	5	85	46	60
Kolar (2161)	Range	5.0–10.2	0.04–1.50	0.0–182.0	9–1144	0.7–141.2	0.04–1.82	0.14–14.4
	Mean	7.0	0.38	20.3	87	7.0	0.34	1.31
	% Deficient		81	31	34	85	87	32
Raichur (1667)	Range	5.1–9.7	0.05–1.48	0.2–169.6	13–1797	0.8–2488	0.04–26.24	0.12–15.24
	Mean	8.3	0.43	11.8	209	46.8	1.17	0.66
	% Deficient			47	4	64	37	78
Tumkur (3041)	Range	5.0–10.0	0.04–2.08	0.1–204.0	11–1470	0.1–128.4	0.03–3.60	0.14–17.26
	Mean	6.6	0.39	5.9	92	5.5	0.33	0.89
	% Deficient			65	34	92	91	50

State total (22867)	Range	5.0–10.5	0.01–3.6	0.1–480	4–3750	0.1–12647	0.02–26.24	0.04–235
	Mean	7.4	0.45	11.4	150	14.4	0.59	0.89
	% Deficient			47	16	83	66	61
Madhya Pradesh								
Badwani (20)	Range	7.6–8.4	0.28–0.76	0.5–18.4	73–299	4.0–40.4	0.18–0.70	0.30–1.14
	Mean	8.1	0.51	4.6	146	11.8	0.42	0.58
	% Deficient			70	0	55	80	75
Dewas (24)	Range	7.0–8.7	0.31–1.00	0.2–10.8	46–456	3.9–9.5	0.12–0.56	0.24–0.82
	Mean	8.0	0.60	2.1	137	6.3	0.24	0.45
	% Deficient			96	4	100	100	96
Guna (38)	Range	7.2–8.5	0.47–1.11	0.1–10.2	86–303	2.7–14.3	0.22–2.20	0.24–1.74
	Mean	8.0	0.65	3.2	158	6.3	0.67	0.51
	% Deficient			79	0	87	50	95
Indore (23)	Range	7.8–8.3	0.43–1.08	0.5–42.2	129–716	5.9–134.4	0.46–1.30	0.56–3.00
	Mean	8.1	0.66	10.4	263	29.7	0.82	1.11
	% Deficient			39	0	9	17	22
Jhabua (22)	Range	6.4–7.4	0.58–1.53	0.2–42.2	88–506	2.7–28.2	0.26–0.76	0.66–3.18
	Mean	7.0	0.88	9.7	216	6.3	0.40	1.54
	% Deficient			45	0	95	91	5
Mandla (21)	Range	5.9–7.2	0.45–1.25	1.0–7.2	82–287	2.0–13.2	0.06–0.80	0.48–1.14
	Mean	6.6	0.68	2.8	143	4.8	0.29	0.79
	% Deficient			90	0	90	86	52
Raisen (20)	Range	7.9–8.4	0.42–0.97	0.5–13.4	118–275	2.9–12.8	0.20–0.74	0.30–0.98
	Mean	8.1	0.58	3.1	199	6.2	0.35	0.49
	% Deficient			90	0	90	90	90

(Continued)

Table 9.2 Continued

District (No.of fields)	Parameter	pH	Organic C (%)	Olsen-P (mg kg⁻¹)	Exch. K (mg kg⁻¹)	Extractable nutrient elements (mg kg⁻¹)		
						S	B	Zn
Rajagarah (30)	Range	6.7–8.3	0.44–1.41	1.6–19.2	51–434	2.9–50.4	0.30–0.92	0.38–3.82
	Mean	7.9	0.78	5.7	203	12.3	0.49	1.14
	% Deficient			60	0	53	73	27
Sagar (32)	Range	6.7–8.0	0.42–2.19	0.5–68.0	149–333	4.2–23.8	0.18–1.22	0.50–3.10
	Mean	7.4	0.72	7.1	265	10.1	0.36	1.04
	% Deficient			78	0	63	91	34
Sehore (19)	Range	7.3–8.4	0.36–0.69	0.5–17.2	48–256	3.0–20.5	0.28–0.62	0.36–0.92
	Mean	8.1	0.50	4.0	167	8.3	0.39	0.53
	% Deficient			84	5	74	95	95
Shajapur (20)	Range	7.1–8.2	0.46–1.15	1.0–25.8	51–249	5.6–42.0	0.18–0.72	0.46–1.42
	Mean	7.7	0.82	8.7	120	17.2	0.43	0.85
	% Deficient			25	0	25	80	40
Vidisha (72)	Range	7.6–8.6	0.31–0.92	0.5–14.1	96–401	1.8–16.6	0.12–0.74	0.10–1.00
	Mean	8.2	0.56	2.8	203	5.5	0.35	0.34
	% Deficient			92	0	96	93	97
State total (341)	Range	5.9–8.7	0.28–2.19	0.1–68	46–716	1.8–134.4	0.06–2.2	0.10–3.82
	Mean	7.8	0.65	5.0	190	9.6	0.43	0.72
	% Deficient			74	1	74	79	66
Rajasthan								
Alwar (30)	Range	7.9–8.8	0.33–0.66	0.5–44.0	53–515	4.5–17.2	0.20–0.68	0.20–2.00
	Mean	8.5	0.46	14.3	128	9.2	0.45	0.56
	% Deficient			10	0	63	87	83
Banswara (30)	Range	6.3–8.1	0.28–1.05	1.0–35.0	31–418	2.4–22.0	0.10–0.54	0.26–2.60
	Mean	7.2	0.56	7.7	107	9.2	0.23	0.70
	% Deficient			50	17	70	100	80
Bhilwara (30)	Range	7.2–8.9	0.32–1.87	0.8–27.0	33–460	4.0–44.9	0.32–1.30	0.16–2.30
	Mean	8.3	0.74	9.2	111	12.8	0.64	0.92
	% Deficient			40	17	43	47	37

Bundi (36)	Range	6.2–8.7	0.18–1.17	0.9–20.1	23–563	3.3–51.0	0.10–0.98	0.20–1.78
	Mean	7.6	0.60	6.2	87	9.2	0.44	0.65
	% Deficient			53	50	72	72	67
Dungarpur (99)	Range	6.2–8.0	0.48–1.99	1.0–28.2	34–240	4.0–31.3	0.28–1.50	0.88–14.10
	Mean	6.9	1.26	6.6	100	9.0	0.70	2.11
	% Deficient			48	8	72	31	0
Jhalawar (30)	Range	8.0–8.6	0.46–1.15	0.9–22.6	51–1358	1.9–78.0	0.22–1.36	0.40–3.40
	Mean	8.4	0.76	10.2	214	8.3	0.49	0.75
	% Deficient			30	0	87	77	60
Sawai Madhopur (44)	Range	7.8–9.4	0.16–0.70	0.2–11.8	44–438	3.1–26.6	0.20–2.18	0.34–28.60
	Mean	8.5	0.38	4.0	137	6.8	0.64	2.54
	% Deficient			73	7	86	52	41
Tonk (78)	Range	6.8–10.2	0.09–1.11	0.2–28.2	14–243	2.3–29.8	0.08–2.46	0.18–14.00
	Mean	8.1	0.36	5.7	83	7.7	0.62	1.61
	% Deficient			55	32	79	64	58
Udaipur (44)	Range	7.3–9.0	0.25–2.37	2.6–41.0	52–288	3.2–274.0	0.22–1.50	0.70–3.92
	Mean	8.2	0.83	15.2	145	26.7	0.83	1.57
	% Deficient			18	0	48	25	5
State total (421)	Range	6.2–10.2	0.09–2.37	0.2–44	14–1358	1.9–274	0.08–2.46	0.16–28.6
	Mean	7.8	0.72	8.1	116	10.6	0.6	1.49
	% Deficient			45	15	71	56	40
Grand total (India) (28270)	Range	5.0–10.5	0.01–3.6	0.1–480	4–3750	0.1–12647	0.02–26.24	0.04–235
	Mean	7.4	0.45	10.9	147	13.6	0.55	0.88
	% Deficient			46	16	82	68	62

[a]Critical limits used in the soil were 5 mg kg^{-1} for Olsen-P; 50 mg kg^{-1} ammonium acetate-extractable K; 8–10 mg kg^{-1} calcium chloride-extractable S; 0.58 mg kg^{-1} hot water-extractable B; and 0.75 mg kg^{-1} DTPA-extractable Zn (see Table 1).

was most prevalent (79% of 341 fields sampled), followed by S (74% fields), Olsen-P (74% fields), and Zn (66% fields) while in Rajasthan, the deficiency of S was most widespread (in 71% of 421 fields sampled), followed by B (56% fields), Zn (40% fields), Olsen-P (45% fields), and K (15% fields) (Table 9.2).

Considering all the four states in the SAT region of India, it can be concluded that the deficiency of S (calcium chloride extractable) was most widespread (on an average 82% of the 28270 farmers' fields sampled were deficient), followed by hot water extractable B (68% of the farmers' fields sampled were deficient), and DTPA extractable Zn (62% of the farmers' fields were deficient), and was indeed most revealing. These results are in accord with those reported earlier with a limited number of soil samples (Rego *et al.*, 2005; Sahrawat *et al.*, 2007, 2010a). On the other hand, K deficiency was not prominent (on an average only 16% of 28270 farmers' fields sampled were deficient) in the rainfed SAT soils (Table 9.2).

These results are significant in showing the widespread nature of the occurrence of the deficiencies of major nutrients such as N and P, but more importantly those of S, B, and Zn in the rainfed production systems of the SAT regions of India. The deficiency levels appear as widespread as those reported from the intensified irrigated systems (Pasricha and Fox 1993; Takkar 1996; Scherer 2001; Fageria *et al.*, 2002; Tandon 2009; Sahrawat *et al.*, 2009, 2010a). In the past, no survey of the nutrient deficiencies in SAT regions has been undertaken and so there are no benchmark results to compare the deficiencies of S and micronutrients in a large number of farmers' fields. But these results demonstrate clearly that in addition to water stress, multiple-nutrient deficiencies have to be managed to unlock the potential of rainfed production systems. The earlier research has mostly concentrated on the major nutrients and the deficiencies of N and P have been reported to be widespread in the rainfed systems (El-Swaify *et al.*, 1985; Burford *et al.*, 1989; Sahrawat *et al.*, 1991, 2001; Rego *et al.*, 2003; Bationo *et al.*, 2008).

9.3 BALANCED NUTRIENT MANAGEMENT: CROP PRODUCTIVITY AND QUALITY

As mentioned earlier, soil fertility management research in the rainfed areas has focused mainly on the management of major nutrients (N, P, and K) and even the amounts of these nutrients is generally inadequate (Rego *et al.*, 2007; Bationo *et al.*, 2008; Sahrawat *et al.*, 2010a). Water stress by erratic and low rainfall is the major bottleneck for farmers to apply adequate amounts of nutrients in the rainfed systems. However, recent work by ICRISAT and its partners and other researchers has shown that for realizing the potential of rainfed systems, both water stress and nutrient deficiencies need to be attended simultaneously (Wani *et al.*, 2003; Ncube *et al.*, 2007; Bationo *et al.*, 2008; Sahrawat *et al.*, 2010b).

For example, during 2002–04, Rego *et al.* (2007) conducted a number of on-farm trials during the rainy season (June–October) in three districts of Andhra Pradesh in the SAT region of India to evaluate crop responses to BN based on soil test results using mung bean (*Vigna radiata*), maize (*Zea mays*), groundnut (*Arachis hypogaea*), castor (*Ricinus communis*), and pigeonpea (*Cajanus cajan*). There were two treatments: (i) control or farmer's nutrient input (FI); and (ii) BN, which consisted of the application of SBZn + NP over FI or FI + SBZn + NP.

Table 9.3 Grain yield of crops in response to fertilization under farmer's nutrient input (FI) and balanced nutrient management (BN) treatments in the semi-arid zone of Andhra Pradesh, India during three (2002 to 2004) rainy seasons[a]

Year	Treatment[b]	Grain yield (kg ha^{-1})				
		Maize	Castor	Mung bean	Groundnut (pod)	Pigeonpea
2002	FI	2730 (20)[c]	590 (8)	770 (9)	1180 (19)	536 (43)
	BN	4560	880	1110	1570	873
	LSD (0.05)	419	143	145	92	156
2003	FI	2790 (24)	690 (17)	900 (6)	830 (30)	720 (12)
	BN	4880	1190	1530	1490	1457
	LSD (0.05)	271	186	160	96.8	220
2004	FI	2430 (19)	990 (6)	740 (12)	1320 (40)	1011 (21)
	BN	4230	1370	1160	1830	1564
	LSD (0.05)	417	285	131	122.5	106

[a]Source: Rego et al. (2007); data on pigeonpea crop are from ICRISAT.
[b]BN = FI + SBZn + NP
[c]The number of farmers' fields on which on-farm trials were conducted is given in parenthesis.

Briefly, for applying nutrients as per BN treatment (FI + SBZn + NP), S, B, and Zn were applied as a mixture, which consisted of 200 kg gypsum (30 kg S ha^{-1}), 5 kg borax (0.5 kg B ha^{-1}), and 50 kg zinc sulfate (10 kg Zn ha^{-1}) ha^{-1}; the mixture was surface broadcast on the plot before the final land preparation. The SBZn + NP treatment consisted of the same amount of S, B, and Zn as in SBZn plus 60 kg N ha^{-1} for maize and castor or 20 kg N ha^{-1} for groundnut and mung bean; and P was added at 30 kg P_2O_5 ha^{-1}. The treatment SBZn was applied along with P plus 20 kg N ha^{-1} as basal to all crops and 40 kg N ha^{-1} was topdressed in the case of maize and castor. In the case of NP treatment, 20 kg N ha^{-1} and 30 kg P_2O_5 ha^{-1} were applied to all crops as basal and 40 kg N ha^{-1} as topdressing for maize and castor. Other nutrient treatments including FI + SBZn, and FI + SBZn + NP or BN were applied as described above (Rego et al., 2007). The grain yields of maize, castor, mung bean, groundnut (pod yield), and pigeonpea were significantly increased under BN treatment with the application of SBZn + NP over the FI treatment in the three seasons (Table 9.3).

A large number of on-farm trials were also conducted in the semi-arid zone of Karnataka during five rainy seasons (2005–09) with maize, finger millet (*Eleusine coracana*), groundnut, and soybean (*Glycine max*) as the test crops. Again, as in the case of trials in Andhra Pradesh, BN treatment significantly increased the grain yields of these crops over the FI treatment (Table 9.4). In another set of trials, conducted during 2005–07 in the semi-arid zone of Karnataka, BN treatment significantly increased maize grain yield and dry matter over the FI treatment; BN also significantly improved the harvest index of the crop during all the three seasons (Rajashekhara Rao et al., 2010) (Table 9.5).

The results of on-farm trials conducted in the SAT zone of Madhya Pradesh with soybean in the rainy season (2008 and 2009) and chickpea in the postrainy season (2008) confirmed the superiority of the BN treatment over the FI treatment and significantly increased soybean and chickpea grain yields (Table 9.6). Similar results were obtained in the on-farm trials conducted during the 2008 rainy season in the semi-arid

Table 9.4 Grain yield of crops in response to fertilization under farmer's nutrient input (FI) and balanced nutrient management (BN) treatments in the semi-arid zone of Karnataka, India during five rainy seasons, 2005–09[a]

Year	Treatment[b]	Grain yield (kg ha^{-1})			
		Maize	Finger millet	Groundnut	Soybean
2005	FI	4000 (6)[c]	2100 (16)	1830 (8)	2030 (6)
	BN	6090	3280	1910	3470
	LSD (0.05)	395	338	91.5	664
2006	FI	4050 (22)	1700 (17)	1080 (17)	1120 (7)
	BN	5400	2170	1450	2650
	LSD (0.05)	240	440	341.4	538
2007	FI	5670 (19)	2000 (27)	1310 (23)	2120 (11)
	BN	8710	2940	2160	3120
	LSD (0.05)	572	230	191.4	262
2008	FI	4400 (27)	1680 (152)	940 (149)	1390 (16)
	BN	6130	2650	1430	1640
	LSD (0.05)	336	125	80.3	249
2009	FP	5460 (90)	1630 (165)	1100 (178)	1770 (36)
	BN	7800	2570	1500	2610
	LSD (0.05)	178	91	49.9	184

[a]Source: Data are from ICRISAT.
[b]BN = FI + SBZn + NP
[c]The number of farmers' fields on which on-farm trials were conducted is given in parenthesis.

Table 9.5 Yield of maize in response to fertilization under farmer's nutrient input (FI) and balanced nutrient management (BN) treatments in on-farm trials in the Haveri district of Karnataka, India, 2005–07[a]

Treatment	Yield (t ha^{-1})		Harvest index (%)[b]
	Grain	Stover	
2005 (9)[c]			
FI	4.00	4.62	46.5
BN	6.09	5.92	50.7
LSD (0.05)	0.49	0.54	1.2
2006 (20)[c]			
FI	3.77	3.80	49.8
BN	5.37	5.12	51.2
LSD (0.05)	0.56	0.52	1.2
2007 (17)[c]			
FI	5.10	4.84	47.2
BN	6.32	5.82	51.3
LSD (0.05)	0.65	0.77	1.6

[a]Source: Adapted from Rajashekhara Rao et al. (2010).
BN = FI + SBZn + NP
The plots under BN treatment received 80 kg N, 30 kg P_2O_5, 30 kg S, 10 kg Zn, and 0.5 kg B ha^{-1}.
[b]Harvest index is Grain wt/(Grain wt + Stover wt) × 100.
[c]Number of participating farmers is given in parenthesis.

Table 9.6 Grain yield of soybean (rainy season) and chickpea (postrainy season) in response to fertilization under farmer's nutrient input (FI) and balanced nutrient management (BN) treatments in Madhya Pradesh, India during 2008 and 2008–09 seasons[a]

Year	Treatment	Grain yield[b] $(kg\,ha^{-1})$	
		Soybean	Chickpea
2008	FI	1490 (117)	1250 (169)
	BN	1840	1440
	LSD (0.05)	56	29
2009	FI	2120 (140)	
	BN	2680	
	LSD (0.05)	95	

[a]Source: Data are from ICRISAT.
BN = FI + SBZn + NP
[b]The number of farmers' fields on which on-farm trials were conducted is given in parenthesis.

Table 9.7 Grain yield of maize and pearl millet in response to fertilization under farmer's nutrient input (FI) and balanced nutrient management (BN) treatments in the semi-arid zone of Rajasthan, India during 2008 rainy season[a]

Treatment	Grain yield[b] $(kg\,ha^{-1})$	
	Maize	Pearl millet
FI	2730 (17)	2310 (16)
BN	2980	2510
LSD (0.05)	55	34.3

[a]Source: Data are from ICRISAT.
BN = FI + SBZn + NP
[b]The number of farmers' fields on which on-farm trials were conducted is given in parenthesis.

zone of Rajasthan with pearl millet (*Pennisetum glaucum*) and maize as the test crops; and the grain yields of these crops were significantly increased in the BN treatment as compared to FI (Table 9.7).

On-farm trials were conducted during the 2006–07 season with a number of vegetable crops in watersheds in Dharwad, Haveri, and Chitradurga districts of Karnataka to study their responses to BN as compared to FI treatment. The results showed an impressive yield response to BN as compared to FI treatment; and the growing of these vegetables under BN was economically viable and remunerative (Srinivasarao *et al.*, 2010) (Tables 9.8 and 9.9).

Balanced plant nutrition is not only important for increasing crop productivity but also critical for enhancing crop quality including grain and stover/straw quality, which has implications for human (grain as food) and animal (straw used as fodder or feed) nutrition. There is a relationship between soil health and food and feed quality, which in turn impacts human and animal health. The importance of mineral nutrition

Table 9.8 Response of vegetables to farmer's nutrient input (FI) and balanced nutrient management (BN) treatments in watersheds in Dharwad and Haveri districts of Karnataka, India[a]

Crop[b]	Fresh fruit yield (kg ha^{-1})		Farm-gate price (₹ kg^{-1})	Additional cost (₹ ha^{-1})	Additional net returns (₹ ha^{-1})	Benefit-cost ratio
	FI	BN				
Ridge gourd (2)	5400	6300	6.0	3050	5700	1.87
Bitter gourd (2)	3000	3900	9.3	3050	8250	2.71
Chili (4)	6000	8500	5.5	3050	13000	4.26
Brinjal (eggplant) (4)	6000	8000	6.8	3050	12770	4.19
Tomato (4)	11200	17100	6.4	3050	34800	11.4

[a]Source: Adapted from Srinivasarao *et al.* (2010).
BN = FI + SBZn + NP
[b]Number of farmers, is given in parenthesis.

Table 9.9 Comparative response of onion to farmer's nutrient input (FI) and balanced nutrient management (BN) treatments in various watersheds in Chitradurga district of Karnataka, India[a]

Watershed	No. of farmers		Onion fresh wt. (t ha^{-1})		Increase in wt. (%)
			FI	BN[b]	
Maradihalli	10	Range	21–30	30–37.5	41
		Mean	24.8	34.5	
Toparamalige	10	Range	22.5–31.5	27.0-38.8	31
		Mean	26.7	34.7	
Belagatta	4	Range	22.5–31.5	27.0-38.8	45
		Mean	27.3	35.6	

[a]Source: Adapted from Srinivasarao *et al.* (2010).
[b]BN = FI + SBZn + NP

of crops along with improved cultivars of crops and crop management cannot be overemphasized for producing nutritious food (Welch *et al.*, 1997; Graham *et al.*, 1998, 2007; Welch and Graham 2002, 2004; Parthasarathy Rao *et al.*, 2006; Sahrawat *et al.*, 2008a) and fodder (Kelly *et al.*, 1996; Sahrawat *et al.*, 2008a; Rattan *et al.*, 2009).

For example, in the on-farm experiments conducted to determine the effects of S, B, and Zn fertilization on the grain and straw quality of sorghum (*Sorghum bicolor*) and maize grown under rainfed conditions in the SAT region of India showed that BN through combined application of S, B, Zn, N, and P as compared to FI increased N, S, and Zn concentrations in the grain and straw of these crops (Sahrawat *et al.*, 2008a) (Tables 9.10 and 9.11). These results stress the importance of balanced mineral nutrition of crops for increased produce quality. For example, the S fertilization of oilseed crops such as soybean (Saha *et al.*, 2001), canola (Brennan and Bolland 2008; Brennan *et al.*, 2010), and sunflower (*Helianthus annuus*) (Usha Rani *et al.*, 2009) is not only required for increasing dry matter and seed yield but also essential for enhancing oil concentration and quality.

Table 9.10 Chemical composition of the grain and straw of sorghum crop as affected by farmer's nutrient input (FI) and balanced nutrient management (BN) treatments in the semi-arid region of India, 2003 rainy season[a]

Treatment[b]	N $(g\,kg^{-1})$	P $(g\,kg^{-1})$	S $(mg\,kg^{-1})$	B $(mg\,kg^{-1})$	Zn $(mg\,kg^{-1})$
Grain					
FI	10.7	2.5	535	0.18	21
BN	13.2	2.8	766	0.22	31
LSD (0.05)	0.9	0.5	46	0.08	5.8
Straw					
FI	2.2	0.7	491	0.7	22
BN	2.6	0.6	537	1.1	31
LSD (0.05)	1.0	0.2	92	0.63	5.7

[a]Source: Adapted from Sahrawat *et al.* (2008a).
[b]BN = FI + SBZn + NP

Table 9.11 Chemical composition of the grain and straw of maize crop as affected by farmer's nutrient input (FI) and balanced nutrient management (BN) treatments in the semi-arid region of India, 2004 rainy season[a]

Treatment[b]	N $(g\,kg^{-1})$	P $(g\,kg^{-1})$	S $(mg\,kg^{-1})$	B $(mg\,kg^{-1})$	Zn $(mg\,kg^{-1})$
Grain					
FI	14.0	3.2	1095	1.4	23
BN	15.1	3.0	1153	1.8	22
LSD (0.05)	1.0	0.4	52	0.67	1.7
Straw					
FI	7.9	1.5	798	5.1	18
BN	6.6	1.3	921	5.9	20
LSD (0.05)	0.7	0.4	193	1.45	4.0

[a]Source: Adapted from Sahrawat *et al.* (2008a).
[b]BN = FI + SBZn + NP

From this discussion on the results obtained in a large number of on-farm trials, it is evident that in the SAT region multiple nutrient deficiencies especially of N, P, S, B, and Zn are holding back the potential of rainfed systems. Also, soil fertility depletion has been recognized as the major biophysical cause of declining food availability in smallholder farms in SSA. Any program aimed at reversing the trend in declining agricultural productivity and food quality, and preserving the environmental quality must begin with soil fertility restoration and maintenance. The decline in productivity is related to decline in soil fertility, which in turn is directly related to decline in soil organic matter status and depletion of the plant nutrient reserves in various production systems with little or no investment in recuperating soil fertility in agroecosystems (Pieri 1989; Sanchez *et al.*, 1997; Izac 2000; Vanlauwe 2004; Bationo *et al.*, 2008; Lal 2008; Stringer 2009; Bekunda *et al.*, 2010).

Soil fertility maintenance is not only a prerequisite for sustainable increase in crop productivity but also equally essential for maintaining crop quality in terms of food, fodder, and feed quality (Kelly *et al.*, 1996; Sahrawat *et al.*, 2008a), especially iron (Fe) and Zn in the grain (Welch and Graham 2004; Graham *et al.*, 2007; Sahrawat *et al.*, 2008a; Rattan *et al.*, 2009). The results from on-farm studies also show that the productivity of the rainfed systems can be enhanced through management of various nutrient deficiencies. Unless the constraints to soil fertility management are alleviated, it would not be possible to achieve the potential productivity of the rainfed systems. Because the area under rainfed production is very large, even a modest increase in yield would contribute in a big way to global food pool, apart from providing source of income and livelihoods to the rural poor.

9.4   SOIL QUALITY AND WATER USE EFFICIENCY

Soil quality is defined for various purposes, but for the purpose of this chapter we use the definition given by Doran and Parkin (1994) and Karlen *et al.* (1997) which relates to the soil's capacity to function, and to perform its agricultural production and environmental functions on a sustainable basis. In the general scientific literature, the terms soil quality and soil health have been interchangeably used, but in the soil science literature the term soil quality is preferred. While soil health refers to the state of soil as a living and dynamic system, soil quality on the other hand emphasizes the soil's capacity to sustain biological productivity and maintain environmental quality. Both soil quality and soil health are functional in nature and soil quality can also be used to cover the soil health too. For detailed discussion on soil quality and soil health, the readers are referred to extensive studies by various researchers (Doran *et al.*, 1996; Freckman and Virginia 1997; Karlen *et al.*, 1997, 2003; Sojka *et al.*, 2003; Sahrawat *et al.*, 2010b).

The productivity in rainfed systems have remained low because of frequent drought due to high variability in both the amount and distribution of rainfall in the growing season, poor soil quality, low use of plant nutrients, small farm holding size, and other farmers' socioeconomic conditions (Pieri 1995; Bationo *et al.*, 2008; Sharma *et al.*, 2009b; Sahrawat *et al.*, 2010b). However, the potential productivity under rainfed condition in the SAT agriculture can be enhanced by improving soil quality by managing plant nutrient disorders (Padwick 1983; Ouédraogo *et al.*, 2001; Tiwari 2008; Scherer 2009; Sahrawat *et al.*, 2010a) and increasing rainfall use efficiency (RUE) (Singh *et al.*, 2009; Wani *et al.*, 2009).

Efficient use of rainwater involves harvesting of extra runoff water (after recharge of the soil profile) in the rainy season and its efficient use for supplemental irrigation wherever the opportunity exists. The use of harvested water for supplemental irrigation of rainfed crops in the SAT regions showed that the benefits of supplemental irrigation in terms of enhancing and stabilizing crop productivity have been excellent even in the areas with relatively assured rainfall areas (Pathak *et al.*, 2009). In the drier areas, supplemental irrigation can make a large difference in crop production and in some instances it can make a difference in having a crop or no crop (Oweis and Hachum 2009). Thus, rainwater management holds the key to successful crop production in the SAT and dry regions (Rockström *et al.*, 2002; Wani *et al.*, 2002, 2008, 2009; Bationo *et al.*, 2008; Pathak *et al.*, 2009). In the light of very impressive responses of crop to supplemental irrigation, it is imperative that most efficient use is made of the scarce

Table 9.12 Effects of micronutrient application on rainfall use efficiency in various field crops in Andhra Pradesh and Madhya Pradesh, India[a]

Crop	Rainfall use efficiency $(kg\,mm^{-1}\,ha^{-1})$	
	Farmer's practice	*Farmer's practice + micronutrients*
Andhra Pradesh		
Maize	5.2	9.2
Groundnut	1.6	2.8
Mung bean	1.7	2.9
Sorghum	1.7	3.7
Madhya Pradesh		
Soybean	1.4	2.7

[a]Source: Adapted from Singh *et al.* (2009).

resource using efficient method of water application at a critical stage of the crop when the response is highest (for review see Oweis and Hachum 2009; Pathak *et al.*, 2009).

For efficient use of water and to increase RUE, soil quality especially the management of various nutrient deficiencies in the production systems is a prerequisite. For example, Singh *et al.* (2009) reported that the application of S, B, and Zn over the FI treatment in on-farm trials in the SAT regions of India (states of Andhra Pradesh and Madhya Pradesh) increased the productivity of rainfed crops, resulting in increased RUE. The RUE of maize for grain production under FI was $5.2\,kg\,mm^{-1}$ water compared to $9.2\,kg\,mm^{-1}$ water with the combined application of S, B, and Zn over the FI treatment (Table 9.12). The best results in terms of RUE for maize and several other crops however, were obtained under the BN treatment when N and P were added along with S, B, and Zn (Singh *et al.*, 2009). These results are in accord with those reported by Rego *et al.* (2007) who found that farmers were applying sub-optimum quantity of major nutrients especially N and P and thus the applications of NP along with SBZn (NP + SBZn) gave the best results in terms of crop yield and biomass production, and nutrient uptake.

In an on-farm study conducted for three seasons (2005-07) in the SAT region of Karnataka, Rajashekhara Rao *et al.* (2010) reported that BN not only increased grain and stover yield of rainfed maize (see results in Table 9.5) but also increased partial factor productivity [Grain yield in fertilized plot = (Grain yield in absolute control + Yield increase due to treatment) × Amount of nutrient applied], agronomic efficiency (the incremental efficiency of applied nutrients over the control), benefit-cost ratio [(Grain yield of fertilized plot × Price of grain) : (Amount of nutrient applied × Price of the applied nutrient inputs)], and RUE (Grain yield/rainfall received during the growing season) for maize production (Table 9.13).

Results from on-farm trials conducted in the SAT regions of Karnataka and Madhya Pradesh in India during rainy season in 2008 and 2009 with maize, finger millet, groundnut, and soybean showed that BN treatment increased the grain yields of these crops and the yield increase was economically attractive and remunerative (Table 9.14).

Table 9.13 Partial factor productivity, agronomic efficiency, benefit-cost ratio and rainfall use efficiency under farmer's nutrient input (FI) and balanced nutrient management (BN) treatments, 2005–07[a]

Production efficiency parameters[b]	FI	BN[c]
Partial factor productivity (kg grain kg^{-1} of nutrients)	40.8	48.4 (18.6)
Agronomic efficiency (kg grain kg^{-1} of nutrients)	12.5	16.0 (28.0)
Benefit-cost ratio	3.2	4.6 (43.8)
Rainfall use efficiency (kg grain mm^{-1} of rainfall)	9.8	14.6 (49.2)

[a]Source: Adapted from Rajashekhara Rao et al. (2010).
[b]Calculated using mean grain yield values in 2005, 2006, and 2007 seasons.
[c]Values in parentheses indicate percent increase or decrease in each parameter over FI treatment.

Table 9.14 Economics of fertilizer use for grain production of crops in on-farm trials conducted during rainy season 2008 and 2009 in the SAT regions of India[a]

State	Crop	Grain yield (kg ha^{-1}) FI[b]	FI + SBZn	Yield increase (kg ha^{-1})	Support price of grain (₹ kg^{-1})	Additional income (₹)	Additional income per rupee invested
2008							
Karnataka	Maize	4400	6130	1730	8.40	14532	7.9
	Finger millet	1680	2650	970	9.15	8876	4.8
	Groundnut (pod)	940	1430	490	21.00	10290	5.6
	Soybean	1390	1640	250	13.90	3475	1.9
Madhya Pradesh	Soybean	1490	1840	350	13.90	4865	2.6
2009							
Karnataka	Maize	5460	7800	2340	8.40	19656	10.6
	Finger millet	1630	2570	940	9.15	8601	4.6
	Groundnut (pod)	1100	1500	400	21.00	8400	4.5
	Soybean	1770	2610	840	13.90	11676	6.3
Madhya Pradesh	Soybean	2120	2680	560	13.90	7784	4.2
Mean							4.4

Fertilizer	Recommended rate (kg ha^{-1})	Cost (₹ kg^{-1})	Total cost (₹ ha^{-1})
Gypsum	200	1.5	300
Borax	5	40	200
Zinc sulfate	50	27	1350
Total cost (ha^{-1})			1850

[a]Source: Data are from ICRISAT.
[b]FI = Farmer's nutrient input
1 US$ = ₹45

Thus, soil quality or health is a major driver of enhanced RUE and productivity in the rainfed systems and needs an implementing strategy in which BN is integrated with soil and water conservation and management (Wani et al., 2009). For maintaining soil health, the changes in soil quality, as impacted by NRM practices, need to be monitored and assessed on a continuing basis as the outcome of such research can offer the valuable opportunity for the implementation of corrective management practices, as and when needed (Mandal *et al.*, 2001; Wander *et al.*, 2002; Sanchez *et al.*, 2003; Andrews *et al.*, 2004; Lilburne *et al.*, 2004; Turner *et al.*, 2007; Wilson *et al.*, 2008; Cotching and Kidd 2010; Sahrawat *et al.*, 2010a).

In the monitoring and assessment of soil health or quality, most of the indices used are chemical and little use is made of the biological fertility indicators in the monitoring program. In a recent study on soil quality evaluation and the interaction with land use and soil order in Tasmania, Australia, Cotching and Kidd (2010) reported that six soil properties [pH, organic C, Olsen-P, aggregate stability, bulk density, and exchangeable sodium percentage (ESP)] were generally responsive to soil order and land use change, although there were differences in their responsiveness to soil order and land use change.

9.5 STRATEGY FOR SCALING-UP THE SOIL TEST-BASED APPROACH FOR ENHANCING AGRICULTURAL PRODUCTIVITY

Low productivity in rainfed systems coupled with water shortage, degraded and marginal soil resource base, and lack of investment in soil fertility maintenance has been marginalizing agriculture and livelihoods in the rainfed areas in much of the SAT regions. To come out of this cycle, there is an urgent need to address the two major constraints to rainfed productivity enhancement, i.e., to simultaneously address the twin problems of water shortage and soil infertility. The watershed approach seems most rational and appropriate to simultaneously implement in an integrated manner both soil and water conservation and management practices with BN at the farm level. The strategy to maintain soil quality and fertility is to use inputs of organic matter and nutrients from both mineral and organic sources (Palm *et al.*, 2001; Bationo *et al.*, 2008). The sources of organic matter inputs should be considered on site-specific basis, but their use is essential for maintaining the physical, chemical, and biological properties of the soil, a prerequisite for the soil to carry out its agricultural production and environmental quality related functions in a sustainable manner (Palm *et al.*, 2001; Sahrawat *et al.*, 2010a, 2010b).

In this section we discuss soil testing as a mechanism for fertility and soil quality management at the farm level. Soil test-based recommendations form the basis for BN to enhance productivity and produce quality. A large body of results presented and discussed in this chapter clearly demonstrates the potential of soil testing for diagnosing and management of the nutrient related disorders in the rainfed agroecosystem (Wani 2008; Subba Rao *et al.*, 2009; Sahrawat *et al.*, 2010a). Since 2002, ICRISAT and its partners have been conducting on-farm trials to develop BN practices to increase agricultural productivity and household incomes in the SAT regions with very impressive results in terms of yields of crops at the farm level. There is an urgent need to scale-up the program so that more and more numbers of farmers are able to benefit from soil test-based nutrient management intervention. This approach will lead to rational and judicious use of purchased inputs of nutrients to enhance and stabilize agricultural productivity in the rainfed areas of the SAT regions.

For scaling-up the soil test-based nutrient management approach at the farm level, a systematic approach is outlined, which has been found useful in our on-farm research (Wani et al., 2002, 2008; Rego et al., 2005, 2007; Sahrawat et al., 2007, 2010a). The first step in this approach is to collect the baseline data on types of soils, dominant crops/cropping systems and their current yield levels, farmer holding size and their socioeconomic status following participatory rural appraisal in the watershed or the cluster of villages to be sampled.

For effective sampling, a watershed or a cluster of villages was divided into three groups based on the position of the fields on a toposequence: top, middle, and bottom, depending on the elevation and drainage of the landscape. Different soil types were separated in each group. For soil sampling, 20% of farmers in each position on the toposequence were randomly selected in proportion to the farm size. Using stratified random sampling methodology (for details see Sahrawat et al., 2008b), 8 to 10 cores of surface soil (0–15 cm deep) were collected to make one composite sample. A farmer participatory approach was used for the collection of soil samples and farmers were trained to collect soil samples from their fields. The soil samples were transported to the ICRISAT laboratory in Patancheru, India for processing and analysis for soil chemical fertility parameters. The samples were air dried and powdered with a wooden hammer to pass through a 2-mm sieve. For organic C analysis, the samples were ground to pass through a 0.25-mm sieve. Standard methods were used for the analyses of soil samples for pH, organic C, and extractable or available major, secondary, and micronutrients (Sahrawat et al., 2007).

The soil test results were shared with farmers during village meetings. They were briefed about the relevance of analysis of results from the stratified soil sampling and how it applies to their fields and recommendations were formulated for the application of BN alongside FI treatment in the on-farm trials. The results of the soil analysis were also disseminated through wall writings in the village. Later, on-farm participatory research and development (PR&D) trials were conducted to determine and compare the crop yields under BN with those in the FI treatment. The results were explained to other farmers during the field days. Various crop, nutrient, and soil management practices are described by Rego et al. (2007) and Sahrawat et al. (2010a).

For example, in the first year of the study, the on-farm trials were conducted in nine nuclear microwatersheds or cluster of villages in a district. During the second year, the nutrient management trials were extended to 5 × 9 watersheds in the same district; and in a period of 3 to 4 years the entire district was covered by the trials. Following the same methodology, the trials cover several districts in a state and eventually the entire state could be covered by BN (Rego et al., 2007). Data from several districts in a state are summarized and interpreted to learn lessons for the extension of such trials in a state or region. The crops covered in these trials include the most important or dominant crops in the district and the results of BN treatment are compared with those of FI treatment. The key to the success of such a program hinges on the participatory nature of the farmers who are involved in planning of the on-farm activities, soil sampling, discussion, and sharing of soil test results for the formulation of recommendations for BN (for details see Rego et al., 2005, 2007; Wani 2008).

The data on yield and additional income earned by farmer households as a result of soil test-based BN, are discussed with farmers and this acts as a catalyst to create awareness and interest among other farmers who had seen good crops in the participating farmers' fields. This has a multiplier effect in the adoption of the technology with adequate support for soil testing, capacity building of all the stakeholders, and implementation of the various practices in the technology at the farm level (Wani 2008; Sahrawat *et al.*, 2010a).

For practical utilization of the soil test-based nutrient management, we have been mapping using the geographical information system (GIS) based extrapolation methodology, the deficiencies of nutrients especially those of S, B, and Zn in various districts in Karnataka. Finally, the soil test-based fertilizer application has been made web-based so that the recommendations can be downloaded and made available nutrient-wise to farmers using color codes depicting the deficiency or sufficiency of a nutrient. Such information can be easily used by smallholder farmers. Typical examples of nutrient mapping for extractable (available) S, B, and Zn, using data from selected districts of Karnataka are shown in Figure 9.1. Such maps can be extended and used by farmers in a cluster of villages to plan the application of deficient plant nutrients to production systems.

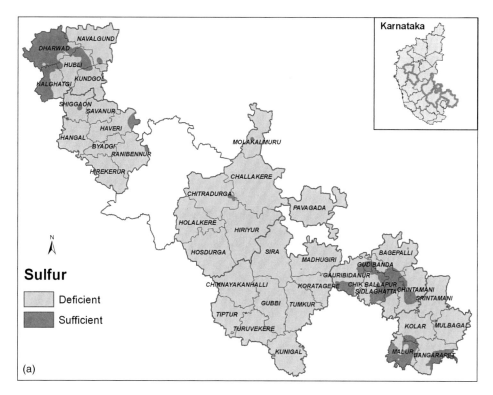

Figure 9.1 Distribution of extractable sulfur, boron, and zinc in soil samples from various districts of Karnataka, India

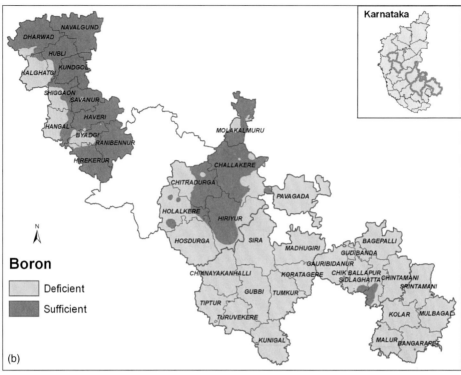

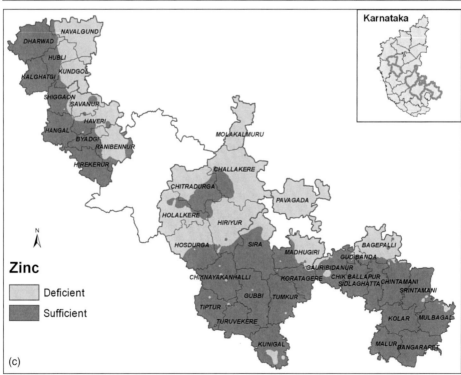

Figure 9.1 Continued

9.6   GENERAL DISCUSSION AND CONCLUSIONS

It is recognized that water shortage related plant stress is the primary constraint to crop production and productivity in the rainfed systems in the SAT regions and consequently the importance of water shortage has globally been rightly emphasized (Molden 2007; Pathak *et al.*, 2009). At the same time, it has been emphasized that the potential of the rainfed systems is much higher than indicated by the current productivity levels (Wani *et al.*, 2009; Rockström *et al.*, 2010). Equally importantly, the water constraint is not always related to absolute water shortage, but is generally caused by a large variability in the availability of water during the cropping season and its improper management. And hence, water management to cover water stress during dry spells can greatly reduce risks (Oweis and Hachum 2009; Pathak *et al.*, 2009; Rockström *et al.*, 2010).

However, apart from water shortage, severe soil infertility is also a problem in the rainfed systems (Black 1993; Rego *et al.*, 2007; Bekunda *et al.*, 2010; Sahrawat *et al.*, 2010a) and managing water stress alone cannot sustainably enhance the productivity of rainfed systems. Hence for achieving sustainable gains in rainfed productivity both water shortage and soil fertility problems need to be simultaneously addressed through effective NRM practices (Wani *et al.*, 2009; Sahrawat *et al.*, 2010b).

For the first time, a large number of farmers' fields in the SAT regions of India were sampled and analyzed for organic C and extractable or available nutrients in an effort to diagnose the prevalence of major and micronutrient deficiencies. Critical limits for various nutrients in the soil from published literature and ICRISAT data were used (Table 9.1) to separate deficient fields from the non-deficient ones (Black 1993; Mills and Jones 1996; Sahrawat 2006; Mahler and Shafii 2009) and for nutrient recommendation for the follow-up on-farm crop response trials. The results on the analyses of 28,270 soil samples (Table 9.2) demonstrate that the soils in rainfed areas are indeed infertile and they are not only deficient in major nutrients especially N (soil organic C status used as an index for available N) and P but are low in organic matter reserve. The most revealing results however, were the widespread acute deficiency of secondary nutrients such as S and micronutrients (especially B and Zn) (Rego *et al.*, 2007; Sahrawat *et al.*, 2007, 2009, 2010a).

A summary of the results on on-farm responses of several field crops to applications of deficient nutrients together with N and P demonstrated that BN has indeed the potential to significantly enhance the productivity of a range of crops (Tables 9.3 to 9.9), improve grain and straw quality (Tables 9.10 and 9.11), enhance RUE (Tables 9.12 and 9.13) and economic gains (Tables 9.13 and 9.14) in the SAT regions under rainfed conditions.

It would appear from these results that soil test-based nutrient management approach can be an important entry point activity and also a mechanism to diagnose and manage soil fertility in practical agriculture. Soil and plant tests have long been used as tools to diagnose and manage soil fertility problems in the intensified irrigated systems and commercial crops including fruit and vegetable crops to maximize productivity (Dahnke and Olson 1990; Black 1993; Mills and Jones 1996; Reuter and Robinson 1997; Subba Rao *et al.*, 2009), but rarely has soil testing been used to diagnose and manage nutrient problems in farmers' fields in the SAT regions at a scale reported in this chapter. The critical limits for P, K, S, B, and Zn in the soil (Table 9.1) seem to provide a fair basis for separating deficient soils from those that are not deficient. Soils below the critical limits of the nutrients evaluated responded to the applications of nutrients; although the overall crop response was regulated by the rainfall received during the cropping season (Rego et al., 2007; Sahrawat et al., 2007, 2010a; Srinivasarao et al., 2008). Soil test-based nutrient application also allows judicious and efficient use of nutrient inputs at the local and regional levels (Black 1993; Subba Rao et al., 2009).

Regarding the source of nutrients, it is recommended that an integrated approach in which both mineral and organic sources of nutrients should be used through the inclusion of legumes in the production systems to supply organic matter as well as nutrients as the organic matter inputs to the soil in any form helps to improve soil physical, chemical, and biological aspects of fertility (Rego and Rao 2000; Aulakh et al., 2001; Bot and Benites 2005; Bationo et al., 2008; Srinivasarao et al., 2009). However, it should be kept in mind that the application of manure alone may not supply enough nutrients to achieve economic yield and the use of organic fertilizers as a complementary nutrient source enhances the contribution of inorganic or chemical fertilizers to yield and soil physical, chemical, and biological properties (Singh et al., 2007; Yan and Gong 2010). Moreover, the nutrient contents of manures vary widely (Lupwayi et al., 2000) and hence it is of critical importance that the rate of application of manure is adjusted based on its nutrient content (Williams et al., 1995; Williams 1999).

For more widespread adoption and use of soil testing for the diagnosis and management of plant nutrient deficiencies in the rainfed systems of the SAT regions, there is a need to strengthen the soil testing facilities at the local and regional levels for science-based management and consider maintenance of soil fertility as a prerequisite for sustainable increase in productivity of the rainfed systems in the SAT (Sahrawat et al., 2007, 2010a). We hope that the research reported in this chapter would stimulate research for widespread use of soil testing as a means for soil fertility management in farmers' fields.

For enhancing the overall agricultural productivity and crop quality of the rainfed systems, the choice of crops and adapted cultivars along with soil, water, and nutrient management practices need to be integrated at the farm level (Wani et al., 2009; Sahrawat et al., 2010a). To achieve this, research and extension support and backstopping along with capacity building of all the stakeholders need to converge (Wani 2008; Sahrawat et al., 2010). Indeed, ICRISAT and its research partners most appropriately advocate the integration of genetics (crops and cultivars) and NRM for technology targeting and greater impact of agricultural research in the SAT regions (Twomlow et al., 2008b). The strategy is based on the use of crop cultivars that are adapted to the harsh conditions of the SAT regions especially water stress and nutrient deficiencies. The soil, water, and nutrient management practices are developed around the adapted cultivars to realize the potential of the cultivars in diverse production systems (Ae et al., 1990; Rego and Rao 2000; Condon et al., 2004; Passioura 2006; Hiradate et al., 2007; Bationo et al., 2008; Sahrawat 2009; Passioura and Angus 2010).

REFERENCES

Ae, N., J. Arihara, K. Okada, T. Yoshihara, and C. Johansen. 1990. Phosphorus uptake by pigeon pea and its role in cropping systems of the Indian subcontinent. *Science* 248: 477–480.

Andrews, S.S., D.L. Karlen, and C.A. Cambardella. 2004. The soil management assessment framework: a quantitative soil quality evaluation method. *Soil Science Society of America Journal* 68:1945–1962.

Aulakh, M.S., T.S. Khera, J.W. Doran, and K.F. Bronson. 2001. Managing crop residue with green manure, urea, and tillage in a rice-wheat rotation. *Soil Science Society of America Journal* 65:820–827.

Bationo, A., J. Kihara, B. Vanlauwe, J. Kimetu, B.S. Waswa, and K.L. Sahrawat. 2008. Integrated nutrient management: concepts and experience from Sub-Saharan Africa. In *Integrated nutrient management for sustainable crop production*, ed. M.S. Aulakh, and C.A. Grant, 467–521. New York, USA: The Haworth Press, Taylor and Francis Group.

Bationo, A., F. Lompo, and S. Koala. 1998. Research on nutrient flows and balances in West Africa: State-of-the-art. *Agriculture, Ecosystems and Environment* 71:19–35.

Bationo, A., and A.U. Mokwunye. 1991. Role of manures and crop residue in alleviating soil fertility constraints to crop production: with special reference to the Sahelian and Sudanian zones of West Africa. *Fertilizer Research* 29:117–125.

Bekunda, M., N. Sanginga, and P.L. Woomer. 2010. Restoring soil fertility in Sub-Saharan Africa. *Advances in Agronomy* 108:183–286.

Bellamy, P.H., P.J. Loveland, R.I. Bradley, R.M. Lark, and G.J.D. Kirk. 2005. Carbon losses from all soils across England and Wales 1978–2003. *Nature* 437:245–248.

Bijay-Singh, Yadvinder-Singh, P. Imas, and J. Xie. 2004. Potassium nutrition of the rice-wheat cropping system. *Advances in Agronomy* 81:203–259.

Black, C.A. 1993. *Soil fertility evaluation and control*. Boca Raton, Florida, USA: Lewis Publishers.

Bossio, D., W. Critchley, K. Geheb, G. van Lynden, and B. Mati. 2007. Conserving land protecting water. In *Water for food, water for life: a comprehensive assessment of water management in agriculture*, ed. D. Molden, 551–584. London, UK: Earthscan; and Colombo, Sri Lanka: International Water Management Institute.

Bot, A., and J. Benites. 2005. *The importance of soil organic matter: Key to drought-resistant soil and sustained food and production*. Rome, Italy: Food and Agriculture Organization of the United Nations.

Brennan, R.F., R.W. Bell, C. Raphael, and H. Eslick. 2010. Sources of sulfur for dry matter, seed yield, and oil concentration of canola grown in sulfur-deficient soils of south-western Australia. *Journal of Plant Nutrition* 33:1180–1194.

Brennan, R.F., and M.D.A. Bolland. 2008. Significant nitrogen by sulfur interactions occurred for canola grain production and oil concentration in grain on sandy soils in the Mediterranean-type climate of south-western Australia. *Journal of Plant Nutrition* 31:1174–1187.

Burford, J.R., K.L. Sahrawat, and R.P. Singh. 1989. Nutrient management in Vertisols in the Indian semi-arid tropics. In *Management of Vertisols for improved agricultural production: proceedings of an IBSRAM Inaugural Workshop, ICRISAT Center, India*, 147–159. Patancheru, India: International Crops Research Institute for the Semi-Arid Tropics.

Carpenter, S.R. 2002. Ecological futures: building an ecology of the long now. *Ecology* 83: 2069–2083.

Condon, A.G., R.A. Richards, G.J. Rebetzke, and G.D. Farquhar. 2004. Breeding for high water-use efficiency. *Journal of Experimental Botany* 55:2447–2460.

Cotching, W.E., and D.B. Kidd. 2010. Soil quality evaluation and the interaction with land use and soil order in Tasmania, Australia. *Agriculture, Ecosystems and Environment* 137: 358–366.

Dahnke, W.C., and R.A. Olson. 1990. Soil test calibration, and the recommendation. In *Soil testing and plant analysis*, ed. R.L. Westerman, 45–71. Third edition, Soil Science Society of America Series 3. Madison, Wisconsin, USA: Soil Science Society of America.

den Biggelaar, C., R. Lal, K. Wiebe, and V. Breneman. 2004a. The global impact of soil erosion on productivity. I: Absolute and relative erosion-induced yield losses. *Advances in Agronomy* 81:1–48.

den Biggelaar, C., R. Lal, K. Wiebe, H. Eswaran, V. Breneman, and P. Reich. 2004b. The global impact of soil erosion on productivity. II: Effects on crop yields and production over time. *Advances in Agronomy* 81:49–95.

Doran, J.W., and T.B. Parkin. 1994. Defining and assessing soil quality. In *Defining soil quality for a sustainable environment*. Soil Science Society of America Special Publication No. 35: 3–21. Madison, Wisconsin, USA: Soil Science Society of America.

Doran, J.W., M. Sarrantonio, and M.A. Liebig. 1996. Soil health and sustainability. *Advances in Agronomy* 56:1–54.

Edmeades, D.C. 2003. The long-term effects of manures and fertilizers on soil productivity and quality: a review. *Nutrient Cycling in Agroecosystems* 66:165–180.

El-Swaify, S.A., P. Pathak, T.J. Rego, and S. Singh. 1985. Soil management for optimized productivity under rainfed conditions in the semi-arid tropics. *Advances in Soil Science* 1: 1–64.

Fageria, N.K., V.C. Baligar, and R.B. Clark. 2002. Micronutrients in crop production. *Advances in Agronomy* 77:185–268.

Freckman, D.W., and R.A. Virginia. 1997. Low-diversity Antarctic soil nematode communities: distribution and response to disturbance. *Ecology* 78:363–369.

Ghosh, K., D.C. Nayak, and N. Ahmed. 2009. Soil organic matter. *Journal of the Indian Society of Soil Science* 57:494–501.

Graham, R.D., D. Senadhira, S.E. Beebe, and C. Iglesias. 1998. A strategy for breeding staple-food with high micronutrient density. *Soil Science and Plant Nutrition* 43:1153–1157.

Graham, R.D., R.M. Welch, D.A. Saunders *et al.* 2007. Nutritious subsistence food systems. *Advances in Agronomy* 92:1–74.

Harris, F. 2002. Management of manure in farming systems in semi-arid West Africa. *Experimental Agriculture* 38:131–148.

Hiradate, S., J.F. Ma, and H. Matsumoto. 2007. Strategies of plants to adapt to mineral stresses in problem soils. *Advances in Agronomy* 96:65–132.

Izac, A.M.N. 2000. What paradigm for linking poverty alleviation to natural resources management. In *Proceedings of an International Workshop on Integrated Natural Resource Management in the CGIAR: Approaches and Lessons*, August 21–25, 2000, Penang, Malaysia.

Kanwar, J.S. 1972. Twenty-five years of research in soil, fertilizer and water management in India. *Indian Farming* 22(5):16–25.

Karlen, D.L., C.A. Ditzler, and S.S. Andrews. 2003. Soil quality: why and how? *Geoderma* 114:145–156.

Karlen, D.L., M.J. Mausbach, J.W. Doran, R.G. Cline, R.F. Harris, and G.E. Schuman. 1997. Soil quality: a concept, definition and framework for evaluation. *Soil Science Society of America Journal* 61:4–10.

Katyal, J.C. 2003. Soil fertility management – a key to prevent desertification. *Journal of the Indian Society of Soil Science* 51:378–387.

Kelly, T.G., P. Parthasarathy Rao, R.E. Weltzien, and M.L. Purohit. 1996. Adoption of improved cultivars of pearl millet in arid environment: straw yield and quality consideration in western Rajasthan. *Experimental Agriculture* 32:161–171.

Lal, R. 1995. Erosion – crop productivity relationships for soils of Africa. *Soil Science Society of America Journal* 59:661–667.

Lal, R. 1997. Degradation and resilience of soils. *Philosophical Transactions of Royal Society, London B* 352:997–1010.

Lal, R. 2007. Anthropogenic influences on world soils and implications to global food security. *Advances in Agronomy* 93:69–93.

Lal, R. 2008. Soils and sustainable agriculture. A review. *Agronomy for Sustainable Development* 28:57–64.

Lilburne, L., G. Sparling, and L. Schipper. 2004. Soil quality monitoring in New Zealand: decomposition of an interpretative framework. *Agriculture, Ecosystems and Environment* 104:535–544.

Lupwayi, N.Z., M. Grima, and I. Haque. 2000. Plant nutrient contents of cattle manures from small-scale farms and experimental stations in the Ethiopian highlands. *Agriculture, Ecosystems and Environment* 78:57–63.

Mahler, R.L., and B. Shafii. 2009. Relationship between soil test boron and pea yields in the Inland Pacific Northwest. *Communications in Soil Science and Plant Analysis* 40:2603–2615.

Mandal, D.K., C. Mandal, and M. Velayutham. 2001. Development of a land quality index for sorghum in Indian semi-arid tropics (SAT). *Agricultural Systems* 70:335–350.

Materechera, S.A. 2010. Utilization and management practices of animal manure for replenishing soil fertility among smallscale crop farmers in semi-arid farming districts of the North West Province, South Africa. *Nutrient Cycling in Agroecosystems* 87:415–428.

Mills, H.A., and J.B. Jones, Jr. 1996. *Plant analysis handbook: A practical sampling, preparation, and interpretation guide.* Athens, Georgia, USA: Micro Macro Publishing, Inc.

Molden, D. 2007. *Water for food, water for life: a comprehensive Assessment of water management in agriculture.* London, UK: Earthscan; and Colombo, Sri Lanka: International Water Management Institute.

Montgomery, D.R. 2007. Soil erosion and agricultural sustainability. *Proceedings of the National Academy of Sciences, USA* 104:13268–13272.

Morris, M., V.A. Kelly, R.J. Kipicki, and D. Byerlee. 2007. *Fertiliser use in African agriculture: lessons learned and good practice guidelines.* Washington, DC, USA: The World Bank.

Ncube, B., J.P. Dimes, S.J. Twomlow, W. Mupangwa, and K.E. Giller. 2007. Participatory on-farm trials to test response of maize to small doses of manure and nitrogen in small-holder farming systems in semi-arid Zimbabwe. *Nutrient Cycling in Agroecosystems* 77:53–67.

Ouédraogo, E., A. Mando, and N.P. Zombré. 2001. Use of compost to improve soil properties and crop productivity under low input agricultural system in West Africa. *Agriculture, Ecosystems and Environment* 84:259–266.

Oweis, T., and A. Hachum. 2009. Water harvesting for improved rainfed agriculture in the dry environments. In *Rainfed agriculture: unlocking the potential,* ed. S.P. Wani, J. Rockström, and T. Oweis, 182–196. Wallingford, UK: CAB International.

Padwick, G.W. 1983. Fifty years of experimental agriculture in the maintenance of soil fertility in tropical Africa: A review. *Experimental Agriculture* 19:293–310.

Palm, C.A., C.N. Gachengo, R.J. Delve, G. Cadisch, and K.E. Giller. 2001. Organic inputs for soil fertility management in tropical ecosystems: Application of an organic resource database. *Agriculture, Ecosystems and Environment* 83:27–42.

Parthasarathy Rao, P., P.S. Birthal, B.V.S. Reddy, K.N. Rai, and S. Ramesh. 2006. Diagnostics of sorghum and pearl millet grains-based nutrition in India. *International Sorghum and Millets Newsletter* 47:93–96.

Pasricha, N.S., and R.L. Fox. 1993. Plant nutrient sulfur in the tropics and subtropics. *Advances in Agronomy* 50:209–269.

Passioura, J.B. 2006. Increasing crop productivity when water is scarce – From breeding to field management. *Agricultural Water Management* 80:176–196.

Passioura, J.B., and J.F. Angus. 2010. Improving productivity of crops in water-limited environments. *Advances in Agronomy* 106:37–74.

Pathak, P., K.L. Sahrawat, T.J. Rego, and S.P. Wani. 2005. Measurable biophysical indicators for impact assessment: changes in soil quality. In *Natural resource management in agriculture: methods for assessing economic and environmental impact*, ed. B. Shiferaw, H.A. Freeman, and S.M. Swinton, 53–74. Wallingford, UK: CAB International.

Pathak, P., K.L. Sahrawat, S.P. Wani, R.C. Sachan, and R. Sudi. 2009. Opportunities for water-harvesting and supplemental irrigation for improving rainfed agriculture in semi-arid areas. In *Rainfed agriculture: unlocking the potential*, ed. S.P. Wani, J. Rockström, and T. Oweis, 197–221. Wallingford, UK: CAB International.

Pieri, C. 1989. Fertilite de terres de savane. Bilan de trente ans de recherché et de developpmenet agricoles au sud du Sahara. Paris, France: CIRAD, Ministere de la cooperation.

Pieri, C. 1995. Long-term soil management experiments in semi-arid Francophone Africa. In *Soil management experimental basis for sustainability and environmental quality. Advances in Soil Science*, ed. R. Lal, and B.A. Stewart, 225–266. Boca Raton, Florida, USA: Lewis Publishers.

Pimentel, D., C. Harvey, P. Resosudarmo *et al.* 1995. Environmental and economic costs of soil erosion and conservation benefits. *Science* 267:1117–1123.

Poch, R.M., and J.A. Martinez-Casanovas. 2006. Degradation. In *Encyclopedia of soil science*, Second edition, ed. R. Lal, 375–378. Philadelphia, Pennsylvania, USA: Taylor and Francis.

Rajashekhara Rao, B.K., K.L. Sahrawat, S.P. Wani, and G. Pardhasaradhi. 2010. Integrated nutrient management to enhance on-farm productivity of rain fed maize in India. *International Journal of Soil Science* 5:216–225.

Rattan, R.K., K.P. Patel, K.M. Manjaiah, and S.P. Datta. 2009. Micronutrients in soil, plant, animal and human health. *Journal of the Indian Society of Soil Science* 57:546–558.

Rego, T.J., and V.N. Rao. 2000. Long-term effects of grain legumes on rainy season sorghum productivity in a semi-arid tropical Vertisol. *Experimental Agriculture* 36:205–221.

Rego, T.J., V.N. Rao, B. Seeling, G. Pardhasaradhi, and J.V.D.K. Kumar Rao. 2003. Nutrient balances – a guide to improving sorghum and groundnut-based dryland cropping systems in semi-arid tropical India. *Field Crops Research* 81:53–68.

Rego, T.J., K.L., Sahrawat, S.P. Wani, and G. Pardhasaradhi. 2007. Widespread deficiencies of sulfur, boron, and zinc in Indian semi-arid tropical soils: on-farm crop responses. *Journal of Plant Nutrition* 30:1569–1583.

Rego, T.J., S.P. Wani, K.L. Sahrawat, and G. Pardhasaradhi. 2005. Macro-benefits from boron, zinc and sulfur application in Indian SAT: a step for gray to green revolution in agriculture. *Global Theme on Agroecosystems Report No. 16*. Patancheru, Andhra Pradesh, India: International Crops Research Institute for the Semi-Arid Tropics.

Reuter, D.J., and J.B. Robinson. 1997. *Plant analysis: An interpretation manual*, Second edition. Australia: CSIRO.

Rockström, J., J. Barron, and P. Fox. 2002. Rainwater management for increased productivity among small-holder farmers in drought prone environments. *Physics and Chemistry of the Earth* 27:949–959.

Rockström J., N. Hatibu, T. Oweis, and S.P. Wani. 2007. Managing water in rainfed agriculture. In *Water for food, water for life: a comprehensive assessment of water management in agriculture*, ed. D. Molden, 315–348. London, UK: Earthscan; and Colombo, Sri Lanka: International Water Management Institute.

Rockström, J., L. Karlberg, S.P. Wani *et al.* 2010. Managing water in rainfed agriculture – The need for a paradigm shift. *Agricultural Water Management* 97:543–550.

Saha, J.K., A.B. Singh, A.N. Ganeshamurthy, S. Kundu, and A.K. Biswas. 2001. Sulfur accumulation in Vertisols due to continuous gypsum application for six years and its effect on yield and biochemical constituents of soybean (*Glycine* max L. Merrill). *Journal of Plant Nutrition and Soil Science* 164:317–320.

Sahrawat, K.L. 2006. Plant nutrients: sufficiency and requirements. In *Encyclopedia of soil science*, Second edition, ed. R. Lal, 1306–1310. Philadelphia, Pennsylvania, USA: Taylor and Francis.

Sahrawat, K.L. 2009. The role of tolerant genotypes and plant nutrients in reducing acid-soil infertility in upland rice ecosystem: an appraisal. *Archives of Agronomy and Soil Science* 55:597–607.

Sahrawat, K.L., M.K. Abekoe, and S. Diatta. 2001. Application of inorganic phosphorus fertilizer. In *Sustaining fertility in West Africa*, ed. G. Tian, F. Ishida, and D. Keatinge, 225–246. Soil Science Society of America Special Publication no. 58. Madison, Wisconsin, USA: Soil Science Society of America and American Society of Agronomy.

Sahrawat, K.L., S.K. Mohanty, and J.R. Burford. 1991. Macronutrient transformations and budgeting in soils. In *Proceedings of the National Symposium on Macronutrients in Soils and Plants*, 198–216. Ludhiana, Punjab, India: Department of Soils, Punjab Agricultural University.

Sahrawat, K.L., K.V.S. Murthy, and S.P. Wani. 2009. Comparative evaluation of Ca chloride and Ca phosphate for extractable sulfur in soils with a wide range in pH. *Journal of Plant Nutrition and Soil Science* 172:404–407.

Sahrawat, K.L., T.J. Rego, S.P. Wani, and G. Pardhasaradhi. 2008a. Sulfur, boron and zinc fertilization effects on grain and straw quality of maize and sorghum grown on farmers' fields in the semi-arid tropical region of India. *Journal of Plant Nutrition* 31:1578–1584.

Sahrawat, K.L., T.J. Rego, S.P. Wani, and G. Pardhasaradhi. 2008b. Stretching soil sampling to watershed: Evaluation of soil-test parameters in a semi-arid tropical watershed. *Communications in Soil Science and Plant Analysis* 39:2950–2960.

Sahrawat, K.L., S.P. Wani, G. Pardhasaradhi, and K.V.S. Murthy. 2010a. Diagnosis of secondary and micronutrient deficiencies and their management in rainfed agroecosystems: Case study from Indian semi-arid tropics. *Communications in Soil Science and Plant Analysis* 41: 346–360.

Sahrawat, K.L., S.P. Wani, P. Pathak, and T.J. Rego. 2010b. Managing natural resources of watersheds in the semi-arid tropics for improved soil and water quality: A review. *Agricultural Water Management* 97:375–381.

Sahrawat, K.L., S.P. Wani, T.J. Rego, G. Pardhasaradhi, and K.V.S. Murthy. 2007. Widespread deficiencies of sulphur, boron and zinc in dryland soils of the Indian semi-arid tropics. *Current Science* 93:1428–1432.

Sanchez, P.A. 2002. Soil fertility and poverty in Africa. *Science* 29:5562–5572.

Sanchez, P.A., C.A. Palm, and S.W. Buol. 2003. Fertility capability classification: a tool to help assess soil quality in the tropics. *Geoderma* 114:157–185.

Sanchez, P.A., K.D. Shepherd, M.J. Soul *et al.* 1997. Soil fertility replenishment in Africa: An investment in natural resource capital. In *Replenishing soil fertility in Africa*, ed. R.J. Buresh, P.A. Sanchez, and F. Calhoun, 1–46. Soil Science Society of America Special Publication no. 51. Madison, Wisconsin, USA: Soil Science Society of America.

Scherer, H.W. 2001. Sulphur in crop production. *European Journal of Agronomy* 14:81–111.

Scherer, H.W. 2009. Sulfur in soils. *Journal of Plant Nutrition and Soil Science* 172:326–335.

Sharma, B.R., K.V. Rao, K.P.R. Vittal, Y.S. Ramakrishna, and U. Amarasinghe. 2010. Estimating the potential of rainfed agriculture in India: Prospects for water productivity improvement. *Agricultural Water Management* 97:23–30.

Sharma, K.L., J.K. Grace, K. Srinivas *et al.* 2009a. Influence of tillage and nutrient sources on yield sustainability and soil quality under sorghum-mung bean system in rainfed semi-arid tropics. *Communications in Soil Science and Plant Analysis* 40:2579–2602.

Sharma, K.L., Y.S. Ramakrishna, J.S. Samra *et al.* 2009b. Strategies for improving the productivity of rainfed farms in India with special emphasis on soil quality improvement. *Journal of Crop Improvement* 23:430–450.

Singh, A.K. 2008. Soil resource management – key to food and health security. *Journal of the Indian Society of Soil Science* 56:348–357.

Singh, G., S.K. Jalota, and Y. Singh. 2007. Manuring and residue management effects on physical properties under rice-wheat system in Punjab, India. *Soil and Tillage Research* 94:229–238.

Singh, P., P. Pathak, S.P. Wani, and K.L. Sahrawat. 2009. Integrated watershed management for increasing productivity and water-use efficiency in semi-arid tropical India. *Journal of Crop Improvement* 23:402–429.

Sojka, R.E., D.R. Upchurch, and N.E. Borlaug. 2003. Quality soil management or soil quality management: performance versus semantics. *Advances in Agronomy* 79:1–68.

Srinivasarao, Ch., K.P.R. Vittal, B. Venkateswarlu *et al.* 2009. Carbon stocks in different soil types under diverse rainfed production systems in tropical soils. *Communications in Soil Science and Plant Analysis* 40:2338–2356.

Srinivasarao, Ch., S.P. Wani, K.L. Sahrawat, K. Krishnappa, and B.K. Rajasekhara Rao. 2010. Effect of balanced nutrition on yield and economics of vegetable crops in participatory watersheds in Karnataka. *Indian Journal of Fertilisers* 6:39–42.

Srinivasarao, Ch., S.P. Wani, K.L. Sahrawat, T.J. Rego, and G. Pardhasaradhi. 2008. Zinc, boron and sulphur deficiencies are holding back the potential of rainfed crops in semi-arid India. *International Journal of Plant Production* 2:89–99.

Sterk, G., L. Herrmann, and A. Bationo. 1996. Wind-blown nutrient transport and soil productivity changes in southwest Niger. *Land Degradation and Development* 7:325–335.

Stoorvogel, J.J., and E.M.A. Smaling. 1998. Research on soil fertility decline in tropical environments: Integration of special scales. *Nutrient Cycling in Agroecosystems* 50:153–160.

Stringer, L.C. 2009. Reviewing the links between desertification and food insecurity: from parallel challenges to synergistic solutions. *Food Security* 1:113–126.

Subba Rao, A., Y. Muralidharudu, B.L. Lakaria, and K.N. Singh. 2009. Soil testing and nutrient recommendations. *Journal of the Indian Society of Soil Science* 57:559–571.

Takkar, P.N. 1996. Micronutrient research and sustainable agricultural productivity. *Journal of the Indian Society of Soil Science* 44:563–581.

Tandon, H.L.S. 2009. *Micronutrient handbook – from research to practical application*. New Delhi, India: Fertiliser Development and Consultation Organisation.

Tiwari, K.N. 2008. Future plant nutrition research in India. *Journal of the Indian Society of Soil Science* 56:327–336.

Turner II, B.L., E.F. Lambin, and A. Reenberg. 2007. The emergence of land change science for global environmental change and sustainability. *Proceedings of the National Academy of Sciences, USA* 104:20666–20671.

Twomlow, S., D. Love, and S. Walker. 2008a. The nexus between integrated natural resources management and integrated water resources management in southern Africa. *Physics and Chemistry of the Earth* 33:889–898.

Twomlow, S., B. Shiferaw, P. Cooper, and J.D.H. Keatinge. 2008b. Integrating genetics and natural resource management for technology targeting and greater impact of agricultural research in the semi-arid tropics. *Experimental Agriculture* 44:235–256.

Usha Rani, K., K.L. Sharma, K. Nagasri *et al.* 2009. Response of sunflower to sources and levels of sulfur under rainfed semi-arid tropical conditions. *Communications in Soil Science and Plant Analysis* 40:2926–2944.

van Asten, P.J.A. 2003. *Soil quality and rice productivity problems in Sahelian irrigation schemes*. Tropical Resources Management Papers No. 46. Wageningen, the Netherlands: Wageningen University and Research Centre, Department of Environmental Sciences.

Vanlauwe, B. 2004. Integrated soil fertility management research at TSBF: The framework, the principles, and their application. In *Managing nutrient cycles to sustain fertility in Sub-Saharan Africa*, ed. A. Bationo, 25–42. Nairobi, Kenya: Academy Science Publishers.

Verhulst, N., B. Govaerts, E. Verachtert *et al.* 2010. Conservation agriculture, improving soil quality for sustainable production systems? In *Advances in soil science: food security and soil quality*, ed. R. Lal, and B.A. Stewart, 137–208. Boca Raton, Florida, USA: CRC Press.

Wander, M.M., G.L. Walter, T.M. Nissen, G.A. Bollero, S.S. Andrews, and D.A. Cavanaugh-Grant. 2002. Soil quality: science and process. *Agronomy Journal* 94:23–32.

Wani, S.P. 2008. Taking soil science to farmers' doorsteps through community watershed management. *Journal of the Indian Society of Soil Science* 56:367–377.

Wani, S.P., P.K. Joshi, K.V. Raju *et al.* 2008. Community watershed as a growth engine for development of dryland areas. A comprehensive assessment of watershed programs in India. *Global Theme on Agroecosystems Report No. 47.* Patancheru, Andhra Pradesh, India: International Crops Research Institute for the Semi-Arid Tropics.

Wani, S.P., P. Pathak, L.S. Jangawad, H. Eswaran, and P. Singh. 2003. Improved management of Vertisols in the semiarid tropics for increased productivity and soil carbon sequestration. *Soil Use and Management* 19:217–222.

Wani, S.P., P. Pathak, H.M. Tam, A. Ramakrishna, P. Singh, and T.K. Sreedevi. 2002. Integrated watershed management for minimizing land degradation and sustaining productivity in Asia. In *Integrated land management in dry areas.* Proceedings of a Joint UNU-CAS International Workshop, 8–13 September 2001, Beijing, China, ed. Z. Adeel, 207–230. Tokyo, Japan: United Nations University.

Wani, S.P., T.K. Sreedevi, J. Rockström, and Y.S. Ramakrishna. 2009. Rainfed agriculture – past trends and future prospects. In *Rainfed agriculture: unlocking the potential*, ed. S.P. Wani, J. Rockström, and T. Oweis, 1–35. Wallingford, UK: CAB International.

Wani, S.P., Ch. Srinivasarao, T.J. Rego, G. Pardhasaradhi, and S. Roy. 2007. On-farm nutrient depletion and buildup in Vertisols under soybean (*Glycine max*) based cropping systems in semi arid central India. *Indian Journal of Dryland Agricultural Research and Development* 22:69–74.

Welch, R.M., G.F. Combs, Jr., and J.M. Duxbury. 1997. Toward a "greener" revolution. *Issues in Science and Technology* 14:50–58.

Welch, R.M., and R.D. Graham. 2002. Breeding crops for enhanced micronutrient content. *Plant and Soil* 245:205–214.

Welch, R.M., and R.D. Graham. 2004. Breeding for micronutrients in staple food crops from a human nutrition perspective. *Journal of Experimental Botany* 55:353–364.

Williams, T.O. 1999. Factors influencing manure utilization by farmers in semi-arid West Africa. *Nutrient Cycling in Agroecosystems* 55:15–22.

Williams, T.O., J.M. Powell, and S.Fernandez-Rivera. 1995. Soil fertility maintenance and crop production in semi-arid West Africa: is reliance on manure a sustainable strategy? *Outlook in Agriculture* 43:43–47.

Wilson, B.R., I. Growns, and J. Lemon. 2008. Land-use effects on soil properties on the northwestern slopes of New South Wales: implications for soil condition assessment. *Australian Journal of Soil Research* 46:359–367.

Yan, X., and W. Gong. 2010. The role of chemical and organic fertilizers on yield, yield variability and carbon sequestration – results of a 19-year experiment. *Plant and Soil* 331:471–480.

Zougmore, R., Z. Zida, and N.F. Kambou. 2003. Role of nutrient amendments in the success of half-moon soil and water conservation practice in semiarid Burkina Faso. *Soil and Tillage Research* 71:143–149.

Chapter 10

Increasing crop productivity and water use efficiency in rainfed agriculture

Piara Singh, Suhas P. Wani, Prabhakar Pathak, K.L. Sahrawat, and A.K. Singh

10.1 INTRODUCTION

Globally rainfed agriculture is very important as 80% of the world's agricultural land area is rainfed and generates 58% of the world's staple foods (SIWI 2001). Most food for poor communities in the developing countries is produced in rainfed areas; for example, in sub-Saharan Africa (SSA) more than 95% of the farm land is rainfed, while the corresponding figure for Latin America is almost 90%, for South Asia about 60%, for East Asia 65%, and for Near East and North Africa 75%. In India, 66% of 142 million ha arable land is rainfed.

Rainfed agriculture in regions characterized by erratic rainfall is subject to large inherent water related risks, which make farmers less likely to invest in production enhancing inputs. If these risks can be lowered through investments in water management techniques to bridge dry spells, farmers' attitude regarding agricultural investments might also change. In rainfed areas, rainfall is the most prominent random parameter beyond farmers' control. Hence, rainfall is both a critical input and a primary source of risk and uncertainty for agricultural production (Rockström *et al.*, 2009). The Comprehensive Assessment of Water Management in Agriculture (Molden *et al.*, 2007) also points out to a large, untapped potential for upgrading rainfed agriculture and calls for increased investments in the sector. On-farm water balance analysis indicates that in semi-arid parts of India only 30–45% of rainfall is used for crop production in the traditional management systems (Wani *et al.*, 2003b). In SSA, less than 30% of rainfall is used as productive transpiration by crops. On severely degraded land, this proportion can be as small as 5% (Rockström and Steiner 2003). Thus, crop failures commonly blamed on "drought" might be prevented in many cases through better farm-level water management.

Current irrigation water withdrawals are already causing stress in many of the world's major river basins (Molle *et al.*, 2007). The world is facing a water crisis with little scope for further expansion of large-scale irrigation. Therefore, it is necessary to improve water management in rainfed agriculture not only to secure the water required for food production (Molden *et al.*, 2007) but also to build resilience for coping with the future water related risks and uncertainties (Rockström *et al.*, 2010). Some experts are predicting further decline in rainfall and amplification of extreme events (IPCC 2007). Thus, the current state-of-affairs and future scenarios underscore the fact that in the future food needs to be met with more efficient use of water resources for providing food and livelihoods for an increasing world population. Many non-water factors also limit production in rainfed agriculture. Production is also limited by labor shortages, insecure land ownership, inadequate access to capital for investments, and limited skills and abilities. As a result, actual production often falls short of potential output.

In this chapter, we briefly describe the concepts of water use efficiency (WUE) and dwell in detail on management options to enhance WUE as a strategy to bridge the yield gaps by following the integrated water resource management (IWRM) framework. The strategies for water harvesting and its use for crop intensification are dealt by Pathak *et al.* and balanced nutrient management strategies for enhancing WUE by Sahrawat *et al.* in this volume; whereas we have discussed in detail the case study results from different semi-arid tropical (SAT) regions to demonstrate the vast scope to bridge the wide existing yield gaps between achievable and current farmers' yields through enhanced WUE.

10.2   WATER USE EFFICIENCY: CONCEPTS AND DEFINITIONS

For increasing and sustaining crop productivity or income, it is important that all the resources input into the production system are efficiently used. Any concept of efficiency is a measure of the output from a given input. There are several definitions of WUE in the literature depending upon the purpose being achieved or the emphasis being placed on the problem being solved. In the biophysical sense, sustainable production refers to maximum economic yield per unit of water being applied or used by the crop, but in the economic sense, it is maximum net income per unit of water applied or used or monetary input to the crop. Some of the definitions of WUE used in the literature are described below.

- WUE_T is the amount of dry matter or marketable yield produced per unit of water taken up (transpiration) by plants. This is also known as transpiration efficiency or transpiration ratio (yield/transpiration).
- WUE_{ET} is the amount of dry matter or marketable yield produced per unit of evapotranspiration (ET) by the crop (yield/ET). ET is the sum of soil evaporation and transpiration by the crop during the season.
- WUE_I is the amount of dry matter or marketable yield produced per unit of irrigation amount applied to the crop (yield/irrigation). Sometimes this is also referred to as water application efficiency (WAE).
- WUE_R is the amount of dry matter or marketable yield produced per unit of rainfall received by the crop or cropping system (yield/rainfall). This is also known as rainfall use efficiency (RUE).
- $WUE_{(ET/R)}$ is the ratio of water used (ET) to the amount of rainfall received by the crop or cropping system during the growing period (ET/rainfall). It is also expressed as percent of rainfall.
- $WUE_{(R+I)}$ is the amount of dry matter or marketable yield produced per unit of rainfall plus irrigation [yield/(rainfall + irrigation)] received by the crop or cropping system during the cropping period.

For a comparative study of WUE of different crops or cropping systems in response to various management practices, equivalent yields of different crops or net income per unit of ET, amount of irrigation, rainfall or rainfall plus irrigation received by the crop or cropping system may be considered. In this chapter, we have considered WUE_{ET}, WUE_R, and $WUE_{(R+I)}$ of crops and cropping systems in terms of economic yield produced or net income per unit of water input or water used.

10.3   WATER BALANCE OF CROPS IN DIFFERENT RAINFED REGIONS

Rainfed regions vary in the amount of rainfall received, its distribution and water balance during the cropping season, thus providing varying opportunities for management of rainfall for enhancing crop yields. For example, total rainfall received during the cropping period in the arid, semi-arid, and subhumid zones of India is about 460, 730, and 980 mm, respectively (Table 10.1). The amount of water used (i.e., ET) by different crops varies with their duration in different zones. Surplus water (runoff + deep drainage) for water harvesting and reuse increases from arid to subhumid zone, thus providing variable opportunities for water management to increase productivity of one crop or to extend the season to grow second or third crop through supplemental irrigation. Thus different agroclimatic zones of rainfed area in India would require different land, water, and crop management practices to enhance overall WUE and crop productivity.

Table 10.1 Average values of water balance components of major rainfed crops in different agroclimatic zones of India[a]

Crop	Agroclimate	Rainfall (mm)	Runoff (mm)	Deep drainage (mm)	Water use (mm)	Soil water change (mm)
Sorghum	Arid	440	50	50	210	130
Pearl millet	Arid	395	55	65	160	115
Soybean	Arid	417	147	0	256	14
Groundnut	Arid	510	147	47	262	54
Pigeonpea	Arid	525	118	43	353	11
Mean		457	103	41	248	65
Sorghum	Semi-arid	795	168	150	337	141
Pearl millet	Semi-arid	671	122	139	248	162
Soybean	Semi-arid	725	195	65	356	108
Groundnut	Semi-arid	687	03	79	325	81
Pigeonpea	Semi-arid	785	183	83	495	24
Mean		733	174	103	352	103
Sorghum	Subhumid	1019	289	253	357	120
Pearl millet	Subhumid	807	230	190	263	123
Soybean	Subhumid	1043	334	205	397	107
Pigeonpea	Subhumid	1052	280	170	581	22
Mean		980	283	205	399	93

[a] Source: Recalculated from the data reported by Bhatia *et al.* (2006) and Murty *et al.* (2007).

10.4 GAPS IN PRODUCTIVITY AND WATER USE EFFICIENCY

In spite of uncertainty in water availability and low crop yields, there exists the potential to increase crop yields enormously in the semi-arid areas (Wani *et al.*, 2003a). Yield gap analyses undertaken by Comprehensive Assessment for major rainfed crops in the semi-arid regions of Asia and Africa and rainfed wheat in West Asia and North Africa (WANA) region revealed large yield gaps. Farmers' yields were lower by a factor 2–4 than achievable yields for major rainfed crops grown in Asia and Africa under water limiting conditions (Singh *et al.*, 2009). In the subhumid and humid tropical zones, agricultural yields in commercial rainfed agriculture exceed $5–6\,t\,ha^{-1}$ (Rockström and Falkenmark 2000; Wani *et al.*, 2003a, 2003b). However, farmers' crop yields oscillate between 0.5 and $2\,t\,ha^{-1}$ in the region with an average of $1\,t\,ha^{-1}$ in SSA and $1–1.5\,t\,ha^{-1}$ in SAT Asia, Central Asia, and WANA for rainfed agriculture (Rockström and Falkenmark 2000; Wani *et al.*, 2003a, 2003b). In India, large yield gaps for all the major rainfed crops have been observed and with the available technologies crop yields can be doubled, demonstrating that in addition to water availability other management factors also hold back the potential of rainfed crops (Table 10.2). The potential to increase productivity of crops increases from the arid to the subhumid agroclimate in the country. Similarly, large gaps exist in the RUE among crops in various agroclimatic zones. In a detailed study, Sharma *et al.* (2010) made a crop-specific assessment of the surplus runoff water available for harvesting across dominant rainfed districts of India. According to their estimates, a surplus rainfall of 114 billion m^3 was available for harvesting from the potential rainfed cropped area (excluding very arid and wet areas) of 28.5 million ha. If only a part of this harvested water is used for providing single supplemental irrigation to rainfed crops under improved management, an average increase of 50% in total production can be expected. Water harvesting and supplemental irrigation were found to be economically viable at the national level. However, the challenge to promote the adoption of technologies that can bridge the gaps in crop yields and WUE remains to be addressed.

Table 10.2 Average value of gap in yield and rainfall use efficiency for major rainfed crops in different agroclimatic zones of India[a]

		Yield (kg ha^{-1})			Rainfall use efficiency (kg ha^{-1} mm^{-1})		
Crop	Agroclimate	Potential	Farmers'	Gap	Potential	Farmers'	Gap
Groundnut	Arid	1480	1050	430	2.9	2.1	0.8
Groundnut	Semi-arid	3138	1088	2050	4.6	1.6	3.0
Pearl millet	Arid	830	605	225	2.1	1.5	0.6
Pearl millet	Semi-arid	2462	1086	1376	3.7	1.6	2.1
Pigeonpea	Semi-arid	1428	573	855	1.8	0.7	1.1
Pigeonpea	Subhumid	1550	770	780	1.5	0.7	0.7
Sorghum	Semi-arid	3195	885	2310	4.0	1.1	2.9
Sorghum	Subhumid	3550	890	2660	3.5	0.9	2.6
Soybean	Semi-arid	1960	1205	755	2.7	1.7	1.0
Soybean	Subhumid	2538	1061	1478	2.4	1.0	1.4

[a]Source: Recalculated from the data reported by Bhatia *et al.* (2006) and Murty *et al.* (2007).

10.5   INTEGRATED APPROACH TO ENHANCE PRODUCTIVITY AND WATER USE EFFICIENCY

To increase agricultural productivity with more efficient use of water, an IWRM framework is needed that can be operationalized through integrated genetic and natural resource management (IGNRM) approach. This approach includes implementation of both scientific and supporting solutions such as enabling policies, institutions, and socioeconomic aspects for enhancing adoption of technologies and practices by farmers and the implementing agencies. An inventory of strategies, purpose, and practices that increase productivity and WUE in the framework of IWRM for rainfed agriculture is given in Table 10.3 and discussed in detail in the following section.

Table 10.3 Inventory of technologies and management practices for increasing water use efficiency in rainfed agriculture

Strategy	Purpose	Practices
Rainfall management to secure water availability		
In-situ soil and water conservation and drainage improvement	Increasing soil water availability and minimizing drought and waterlogging stresses to crops	• Land surface management: Broad-bed and furrow (BBF), ridges and furrows, micro-basins, dead furrows, staggered trenches, contour farming, contour bunds, conservation furrows, and terraces • Tillage practices and conservation agriculture • Providing green cover to reduce runoff
Ex-situ water conservation and groundwater recharge	Conserving surplus water for supplemental irrigation to mitigate dry spells and to extend the cropping season	• Surface ponds: On-farm ponds, surface micro-dams, percolation ponds, check-dams, etc. • Groundwater recharging: Percolation ponds, check-dams, gully plugs, groundwater recharging structures, and subsurface ponds • Recharging of open wells
Increasing water use and water use efficiency		
Efficient supplemental irrigation	Mitigate dry spells, extend the cropping season, crop intensification and diversification	• Efficient water conveyance and application methods • Irrigation scheduling and deficit irrigation, conjunctive use of rainfall and irrigation • Intensification and diversification with high-value crops
Increasing soil water uptake	Increasing productivity and reducing water stress	• Improved crop agronomy: Early sowing, dry planting, seeding rate, plant geometry, crop choice • Balanced plant nutrition: Integrated nutrient management, water conservation and nutrient management • Crop protection: Integrated pest/disease management practices • Intercropping, crop rotations, crop diversification • Crop intensification: Intensification of rainy season fallows and rice fallows • Contingency and dynamic cropping

(Continued)

Table 10.3 Continued

Strategy	Purpose	Practices
Reducing soil evaporation	Minimizing unproductive losses	• Mulching (plastic, straw, or stone) and microclimate modification • Conservation agriculture
Increasing plant productivity per unit of water uptake	Increasing productivity and income per unit of water used by the crop	• Breeding high-yielding and drought tolerant varieties to increase water productivity
Promoting adoption of technologies		
Enabling policies	To enhance productivity, income, and efficient water use	• Greater investments in rainfed agriculture, sustained access to resources and inputs, financial support and selective incentives for rainfed and water efficient crops, water and electricity pricing, crop insurance • Efficient markets and infrastructure
Building institutions		• Enhancing participation of rural communities and bottom-up participatory approach • Building and strengthening community-based organizations • Collective and participatory water management • Consortium partners and efficient implementing agencies
Increasing awareness and capacity building	To increase knowledge and skills and provide options for efficient use of natural resources	• Efficient knowledge sharing • Building awareness about national and international policies • Building human capital particularly empowerment of women and underprivileged groups and institutional capacities

10.6 RAINFALL MANAGEMENT TO SECURE WATER AVAILABILITY

In the semi-arid and dry subhumid zone, it is not always the amount of rainfall that is the limiting factor for production (Klaij and Vachaud 1992; Hatibu *et al.* 2003; Wani *et al.* 2003b), it is rather the extreme variability of rainfall, with high rainfall intensities, fewer rain events, and poor spatial and temporal distribution of the rainfall. By contrast, in the arid zone, crop water needs often exceed the total rainfall, causing absolute water scarcity. In the semi-arid and subhumid agroecosystems, dry spells as short periods of drought during critical growth stages occur in almost every rainy season (Barron *et al.*, 2003; Rao *et al.*, 2003). By contrast, the meteorological droughts occur on average once or twice every decade. Frequencies of both meteorological droughts and dry spells are predicted to increase with climate change (IPCC 2007). While dry spells can be bridged through investments in appropriate water management techniques, crop yields cannot be sustained during a meteorological drought and different coping mechanisms are required. Some of the available options to enhance water availability are described below.

Figure 10.1 BBF system of soil and water conservation on a Vertic Inceptisol watershed (BW7 watershed) at ICRISAT, Patancheru, India (Source: Singh *et al.*, 2009)

10.6.1 In-situ soil and water conservation

Rainfed crop production, which uses infiltrated rainfall that forms soil moisture in the root zone, accounts for most of the crop water consumption in agriculture. Soil and water conservation, or in-situ water harvesting, has been the focus of most of the investment in water management in rainfed agriculture during the past 50 years. As in-situ water harvesting can be applied on any piece of land and is affordable by most smallholder farmers, the farmers can adopt these practices with little training (Wani *et al.*, 2003b; Sreedevi *et al.*, 2004). These management systems need to be in place prior to investing in ex-situ water harvesting options. Their implementation in the field depends on the characteristics of the soil, climate, farm size, capital, and availability of human and traction power resources. Some of the in-situ water conservation practices that can be implemented for increasing soil water availability are described.

10.6.1.1 Land surface management

Land smoothening and forming field drains are basic components of land and water management for conservation and safe removal of excess water. Broad-bed and furrow (BBF) system is an improved in-situ soil and water conservation and drainage technology for the Vertisols. This system is useful for clayey soils with low infiltration capacity as soil profile gets saturated and waterlogged with the progression of rainy season. The system consists of raised bed approximately 100 cm wide and shallow furrow about 50 cm wide laid out in the field with a slope of 0.4 to 0.8% (Figure 10.1). The BBF system helps in the safe disposal of excess water through furrows when there is high intensity rainfall with minimal soil erosion, at the same time it serves as land surface treatment for in-situ moisture conservation. Contour farming is practiced on lands having medium slope (0.5–2.0%) and permeable soils, where farming operations such as plowing and sowing are carried out along the contour. The system helps to reduce

Table 10.4 Effect of land configuration on productivity of soybean- and maize-based system in the watersheds of Madhya Pradesh, India, 2001–05[a]

Watershed location	Crop	Grain yield (t ha^{-1})		Increase in yield (%)
		Farmers' practice	*BBF system*	
Vidisha and Guna	Soybean	1.27	1.72	35
	Chickpea	0.80	1.01	21
Bhopal	Maize	2.81	3.65	30
	Wheat	3.30	3.25	16

[a]Source: Singh *et al.* (2009).

Table 10.5 Rainfall use efficiency of different cropping systems under improved land management practices in Bhopal, Madhya Pradesh, India[a]

Cropping system	Rainfall use efficiency (kg ha^{-1} mm^{-1})	
	Flat-on-grade	*BBF system*
Soybean–chickpea (sequential)	8.2	11.6
Maize–chickpea (sequential)	8.9	11.6
Soybean/maize–chickpea (intercrop and sequential)	8.9	10.9

[a]Source: Singh *et al.* (2009).

the velocity of runoff by impounding water in series of depressions and thus decrease the chance of developing rills in the fields. Conservation furrows is another promising technology for Alfisols having moderate slope (0.2–0.4%) and receiving seasonal rainfall of 500–600 mm. It comprises a series of dead furrows across the slope at 3 to 5 m intervals, where the size of furrows is about 20 cm wide and 15 cm deep. Contour bunding is recommended for medium to low rainfall areas (<700 mm) on permeable soils with less than 6% slope. It consists of a series of narrow trapezoidal embankments along the contour to reduce and store runoff in the fields. The BBF system and contour bunds must be in place before sowing, while conservation furrows and other operations along the contour can be carried out at sowing or later during the crop growing season.

On-farm trials on land management of Vertisols of Central India revealed that BBF system resulted in 35% yield increase in soybean during the rainy season and yield advantage of 21% in chickpea during the postrainy season when compared with farmers' practice. Similar yield advantage was recorded in maize and wheat rotation under BBF system (Table 10.4). Yield advantage in terms of RUE was also reflected in the cropping system involving soybean-chickpea, maize-chickpea, and soybean/maize-chickpea under improved land management systems. The RUE ranged from 10.9 to 11.6 kg ha^{-1} mm^{-1} across cropping systems in BBF system compared to 8.2 to 8.9 kg ha^{-1} mm^{-1} in flat-on-grade system of cultivation on Vertisols (Table 10.5). The benefits due to conservation furrow landform treatment were also evaluated on Alfisols in the Haveri, Dharwad, and Tumkur districts of Karnataka, India. Yield advantage of 15 to 20% was recorded in maize, soybean, and groundnut with conservation furrows over farmers' practices (Table 10.6).

Table 10.6 Effect of improved land and water management on crop productivity in the Sujala watersheds of Karnataka, India during 2006–07[a]

| Watershed | Crop | Grain yield (t ha^{-1}) | | Increase in yield (%) |
		Farmers' practice	Conservation furrows	
Haveri	Maize	3.57	4.10	15
Dharwad	Soybean	1.50	1.80	20
Kolar	Groundnut	1.05	1.22	16
Tumkur	Groundnut	1.29	1.49	15

[a]Source: Singh et al. (2009).

Table 10.7 Effect of summer plowing and other agronomic practices on yield and water use efficiency (WUE) of pearl millet[a]

| Treatment | Grain yield (kg ha^{-1}) | | WUE_{ET} for grain yield (kg ha^{-1} mm^{-1}) | |
	1997	1998	1997	1998
No summer plowing	1880	1912	7.34	7.00
Summer plowing	2173	2292	8.41	7.96
Summer plowing + farmyard manure + insecticide + herbicide	2270	2509	8.73	8.50
CD ($P = 0.05$)	190	155		

[a]Source: Jat and Gautam (2001).

10.6.1.2 Tillage

Tillage roughens the soil surface and breaks apart any soil crust or compaction. This leads to increased water storage by increased infiltration into the soil as well as increased water loss by evaporation compared with residue-covered surface. After initial water loss, tilled surface soil also acts as soil mulch and reduces loss of water from the subsoil because of break of continuity of capillaries. More aggressive and frequent tillage also damages the soil structure, reduces macro porosity and reduces rainwater infiltration into the soil through the effect on hydraulic conductivity (Hatfield et al., 2001).

Jat and Gautam (2001) studied the productivity and water use of rainfed pearl millet as influenced by summer plowing and in-situ moisture conservation practices under the semi-arid conditions of New Delhi, India. Summer plowing alone or in combination with soil fertility management and crop protection practices increased productivity and WUE of pearl millet than no summer plowing (Table 10.7). Jat et al. (2006) studied the effects of tillage practices on the productivity and WUE of maize on a sandy loam in the Bhilwara region of western India. Tillage practice with summer disc plow, followed by cultivator was more beneficial to the farmer in terms of increased maize yield and higher net returns despite the higher cost of cultivation. This practice also reduced runoff by 32.9%, soil loss by 66.4%, and increased WUE by 85.7% over the practice of tilling the soil using cultivator two times at the time of sowing.

These studies indicated that summer plow on sandy loam soils of North India increases productivity and WUE of dryland crops.

Oswal and Dakshinamurti (1976) investigated the effects of subsoiling plus 2 disking, chisel-plow plus 2 disking, moldboard plowing plus 2 disking, three surface cultivations or fallowing on the yield and WUE of pearl millet and mustard; WUE was highest with subsoiling. Jin *et al.* (2007) evaluated various tillage practices on the silt loam soils of the loess plateau in China. Four years of no till followed by one subsoiling with soil cover reduced soil compaction, increased WUE (+10.5%) and yield (+12.9%) of maize and wheat as compared to traditional tillage methods, and also provided 49% economic benefit for maize and 209% for the wheat crop. The above studies indicate that tillage practices increase infiltration, reduce soil evaporation, enhance root penetration and extraction of water and nutrients from the soil profile, and increase productivity and WUE.

10.6.1.3 Conservation agriculture

The three basic elements of conservation agriculture are: (1) No or minimum tillage without significant soil inversion; (2) Retention of crop residues on the soil surface; and (3) Growing crops in rotations appropriate to the soil-climate environment and socioeconomic conditions of the region. This practice promotes in-situ conservation of rainfall, reduces soil evaporation, moderates soil temperature, improves crop productivity and soil quality through reduced soil erosion, and improves soil organic matter and other soil physical, chemical, and biological properties (Rockström and Steiner 2003; Rockström *et al.*, 2009).

Some form of conservation agriculture is practiced on 40% of the rainfed farm land in the United States and has generated an agricultural revolution in several countries in Latin America (Derpsch 2005; Landers *et al.*, 2001). Examples from SSA show that converting from plow to conservation agriculture results in yield improvements ranging between 20% and 120%, with water productivity improving from 10% to 40% (Table 10.8) (Rockström *et al.*, 2009). In northern China on the loess plateau conservation tillage (no tillage and straw management) increased wheat crop productivity and WUE by up to 35% compared to conventional tillage, especially in the low rainfall years. Conservation tillage is a more sustainable farming system in terms of increased productivity, improved soil structure, and positive environmental impacts in the dry farming areas in northern China (Li HongWen *et al.*, 2007; Wang *et al.*, 2007). Other advantages of non-inversion tillage systems include saving in labor needed for plowing. The potential disadvantages include higher costs of pest and weed control, the cost of acquiring new management skills, and investments in new planting equipment. Conservation agriculture can be practiced on all soils, especially light soils and does not require water harvesting structures. It increases productivity, sustainability, and efficient use of natural resources (Rockström *et al.*, 2009).

10.6.2 Water harvesting and groundwater recharge

In medium to high rainfall areas, despite following in-situ moisture conservation practices, rainfall runoff occurs due to high intensity storms or water surplus opportunities after filling up the soil profile. This excess water should be harvested in surface ponds

Table 10.8 Average maize grain yield under various tillage and conservation farming systems in Ethiopia, 1999–2003[a]

Treatment	Fertilized Mean yield (kg ha^{-1})	n	Non-fertilized Mean yield (kg ha^{-1})	n
Ripping + ridging	1775 (111)[a]	32	1462 (133)[bc]	19
Ripping + wing-plow	1609 (128)[ab]	19	1403 (179)[ab]	9
Ripping + subsoiling	1540 (127)[abc]	25	1266 (141)[bc]	19
Conventional/Maresha	1458 (100)[bc]	32	1258 (131)[c]	18

[a]Standard error (SE) is given in paranthesis. Values are significantly different at $P < 0.05$.
Source: Rockström *et al.* (2009).

Figure 10.2 Water harvesting structure in Wang Chai watershed in Thailand (Source: ADB 2006)

for recycling through supplemental irrigation or the groundwater should be recharged for later use in the postrainy season. The size and shape of the water harvesting structure and its location in the landscape depend upon the topography, amount of runoff expected, supplemental irrigation needs, socioeconomic condition of the farmers, and the equity concerns. Various types of water harvesting structures were tried in the Adarsha watershed in Kothapally village in Andhra Pradesh, India, Tad Fa watershed in Thailand, and Thanh Ha watershed in Vietnam with the participation of farmers (Figure 10.2). Water harvesting in these structures resulted in increase in groundwater levels (Figure 10.3). Additional water resource thus created was used by the farmers to provide supplemental irrigation to the crops especially to postrainy season crops such as chickpea or to grow high-value crops such as vegetables in these watersheds. Small, low-cost, and well distributed water harvesting structures throughout the toposequence in the watershed area provided equity and benefited more number of farmers than the large size structures which benefit only a few selected farmers (Wani *et al.*, 2003c, 2008; Pathak *et al.*, 2009).

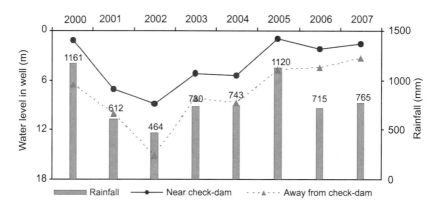

Figure 10.3 Influence of water harvesting structures on groundwater levels in Adarsha watershed, Kothapally, India (Source: Pathak *et al.*, 2009)

10.7   INCREASING WATER USE AND WATER USE EFFICIENCY

10.7.1   Efficient supplemental irrigation

In the semi-arid and subhumid agroecosystems, dry spells occur in almost every season. These dry spells need to be mitigated to save the crop from drought and minimize the climate risks to crop production in rainfed systems. Supplemental irrigation is also used to secure harvests or to provide irrigation to the second crop during the postrainy season. Supplemental irrigation systems are ex-situ water harvesting systems comprising surface ponds or recharged groundwater. Efficient use of water involves both the timing of irrigation to the crop and efficient water application methods. Broadly, the methods used for application of irrigation water can be divided into two types, viz., surface irrigation systems (border, basin, and furrow) and pressurized irrigation systems (sprinkler and drip). In the surface irrigation system, the application of irrigation water can be divided into two parts: (1) Conveyance of water from its source to the field; and (2) Application of water in the field.

10.7.1.1   Conveyance of water to the field

In most SAT areas, the water is carried to cultivated fields through open channels, which are usually unlined and therefore, a large amount of water is lost through seepage. On the SAT Vertisols, generally there is no need of lining the open field channels as the seepage losses in these soils are low mainly due to very low saturated hydraulic conductivity in the range of 0.3 to 1.2 mm h^{-1} (El-Swaify *et al.*, 1985). On Alfisols and other sandy soils having more than 75% sand, the lining of open field channel or use of irrigation pipes is necessary to reduce the high seepage water losses. The uses of closed conduits (plastic, rubber, metallic, and cement pipes) are getting popular especially with farmers growing high-value crops, viz., vegetables and horticultural crops (Pathak *et al.*, 2009).

Table 10.9 Grain yield of chickpea in different treatments on Vertisols at ICRISAT, Patancheru, Andhra Pradesh, India[a]

Treatment	Mean depth of water application (cm)	Grain yield (kg ha^{-1})
No supplemental irrigation	0	690
One supplemental irrigation on uncultivated furrows	6.3	920
One supplemental irrigation on cultivated furrow	4.6	912
SEM		19
CV (%)		5.55

[a]Source: Pathak *et al.* (2009).

10.7.1.2 Methods of application of supplemental water on SAT Vertisols

Formation of deep and wide cracks during soil drying is a common feature of the SAT Vertisols. The abundance of cracks is responsible for high initial infiltration rates (as high as $100 \, \text{mm h}^{-1}$) in dry Vertisols (El-Swaify *et al.*, 1985). This specific feature of Vertisols makes efficient application of limited supplemental water to the entire field a difficult task. Among the various systems studied at the International Crops Research Institute for the Semi-Arid Tropics (ICRISAT), Patancheru, India, the BBF system was found to be most appropriate for applying irrigation water on Vertisols. As compared to narrow ridge and furrow, the BBF system saved 45% of the water without affecting crop yields. Compared to narrow ridge and furrow and flat systems, the BBF system had higher WAE, water distribution uniformity, and better soil wetting pattern (Pathak *et al.*, 2009). Studies conducted to evaluate the effect of shallow cultivation in furrow on efficiency of water application showed that the rate of water advance was substantially higher in cultivated furrows as compared to that in uncultivated furrows. Shallow cultivation in moderately cracked furrows before the application of irrigation water, reduced the water required by about 27% with no significant difference in chickpea yields (Table 10.9).

10.7.1.3 Efficient application of supplemental water on SAT Alfisols

On Alfisols, surface irrigation on flat cultivated fields results in very poor distribution of water and high water loss. The wave-shaped BBF system, with checks at every 20 m length along the furrows, was found to be most appropriate for efficient application of supplemental water and increasing crop yields. The moisture distribution across the beds was uniform in the wave-shaped BBF system with checks compared to normal BBF system (Pathak *et al.*, 2009). Sorghum yield in wave-shaped BBF system with checks was higher at every length of run compared to normal BBF (Table 10.10). When irrigation water was applied in normal BBF system on Alfisols, the center of the broad-bed remained dry. The center row crop did not get sufficient irrigation water, resulting in poor crop yields. In another experiment on Alfisols, normal BBF system (150 cm wide) was compared with narrow ridge and furrow system (75 cm wide). The narrow ridge and furrow system performed better than BBF system both in terms of uniform water application and higher crop yields. Therefore, for Alfisols, the

Table 10.10 Sorghum grain yield as affected by water distribution in different surface irrigation systems on Alfisols[a]

	Grain yield (t ha^{-1})	
Length of run (m)	Normal BBF	Wave-shaped broad-bed with check in furrow
0	2.07	2.52
20	2.38	3.91
40	2.56	4.42
60	3.06	4.54
80	3.26	4.53
100	3.08	4.42

[a]Source: Pathak *et al.* (2009).

wave-shaped broad-bed with check in the furrow is the most appropriate land surface configuration for efficient application of supplemental irrigation water, followed by narrow ridge and furrow system (Pathak *et al.*, 2009).

The improved surge flow irrigation method can also be used for improving the performance of furrow irrigation. This system saves water, uses less energy, and improves water productivity. With proper planning and design surge flow system can be extensively used for efficiently irrigating high-value crops grown using the ridge and furrow landform (Singh 2007). Modern irrigation methods, viz., sprinklers and drip irrigation, can play vital roles in improving water productivity. These irrigation systems are highly efficient in water application and have opened up opportunities to cultivate light-textured soils with very low water-holding capacity and in irrigating undulating farm lands. The technology has also enabled regions facing limited water supplies to shift from low-value crops with high water requirements such as cereals to high-value crops with moderate water requirements such as fruits and vegetables (Sharma and Sharma 2007).

Burney *et al.* (2010) studied the role of solar-powered drip irrigation systems in enhancing food security in the Sudano-Sahelian region of Africa and concluded that the system can provide substantial economic, nutritional, and environmental benefits to the population. Implementation of these improved irrigation techniques can be used to save water and energy, and increase crop yields. However, currently the use of these improved irrigation methods are limited, primarily due to the high initial cost. Favorable government policies, availability of credit, institutional support, and training of farmers are essential for popularizing these irrigation methods.

10.7.1.4 Scheduling of irrigation and deficit irrigation

Srivastava *et al.* (1985) studied the response of postrainy season crops to supplemental irrigation grown after maize or mung bean on a Vertisol. The highest WAE was recorded for chickpea (5.6 kg mm^{-1} ha^{-1}), followed by chili (4.1 kg mm^{-1} ha^{-1}), and safflower (2.1 kg mm^{-1} ha^{-1}) (Table 10.11). It was concluded that a single pre-sowing irrigation to the sequential crops of chickpea and chili was profitable on Vertisols. Average additional gross returns due to supplemental irrigation were about ₹1630 ha^{-1} for safflower, ₹7900 ha^{-1} for chickpea, and ₹14600 ha^{-1} for chili.

Table 10.11 Response of sequential crops in the postrainy season to supplemental irrigation on a Vertisol watershed at ICRISAT, Patancheru, Andhra Pradesh, India, 1981–85[a]

Cropping system (sequential)	Yield (kg ha⁻¹)		Water application efficiency (kg mm⁻¹ ha⁻¹)
	Irrigated	Increase due to irrigation	
Maize-chickpea	1540	493	5.6
Mung bean-chili	1333	325	4.1
Maize-safflower	1238	165	2.1

[a]Source: Pathak *et al.* (2009).

Table 10.12 Grain yield response of cropping systems to supplemental irrigation on an Alfisol watershed at ICRISAT, Patancheru, Andhra Pradesh, India, 1981–82[a]

Yield with irrigation (kg ha⁻¹)	Yield increase (kg ha⁻¹)	WAE (kg ha⁻¹ mm⁻¹)	Yield with irrigation (kg ha⁻¹)	Yield increase (kg ha⁻¹)	WAE (kg ha⁻¹ mm⁻¹)	Combined WAE (kg ha⁻¹ mm⁻¹)
Intercropping system						
Pearl millet			Pigeonpea			
2353	403	10.0	1197	423	5.3	6.8
Sorghum			Pigeonpea			
3155	595	14.9	1220	535	6.7	9.4
Sequential cropping system						
Pearl millet			Cowpea			
2577	407	10.2	735	425	5.3	6.9
Pearl millet			Tomato			
2215	350	8.8	26250	14900	186.3	127.1

[a]Source: Pathak and Laryea (1991).
Irrigation of 40 mm each was applied.

Impressive benefits have also been reported from supplemental irrigation of rainy and postrainy season crops on Alfisols at ICRISAT, Patancheru, India (El-Swaify *et al.*, 1985; Pathak and Laryea 1991). The average WAE for sorghum (14.9 kg mm⁻¹ ha⁻¹) was more than that for pearl millet (8.8 to 10.2 kg mm⁻¹ ha⁻¹) (Table 10.12). An inter-cropped pigeonpea responded less to irrigation and its average WAE ranged from 5.3 to 6.7 kg mm⁻¹ ha⁻¹ for both sorghum/pigeonpea and pearl millet/pigeonpea intercrop systems. Tomato responded very well to water application with an average WAE of 186.3 kg mm⁻¹ ha⁻¹ (Table 10.12).

For the sorghum/pigeonpea intercrop, two irrigations of 40 mm each, gave an additional gross return of ₹9750 ha⁻¹. The highest additional gross return from supplemental irrigation was obtained by growing tomato (₹58300 ha⁻¹). These results indicate that on Alfisols, significant returns can be obtained from relatively small quantities of supplemental water.

The above studies indicate that on Alfisols, the best results from limited supplemental irrigation were obtained during the rainy season. On Vertisols in medium to high rainfall areas, pre-sowing irrigation for postrainy season crops was found to be the

Table 10.13 Effect of irrigation on sorghum (CSH6) yield (kg ha^{-1}) on different sections of the slope on Alfisols at ICRISAT, Patancheru, India, 1985–1986[a]

Treatment	Grain yield (kg ha^{-1})								WAE[b] (kg mm^{-1} ha^{-1})	
	Upper section (0–20 m)		Middle section (20–40 m)		Lower section (40–60 m)		Average			
	1985	1986	1985	1986	1985	1986	1985	1986	1985	1986
Rainfed	1058	2220	1618	2110	1710	2140	1659	2150	–	–
Full irrigation[c]	3716	3404	3516	3200	2960	3458	3390	3352	6.9	7.5
LID system	3413	3090	2600	2710	2000	2110	2671	2636	12.1	9.2

[a]Source: Pathak et al. (2009).
[b] Water application efficiency (WAE) = Increase in yield due to irrigation/Depth of irrigation
[c]Five irrigations totalling 250 mm and 4 irrigations totalling 130 mm were applied during 1985 and 1986 respectively on full irrigation and LID (upper section) treatments on area basis.

most beneficial. The best responses to supplemental irrigation were obtained when irrigation water was applied at critical stages of the crop. To get the maximum benefit from the available water, growing high-value crops (viz., vegetables and horticultural crops) is becoming popular even with poor farmers (Pathak *et al.*, 2009). According to Oweis (1997), supplemental irrigation of 50-200 mm can bridge critical dry spells and stabilize yields in arid to dry subhumid regions. The potential yield increase in supplemental irrigation varies with rainfall. An example from Syria illustrates that improvements in yields can be more than 400% in arid regions (Oweis 1997). Several studies indicate that supplemental irrigation systems are affordable by small-scale farmers (Fan *et al.*, 2000; Fox *et al.*, 2005). However, policy framework, institutional structure, and human capacity similar to those for full irrigation infrastructure are required to successfully apply supplemental irrigation in rainfed agriculture.

10.7.1.5 Conjunctive use of rainfall and limited irrigation water

Stewart *et al.* (1983) developed a limited irrigation dryland (LID) system for efficient use of limited irrigation water for crop production. The objective of the LID system concept was to maximize the combined use of growing-season rainfall, which varies for any given year, with a limited supply of irrigation water. This system was studied at ICRISAT, Patancheru, India for rainy season sorghum on Alfisols. Results demonstrated the usefulness of LID system in the application of limited water under uncertain and erratic rainfall conditions. The LID system increased both crop yields and WAE during the two years of study (Table 10.13).

10.7.1.6 Supplemental irrigation and crop intensification or diversification

The primary constraints for food security in developing countries are low productivity per unit area, shrinking land and water sources available for cropping, and escalating costs of crop production. Under these circumstances, crop diversification can be useful to increase crop output under different conditions of available resources either

Table 10.14 Crop diversification with high-value crops under supplemental irrigation in the Ringnodia watershed, Madhya Pradesh, India[a]

Crop	Area covered (ha)	Yield (t ha^{-1})	Net income (₹ ha^{-1})
Potato	8.3	17.5	29130
Onion	1.0	25.2	42000
Garlic	1.5	7.6	15750
Hybrid tomato	1.5	66.8	55000
Coriander	2.9	6.1	12700

[a]Source: Singh *et al.* (2009).

through broadening the base of the system by adding more crops coupled with efficient management practices or replacing the traditional crops by high-value crops. Crop diversification allows realization of the real value of improved water availability through watershed programs either through growing high-value crops like vegetables or more number of crops with supplemental irrigation. However, crop diversification takes place automatically from traditional agriculture to high-value/commercial agriculture at the field level once the water availability is improved. On-farm survey in Ringnodia watershed in Madhya Pradesh revealed the spread of high-value crops like potato, coriander, garlic, etc. and increase in net income from farming activities once the scope for supplemental irrigation was established in the watershed (Table 10.14).

10.7.2 Increasing soil water uptake

10.7.2.1 Improved crop agronomy

Many studies have clearly shown that delayed planting after the site-specific optimum date often results in grain yield losses of 4 to 7% per week. Yield reductions in late-sown wheat is often attributed to inadequate tillering and reduced transpiration late in the season (Doyle and Fischer 1979). High seeding rates can offset much of the adverse effect of late seeding (Khalifa *et al.*, 1977; Doyle and Fischer 1979). For other situations the cause for low yields may be related to occurrence of pests and diseases associated with the sowing date. Off-season tillage and early bed preparation can be of considerable benefit for timely sowing. In southern India below 18°N latitude, advancing the sowing of postrainy season crops is a simple and effective practice of increasing WUE.

In dryland agriculture, the adjustments in plant population and row spacing are often needed for optimizing the use of light and water and to achieve high harvest index of crops. However, these practices are crop, season, and site specific considering the water availability environment. A crop with high plant density uses soil moisture early in the season resulting in low grain yield. On the other hand, a crop with low plant density does not fully extract the available soil moisture and thus gives reduced yields. Steiner (1986) showed that high plant population of dryland sorghum significantly reduced grain yield because of severe decrease in harvest index. However, there was no difference in the grain yield under low and medium plant populations. These results show that harvest index and amount of water extracted can be affected by planting

Table 10.15 Effects of sulfur and micronutrient amendments on yields of selected field crops in Andhra Pradesh, India[a]

Crop	Yield (kg ha^{-1})		Increase (%) over control
	Control	Sulfur + micronutrients	
Maize	2800	4560	79
Mung bean	770	1110	51
Castor	470	760	61
Groundnut (pod)	1430	1825	28

[a]Source: Singh et al. (2009).

geometry, but the range is very wide before grain yield is severely affected. Therefore, the best strategy is to select a combination of moderate plant population and row width for higher yields and higher WUE. And these combinations are determined by the crop or variety, season, and the site where the crop is grown.

10.7.2.2   Balanced plant nutrition

Besides water scarcity, low fertility is one of the major causes for low productivity under rainfed system. The deficiency of nitrogen (N) and phosphorus (P) among the nutrients is considered an important issue in soil fertility management programs. However, the ICRISAT-led watershed program across the Indian subcontinent provided the opportunity to diagnose and understand the widespread deficiencies of secondary nutrients such as sulfur (S) and micronutrients such as boron (B) and zinc (Zn) in the soils of rainfed areas. On-farm survey across various states in India, revealed that out of 1926 farmers' fields, 88 to 100% was deficient in available S, 72–100% in available B, and 67–100% in available Zn (Sahrawat et al., 2007).

On-farm trials evaluated the response of crops to the application of S and micronutrients at the rate of 30 kg S, 0.5 kg B, and 10 kg Zn ha^{-1}. The results revealed 79% yield advantage in maize, 61% in castor, 51% in mung bean (green gram), and 28% in groundnut compared to the yield levels without the application of S and micronutrients (Table 10.15). Impressive economic gains due to improved soil fertility management to the extent of ₹5948 ha^{-1} in maize and ₹4333 ha^{-1} in groundnut were also reported from trials conducted under the ICRISAT-led watershed program in Andhra Pradesh (Table 10.16). Addition of micronutrients and S substantially increased productivity of crops and this resulted in increased RUE. The RUE of maize for grain yield under farmer nutrient inputs was 5.2 kg mm^{-1} compared to 9.2 kg mm^{-1} with S, B, and Zn application and farmer nutrient inputs; respective values in the same order of treatment were 1.6 kg mm^{-1} and 2.8 kg mm^{-1} for groundnut and 1.7 kg mm^{-1} and 2.9 kg mm^{-1} in mung bean (Table 10.17). However, addition of recommended dose of N and P along with S, B, and Zn in legumes further increased agricultural productivity, RUE, and incomes of the farmers.

Deshpande et al. (2007) investigated the effects of application of combination of mineral fertilizer (urea) and different organic sources (crop residues, sorghum waste, farmyard manure, and Leucaena loppings) on the productivity and WUE of sorghum

Table 10.16 Yield and economic returns in response to the application of nutrients in maize and groundnut in Andhra Pradesh, India[a]

	Maize		Groundnut	
Treatment	Yield increase (%) over FP	Economic returns (₹ ha^{-1})	Yield increase (%) over FP	Economic returns (₹ ha^{-1})
Farmers' practice (FP)	–	13931	–	12490
FP + S	26	17228	12	13660
FP + Zn	33	17479	27	14780
FP + B	33	18354	20	14850
FP + S + B + Zn	49	19429	48	16830
FP + S + B + Zn + N + P	75	21766	78	19520

[a]Source: Singh *et al.* (2009).

Table 10.17 Effect of micronutrient application on rainfall use efficiency in various field crops in Andhra Pradesh and Madhya Pradesh, India[a]

	Rainfall use efficiency (kg mm^{-1} ha^{-1})	
Crop	Farmers' practice	Farmers' practice + micronutrients
Andhra Pradesh		
Maize	5.2	9.2
Groundnut	1.6	2.8
Mung bean	1.7	2.9
Sorghum	1.7	3.7
Madhya Pradesh		
Soybean	1.4	2.7

[a]Source: Singh *et al.* (2009).

over five years. Higher consumptive use of water was recorded with the application of 50 kg N ha^{-1} through urea. Higher WUE was recorded with the application of 25 kg N ha^{-1} through crop residues +25 kg N ha^{-1} *Leucaena* loppings. The beneficial effects of organic material incorporation are attributed to improvements in the physical, chemical, and biological properties of the soil, e.g., infiltration rate, soil organic matter, and nutrient availability. Barros *et al.* (2007) reported that application of fertilizers (NPK) and lime to maize/cowpea intercrop on acid Acrisols of semi-arid northeastern Brazil increased biomass production and grain yield of the intercrop up to 400% and 550%, respectively. Improved crop growth with balanced nutrition reduced deep percolation and soil evaporation of rainfall, improved root development, and increased productive transpiration flow leading to overall increase in WUE. The omission of lime showed only minor effects on the evaporation and transpiration WUE. Nevertheless, the gross WUE was reduced up to 58% when lime was omitted and NPK applied at high inputs.

Table 10.18 Performance of improved varieties of finger millet and groundnut under different levels of management in Kolar and Tumkur districts, Karnataka during 2005 rainy season[a]

Finger millet			Groundnut		
	Yield (t ha^{-1})			Yield (t ha^{-1})	
Variety	Farmers' practice	Improved management	Variety	Farmers' practice	Improved management
Local	1.97	–	TMV 2 (local)	1.38	1.74
GPU 28	3.00	3.68	JL 24	1.92	2.80
MR 1	2.83	3.93	ICGV 91114	2.32	3.03
HR 911	2.90	3.66			
L 5	3.20	4.65			
Mean	3.00	4.00		1.88	2.52
Increase (%) over local variety	52	103		36	83

[a]Source: Singh *et al.* (2009).

10.7.2.3 Improved crop varieties and nutrient management

The adoption of improved varieties always generates significant field level impact on crop yield and stability. The yield advantage through the adoption of improved varieties has been recognized undoubtedly in farmer participatory trials across India under rainfed systems. Recent trials during the rainy season conducted across the Kolar and Tumkur districts of Karnataka, India revealed that mean yield advantage of 52% in finger millet was achieved with the use of high-yielding varieties like GPU 28, MR 1, HR 911, and L 5 under farmer nutrient inputs traditional management compared with use of local variety and farmer management (Table 10.18). These results showed that the efficient use of available resources by the improved varieties reflected in grain yields under given situations. However, yield advantage of 103% was reported in finger millet due to improved varieties under best-bet management practices (balanced nutrition including the application of Zn, B, and S and crop protection). Similarly, the use of improved groundnut variety ICGV 91114 resulted in pod yield of 2.32 t ha^{-1} under farmer management compared to the local variety under similar inputs. The yields of improved varieties further improved by 83% over the local variety with improved management that included balanced nutrient application (Singh *et al.*, 2009).

10.7.2.4 Water conservation practices and nutrient management

Rao *et al.* (2003) reported that the soils (Vertisols, Alfisols, Inceptisols, Entisols, and Aridisols) in rainfed areas are generally deficient in one or more nutrients. Balanced nutrition increased the amount of vegetative cover, which has a key role in reducing runoff and increasing the water infiltration. In view of the multi-nutrient deficiencies including those of major and micronutrients, the addition of optimum nutrients acts as an insurance against drought for the dryland crops. The supply of nutrients in the form of organic manures helps in retaining more moisture and increasing the water storage capacity and thereby increases water and nutrient use efficiency in drylands.

Combining the in-situ soil moisture conservation and balanced nutrient supply could boost the productivity levels in dryland agriculture (Rao *et al.*, 2003).

Degraded soils in the sub-Saharan zone are often unproductive because of nutrient imbalance and an inadequate water supply. Zougmoré *et al.* (2004) studied the effect of integrated local water and nutrient management practices on soil water balance, sorghum yield, and WUE on a Ferric Lixisol with 1.5% slope in the northern Sudanian zone of Burkina Faso. The treatments evaluated were soil and water conservation measures (stone rows, grass strips) and application of organic or mineral N-inputs (compost, manure, and urea) alone or in combination; and compared to a control treatment without N-input and soil and water conservation. The application of compost improved soil water storage in the rooting zone (0–80 cm) when combined with stone rows or grass strips and when the season had well-distributed rainfall. However, during an erratic rainy season, there was less soil water storage in the organic treatments than in the mineral source treatment. The authors concluded that the synergistic effect of water harvesting practices and the supply of organic or mineral resources increased WUE. It seems that an optimum combination of organic resources and fertilizers could improve the WUE (i.e., reduce runoff and drainage losses) and the productivity of Sahelian rainfed agriculture.

Oweis *et al.* (2003) has shown that substantial and sustainable improvements in water productivity can only be achieved through integrated farm-resources management. On-farm water-productive techniques coupled with improved irrigation management options, better crop selection and appropriate cultural practices, improved genetic make-up, and timely socioeconomic interventions help to achieve this objective. Conventional water management guidelines, designed to maximize yield per unit area, need to be revised for achieving maximum water productivity instead. A case study from Syria showed that when water is scarce, higher farm incomes can be obtained by maximizing water productivity than by maximizing land productivity.

10.7.2.5 Crop protection

Integrated pest management (IPM) is an effective and environmentally sensitive approach to pest management that relies on a combination of available pest suppression techniques to keep the pest populations below the economic thresholds. In other words, IPM is a sustainable approach to managing pests by combining biological, cultural, physical, and chemical tools in a way that minimizes economic, health, and environmental risks. New IPM products and methods are developed and extended to producers to maximize yields. On-farm trials on IPM were evaluated in the Bundi watershed, Rajasthan and Kothapally watershed in Andhra Pradesh and the results clearly demonstrated that IPM comprising use of suitable varieties, clean cultivation, scouting through pheromone traps, use of NPV (nuclear polyhedrosis virus) against lepidopteron pests, and installing bird perches resulted in yield advantage ranging from 18 to 56% for different crops. IPM practices also reduced the cost of pest management and provided stability in production as compared to farmers' practice of chemical control alone (Table 10.19). Beneficial effects on health and environment are additional bonus to the farmer and the society. Thus, IPM practices also contribute to increase in WUE through the increase in productivity per unit of rainfall or water used by the crops.

Table 10.19 Effect of IPM on the productivity of crops in Bundi and Kothapally watersheds in India[a]

Technology	Crop	Cost of pest management (₹ ha^{-1})	Yield (t ha^{-1})	Increase (%) in yield
Bundi Watershed, Madhya Pradesh				
Farmers' practice	Green peas	1800	3.53	
IPM		1080	4.16	18
Kothapally Watershed, Andhra Pradesh				
Farmers' practice	Tomato	2057	2.45	
IPM		2637	3.82	56
Farmers' practice	Cotton		1.31	
IPM			1.64	25
Farmers' practice	Pigeonpea		0.52	
IPM			0.75	44
Farmers' practice	Chickpea		0.71	
IPM			0.84	18

[a]Source: GV Ranga Rao, ICRISAT, Personal communication.

10.7.2.6   Crop intensification (double cropping)

Evidence from long-term experiments at ICRISAT, Patancheru, India since 1976, demonstrated the virtuous cycle of persistent increase in yield and RUE through improved land, water, and nutrient management in rainfed agriculture. Improved systems of sorghum/pigeonpea intercrops produced higher mean grain yields (5.1 t ha^{-1} per yr) compared to, average yield of sole sorghum (1.1 t ha^{-1} per yr) in the traditional postrainy system (farmers' practice) where crops are grown on stored soil moisture with 5 t ha^{-1} farmyard manure once in two years. The annual gain in grain yield in the improved system was 82 kg ha^{-1} compared with 23 kg ha^{-1} in the traditional system. The large gaps in yield and RUE show that a large potential of rainfed agriculture in terms of enhancing crop yields and RUE remains to be tapped (Figure 10.4). Moreover, the improved management system is still gaining in productivity as well as improved soil quality (physical, chemical, and biological parameters) along with increased carbon (C) sequestration of 330 kg C ha^{-1} per year (Wani *et al.*, 2003a, 2009).

The practice of fallowing Vertisols and associated soils during the rainy season in Madhya Pradesh has decreased after the introduction of soybean. However, it is estimated that about 2.02 million ha of cultivable land is still kept fallow in Central India, where there is a vast potential for having crop during *kharif* (rainy season) (Wani *et al.*, 2002). A recent survey of farmers' fields revealed that the introduction of rainy season crop delays sowing of the postrainy season crop and frequent waterlogging of crops during the *kharif* season forces farmers to keep the cultivable lands fallow. Under such situations, ICRISAT research demonstrated the avoidance of waterlogging during the initial crop growth period on Vertisols by preparing the fields in BBF landform along with grassed waterways. Hence, timely sowing with short-duration soybean genotypes would pave the way for successful postrainy season crop where the moisture carrying capacity is sufficiently high to support successful performance of the postrainy season crop. Yield maximization and alternate crops can be tried in the postrainy season as there is assured moisture availability in the Vertisols of the region. On-station research

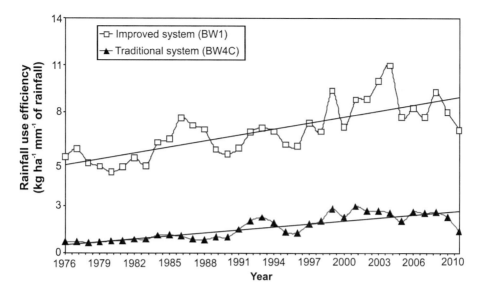

Figure 10.4 Three-year moving average of rainfall use efficiency in improved and traditional management systems during 1976–2010 at ICRISAT, Patancheru, India (Source: Recalculated from the data presented by Wani *et al.*, 2009)

was initiated with Indian Institute of Soil Science (IISS), Bhopal to address issues related to soil, water, and nutrient management practices for sustaining the productivity of soybean-based cropping systems in Madhya Pradesh. Then, the conceptual best-bet options were scaled-up in farmers' fields and yield advantages of 30 to 40% over the traditional system were recorded.

On-farm trials on soybean conducted by ICRISAT and partners to test improved land configuration (BBF system) and short-duration soybean varieties along with fertilizer application (including micronutrients) showed yield increase of 1300 to 2070 kg ha^{-1} compared to 790 to 1150 kg ha^{-1} in Guna, Vidisha, and Indore districts of Madhya Pradesh. The soybean varieties Samrat, MAUS 47, NRC 12, Pusa 16, NRC 37, JS 335, and PK 1024 were evaluated and the performance of JS 335 was better in Guna watershed than in Vidisha and Indore. Combination of improved technologies (land management, new varieties, and improved crop agronomy) increased crop yields (40–200%) and incomes (up to 100%) (Wani *et al.*, 2008).

10.7.2.7   Crop diversification with chickpea in rice fallows

It is estimated that about 11.4 million ha of rice fallows are available in India. The amount of soil moisture remaining in the postrainy season after harvest of the rice crop is usually adequate for raising a short-duration legume crop. Despite low yields of legumes grown after rice due to progressively increasing biophysical stresses, their low-cost of production and higher market prices often results in greater returns to the farmer. Thus the twin benefits of income and nutrition could be realized from legumes rather than from rice in spite of moderate yields of legumes. Introduction

Table 10.20 Productivity and water use efficiency of rice-chickpea system compared with rice-fallow system in the two districts of Chhatisgarh during 2008–09[a]

District	Rainfall + irrigation (mm)	Cropping system	Total seed yield (kg ha^{-1})	Total net income (₹ ha^{-1})	$WUE_{(R+I)}$ (kg ha^{-1} mm^{-1})	$WUE_{(R+I)}$ (₹ ha^{-1} mm^{-1})
Kanker	350	Rice-fallow	6090	55320	17.4	158
	390	Rice-chickpea	8510	84080	21.6	214
Bastar	350	Rice-fallow	3910	37300	11.1	106
	400	Rice-chickpea	5600	51480	13.9	127

[a] Source: Project Completion Report, Ministry of Water Resources, India.

of early maturing cool season chickpea in the rice fallows by addressing the crop establishment constraints will certainly improve cropping intensity and sustainability of the system. The main constraints to production of legumes in rice fallows are low P in the soil, poor plant establishment, low or absence of native Rhizobial population, root rot, and terminal drought. On-farm trials in eastern states of India on growing of early maturing chickpea in rice fallows with suitable best-bet management practices revealed that chickpea grain yields of 800-850 kg ha^{-1} can be obtained (Kumar Rao *et al.*, 2008). On-farm trials conducted in the two districts of Chhatisgarh state revealed that both productivity and WUE can be significantly increased by growing chickpea after rice. Because of high market value of chickpea, substantial improvements were recorded in the income of farmers per unit of water received by the crops (Table 10.20).

10.7.2.8 Contingent and dynamic cropping

Sadras *et al.* (2003) tested the hypothesis for the Mallee region of southeastern Australia that whole-farm profitability could be enhanced by the adoption of a dynamic cropping strategy shifting from a cereal-only, conservative strategy in dry years, to a more risky strategy involving both cereals and canola in wet years. To test this hypothesis, they used 40-years rainfall series to: (i) investigate rainfall features in 11 locations in the Mallee region, (ii) test the skill of simple rules to predict seasonal rainfall, as developed by local farmers, and (iii) calculate whole-farm profit for conservative, risky, and dynamic cropping strategies. Rainfall and profit were linked with a whole-farm model that estimates crop yield as a function of seasonal rainfall (i.e., rainfall from April to October) and WUE. Among locations, annual rainfall ranged from 259 to 358 mm. For each location, two types of seasons were defined: likely wet, when April rain was above the median and likely dry otherwise. The strength of the association between April and seasonal rain varied widely among sites; it was stronger in locations with more marked rainfall seasonality. Contrasting whole-farm profit responses to cropping strategies were found in locations with annual rainfall below or above a threshold around 300 mm. For wetter locations (annual rain above the threshold), the more risky cropping strategy including canola was generally more profitable than the more conservative strategy. For farms in drier areas, the cereal-based conservative strategy outperformed the more risky strategy in seasons predicted to be dry, but was less profitable in wet seasons. The dynamic cropping strategy had a substantial effect on extreme years, alleviating economic losses associated with the risky strategy in dry seasons, while being able to capture the benefits of more favorable seasons. Analysis of rainfall patterns, development of a rainfall forecasting procedure and quantification of whole-farm profit in response to cropping strategies, all highlighted the need for decision support tools that account for small-scale variation in rainfall characteristics.

10.7.3 Reducing soil evaporation

10.7.3.1 Mulches

In the semi-arid areas up to 50% of the rainfall is lost from the fields as non-productive soil evaporation. Converting some of that water to productive transpiration through evaporation management will increase water productivity in the arid, semi-arid, and dry subhumid regions. Options to reduce soil evaporation include dry planting, conservation agriculture, and mulching. Higher water productivity is achieved also by improving crop yields. When yields are low (between 1 and $2 \, t \, ha^{-1}$), even small improvements in yield will generate large gains in water productivity. This non-linear relationship between water productivity and yield is due to the shading of the soil when the crop canopy becomes denser with higher yield, thus changing the ratio between productive transpiration and non-productive evaporation. Hence efforts to improve crop yields are beneficial from both water saving and income enhancing perspectives.

Dang TingHui et al. (2008) studied the effects of different straw and plastic film mulching modes and N fertilizer on yield of dryland wheat from 1998 to 2003. Under the dual mulching mode of plastic film and straw, grain yield increased by 12.11–17.65%, WUE increased by 7.2–30.8%, water content in the arable layer increased to 12–16%, and the nitrate N content in the arable layer increased to $4.70–10.17 \, mg \, kg^{-1}$. The dual mulching mode of plastic film and straw significantly increased crop yield and WUE; thus nitrate leaching and accumulation in the soil profile was alleviated. Similar results on increase in productivity and WUE for maize have been reported (Mai ZiZhen et al., 2007).

10.7.3.2 Microclimate modifications

The presence of windbreaks usually reduces ET by the crop. For this reason, the windbreak barrier is included among the agro-techniques specific for the dry farming systems. Campi et al. (2009) studied the effect of windbreaks on crop water requirements and yield on durum wheat growing in open field, in a typical Mediterranean environment. A windbreak of Cupressus arizonica (3 m in height) bordered at north of the experimental field. The analysis of the microclimatic observations showed that when wind blew from the North, the windbreak influenced the wind speed until the distance $12.7H$ (H is the windbreak height) and temperature increased in a distance of $4.7H$ from the barrier. On the basis of the soil water content, continuously measured by Time Domain Reflectometry (TDR) technique, ET was daily determined and season ET calculated. Windbreaks mitigated ET for a distance of 12.7 times the windbreak height. Outside this area, the ET was 16% higher than the ET measured near the windbreak belt ($<4.7H$). Yield performances changed according to the distance from the windbreak. Within the distance of 18 times the windbreak height, wheat production was higher than that obtained in the zone not influenced by the windbreak. Within

Table 10.21 Changes in water productivity from the adoption of sustainable agricultural practices in 144 projects by crop type (kg of produce per m³ of water used)[a]

Crop	Before intervention	After intervention	Gain	Increase (%)
Irrigated agriculture				
Rice (18 projects)	1.03 (±0.52)	1.19 (±0.49)	0.16 (±0.16)	15.5
Cotton (8 projects)	0.17 (±0.10)	0.22 (±0.13)	0.05 (±0.05)	29.4
Rainfed agriculture				
Cereals (80 projects)	0.47 (±0.51)	0.80 (±0.81)	0.33 (±0.45)	70.2
Legumes (19 projects)	0.43 (±0.29)	0.87 (±0.68)	0.44 (±0.47)	102.3
Roots and tubers (14 projects)	2.79 (±2.72)	5.79 (±4.04)	3.00 (±2.43)	107.5

[a]Figures in parentheses are standard errors. Source: Pretty *et al.* (2006).

the protected area, wheat WUE (calculated as the ratio between yield and seasonal ET) attained the maximum value of 1.15; outside the area of windbreak protection, WUE was $0.70 \, kg \, m^{-3}$. Since windbreaks reduce ET, farms of the Mediterranean environments should be redesigned in order to consider windbreaks as a possible issue of sustainability.

10.7.3.3   Land degradation, conservation agriculture, and water use efficiency

Land degradation reduces WUE at field and landscape scales and affects water availability, quality, and storage. Because of this strong link between land and water productivity, improving water management in agriculture requires that land degradation be mitigated or prevented. Bossio *et al.* (2010) reviewed the global experiences relating to land degradation and highlighted the important degradation processes (loss of soil organic matter, soil physical degradation, nutrient depletion, chemical degradation, soil erosion and sedimentation, and degradation of landscape functions) that are closely linked to water use and management. Investing in improved land management, such as resources-conserving technologies, can considerably improve on-farm WUE in both rainfed and irrigated agriculture (Table 10.21) (Bossio *et al.*, 2008). Resources-conserving technologies cover a broad range of systems which have the potential to improve WUE and water management in various ways. For example, soil management practices (such as zero till) to improve infiltration and soil water storage can boost WUE by an estimated 25–40%, while nutrient management can boost WUE by 15–25% (Hatfield *et al.*, 2001). Water productivity improvements can range from 70 to 100% in rainfed systems using resources-conserving technologies that enhance soil fertility and reduce water evaporation (Pretty *et al.*, 2006).

10.7.4   Crop breeding for increased water productivity

Improved crop varieties that produce more biomass and economic yield per unit of water uptake in water stressed environments would enhance the overall WUE of the production system. Gowda *et al.* (2009) reviewed genetic enhancement of dryland crops for improving crop water productivity. A combination of approaches has been employed to enhance the adaptation of crops to varying water availability environments. As a result of these approaches, several genetically enhanced products have been developed, some of which have reached the farmers' fields. Products of marker-assisted selection in pearl millet and maize have shown superior performance under severe drought conditions. There are several other successful plant breeding efforts that have improved plant water productivity (Dingkuhn *et al.*, 2006; Richards 2006).

10.8 PROMOTING ADOPTION OF TECHNOLOGIES

10.8.1 Enabling policies

In spite of greater investments in irrigated agriculture, crop yields have reached a plateau in most of the irrigated areas that had brought green revolution in Asia and other parts of the globe. The reasons for this are many and it is becoming difficult to enhance productivity and WUE with additional incremental input. Greater investments need to be made by governments in rainfed agriculture to close the yield gaps for efficient use of natural resources, particularly water. Greater national and local-level enabling policies are needed for the adoption of better practices by farmers. This would include sufficient and sustained access to natural resources (e.g., land and water), agricultural inputs, and credit to the farmers. Currently, these are insufficient and not at the desired level and need favorable policy changes in terms of enhancing their availability and easy access. For optimal water management, policies are needed to promote low-cost soil and water conservation measures and structures that have proved more useful and economical and more equitable than the large dams, which require heavy investments in the irrigated command area. In spite of increased water harvesting in rainfed areas, the groundwater levels in the rainfed areas are declining because of increased extraction of water by relatively rich farmers using bore wells. Water sharing mechanisms and water markets need to be developed for more efficient use of water.

10.8.2 Building institutions

To facilitate adoption of technologies and practices that enhance productivity and efficient use of resources, various institutions both at the local and state level need to be in place. First at the local level, all developmental approaches need to be farmer participatory, demand driven, and must provide tangible benefits to the farmers. Factors that promote collective action by the community, especially for the management and use of water resources, need to be promoted and practices that promote equity, equal partnership, shared vision, and trust should be encouraged. A consortium of institutions for technical guidance on the management of watersheds and water resources need to be fully operational. Institutional mechanisms at the district, state, and national level that promote adoption of technologies need to be more efficient in delivery to the farmers (Wani *et al.*, 2009). Water management institutions at the local level for participatory groundwater management need to be supported and guided for more efficient use of water resources. Various institutional structures such as market, finance, and risk management need to be in place for transaction of agricultural inputs and outputs by farmers and to cover the risks associated with farming in rainfed areas.

10.8.3 Raising awareness and capacity building

Farming communities have traditional knowledge, which they have been applying over years for the management and use of natural resources, while the scientific community and the government implementing agencies are more equipped with the new knowledge gathered through scientific developments. In many situations the traditional knowledge needs to be blended with new developments for upgrading the practices for sustainable agricultural development. There is a need to empower communities and village-level institutions for adopting new technologies that enhance productivity and efficient use of natural resources that enhance productivity and a strong social and human capital will enhance sustainability of the developmental programs. Often the poor and women take care of agriculture when the men migrate for alternative livelihood. Their participation in water user groups and capacity building could prove more meaningful and effective in management of natural resources. As the extension services are not able to cope with the growing needs of the farmers, new ICT (information and communications technology) based knowledge transfer systems need to be put in place at village levels for serving the needs of farmers (Sreedevi and Wani 2009).

10.9 SUMMARY AND CONCLUSIONS

Globally, rainfed agriculture is very important as it generates 58% of the world's staples; and most food of the poor communities is produced in the rainfed areas of developing countries. Underdevelopment, rapid population increase, land degradation, climate uncertainty, water scarcity, and unfavorable government policies are the major bottlenecks to achieving higher agricultural production and improved rural livelihoods. The farming community is facing water crisis due to excessive groundwater withdrawal and climate change with little scope for expansion of large-scale irrigation systems. Therefore, the available water resources need to be used more efficiently not only for increasing crop production, but also to build resilience for coping with the climate related risks and uncertainties. An assessment of potential yields and water use by crops have indicated that there are large gaps in productivity and WUE of rainfed crops in different agroclimatic regions. The potential to increase productivity and WUE of crops increases from the arid to subhumid agroclimate where more surplus water is available for increasing total productivity per unit of land. Therefore, an IWRM approach is needed, which involves efficient management of all the components of water cycle for enhancing crop productivity to meet the current and future food security of the nations. A large number of rainwater conservation and management and productivity enhancing technologies have been successfully demonstrated by various workers at research stations and under on-farm situations in farmers' fields. The challenge is how to further scale-up the adaptation and adoption of these technologies by large number of farmers in different agroclimatic zones of rainfed areas to improve productivity of water. Climate change will have both direct and indirect adverse impact on productivity and WUE of production systems. The currently available technologies will have to be further fine tuned or newer ones may have to be developed to make the production systems more sustainable and resilient to the adverse impacts of climate change. Production is also limited by various non-water factors such as labor shortage, insecure land ownership, inadequate access to capital for investments, and limited skills and abilities. Therefore, for increased adoption of crop productivity and WUE, enhancing technologies by rural communities, appropriate institutional and policy support, and increased awareness and capacity building of stakeholders at different levels are essential to achieve the overall goal of food security and resilience in the rainfed areas.

REFERENCES

ADB. 2006. Participatory watershed management for reducing poverty and land degradation in SAT Asia. Technical Assistance (RETA 6067) Completion Report (Jan 2003–June 2006). Asian Development Bank.

Barron, J., J. Rockström, F. Gichuki, and N. Hatibu. 2003. Dry spell analysis and maize yields for two semi-arid locations in East Africa. *Agricultural Forest Meteorology* 117(1–2):23–37.

Barros, I. De, T. Gaiser, F.M. Lange, and V. Römheld. 2007. Mineral nutrition and water use patterns of a maize/cowpea intercrop on a highly acidic soil of the tropic semiarid. *Field Crops Research* 101(1):26–36.

Bhatia, V.S., Piara Singh, S.P. Wani, A.V.R. Kesava Rao, and K. Srinivas. 2006. Yield gap analysis of soybean, groundnut, pigeonpea and chickpea in India using simulation modeling. *Global Theme on Agroecosystems Report No. 31*. Patancheru, Andhra Pradesh, India: International Crops Research Institute for the Semi-Arid Tropics.

Bossio, D., K. Geheb, and W. Critchley. 2010. Managing water by managing land: addressing land degradation to improve water productivity and rural livelihoods. *Agricultural Water Management* 97(4):536–542.

Bossio, D., A. Noble, D. Molden, and V. Nangia. 2008. Land degradation and water productivity in agricultural landscapes. In *Conserving land, protecting water,* ed. D. Bossio, and K. Geheb, 20–32. Wallingford, UK: CABI Publishing; and Colombo, Sri Lanka: International Water Management Institute.

Burney, J.L., M. Woltering, M. Burke, R. Naylor, and Dov Pasternak. 2010. Solar-powered drip irrigation enhances food security in the Sudano-Sahel. *PNAS* 107(55):1848–1853.

Campi, P., A.D. Palumbo, and M. Mastrorilli. 2009. Effects of tree windbreak on microclimate and wheat productivity in a Mediterranean environment. *European Journal of Agronomy* 30(3):220–227.

Dang TingHui, Guo Dong, and Qi LongHai. 2008. Effects of wheat yield and water use under dual-mulching mode of plastic film and straw in the dryland farming. *Transactions of the Chinese Society of Agricultural Engineering* 24(10):20–24.

Derpsch, R. 2005. The extent of conservation agriculture adoption worldwide: implications and impact. Keynote paper at the 3rd World Congress on Conservation Agriculture, Regional Land Management Unit, World Agroforestry Centre, 3–7 October, Nairobi, Kenya.

Deshpande, A.N., P.B. Jagtap, B.G. Gaikwad, and A.L. Pharande. 2007. Effect of long term application of organic materials in soil and its effect on soil properties and performance of sorghum under dryland conditions during post-rainy season. *Indian Journal of Dryland Agricultural Research and Development* 22(1):41–47.

Dingkuhn, M., B.B. Singh, B. Clerget, J. Chanterau, and B. Sultan. 2006. Past, present and future criteria to breed crops for water-limiting environments in West Africa. *Agricultural Water Management* 80(1–3):241–261.

Doyle, A.D. and R.A. Fischer. 1979. Dry matter accumulation and water use relationships in wheat crops. *Australian Journal of Agricultural Research* 30:815–829.

El-Swaify, S.A., P. Pathak, T.J. Rego and S. Singh. 1985. Soil management for optimized productivity under rainfed conditions in the semi-arid tropics. *Advances in Soil Science* 1:1–64.

Fan, S., P. Hazell, and P. Haque. 2000. Targeting public investments by agro-ecological zone to achieve growth and poverty alleviation goals in rural India. *Food Policy* 25(4):411–428.

Fox, P., J. Rockström, and J. Barron. 2005. Risk analysis and economic viability of water harvesting for supplemental irrigation in semi-arid Burkino Faso and Kenya. *Agricultural Systems* 83(3):231–250.

Gowda, C.L.L., R. Serraj, G. Srinivasan *et al.* 2009. Opportunities for improving crop water productivity through genetic enhancement of dryland crops. In *Rainfed agriculture: unlocking the potential,* ed. S.P. Wani., J. Rockström, and T. Oweis. 133–163. Comprehensive Assessment of Water Management Series. Wallingford, UK: CAB International.

Hatfield, J.L., T.J. Sauer, and J.H. Prueger. 2001. Managing soils to achieve greater water use efficiency: a review. *Agronomy Journal* 93(2):271–280.

Hatibu, N., M.D.B. Young, J.W. Gowing, H.F. Mahoo, and O.B. Mzirai. 2003. Developing improved dryland cropping systems for maize in semi-arid Tanzania. Part 1: Experimental evidence of the benefits of rainwater harvesting. *Journal of Experimental Agriculture* 39(3):279–292.

IPCC. 2007. Climate change–impacts, adaptation and vulnerability. Technical Summary of Working Group II. In *Fourth Assessment Report of Inter-governmental Panel on Climate Change,* ed. M.L. Parry, O.F. Canziani, J.P. Paultikof *et al.,* 23–78. Cambridge, UK: Cambridge University Press.

Jat, M.L., and R.C. Gautam. 2001. Productivity and water use of rainfed pearl millet (*Pennisetum glaucum*) as influenced by summer plowing and *in-situ* moisture-conservations of north-west India. *Indian Journal of Agronomy* 46(2):266–272.

Jat, M.L., P. Singh, S.K. Sharma, J.K. Balyan, R.K. Sharma, and L.K. Jain. 2006. Energetics and profitability of tillage practices for maize (*Zea mays* L.) cultivation in semi arid tropics. *Current Agriculture* 30(1/2):11–21.

Jin, H., Li Hong Wen, Wang Xiao Yan, A.D. McHugh, Li Wen Ying, Gao Huan Wen, and N.J. Kuhn. 2007. The adoption of annual subsoiling as conservation tillage in dryland maize and wheat cultivation in northern China. *Soil and Tillage Research* 94(2):493–502.

Khalifa, M.A., M.H. Akasha, and M.B. Said. 1977. Growth and N-uptake by wheat as affected by sowing date and nitrogen in irrigated semi-arid conditions. *Journal of Agricultural Science* 89:35–42.

Klaij, M.C., and G. Vachaud. 1992. Seasonal water balance of a sandy soil in Niger cropped with pearl millet, based on profile moisture measurements. *Agricultural Water Management* 21(4):313–330.

Kumar Rao, J.V.D.K., D. Harris, M. Kankal, and B. Gupta. 2008. Extending *rabi* cropping in rice fallows of eastern India. In *Improving agricultural productivity in rice-based systems of the High Barind Tract of Bangladesh,* ed. C.R. Riches, D. Harris, D.E. Johnson, and B. Hardy, 193–200. Los Banos, Philippines: International Rice Research Institute.

Landers, J.N., H. Mattana Saturnio, P.L. de Freitas, and R. Trecenti. 2001. Experiences with farmer clubs in dissemination of zero tillage in tropical Brazil. In *Conservation agriculture, a worldwide challenge,* ed. L. Garcia-Torres, J. Benites, and A. Martinez-Vilela. Rome, Italy: Food and Agriculture Organization of the United Nations.

Li HongWen., HuanWen Gao, HongDan Wu, Li Weng Ying, Hizo Yan Wang, and Jin He. 2007. Effects of 15 years of conservation tillage on soil structure and productivity of wheat cultivation in northern China. *Australian Journal of Soil Research* 45(5):344–350.

Mai ZiZhen, Luo ShiWu, Cheng BingWen, and Wang Yong. 2007. Soil water content dynamics and water use efficiency under plastic film and straw dual-mulching in maize fields. *Chinese Journal of Eco-Agriculture* 15(3):68–70.

Molden, D., K. Frenken, R. Barker *et al.* 2007. Trends in water and agricultural development. In *Water for food, water for life: a comprehensive assessment of water management in agriculture,* ed. D. Molden, 57–89. London, UK: Earthscan; and Colombo, Sri Lanka: International Water Management Institute.

Molle, F., P. Wester, P. Hirsch *et al.* 2007. River basin development and management. In *Water for food, water for life. a comprehensive assessment of water management in agriculture*, ed. D. Molden, 585–625. London, UK: Earthscan; and Colombo, Sri Lanka: International Water Management Institute.

Murty, M.V.R., Piara Singh, S.P. Wani, I.S. Khairwal, and K. Srinivas. 2007. Yield gap analysis of sorghum and pearl millet in India using simulation modeling. *Global Theme on Agro-ecosystems Report No. 37*. Patancheru, Andhra Pradesh, India: International Crops Research Institute for the Semi-Arid Tropics.

Oswal, M.C., and C. Dakshinamurti. 1976. Effect of different tillage practices on water-use efficiency of pearl-millet and mustard under dry-farming conditions. *Indian Journal of Agricultural Sciences* 45(6):264–269.

Oweis, T. 1997. *Supplemental irrigation: A highly efficient water-use practice*. Aleppo. Syria: International Center for Agricultural Research in the Dry Areas.

Oweis, T.Y., A.Y. Hachum, J.W. Kijne, R Barker, and D. Molden. 2003. Improving water productivity in the dry areas of West Asia and North Africa. *Water productivity in agriculture: limits and opportunities for improvement*, 179–198. Wallingford, UK: CABI Publishing.

Pathak, P., and K.B. Laryea. 1991. Prospects of water harvesting and its utilization for agriculture in the semi-arid tropics. In *Proceedings of the symposium of the SADCC land and water management research program scientific conference, 8–10 Oct 1990, Gaborone, Botswana*, 253–268.

Pathak, P., K.L. Sahrawat, S.P. Wani, R.C. Sachan, and R. Sudi. 2009. Opportunities for water harvesting and supplemental irrigation for improving rainfed agriculture in semi-arid areas. In *Rainfed agriculture: unlocking the potential*, ed. S.P. Wani, J. Rockström, and T. Oweis, 197–221. Wallingford, UK: CAB International.

Pretty, J., A. Noble, D. Bossio *et al.* 2006. Resource-conserving agriculture increases yields in developing countries. *Environmental Science and Technology* 40(4):1114–1119.

Rao, C.S., J.V.N.S. Prasad, K.P.R. Vittal, B. Venkateswarlu, and K.L. Sharma. 2003. Role of optimum plant nutrition in drought management in rainfed agriculture. *Fertiliser News* 48(12):105–114.

Richards, R.A. 2006. Physical traits used in the breeding of new cultivars for water-scarce environments. *Agricultural Water Management* 80(1–3):197–211.

Rockström, J. and M. Falkenmark. 2000. Semiarid crop production from a hydrological perspective: gap between potential and actual yields. *Critical Reviews in Plant Science* 19(4): 319–346.

Rockström, J., L. Karlberg, S.P. Wani *et al.* 2010. Managing water in rainfed agriculture – The need for a paradigm shift. *Agricultural Water Management* 97:543–550.

Rockström, J., P. Kaumbutho, J. Mwalley *et al.* 2009. Conservation farming strategies in East and Southern Africa: Yields and rain water productivity from on-farm action research. *Soils and Tillage Research* 103:23–32.

Rockström, J., and K. Steiner. 2003. Conservation farming–a strategy for improved agricultural and water productivity among small-holder farmers in drought prone environments. In *Water conservation technologies for sustainable dryland agriculture in sub-Saharan Africa*, ed. D. Beukes, M. De Villiers, S. Mkhize *et al.* Proceedings of the International Water Conservation Technologies Workshop, Blomfontein, South Africa, April 8–11.

Sadras, V., D. Roget and M. Krause. 2003. Dynamic cropping strategies for risk management in dry-land farming systems. *Agricultural Systems* 76(3):929–948.

Sahrawat, K.L., S.P. Wani, T.J. Rego, G. Pardhasaradhi, and K.V.S. Murthy. 2007. Widespread deficiencies of sulphur, boron and zinc in dryland soils of the Indian semi-arid tropics. *Current Science* 93:1428–1432.

Sharma, B.R., K.V. Rao, K.P.R. Vittal, Y.S. Ramakrishna, and U. Amarasinghe. 2010. Estimating the potential of rainfed agriculture in India: Prospects for water productivity improvements. *Agricultural Water Management* 97:23–30.

Sharma, K.D., and Sharma Anupama. 2007. Strategies for optimization of groundwater use for irrigation. In *Ensuring water and environment for prosperity and posterity. Souvenir.* 10th Inter-regional Conference on Water and Environment (ENVIROWAT 2007), 17–20 October 2007, organized by Indian Society of Water Management in collaboration with Indian Society of Agricultural Engineers and International Commission on Agricultural Engineering. pp. 52–58.

Singh, H.P. 2007. Enhancing water productivity in horticultural crops. In *Ensuring water and environment for prosperity and posterity. Souvenir.* 10th Inter-regional Conference on Water and Environment (ENVIROWAT 2007), 17–20 October 2007 organized by Indian Society of Water Management in collaboration with Indian Society of Agricultural Engineers and International Commission on Agricultural Engineering pp. 40–48.

Singh, P., P. Pathak, S.P. Wani, and K.L. Sahrawat. 2009. Integrated watershed management for increasing productivity and water use efficiency in semi-arid tropical India. *Journal of Crop Improvement* 23(4):402–429.

SIWI. 2001. Water harvesting for upgrading of rainfed agriculture. Policy analysis and research needs. SIWI Report II. Stockholm, Sweden: Stockholm International Water Institute.

Sreedevi, T.K., B. Shiferaw, and S.P. Wani. 2004. Adarsha watershed in Kothapally: understanding the drivers of higher impact. *Global Theme on Agroecosystems Report No. 10.* Patancheru, Andhra Pradesh, India International Crops Research Institute for the Semi-Arid Tropics.

Sreedevi, T.K., and S.P. Wani 2009. Integrated farm management practices and upscaling the impact for increased productivity of rainfed systems. In *Rainfed agriculture: unlocking the potential,* ed. S.P. Wani, J. Rockström, and T. Oweis, 222–257. Comprehensive Assessment of Water Management in Agriculture Series. Wallingford, UK: CAB International.

Srivastava, K.L., P. Pathak, J.S. Kanwar, and R.P. Singh. 1985. Watershed-based soil and rain water management with special reference to Vertisols and Alfisols. Presented at the National Seminar on Soil Conservation and Watershed Management, 5–7 Sep 1985, New Delhi, India.

Steiner, J.L. 1986. Dryland grain sorghum water use. Light interception and growth responses to planting geometry. *Agronomy Journal* 78:720–726.

Stewart, B.A., J.T. Musick, and D.A. Dusek. 1983. Yield and water use efficiency of grain sorghum in a limited irrigation-dryland farming system. *Agronomy Journal* 75(4): 629–634.

Wang, X.B., D.X. Cai, W.B. Hoogmoed, O. Oenema, and U.D. Perdok. 2007. Developments in conservation tillage in rainfed regions of North China. *Soil and Tillage Research* 93(2):239–250.

Wani, S.P., R.S. Dwivedi, K.V. Ramana, A. Vadivelu, R.R. Navalgund, and A.B. Pande. 2002. Spatial distribution of rainy season fallows in Madhya Pradesh: Potential for increasing productivity and minimizing land degradation. *GT3: Water, Soil and Agrobiodiversity Management for Ecosystem Health. Report No. 3.* Patancheru, Andhra Pradesh, India: International Crops Research Institute for the Semi-Arid Tropics.

Wani, S.P., P.K. Joshi, Y.S. Ramakrishna *et al.* 2008. A new paradigm in watershed management: a must for development of rainfed areas for inclusive growth. In *Conservation farming: enhancing productivity and profitability of rainfed areas,* ed. Anand Swarup, Suraj Bhan, and J.S. Bali, 163–178. New Delhi, India: Soil Conservation Society of India.

Wani, S.P., P. Pathak, L.S. Jangawad, H. Eswaran, and P. Singh. 2003a. Improved management of Vertisols in the semiarid tropics for increased productivity and soil carbon sequestration. *Soil Use and Management* 19(3):217–222.

Wani, S.P., P. Pathak, T.K. Sreedevi, H.P. Singh, and P. Singh. 2003b. Efficient management of rainwater for increased crop productivity and groundwater recharge in Asia. In *Water productivity in agriculture: limits and opportunities for improvement,* ed. J.W. Kijne, R. Barker, and D. Molden. Wallingford, UK: CABI Publishing; and Colombo, Sri Lanka: International Water Management Institute.

Wani, S.P., A. Ramakrishna, T.J. Rego, T.K. Sreedevi, P. Singh, and P. Pathak. 2003c. Combating land degradation for better livelihoods: The Integrated Watershed Approach. *Journal of Arid Land Studies* 14:115–118.

Wani, S.P., T.K. Sreedevi, J. Rockström, and Y.S. Ramakrishna. 2009. Rainfed agriculture – past trends and future prospects. In *Rainfed agriculture: unlocking the potential,* ed. S.P. Wani., J. Rockström, and T. Oweis, 1–35. Comprehensive Assessment of Water Management in Agriculture Series. Wallingford, UK: CAB International.

Zougmoré, R., A. Mando, and L. Stroosnijder. 2004. Effect of soil and water conservation and nutrient management on the soil-plant water balance in semi-arid Burkina Faso. *Agricultural Water Management* 65(2):103–120.

Chapter 11

Impact of watershed projects in India: Application of various approaches and methods

K. Palanisami, Bekele Shiferaw, P.K. Joshi, S. Nedumaran, and Suhas P. Wani

11.1 INTRODUCTION

Governments in developing countries like India actively pursue various forms of policy instruments like the implementation of development programs to achieve desired economic growth. The objective of such development programs is to transform a set of resources into desired results for upscaling. This is particularly so for the policies designed to alleviate rural poverty and foster economic growth in the agricultural sector of developing economies. For achieving these goals, understanding the nature, objectives, and scope of the development program and the responsiveness of target groups is imperative for all those engaged in developmental work including staff and policy makers. This applies to watershed program staff engaged in the development and implementation of technologies for enhancing food, fodder, and fuel productivity and ensure livelihood security for those below the poverty line. Thus, a systematic feedback from the project areas and beneficiaries is of crucial importance. Evaluation and monitoring studies provide the needed information for upscaling the interventions by implementing agencies. The objective of this chapter is to discuss in detail various methodologies employed in evaluating the performances of the watershed development programs in India. Given the significance of the watershed programs in meeting the challenges especially in rainfed agriculture, it is important to see how the issues facing the watershed evaluation could be addressed through the review of the different evaluation criteria and methods that have been field tested already. This chapter aims to derive the messages from the past studies focusing on the measurement methodologies in watershed evaluation. The chapter is organized into four sections. Section 1 deals with the introduction and an overview of the watershed development programs in India, section 2 outlines the various approaches used in watershed evaluation, section 3 applies the various methodologies with examples from the fields, and section 4 gives the conclusions and policy recommendations.

11.1.1 An overview of watershed development programs in India

The concept and history of watershed development in India started way back in 1880 with the Famine Commission and then in the Royal Commission of Agriculture in 1928. Both Commissions laid the foundation for organized research in a watershed framework. Small-scale watershed development programs to conserve soil and water and prevent land degradation began during the early twentieth century, e.g., Lingajat

Peetadhipathi, near Bijapur in Karnataka. The activities included construction of bunds in the then Bombay Provinces for rural employment during drought relief operations. In this sequence, Bombay Land Development Act, 1943, provided a model for other states enlightening watershed development. Realizing the importance of the watershed programs for land reclamation, a multidisciplinary Soil Conservation Department was set up at Hazaribagh under the Damodar Valley Corporation. Then the Government of India supported program started in the mid-1950s and the focus on watershed programs was sharpened with the establishment of the Soil Conservation Research, Demonstration and Training Center at eight locations, namely Dehradun, Chandigarh, Agra, Valsad, Kota, Hyderabad, Bellary, and Ootacamund, which in turn established as Central Soil and Water Conservation Research and Training Institute (CSWCRTI) by linking all the eight centers in 1956. The center started watershed activities in 42 locations mainly at a small scale to understand the technical processes of soil degradation and options for soil conservation (for review see Joshi et al., 2004).

The first large-scale government supported watershed program was launched in 1962–63 to monitor the siltation of the multipurpose reservoirs as "Soil Conservation Works in the Catchments of River Valley Projects (RVP)". This was followed by another mega-project, the Drought Prone Area Programme (DPAP) in 1972–73, which aimed at mitigating the impact of drought in vulnerable areas. On similar lines, the Desert Development Programme (DDP) was added for the development of desert areas and for drought management in the fragile, marginal, and rainfed areas. These schemes were implemented in 45 catchments spread over 20 states covering about 96.1 million ha area (Government of India 2001a).

Several programs were launched under the Operational Research Program (ORP) of CSWCRTI and Central Research Institute for Dryland Agriculture (CRIDA) and 41 model watersheds under the framework of the Integrated Watershed Development Program, which includes a system combining erosion and runoff, and improved land management (i.e., through vegetative cover, bunds, check-dams, and small percolation tanks) with irrigation wells for lifting groundwater on a sustainable basis so that the amount of water withdrawn is less than or equal to the annual recharge of groundwater. The system was an extension of the idea of water harvesting by which runoff water is collected in small ponds directly through gravity flows (Rajagopalan 1991). The program was organized as multidisciplinary and multi-agency and functionally participatory with the active involvement of farmers of the watershed. The key for the success of the integrated watershed development program was participatory planning and implementation by government agencies and non-government organizations (NGOs). The impact was documented in terms of increased crop productivity, increased employment, better crops and cropping systems, which ensure higher and regular cash flow, additional area under sustained irrigation and cropping, and reduced production risks (Joshi et al., 2004).

The severe drought during 1987 forced the Government of India to give more thrust to agriculture in the rainfed areas. Hence, a committee was constituted to examine the effectiveness of watershed-based programs in the rainfed areas. The committee recommended that the watershed development programs in the rainfed areas should optimize the production of rainfed crops (like pulses, oilseeds, coarse cereals, cotton, etc.), which improve the livelihood of the poor farmers along with soil and water conservation. The recommendations of the committee led to the formation of National

Watershed Development Project for Rainfed Areas (NWDPRA) in 1990–91. Then the Ministry of Agriculture terminated all the earlier watershed programs during the VII Five Year Plan and started new programs to cover both arable and non-arable areas and give more thrust for area-based approach for watershed development under NWDPRA. During the VIII Five Year Plan, an area of 4.23 million ha covering 2554 watersheds in 350 districts located over 25 states and two union territories were treated and developed with an expenditure of ₹9679 million. In the IX Five Year Plan, an outlay was raised to ₹10200 million to treat 2.25 million ha, which was slightly more than half of the area treated in the VIII Five Year Plan (Joshi *et al.*, 2004).

All the government-sanctioned programs in the 1980s paid more focus on soil and water conservation and attention to poverty alleviation as they operated in relatively poor and degraded areas. Economic improvement in these agricultural-dependent areas required making the land more productive, so poverty alleviation benefits were implicit. The programs also employed very poor people to carry out watershed work. They all adopted the technological approaches used in the model watersheds and none of them incorporated lessons learnt regarding institutional arrangements (World Bank 1990; Government of India 1994a). In earlier programs, the benefits and costs of watershed were unevenly distributed among all the stakeholders and programs made little or no effort to organize communities in the watersheds to solve the problems collectively. In the earlier watershed programs where village-level participation was attempted, it typically involved one or two key persons, such as the village *sarpanch* (leader) in the ICAR (Indian Council of Agricultural Research) watersheds or a trained technician in the NWDPRA (Government of India 1990).

The impact of these watershed programs showed disappointing results associated with the top-down implementation and management, inflexible or lack of site specific technology, and lack of attention to institutional arrangements (Shah 2000). Some of these programs showed good technical and economic performance in the early years, especially while project staff were still in place and the work was heavily subsidized (IJAE 1991). The benefits were not sustained for long beyond the project period in many cases (World Bank 1990; Government of India 1994a; Farrington *et al.*, 1999; Reddy 2000).

In the late 1980s, many NGOs introduced watershed development activities along with their other activities, and were better able to target the poorest people's needs. MYRADA in Karnataka, the Aga Khan Rural Support Programme (AKRSP) in Gujarat, and Social Centre in Maharashtra, all provided excellent examples of such approaches (Farrington and Lobo 1997; Hinchcliffe *et al.*, 1999). These organizations devoted much attention to organizing politically and economically weaker groups to initiate self-help activities such as thrift and credit associations and build their organizational skills, which give confidence to demand better services from the government agencies. This approach was used in the NGO-implemented watersheds to encourage people participation and sharing net benefits from watershed development (Fernandez 1994).

In the 1990s several European bilateral agencies established major watershed initiatives. Generally, these projects aimed to promote collaboration between government and NGO projects to draw on the strengths of each and to make government agencies more sensitive to the institutional issues. Some of the projects, including Indo-German in Maharashtra and Indo-British in Karnataka, drew on some NGOs' approaches to promote benefit sharing, and they tried to implement on large scale the associated institutional approaches (Farrington and Lobo 1997; Nanan 1998). Nanan (1998), however, found that despite a common focus on poverty alleviation in projects sponsored by the European Union, Danida, and the German Development Bank, benefits tended to favor landowners, whereas the landless benefited only marginally.

All these programs had their own guidelines, norms, funding patterns, and technical components based on their respective and specific aims (Government of India 1994b). In 1994 the Ministry of Rural Development introduced new comprehensive guidelines for all its projects that bypassed the state-level bureaucracy, giving unprecedented autonomy to village-level organizations to choose their own watershed technology and obtain assistance from NGOs rather than government line departments (Government of India 1994a, 1994b). These guidelines were used by the centrally sponsored schemes for watershed development under the Ministry of Rural Development and the Ministry of Agriculture.

The 1994 guidelines were in operation for five years. The guidelines were revolutionary in the extent to which they devolved power, promoted indigenous technology, and created a role for NGOs. This period has seen many successes as well as some failures in watershed development. Shah (2001) reviewed the performance of projects under the new guidelines in Gujarat state and found that benefits were heavily skewed towards wealthier households. Hence greater flexibility of the guidelines was essential to enhance the robustness of the response to the regionally differentiated demands that characterize rural India. Since different ministries were involved in the watershed development, it was decided to develop common guidelines. The Ministries of Agriculture and Rural Development jointly developed the 'Common Approach/Principles of Watershed Development' in 2000 (Government of India 2000). The Ministry of Agriculture brought out the new guidelines based on the 'Common Approach' in 2000 for NWD-PRA as Watershed Areas for Rainfed Agriculture System Approach (WARASA) or *Jan Sahbhagita*. The approach allows decentralization of procedures, flexibility in choice of technology, and provisions for active involvement of the watershed community in planning, execution, and evaluation of the program.

In 2001 the Ministry of Rural Development prepared a document of revised guidelines (Guidelines for Watershed Development) based on the common principles (Government of India 2001b). The new guidelines give more flexibility that was needed at village/watershed level. These guidelines, inter alia, envisage the convergence of different programs of the Ministry of Rural Development, Ministry of Agriculture, and other Ministries and Departments. Following the 73rd and 74th Amendments to the Constitutions of India in the early 1990s, the Panchayat Raj Institutions (PRIs) have been mandated with enlarged role in the implementation of developmental programs at the grassroots level, and accordingly their role has been more clearly brought out. The 1994 guidelines were made more flexible, and workable with greater participation of the community. The new guidelines lay greater emphasis on local capacity building through various training activities and empowering community organization.

To further simplify the procedures and involve the PRIs more meaningfully in the planning, implementation, and management of economic development activities in rural areas, the new guidelines called Guidelines for Hariyali were documented in 2003 by the Ministry of Rural Development (Government of India 2003). All the new projects under the area development programs have been implemented in accordance with the Guidelines for Hariyali with effect from 1 April 2003. This committee should oversee the implementation of watershed activities concerning drinking water security.

The Watershed Development Fund (WDF) was established by the National Bank for Agriculture and Rural Development (NABARD) during 1990–91, to integrate all the watershed programs in 100 priority districts in different states of the country. A total of ₹2000 million, which includes ₹1000 million by NABARD and a matching fund by the Ministry of Agriculture, was made available under the fund. The WDF was set up on the lines of the Rural Infrastructure Development Fund (RIDF) to help the state governments augment their watershed development programs (Sharma 2001). The main purpose of the fund was to create the framework conditions to replicate and consolidate the isolated successful initiatives under the different watershed development programs.

11.1.2 Synthesis of past experience of watershed development in India

To provide useful insights on the performance of numerous watershed development programs and to examine conditions for the success of the watershed programs across different geographical regions of India, a study was carried out by Joshi *et al.* (2005). The purpose of the study was to provide insights into the importance of economic, policy, and institutional issues and constraints and suggest options for the watershed management and also identify the areas of future research.

The study concluded that even though there are some visible gains from the various watershed development programs, the sustainability of the investments undertaken by the different agencies has not been ensured mainly because of insufficient participation of the local communities. The first generation watershed programs suffered from a top-down approach and technical focus on soil and water conservation without sufficient emphasis on livelihood benefits to the rural poor. Along with several socioeconomic studies, which documented the weaknesses of various watershed management approaches, experience has shown the difficulties of the top-down approach to natural resource management (NRM). This has led to the development of new policies and guidelines for a common approach to watershed management across the different implementing agencies in the country. These policies combine the technical strengths of the older programs along with the lessons learned about the role of community participation. Even after the new policies have been issued, the watershed development program suffers from second and third generation problems. The review of literature on the policy and institutional issues for watershed management and major lessons from the case studies examined in this study indicate the few critical areas that continue to affect the success of participatory community watershed management in the country. These are mainly related to profitability of the interventions, problems of collective action and active participation by the community, cost-sharing between individual farmers and the community/state, distribution of the gains from watershed management (equity), and negative externalities (e.g., upstream-downstream tradeoffs).

These challenges are made more complex by the lack of supportive policies and legislations that encourage cost-sharing and private and collective action in watershed programs. The landless households and marginalized groups are especially vulnerable to exclusion from accessing the benefits of the programs. The high subsidies provided for the program, including soil and water conservation investments on private lands, make it difficult to effectively assess the real farmer and community demand for the programs.

Further, it is essential to overcome the conflicting objectives and share benefits and cost evenly in the heterogeneous rural setting. Given the diversity of the rural social structure, different groups and individual farmers have different and often conflicting interests. The conflicting objectives are to be minimized by evolving appropriate policies and institutional arrangements. The case studies assessed in the synthesis study have clearly shown that the success in attaining the stated objectives is associated with an integrated approach where availability of profitable technologies for resource conservation and access to local markets encourage people's participation in the watershed programs. Depending on the focus given to this combination of technical support, social organization, and market access, the review of diverse development experiences indicates that most of the government-managed watershed programs performed poorly, while those managed by research institutions and some NGOs were quite successful. Lack of capacity in these important aspects is the principal reason for poor performance and failure of many watershed development programs. Careful integration of these components in the future policies and programs would help transform subsistence agriculture in rainfed areas while also protecting the vital resource base. Periodic monitoring and evaluation of the effectiveness and efficiency of the interventions and approaches as well as assessment of the multi-faceted impacts of the new generation of watershed programs implemented under the new guidelines would be useful to generate needed data and lessons for scaling-up successful approaches.

11.1.3   Need for economic impact assessment of watershed

The watershed development programs involve the entire community and natural resources and influence: (i) productivity and production of crops, changes in land use and cropping pattern, adoption of modern technologies, increase in milk production, etc.; (ii) attitude of the community towards project activities and their participation in different stages of the project; (iii) socioeconomic conditions of the people such as income, employment, assets, health, education, and energy use; (iv) environment; (v) use of land, water, human, and livestock resources; (vi) development of institutions for implementation of watershed development activities; and (vii) sustainability of improvement. It is thus clear that watershed development is a key to sustainable production of food, fodder, fuel wood, and for meaningfully addressing the social, economical, and cultural conditions of the rural community. By virtue of its nature, watershed is an area based technology cutting across villages comprising both private and public lands. The benefits from watershed development activities are not only limited to the users/beneficiaries, but also the non-participating farmers.

Experience shows that various watershed development programs brought significant positive impact. There has been a marked improvement in the access to drinking water due to groundwater recharges in the project area (Kerr *et al.*, 2000; Reddy *et al.*, 2001; Kakade *et al.*, 2001), increase in crop yields and substantial increase in the cropped area (Erappa 1998; Wani *et al.*, 2002), rise in employment and reduction in migration of labor (Deshpande and Ratna Reddy 1991; Kerr *et al.*, 2000).

Availability of fodder has also improved, leading to a rise in the yield of milk. The most important factor accounting for the positive impact of watershed development programs is community participation and decentralization of program administration. Experience from Maharashtra shows that the encouraging performance is attributable largely to the positive response from the people, especially in the tribal areas, owing to their traditions of community participation and to political and administrative will for decentralizing administration and strengthening of the PRIs (Hanumantha Rao 2000).

A program such as watershed development, which involves a hierarchy of administration and communities at the grassroots level in highly varying agroclimatic and socioeconomic conditions, invariably requires periodical assessment for achieving developmental objectives. Typically, an implementing agency would see a greater value in spending an extra few crores of rupees for undertaking works in the field rather than spending this money for monitoring and evaluation. However, according to some observers, mid-course corrections can improve the program benefits substantially, in some cases up to 100%. But even if we consider the improvement to be very modest, say, 10%, then a one per cent of program outlay on meaningful monitoring and evaluation would have a very high payoff in terms of achieving the program objectives. It is of utmost importance therefore, to put in place institutional mechanism for research and monitoring and evaluation in the field of watershed development by involving reputed institutions in the country for upgrading the quality of evaluation.

Information generated by impact evaluations of watershed development informs decisions on whether to expand, modify, or eliminate a particular policy, and can also be used in prioritizing public actions. In addition, impact evaluation contributes to improve the effectiveness of the policies and programs by addressing the questions such as:

- Does the program achieve the intended goal?
- Can the changes in outcomes be explained by the program, or are they the result of some other factors occurring simultaneously?
- Do program impacts vary across different groups of intended beneficiaries (males, females, and indigenous people), regions, and over time?
- Are there any unintended effects of the program, either positive or negative?
- How effective is the program in comparison with alternative interventions?
- Is the program worth the resources it costs?

11.1.4 Challenges in impact assessment of watershed development

Impact analysis of an area based program like watershed development has inherent difficulties. Apart from the benefits accrued from different technologies, the impact of watershed development should be looked into three major dimensions, viz., scales (household level, farm level, and watershed level), temporal, and spatial. The dimensions of impact of watershed technologies further complicate the impact assessment.

The problem of impact assessment of watershed development project lies in the following: (i) Developing a framework to identify what impacts to assess, where to look for these impacts, and selecting appropriate indicators to assess the impacts; and (ii) Developing a framework to look after the indicators together and assessing the overall impact of the project. The nature of watershed technologies and its impact on different sectors pose challenges to project monitoring and evaluating agencies, economists, researchers, and policy makers. More specifically, major challenges include: (i) the choice of methodologies, (ii) selection of indicators, and (iii) choice of discount rate.

11.1.4.1 Methods of impact assessment

Choosing appropriate methodology for impact assessment is essential. Different methodologies have been used in the evaluation literature, mainly qualitative and quantitative methods. The quantitative methods such as experimental or randomized control designs are being widely used. Some other quasi-experimental designs are also widely used (Baker 2000). The non-experimental or quasi-experimental designs such as matching methods or constructed controls, double difference, instrumental variables or statistical control methods, and reflexive comparisons are being used by the evaluating agency. Qualitative techniques are also used for carrying out impact evaluation with the intent to determine impact by the reliance on something other than the counterfactual to a causal inference (Mohr 2000). The qualitative approach uses relatively open-ended methods during design, collection of data, and analysis. The benefits of qualitative assessments are that they are flexible, can be specifically tailored to the needs of the evaluation using open-ended approaches, can be carried quickly using rapid techniques, and can greatly enhance the findings of an impact evaluation through providing a better understanding of stakeholders' perceptions and priorities and the conditions and processes that may have affected program impact (Baker 2000). The qualitative methods are not exempted from limitations. Limitations like subjectivity involved in data collection, the lack of comparison group, and the lack of statistical robustness, given mainly small sample sizes, all of which make it difficult to generalize to a larger, representative population. Also, the validity and reliability of data from qualitative analysis are highly dependent on the methodological skill, sensitivity, and training of the evaluator.

11.1.4.2 Approaches of impact assessment

One dominant perspective in impact assessment literature is to view natural resources development projects as constituting a set of inputs that are transformed through activities into a set of outputs and the impact of these projects on people are through the changes in output and through activities that produce these outputs (Gregerson and Contreras 1992). These impacts are of main concern in economic approaches. The other approach, resulting from a change in the basic conception of development, sees projects more in terms of process pursuing multiple objectives: social, economic, environmental, and institutional (e.g., equity, efficiency, sustainability, community organizations, etc.). Project goals and objectives, and assessment of achievements and impacts have become the central concerns of this approach. Many studies using or proposing this approach implicitly or explicitly use variants of a Logical Framework Approach as a basis. These approaches build the evaluation function within the management systems of the project cycle. The third approach is participatory evaluation where evaluation systems are designed and implemented in partnership mode with the people involved in the projects.

11.1.4.3   Scale or time lags

Being a common property resource, treatments in watersheds generate externalities. Conflicts arise between downstream and upstream farmers in sharing benefits and making investments. As watersheds include private and common lands, the impact of various watershed treatment activities on different scale of dimensions such as farm level, household level, and watershed level is crucial in impact assessment. Time is an extremely important element in NRM particularly watershed development projects where the benefits and costs of development activities rarely occur the same time. For instance, investments on construction of rainwater harvesting structures occur in the early years, but the benefits occur during later part, resulting in a large time gap between investment and receipt of revenues. Time also complicates comparing investments with different timings and magnitude of benefits and costs.

11.1.4.4   Samples for the study

Another important issue faced by the evaluators is the choice of methodology for selecting sample respondents for the impact assessment. Should the researcher study the samples from the watershed area itself employing before/after approach or should he/she study samples both in the treated and control villages employing with/without approach? Also, case studies raise a number of methodological issues in impact assessment of watershed development activities. For instance, issue arises in relation to sampling, i.e., should the researcher use random sampling or purposive sampling in selecting among watersheds to assess the impacts? Each approach has its own pros and cons and no clear consensus seems to have emerged.

11.1.4.5   Selection of indicators

There are various indicators of impact. Changes in economic welfare are an obvious one and changes in distributional outcomes are another. It is difficult to derive appropriate indicators in assessing the program impacts. Assessing the economic value of the increased outputs in the watersheds as a result of various treatment activities is a valid measure of its impacts.

Development of indicators for impact assessment forms crucial aspects in impact assessment of watershed development programs, where the impact of different activities on different development domains is complex. Although several studies list a good number of indicators, there is little effort in developing a comprehensive framework for the identification, analysis, and usage of appropriate indicators in watershed development projects. They can be obtained either by synthesis (a range of information obtained from primary or secondary data is combined to form the indicator) or selection (from primary or analyzed data). It is important to identify data requirements, generate data, and update the database at regular intervals. In using the indicators, there are many problems such as: (i) establishing causal links between indicators and the actual changes they are supposed to reflect; (ii) different indicators may give conflicting signals for the same results; (iii) establishing the relative importance of the changes in different indicators (as a common denominator like price/money value is lacking); and (iv) lack of or problem of arriving at a rational method to assess the significance of quantum of change. Another such problem lies in the inter-comparison of projects.

As the impact of the watershed developmental activities is multifaceted and complex, it may not always be possible to measure the results that have been achieved because they may be intangible or it may be too costly to measure them effectively. In such cases indications that success is being achieved will make good proxies. Such indicators, however, must be chosen carefully so that they are reliable substitutes to direct measurement and are easy to measure in terms of time and effort. The choice of indicators is determined by who the end-user is. These issues pose challenges in impact assessment of watershed or activities therein.

11.1.4.6   Choosing the discount rate

There has been much discussion and debate on natural resources economics on the determination of methodology to use in discounting and selection of a discount rate. If the economy is optimal and all of society's wishes are reflected in financial markets, the determination of a discount rate would be straight forward. It would be related to some financial rate such as interest on bank deposits. However, the economy is non-optimal or second best. Furthermore, determining society's preferences and how these are reflected through government spending is difficult. Problems centered on whether discounting should occur at the social rate of time preference (the social discount rate) or at a marginal rate for private investment (the private discount rate). It is generally argued that society is more concerned with the future, especially with negative natural resource and environmental consequences than the individual or private firms. Consequently, the social discount rate will be lower; however, some support the notion that private and social rates do not differ. Most economists suggest using an opportunity cost approach for evaluating government projects as it is the most efficient and easiest to implement.

One big debateable issue in the field of natural resources evaluation is the choice of discount rate to be used in either economic analysis or financial analysis of project impact assessment. Impact assessment of watershed development is not an exception to this. As watershed development involves development of both common and private lands, it generates many positive externalities and leads to spill over effects. Moreover, as it involves huge government spending, the selection of a 'discount rate' is a crucial one.

11.1.5   Indicators for evaluation of watershed development projects

The problem of developing an evaluation framework for any watershed development project lies in the following:

- Developing a framework to identify what impacts to assess, where to look for these impacts, and selecting appropriate indicators to assess the impacts.
- Developing a framework to look after the indicators together and assessing the overall impact of the project.

Evaluation is a periodic assessment of the relevance, performance, efficiency, and impact of the project in the context of its stated objectives. Several types of evaluation were used in different studies. Some useful typologies are reviewed here. Based on the objectives of the project, the evaluation system may be defined as:

- Validation evaluation to evaluate the assumptions used in the project formulations.
- Effectiveness evaluation to evaluate progress towards stated physical and financial goals.
- Achievement evaluation to evaluate changes in living standards or in hydrologic and environmental conditions brought about by the project.

Based on the stage in the project cycle at which evaluation is conducted, evaluation systems are classified as:

- Baseline: Pre-project assessment to analyze viability of the project.
- Ongoing or intermediate to check the effectiveness of each individual activity conducted throughout the project's life cycle.
- Terminal evaluation – conducted at the end of the project to evaluate the efficiency of project implementation.
- Post-terminal evaluation – to evaluate the long-term project accomplishments conducted several years after the project completion.

In general, the evaluating agency will evaluate the project either during the project implementation phase, i.e., mid-term or ongoing evaluation or after the project period is over, i.e., ex-post evaluation. Ongoing evaluation is a series of periodic 'breaks' to analyze the monitored information to probe further the signals received and assess how things are moving. Some important questions are raised: Are activities being accomplished on time? Is progress towards achievement of objectives satisfactory? Throughout the ongoing evaluation of a program emphasis is placed on delivering information, which is modest in both scale and scope but sharply focused on the practical implications for management. The very purpose of ongoing evaluation is to assess continuing relevance, present and future outputs, and effectiveness during implementation. The main focus is on assessing the validity of the project design and targets, assessment of effects and review of cost effectiveness. Ongoing evaluation is target oriented. Terminal/ex-post evaluation is usually done after completion of the project mainly to assess the impact of the project, i.e., assess success/failure of the project. The purpose of ex-post evaluation is to assess output, effect, and impact and drawing lessons for future planning and development. This type of valuation is beneficiary oriented. The performance indicators used for the watershed impact studies are given Table 11.1.

11.2 APPROACHES

11.2.1 Before and after

Project parameters compared to the 'pre-project' situation gives the incremental benefits due to the project. But these increments in the parameters intrinsically include the changes due to state-of-the-art of technology. This approach would be viable when the benchmark information is available. But in reality, most of the watershed development

Table 11.1 Performance indicators used for watershed impact evaluation

Performance criteria	Indicators	Measures
Groundwater recharge and water resource potential	Measurement of groundwater levels, climate variation, and pumping volume	• Duration of water availability • Water table of wells • Surface water storage capacity • Hydrological Index • Index of water conservation practices • Difference in number of wells • Number of wells recharged/defunct • Difference in irrigated area • Difference in number of seasons irrigated • Difference in village-level drinking water adequacy • Difference in irrigation intensity
Agricultural productivity/profits	Agricultural productivity and net returns at plot level	• Agricultural Productivity Index (API) • Crop Yield Index (CYI) • Crop Diversification Index (CDI) • Cropping System Index (CSI) • Index of Agroforestry Practices (IAP) • Difference in cropping pattern • Difference in cropping intensity • Difference in yield of crops • Farm profit
Household welfare	Household income and wealth	• Difference in per capita income • Difference in employment • Difference in household income • Difference in persons migrated
	Nutritional status	• Food security index (FSI) • Child nutrition and health
Socioeconomic indicators	Development of infrastructure	• Infrastructure Development Index (IDI)
	Impact on women (decision-making, health, life style and awareness)	• Women's Participation Index (WPI)
	People's participation	• Index of Social Affiliation (ISA)
	Institutions	• Difference in number of institutions
	Ownership rights	• Difference in number of agricultural laborers • Difference in number of landless laborers • Difference in farm households by size groups
Overall impact	Economic returns to investment	• Net present value, benefit-cost ratio, and internal rate of return
	Extent of green cover	• Forest Eco Index

[a]Source: Palanisami *et al.* (2002a).

programs are implemented without collecting full set of benchmark information. Thus sometimes, the benefits may be exaggerated.

11.2.2 With and without

A comparison between the 'project parameters' with the 'non-project control region' is used for evaluation. This method automatically incorporates the correction for the impact of technology in the absence of the project. But this approach also has limitations. Though the watershed treated and control regions fall within the same agroclimatic conditions, the differences in hydro-geological profile vary within a village/even across plots in the farm. Thus, this approach can be only used to compare the villages having homogeneous conditions.

11.2.3 Combination of with and without using double difference method

When the time span is too long, economists adopt *a combination of both with and without and before and after approaches,* where they compare pre- and post-project period and with the control village as well so as to get a holistic picture on impact of watershed development activities. The double difference method explained below can be applied.

Data may be collected for both watershed treated villages and control villages before and after watershed development intervention. This enables the use of the double difference method to study the impacts due to watershed development intervention. The framework was adopted from the program evaluation literature (see Figure 11.1) (Maluccio and Flores 2005).

11.3 METHODOLOGIES: APPLICATION OF WATERSHED EVALUATION METHODS

11.3.1 Conventional benefit-cost analysis

The conventional analysis primarily includes:

- Net present value (NPV)
- Benefit-cost ratio (BCR)
- Internal rate of return (IRR)

The limitations and complexities associated with measuring, monitoring, and valuing social costs and benefits associated with NRM interventions require more innovative assessment methods. An important factor that needs to be considered in the selection of appropriate methods is the capacity for simultaneous integration of both economic and biophysical factors and ability to account for non-monetary impacts that NRM interventions generate in terms of changes in the flow of resource and environmental services that affect economic welfare, sustainability, and ecosystem health. Hence a mix of qualitative and quantitative methods is the optimal approach for capturing on-site and off-site economic welfare and sustainability impacts (Freeman *et al.*, 2005).

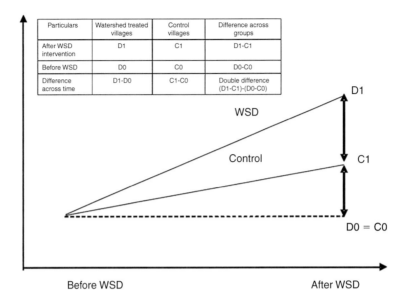

Particulars	Watershed treated villages	Control villages	Difference across groups
After WSD intervention	D1	C1	D1-C1
Before WSD	D0	C0	D0-C0
Difference across time	D1-D0	C1-C0	Double difference (D1-C1)-(D0-C0)

Figure 11.1 Illustration of impact of watershed development (WSD) intervention by double difference method (Source: Maluccio and Flores 2005)

The approaches that have been developed recently for evaluating the impacts of agricultural and NRM interventions are presented.

11.3.2 Econometric methods (Economic surplus approach)

The economic surplus approach to impact assessment is rooted in the microeconomics of supply and demand (Bantilan *et al.*, 2005). The basic idea is simple and is illustrated in Figure 11.2. Consumer demand can be described by downward sloping demand curve illustrating that some consumers are willing to pay more than others for a given commodity. At a market-clearing equilibrium price, P*, those consumers who were willing to pay more than P* realize benefits by getting the product for less money than they were willing to pay. Across all consumers, the area beneath the demand curve, D, and above the equilibrium price, P*, measures the total value of consumer surplus.

Producer supply can be described by an upward sloping curve that illustrates that some producers can supply a product for a lower price than others. At a market-clearing equilibrium price, P*, those producers who could supply the products at a lower price obtain extra benefits. The aggregate benefits described by the area above the supply curve, S, and below the equilibrium price, P*, measure the total producer surplus. Economic surplus is the sum of consumer surplus and producer surplus.

This is the most commonly used method for assessing the impact of agricultural research investment, particularly those related to crop improvement. This approach estimates the benefits of research in terms of change in consumer surplus and producer surplus, resulting from a shift in the supply curve by introduction of new technology. Thus, the economic surplus (sum of producer surplus and consumer surplus) is

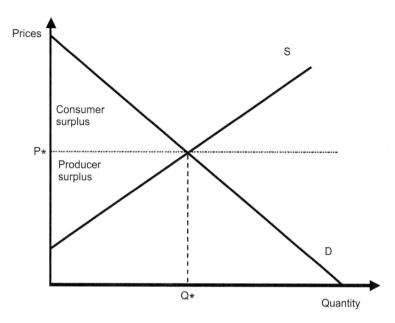

Figure 11.2 Economic surplus divided between consumer and producer surplus (Note: P* = equilibrium price; Q* = equilibrium quantity; S = supply curve; D = demand curve) (Source: Bantilan *et al.*, 2005)

taken as a measure of the gross benefit from research investment in a given year. The major challenge is to make a plausible link between changes in NRM practices and the supply of economic goods and services. The presence of non-marketed externalities further complicates the approach, although in theory, the social marginal cost of production could be used to internalize the externalities. New methods (e.g., benefit transfer function) are developed to extend the economic surplus approach for assessment of non-marketed social gains from improved NRM technologies. Bantilan *et al.* (2005) used the economic surplus approach to estimate empirically the economic and environmental impact of groundnut production technology in Maharashtra.

The econometric approach is also used to link measures of output, costs, and profits directly to past watershed development investments. The econometric approach uses regression models [like probit, logit, tobit, and two stage least squares (2SLS) regressions] to explain variations in agroecosystem services through changes in NRM pattern. This approach uses the changes in biophysical, economic, and environmental indicators as proximate indicators of the impact of the NRM technologies. The indicators include changes in land productivity; total factor productivity; reduction in costs (e.g., reduced use of fertilizers and pesticides); reduced risk and vulnerability to drought and flooding; and improved net farm income and change in poverty levels (e.g., head count ratio). However, there are some limitations in this approach related to data availability and measurement errors, and problem in internalizing externalities and inter-temporal effects. For example, the time-varying nature of the impacts of NRM practices require time-series data, ideally panel data with repeated observations from the same households and plots over a period of many years so that the dynamics of these impacts and their feedback effected on household endowments and subsequent NRM decisions are adequately assessed (Pender 2005).

Unfortunately, household and plot-level panel data sets with information on both NRM practices and causal factors and outcomes are quite rare. In the absence of such data, inferences about NRM impacts will remain limited to those possible based on the available short-term experimental data and cross-sectional econometric studies. These can provide information on near-term impacts, for example, on current production, income, and current rates of resource degradation or improvement, but do not reveal feedback effects such as how the changes in income or resource conditions may lead to changes in future adoption, adaptation, or non-adoption of NRM practices (Barrett et al., 2002; Pender 2005).

Assessing the multiple and complex mechanisms by which NRM (and other factors) may affect outcomes is an important issue, and one that is more difficult to address when limited dependent variable models (such as the probit, ordered probit, and tobit models) or other non-linear models are estimated. In linear system of structural equations, the total impacts of any variable on the outcomes can be determined by total differentiation of the system and by adding up the partial effects (Fan et al., 1999). But with limited dependent variable models or other non-linear models, this approach does not work. There will be no simple general relationship between the estimated coefficients of the structural model and the total impact; these relationships all depend on the level of each variable in non-linear models.

Pender (2005) applied an alternative approach to estimate total effects in non-linear models by using predictions from the estimated model to simulate both indirect and direct impacts of changes in the explanatory variables. Even though econometric models are useful in assessing the NRM impacts, they are not without problems and limitations. The most important problems are those of endogeneity of NRM practices and omitted variable bias, which can be addressed through careful data collection and use of instrumental variables estimators.

Kerr and Chung (2005) also applied the econometric approach to assess the impact of the watershed program in the semi-arid tropical India. In this study they used instrumental variables approach for evaluation because of inadequate data on baseline conditions and lack of hydrological data (such as groundwater level, runoff, soil erosion, etc.). The study found that the projects involving greater degree of participation were more successful in protecting upper catchments to promote water harvesting. On the other hand, often protection of upper catchments came at the expense of landless people whose livelihood relied heavily on them.

11.3.2.1   Application of economic surplus method to watershed evaluation

Watershed programs play a dual role of safeguarding the interest of the producers as well as consumers, as in several locations, the drought-proofing aspects of the watershed programs are easily felt (Palanisami and Suresh Kumar 2007). In the case of producers, they can change the crop pattern due to increased water levels in their wells, enhance moisture conservation in the soil, increase water use for the existing crops, and increase the number of livestock and fodder production. There is also a change in the cost of production of the commodities in the watershed. Over the years, there is an increase in technology adoption due to the watershed programs. In the case of consumers, the increased crop production in the watershed results in the availability of produce at lower prices. Consumption levels also get increased among the consumers. Labor employment is increased due to increased land and crop production and processing activities in the watershed. Evidence shows that the production levels have increased as a result of watershed interventions and the consumers have started enjoying the benefits of the localized production in the regions. Hence, for the purpose of the analysis, it was assumed that the output supply curve shifts gradually over time when the benefits from the watershed developmental activities start benefiting the agricultural sector through water resource enhancement. The supply shift factor due to technological change, in our case, watershed intervention, is known as K. This factor varies in time depending on the dynamics of the rainfall, adoption, dissemination of soil and moisture conservation technologies, and the repair and maintenance activities undertaken in the watershed. The supply shift factor (K) can be interpreted as a reduction of absolute costs for each production level, or as an increase in production for each price level (Libardo *et al.*, 1999).

Micro economic theory defines consumer surplus (individual or aggregated) as the area under the (individual or aggregated) demand curve and above a horizontal line at the actual price (in the aggregated case: the equilibrium price). The demand curve is assumed to be log-linear with constant elasticity. Thus, the equation for this demand function can be written as:

$$P = gQ^{\eta} \tag{1}$$

where, η is the elasticity and g is a constant. Once, the parameters η and g are estimated, then consumer surplus could be estimated by equation 2:

$$CS = \int_{Q_0}^{Q_1} gQ^{\eta} dQ - (Q_1 - Q_0)P_1 \tag{2}$$

Combined, the consumer surplus and the producer surplus make up the total surplus. The estimation of benefits is given in Box 1.

11.3.2.2   Cost of project

The cost involves the watershed development investment during the project period and maintenance expenditure incurred in the project. For watershed development projects with multiple technologies or crops, incremental benefits from each technology and crop were added to compile the total benefits.

11.3.2.3   Results of the economic surplus method

This section presents the key indicators from the field experience on impact assessment of watershed programs implemented under DPAP in Coimbatore district of Tamil Nadu. The general characteristics of the sample farm households in the study watershed were analyzed (Table 11.2). The average size of the holding was 1.28 ha and 1.75 ha, respectively for the watershed and control villages. It is evident from the analysis that

---

Box 1. Estimation of Benefits

Following the theory of demand and supply equilibrium, economic surplus (benefits) as a result of watershed development intervention is measured by equation (3):

$$B = K^* P_0^* A_0^* Y_0^* (1 + 0.5 Z^* \varepsilon_d) \tag{3}$$

where, K is the supply shift due to watershed intervention.

The supply shift due to watershed intervention can be mathematically represented by equation (4):

$$K = \forall^* \rho^* \psi^* \Omega \tag{4}$$

where, K represents the vertical shift of supply due to intervention of watershed development technologies and is expressed as a proportion of initial price. $\forall$ is net cost change defined as the difference between reduction in marginal cost and reduction in unit cost. The reduction in marginal cost is defined as the ratio of relative change in yield to price elasticity of supply (ε_s). Reduction in unit cost is defined as the ratio of change in cost of inputs per hectare to $(1 + \text{change in yield})$. ρ is the probability of success in watershed development implementation. ψ represents adoption rate of technologies and Ω is the depreciation rate of technologies.

Z represents the change in price due to watershed interventions. Mathematically, Z can be defined by equation (5):

$$Z = K^* \frac{\varepsilon_s}{(\varepsilon_d + \varepsilon_s)} \tag{5}$$

where, P_0, A_0, and Y_0 represent prices of output, area, and yield of different crops in the watershed before implementation of the watershed development program. If we use with and without approach, then these represent area, yield, and price of crops in control village.

Table 11.2 General characteristics of sample farm households[a]

Particulars	Watershed village	Control village
Farm size (ha)	1.28	1.75
Household size	3.31	3.34
Land value (₹ ha^{-1})	230657	153452
No. of wells owned	1.35	1.20
Average area irrigated by wells (ha)	1.48	1.80
Value of household assets (₹)	261564*	184385
No. of persons in the household	4.07	4.2
No. of workers	2.5	2.1
Labor-force participation (%)	61.48	50.79

[a],*indicates that value was significantly different at 10% level from the corresponding values of control village.
Source: Palanisami and Suresh Kumar (2006).

the average number of workers was 2.5 and 2.1 out of 4.07 and 4.2 for the watershed and control villages.

The labor force participation rate was 61.48% and 50.79%. The higher labor force participation was due to better scope for agricultural production, livestock activities, and other off-farm and non-farm economic activities. Results from the analysis showed that the labor force participation rate among farmers in watershed villages was higher, implying that the enhanced agricultural production was due to watershed treatment activities. Construction of new percolation ponds, major and minor check-dams, and the rejuvenation of existing ponds/tanks had enhanced the available storage capacity in the watersheds to store the runoff water for surface water use and groundwater recharge. The additional surface water storage capacity created in the watersheds ranged from $9299\,m^3$ to $12943\,m^3$. This additional storage capacity further helped in improving the groundwater recharge and water availability for livestock and other non-domestic uses in the village. On the basis of the data collected from the sample farmers, it was found that the water level in the open dug-wells had risen in the range of 0.5–1.0 m in watershed villages. The depth of the water column in the few sample wells was recorded both in watershed and control villages for comparison. The depth of the water column in the wells was found to be higher in the watershed villages than in control villages. For instance, depth of the water column in the wells in Kattampatti watershed village was 3.53 m compared to 2.16 m in the control village, leading to a difference of 63.43%.

Information related to duration of pumping hours before well went dry (or water level depressed to a certain level) and time it took to recuperate to the same level were collected for the sample farmers across villages. Due to watershed treatment activities, groundwater recuperation in the nearby wells had increased. The increase in recuperation rate varied from 0.1 to $0.3\,m^3\,h^{-1}$. It was also observed that the recharge to wells decreased with their distance from the percolation ponds and check-dams and the maximum distance where the recharge to the wells had occurred was observed to be 500–600 m from the percolation ponds.

The area irrigated in the watershed village registered a moderate increase after the watershed development activities in most of the watersheds, whereas in the control village it declined slightly over the period. The irrigation intensity was found higher in watershed treated village than in the untreated village. The watershed developmental activities helped increase the water resource potential of a region through enhanced groundwater resources coupled with soil and moisture conservation activities. In the case of control village, the water table in the wells had declined due to continuous pumping. It is one of the reasons why most farmers demand watershed program in their villages. The analysis also revealed increase in net cropped area, gross cropped area, and cropping intensity in both the watersheds (Table 11.3). For example, the cropping intensity worked was 146.88% in the watershed village, which is higher than in the control village (133.33%). The composite entrophy index (CEI) was used to compare diversification across situations having different and large number of activities. The CEI has two components, viz., distribution and number of crops or diversity. The value of crop diversification index (CDI) increases with the decrease in concentration and rises with the number of crops/activities. In general, CDI is higher in the case of watershed treated villages than control villages, confirming that watershed treatment activities help diversification in crop and farm activities.

Table 11.3 Impact of watershed activities on cropped area, cropping intensity, and crop diversification[a]

Particulars	Watershed villages		Control villages	
	Before	After	Before	After
Net area irrigated (ha)	1.08	1.10***	1.68	1.62
Gross area irrigated (ha)	1.25	1.35**	1.84	1.62
Irrigation intensity (%)	115.74	122.73**	109.52	100.00
Net cropped area (ha)	1.15	1.28**	1.78	1.62
Gross cropped area (ha)	1.38	1.88**	2.43	2.16
Cropping intensity (%)	120.00	146.88	136.52	133.33
Crop diversification index (CDI)	1.0		0.97	

[a],** and *** indicate that values were significantly different at 1 and 5% levels from the corresponding values of control village.
Source: Palanisami and Suresh Kumar (2006).

Table 11.4 Livestock maintained in watershed and control villages[a]

Particulars	Watershed village	Control village
% of households maintained livestock	46.67	93.33
Livestock (number per household)	2.57	2.64
Livestock (number per hectare of gross cropped area)	2.01	1.63

[a]Source: Palanisami and Suresh Kumar (2006).

Based on the proportion of different crops in the farm, *CEI* for crop diversification was estimated as:

$$CEI = -\left(\sum_{i=1}^{N} P_i \cdot \log_N P_i\right)^{*} \{1 - (1/N)\} \qquad (6)$$

where,
P_i = Acreage proportion of ith crop in total cropped area, and
N = Total number of crops.

The livestock income has been a reliable source of income for the livelihood of the resource-poor farmer households. Cattle, sheep, and goats were maintained as important sources of manure and were the liquid capital resource. Nearly 46.67% and 93.33% of the households in watershed and control villages respectively maintained cattle (Table 11.4).

Access to grazing land and fodder had made the farm households to maintain livestock in their farms to derive additional income. But the analysis revealed that relatively greater number of households in the control village maintained livestock. It was mainly due to the fact that inadequate grazing land and poor resource-base for stall feeding persuaded them to feed their livestock with green leaves and fodder obtained from crops and crop residues. The farm households in the control village maintained mainly milch animals to derive additional income for their livelihood.

Table 11.5 Impact of watershed development intervention on crop yield and cost[a]

Crop/Commodity	Change in yield (%)	Reduction in marginal cost (%)	Reduction in unit cost (%)	Net cost change (%)
Sorghum	33	63.6	3.76	59.8
Maize	31	39.9	2.29	37.6
Pulses	36	41.0	1.47	39.6
Vegetables	32	32.8	0.76	31.9
Milk	28	27.3	7.81	19.5

[a]The reduction in marginal cost (C_m) was the ratio of relative change in yield to price elasticity of supply (ε_s). Reduction in unit cost (C_u) was the ratio of change in cost of inputs per hectare to (1 + change in yield). C_i was the input cost change per hectare, i.e., $C_u = C_i/(1 + \text{Change in yield})$. The net cost change ($\forall$) was the difference between reduction in marginal cost and reduction in unit cost, i.e., $\forall = C_m - C_u$.
Source: Palanisami *et al.* (2009).

Table 11.6 Impact of watershed development activities on the village economy

Crop/Commodity	Total benefits due to watershed intervention[a]		
	Change in total surplus (ΔTS)	Change in consumer surplus (ΔCS)	Change in producer surplus (ΔPS)
Sorghum	293177.3 (100.0)[b]	113636.3 (38.8)	179541.0 (61.2)
Maize	177774.2 (100.0)	85424.0 (48.1)	92350.2 (51.9)
Pulses	25777.5 (100.0)	12580.3 (48.8)	13197.2 (51.2)
Vegetables	29663.6 (100.0)	10627.5 (35.8)	19036.1 (64.2)
Milk	176878.5 (100.0)	105974.1 (59.9)	70904.4 (40.1)

[a]The change in total surplus in the village economy due to watershed intervention was decomposed into change in consumer surplus and change in producer surplus. The decomposition of total surplus was as follows:
$\Delta TS = \Delta CS + \Delta PS = P_0 Q_0 K(1 + 0.5Z\eta)$
$\Delta CS = P_0 Q_0 Z(1 + 0.5Z\eta)$
$\Delta PS = P_0 Q_0 (K - Z)(1 + 0.5Z\eta)$
[b]Percentage values are given in parentheses.

The impact of watershed development activities on yield of crops and hence the cost was estimated (Table 11.5). The change in yield due to watershed intervention across crops varied from 31% in maize to 36% in pulses. The change in yield was maximum due to watershed intervention. Reduction in marginal cost due to supply shift ranged from 32.8% in vegetables to 63.6% in sorghum. Net cost change varied from 32% in vegetables to 59.8% in sorghum. The change in total surplus was higher in sorghum and maize than crops like pulses and vegetables (Table 11.6). Being the major rainfed crops, these two crops benefited more from the watershed interventions.

Table 11.7 Results of economic analysis employing economic surplus method[a]

Particulars	Economic surplus method	Conventional method
Benefit-cost ratio	1.93	1.23
Internal rate of return (%)	25	14
Net present value (₹)	2271021	567912

[a]Conventional method refers to the crop production related costs and benefits.
Source: Palanisami *et al.* (2009).

The change in total surplus due to watershed intervention was decomposed into change in consumer surplus and change in producer surplus. It was evident that the producer surplus was higher than the consumer surplus in all the crops. For instance, in sorghum, the producer surplus was 61.2%, whereas the consumer surplus was only 38.8%. Watershed development activities benefited the agricultural producers more. It was interesting to note that unlike in the crop sector, milk production had different impacts on the society. The decomposition analysis revealed that watershed development activities generated more consumer surplus in milk production.

The overall impact of different watershed treatment activities was assessed in terms of BCR, and IRR using the economic surplus methodology assuming 10% discount rate and 15 years life period. The BCR was more than one, implying that the returns to public investment such as watershed development activities were feasible. Similarly, the IRR was 25%, which is higher than the long-term loan interest rate by commercial banks indicating the worthiness of the government investment on watershed development (Table 11.7). The NPV per hectare was ₹4542 (where the total area treated was 500 ha), which implied that the benefits from watershed development were higher than the cost of investment of the watershed development programs of ₹4000 per ha. However, recently the watersheds in India have been allotted a budget of approximately ₹6000 per ha. Thus, a watershed with a total area of 500 hectares receives ₹3 million for a five-year period. The bulk of this money (80%) is meant for development/treatment and construction activities. According to the new Common Guidelines 2008, the budgetary allocation is of ₹12000 per ha.

11.3.3 Bioeconomic modeling approach

Even though the economic surplus method could incorporate both the consumer and producer benefits, improvements could be made while accounting for the watershed related impacts. Further, the individual impacts of various technologies are known but there is little information on their combined impact or on the role of policy and institutional arrangements in conditioning their outcomes (Okumu *et al.*, 2000). In addition, past research seldom included the biophysical factors (like soil erosion, nutrient depletion, water conservation, etc.) in their studies, which have a direct effect in the productivity of the numerous enterprises (like crop production, livestock production, forestry, pasture development). In the recent past, the methodologies that are capable of simultaneously addressing the various dimensions of agriculture and NRM technology changes and the resulting tradeoffs among economic, sustainability, and environmental objectives have been developed (e.g., Barbier 1998; Barbier and Bergerson 2001; Holden and Shiferaw 2004; Holden *et al.*, 2004). The main innovation in the development of such methodologies is the integration of biophysical and economic information into a single integrated bioeconomic model. Bioeconomic models link economic behavioral models with biophysical data to evaluate potential effects of new technologies, policies, and market incentives on human welfare and the sustainability of the environment or natural resources (Shiferaw and Freeman 2003). So it helps the researchers in the selection of technologies that may improve the farmers' economic efficiency and welfare as well as the condition of the natural resource base over time. The models can also be used to account for externalities if the generation of externalities can be linked with NRM and economic factors. Bioeconomic models have been applied at the level of the household (e.g., Holden and Shiferaw 2004; Holden *et al.*, 2004; Holden 2005), village and watershed levels (e.g., Barbier 1998; Barbier and Bergerson 2001; Sankhayan and Hofstad 2001; Okumu *et al.*, 2002) and for agricultural sector (e.g., Schipper 1996).

11.3.3.1 Advantages of bioeconomic modeling in impact assessment studies

Bioeconomic models are used to incorporate changes in the biophysical conditions of the natural resource use within the economic behavioral models with the purpose of exploring or understanding the two way interaction (i.e., how changes in biophysical conditions affect welfare and vice versa). Such models are useful to evaluate the potential effects of new agricultural and NRM technologies, policies and market incentives on human welfare as well as the quality of the resource base and the environment. Possibilities to address dynamic issues and linking changes in biophysical indicators with economic models are important advantages of this method. The integrated framework allows a consistent analysis of technology impacts within a given socioeconomic and policy setting. According to Holden (2005) the main advantages of using bioeconomic models for NRM technologies and policy impact assessment are:

- They allow consistent treatment of complex biophysical and socioeconomic variables, providing a suitable tool for interdisciplinary analysis.
- They allow sequential and simultaneous interactions between biophysical and socioeconomic variables.
- They can be used to assess the potential impacts of new technologies and policies (*ex ante* impact assessment).
- They allow disturbing variation to be controlled (*ceteris paribus* conditions) for evaluation of impacts of certain interactions by isolating effects from other influences.
- They can capture both direct and indirect effects (i.e., the total effect of technology or policy change can be estimated).
- They can be used to carry out sensitivity analyses in relation to various types of uncertainties.

11.3.3.2 Application of bioeconomic model for impact evaluation of watershed development program in semi-arid tropics of India

Even though there have been several case studies of successful watershed development in India (e.g., Kerr *et al.*, 2000; Wani *et al.*, 2002), the impact of the approach in improving the welfare of the poor and the natural resource condition in the semi-arid tropical areas is not fully known. A study was carried to assess the inter-temporal impacts of key integrated watershed management technologies (e.g., high-yielding varieties and soil and water conservation structures) on household income, food security, soil erosion, and nutrient mining in selected micro-watersheds.

Based on the lessons learnt from the success of on-station soil, water, and nutrient management (SWNM) research in watershed, the International Crops Research Institute for the Semi-Arid Tropics (ICRISAT) developed a new Integrated Genetic and Natural Resource Management (IGNRM) model. In one of the on-farm watersheds in India (Adarsha watershed, Kothapally), a participatory community watershed management program was initiated in collaboration with DPAP of Government of India. Along with ICRISAT, a consortium of NGOs and national research institutes have been testing and developing technological, policy, and institutional options for integrated watershed management in the village (Wani *et al.*, 2002; Shiferaw *et al.*, 2003). A package of IGNRM practices were evaluated on farmers' fields including soil and water conservation, new high-yielding varieties, integrated pest management and integrated nutrient management through participatory approaches.

11.3.3.3 Biophysical and socioeconomic data

ICRISAT has installed an automatic weather station in Kothapally village, which allows regular collection of weather parameters (e.g., temperature, rainfall, sunshine, wind speed and direction, etc.). In 2001, ICRISAT conducted census of all households in Kothapally village and five adjoining villages, i.e., non-watershed/control villages (namely Husainpura, Masaniguda, Oorella, Yankepally, and Yarveguda) located outside the watershed with comparable biophysical (rainfall, soil, and climate) and socioeconomic conditions. Based on the information from the census analysis, a random sample of 60 households from watershed village (Kothapally) and another 60 households from non-watershed villages were selected for detailed survey. Along with other standard socioeconomic data, detailed plot-wise and crop-wise input and output data were collected immediately after harvest from the operational holdings of all the sample households.

11.3.3.4 Bioeconomic modeling

Bioeconomic model combines both socioeconomic factors influencing farmers' decision-making with biophysical factors affecting crop production and natural resource conditions (Barbier 1998; Woelcke 2006). The model consists of three components: (i) a mathematical programming model that reflects the farm household decision-making process under certain constraints; (ii) estimation of crop yield response to soil depth; and (iii) nutrient balances as a sustainability indicator. The results of the marginal yield response for soil depth and estimation of soil erosion under different cropping systems are then incorporated into programming model (for the detailed description of the bioeconmic model refer Nedumaran 2007, 2009).

11.3.3.5 Validation of the bioeconomic model

The challenge in the development of bioeconomic models is to ensure that the results can be trusted and that the model can be reused in similar other settings. The validation of the complex models like bioeconomic models is much debated; for example, Janssen and van Ittersum (2007) reviewed 48 bioeconomic models and found that only 23 studies validated their results using observed qualitative and quantitative data.

Based on McCarl and Apland (1986), the *ex-ante* bioeconomic model was validated by conducting regression analysis between observed and simulated land use values. A regression line was fitted through the origin for the observed land use in 2003 and first year of simulated land use of major seven crops expressed in percentage of total area of these crops in the total cultivated area in the watershed. The comparison was done at watershed level. Figure 11.3 compares the observed with the simulated land use at the watershed level. The parameter coefficients are close to unity at watershed level with an explained variance of 97%, which indicates the model results are almost identical with the 2003 land use trend in the Kothapally watershed.

11.3.3.6   Impact of change in yield of dryland crops

The simulation results showed that the per capita income of all three household groups were above the baseline level when the yields of dryland crops increased (Table 11.8). Increase in area of the dryland crops (sorghum and maize) in the watershed increases fodder production, which in turn enhances the carrying capacity of livestock in the watershed. This increased livestock production and also increased the income from livestock gradually for all the household groups.

The soil erosion under the scenario of increased yield of dryland crops was higher than the baseline level at the initial years and started declining from the fifth year of simulation. The increase in area of dryland crops increased the demand for on-farm labor in the initial years, which reduced the incentive to use the labor for conservation measures and this caused higher soil erosion in the initial year of simulation. However, the population growth in the watershed over the years drove the farmers to use more labor for conservation measures in the field, which declined the soil erosion towards the end of the simulation period. The results revealed that the decline in soil erosion was 6% compared to the baseline in the final year of simulation. Under the decreased dryland crop yield scenario, the soil erosion had not changed much compared to the baseline scenario.

The increase in area under sorghum and maize and decline in the area of high nutrient mining crops like cotton and sunflower under the scenario of increased yields of dryland crops had reduced soil nutrient mining by 4, 1, and 3% of nitrogen, phosphorus, and potassium respectively compared to baseline level. If the yield of dryland crops had decreased by 10%, the results showed that nutrient balances in the watershed were similar to baseline level.

11.3.3.7   Impact of change in irrigated area in the watershed

One of the important objectives of the watershed development program is to conserve rainwater by reducing outflows from the watershed by constructing check-dams and other in-situ soil and water conservation systems. The stored water improves the groundwater table, which in turn helps to increase the area under irrigation in the watershed. In this context, simulation was carried out to assess the impact of changes in irrigated area resulting from adoption of the soil and water conservation measures on household welfare, soil loss, and nutrient balance in the watershed. Hence, the

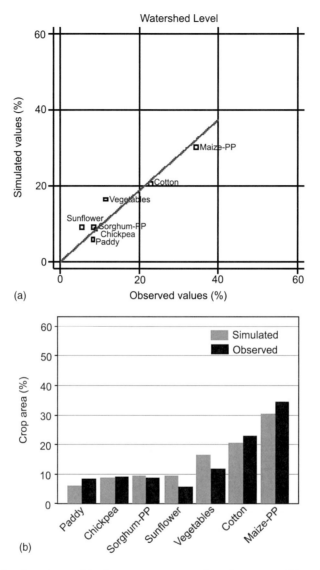

Figure 11.3 Simulated vs observed land use as percentage of total crop area (Note: PP = pigeonpea); Regression line fit: Co-eff = 0.93; SE = 0.51; R^2 = 0.97 (Source: Nedumaran 2009)

baseline scenario in the watershed was compared with two alternative scenarios: (1) increasing irrigated area by 25%; and (2) reducing the area under irrigation by 25%. These changes were simulated through comparative adjustments in dryland area so that the total cultivable area in the watershed remained unchanged.

The results revealed that if irrigated area increased, the per capita income of all the three household groups were more than the baseline level (Table 11.8). This was due to higher productivity of crops like cotton, vegetables, and sunflower under irrigation and increasing the area of these crops under irrigation resulted in increased production

Table 11.8 Impact of change in irrigated area in the watershed[a]

Scenario	Per capita income (₹)			Soil loss (t ha⁻¹)	Conservation labor (person-days)	Nutrient balance (t)		
	Small	Medium	Large			N	P	K
Baseline	5080	9110	16160	4.04	4092.2	−11.74	12.25	−94.79
Irrigated area (+25%)	5160	9500	17810	4.13	4374.18	−14.38	11.37	−98.94
Irrigated area (−25%)	4730	8700	16720	3.92	3600.95	−9.2	14.46	−88.98

[a]Average of 10 years simulation.
Source: Nedumaran (2009).

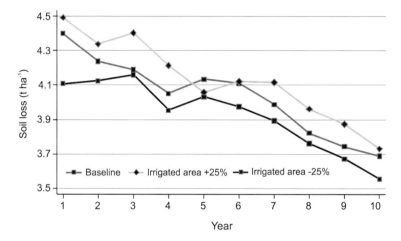

Figure 11.4 Simulated soil loss in the watershed indicating an alternative scenario for change in irrigated area (Source: Nedumaran 2009)

in the watershed. The increased marketable surplus of these crops increased the income of the household groups. The scenario of decreasing the irrigated area by 25% led to reduction in the per capita income for small and medium farm household because the area under commercial crops like vegetables and cotton decreased. The per capita income of the large farmers had not changed much because these farmers were not constrained by irrigated land.

Soil erosion was higher when irrigated area increased in the watershed compared to the baseline level (Figure 11.4). The area under irrigated cotton, sunflower, and vegetables increased because of expanding irrigated land. The increase in the area of erosive crops (wide spaced crops) like cotton and vegetables resulted in higher erosion by 2% compared to baseline level. On the contrary, reduction in irrigated land in the watershed increased the area under less erosive dryland crops like maize and sorghum, reducing the soil erosion by about 7%.

When irrigated area increased by 25%, the labor used for conservation measures was less than the baseline level in the initial years and increased above the baseline

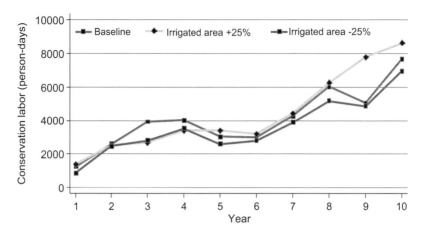

Figure 11.5 Simulated labor used for conservation measures indicating an alternative scenario for change in irrigated area (Source: Nedumaran 2009)

level towards the end of simulation (Figure 11.5). When the irrigated area decreased by 25%, soil erosion was below the baseline level, even though the total labor used for soil and water conservation was lower than the baseline level. This was mainly due to change in cropping pattern, where area under less erosive dryland crops like maize and sorghum increased in the watershed.

The soil nutrient balance indicated that nutrient mining was higher compared to the baseline level when irrigated area increased by 25% (Table 11.8). This was due to increase in the area of high nutrient extraction by irrigated crops like vegetables, cotton, and sunflower compared to baseline level. The reduction in irrigated area increased the area under cereal-legume cropping systems like maize/pigeonpea and sorghum/pigeonpea which removed comparatively less nutrients from the soil and also improved the soil nutrient status through biological nitrogen fixation. Although increase in irrigated area in the watershed improved the welfare of the farmers and the cropping pattern, it caused negative effects on the environment by increasing the erosion level and soil nutrient mining.

Bioeconomic modeling indicated that the introduction of high-yielding varieties and cereal-legume intercropping systems helped to improve the welfare of smallholder farmers by increasing the income and enhancing the sustainability of the natural resource base. It also stimulates sustainable intensification of production by controlling soil erosion and nutrient mining through investment in conservation and adoption of better land use patterns in the watershed. So, it is important to focus more on crop-specific research to develop drought tolerant high-yielding varieties of dryland crops, which are also resistant to pests and diseases. The increase in irrigated area under cotton, vegetables, and sunflower due to the availability of water from community and in-situ soil and water conservation in the watershed improved farmers' income. The erosion level and nutrient mining in the watershed however, increased because of increase in the area under soil erosive and nutrient mining crops like cotton and vegetables. It is important to promote irrigated cereal crops in the watershed, so that erosion level will be minimized and fodder production enhanced to create complementarities with livestock production, leading to increased manure availability for use to replenish soil fertility. The results clearly indicated that care should be taken while developing technologies for watershed development to avoid the promotion of conflicting technologies.

11.3.4 Meta analysis

The economic surplus method and bioeconomic models have demonstrated clearly the use of improved measurement methodologies in watershed evaluation. However, it is also important to examine how in the long run such methods could be applied if the present level of watershed development works is carried out.

Earlier meta analysis was applied to assess the returns on investment in education and understand the implications of certain medical treatments on offspring and the returns to research investment at the global level. Ordinary least square (OLS) approach was employed to estimate the regression equation:

$$BCR = f(L, S, F, R, I, P, T, A, SL) \tag{7}$$

where,
BCR = Benefit-cost ratio;
$\quad L$ = Geographical location of watershed;
$\quad S$ = Size of watershed;
$\quad F$ = Focus of watershed;
$\quad R$ = Rainfall in the watershed area;
$\quad I$ = Implementing agency of the watershed;
$\quad P$ = Peoples' participation;
$\quad T$ = Time gap between project implementation and evaluation;
$\quad A$ = Various activities performed in the watershed area; and
$\quad SL$ = Type of soil in the watershed area as explanatory variables.

Meta analysis has become popular among economists to assess the impacts at macro level. The purpose is to collate research findings from previous studies, and distil them for broad conclusions. The approach is popularly known as analysis of the analyses. Meta analysis can be helpful for policymakers, who may be confronted by numerous conflicting conclusions (Joshi *et al.*, 2005).

11.3.4.1   Review of studies on meta analysis

This section is mainly drawn from the recent study made by the ICRISAT-led consortium team (Wani *et al.*, 2008). Reddy (2000) reviewed 22 impact assessment studies conducted across the country from 1967 to 1997. The impact of watershed development projects showed positive impacts on crop yields, cropping intensity, and cropping pattern changes. However, there was a large variation in the magnitude of the impact across regions and crops. The magnitude of the impact is dependent on the nature of activities undertaken in the watershed (i.e., higher the agricultural and livestock interventions, higher will be the overall benefits from the watershed program). In general, net income increase had favorable BCR. The BCR is stable at 1.75, implying positive impacts by the watershed development programs in the country.

Many other studies (Palanisami *et al.*, 2002a; Ramaswamy and Palanisami 2002; Sastry *et al.*, 2002; Sreedharan 2002; Palanisami and Suresh Kumar 2006) employed before and after approaches to assess the impact of watershed development activities. Others (Lokesh *et al.*, 2006; Ramakrishna *et al.*, 2006) adopted with and without approach to asses the impact.

These studies focused on the impact of watershed activities on various impact domains like soil and moisture conservation, water resources development, impact of cropping pattern and yield, and overall economic impacts. These studies found that there is significant impact on soil erosion control, soil moisture conservation, water resources development, and increased crop yields. The watershed development has also produced desired results in terms of improvement in socioeconomic conditions, and the environment. Experiences of most of the impact assessment studies report that watershed development interventions have produced desired positive impacts. But the magnitude of the impact was found to vary across regions and impact domains.

The impacts of various watershed development activities are discussed under different domains with various indicators. The watershed development activities are expected to influence various biophysical aspects such as soil fertility, expansion in cropped area, cropping intensity, and productivity of crops; socioeconomic aspects such as employment, food security, income of the households, migration, and people's participation; economic aspects such as overall impacts on the rural economy; environmental aspects such as water table in the wells, irrigated area, soil loss, runoff, and water pollution; expansion in production of high-value agricultural commodities; and non-farm ancillary activities. These impacts on different domains are discussed hereunder.

11.3.4.2 Biophysical impacts

The watershed development activities have significant positive impacts on various biophysical aspects such as investment on soil and water conservation measures, soil fertility status, soil and water erosion, expansion in cropped area, changes in cropping pattern, cropping intensity, production and productivity of crops (Figures 11.6 and 11.7). These include improved conservation of soil and moisture, improvement and maintenance of fertility status of the soil (Sikka *et al.*, 2000; AFC 2001; Ramasamy and Palanisami 2002; Sastry *et al.*, 2002) and reduced soil and water erosion. The organic carbon increased by 37% due to watershed intervention (Sikka *et al.*, 2000). Significant reduction in soil and water erosion (77.78% reduction) was observed by Milkesha Wakjira (2003).

Impact and evaluation study of the soil conservation scheme under DPAP indicates that only marginal impact was realized in land use, cropping pattern, and yield (Evaluation and Applied Research Department 1981). Evidences show that soil conservation improved moisture retention, reduced soil erosion, changed land use pattern, and increased crop yield. Soil loss reduced from $18758\,kg\,ha^{-1}$ in 1988 to $6764\,kg\,ha^{-1}$ in 1989. Between 1985–86 and 1989–90, crop yield had increased at annual compound growth rate (CGR) of 3.94% to 16.40% (Evaluation and Applied Research Department 1991).

Improvement in soil fertility coupled with increased water resources in the watershed area led to expansion in cropped area, cropping intensity, and increase in

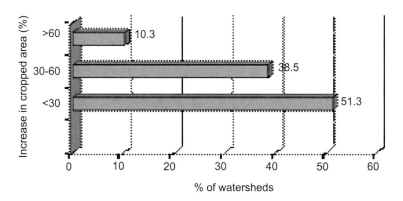

Figure 11.6 Distribution (%) of watersheds by increase in cropped area

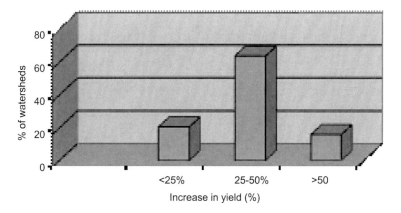

Figure 11.7 Distribution (%) of watersheds by increase in yield

production and productivity of crops. Most of the studies indicated significant increase in cropped area, which ranged from 6.84% (Sreedharan 2002) to 52% (Sastry *et al.*, 2002). The increase in cropped area helped increase production and productivity. The productivity enhancement due to watershed development is a common phenomenon in most of the watersheds (Figure 11.7). The increase in yield of crops ranged from 5% (Shobha Rani 2001) to 91.11% in Karnataka (Milkesha Wakjira 2003).

The cropping pattern changes have taken place both in additional area brought under well irrigation from the fallow lands and in the area under rainfed cultivation. The area under high water consuming crops increased by 25.3% in first crop and 29.4% in second crop period. Similarly, cropping intensity increased from 120% to 146.88% in Kattampatti watershed and from 102.14% to 112.08% in Kodangipalayam watershed (Palanisami and Suresh Kumar 2004). Increase in Crop Productivity Index, Fertilizer Application Index, and Crop Diversification Index was also observed (Sikka *et al.*, 2000, 2001).

It is lucid from the analysis that though there are differences in impacts, the watershed development activities have made significant positive impacts on the bio-physical aspects leading to increased soil fertility, cropping pattern changes, and crop production and productivity.

11.3.4.3 Socioeconomic impacts

The watershed development technologies aimed at not only conserving the natural resources but also improving the socioeconomic conditions of the rural people who depend upon these for their livelihood. The impacts of various watershed treatments is however widespread. The changes in various biophysical, and environmental aspects impacts socioeconomic conditions of the people. Watershed development programs are designed to influence the biophysical aspects and environmental aspects and thereby bringing changes in socioeconomic conditions (Deshpande and Rajasekaran 1997).

The socioeconomic indicators like changes in household income, changes in per capita income, consumption expenditure, differences in employment, changes in persons migrated, peoples' participation, changes in household assets, and changes in wage rate at village level were considered for the impact assessment. The watershed intervention helped the rural farm and non-farm households to enhance their income level. Evidences show that the rural labor households in the treated villages derive ₹28732 when compared to ₹22320 in control village, which is 28.73% higher in Kattampatti watershed. Similarly, the per capita income was also relatively higher among households of watershed treated villages. The percentage difference among households across villages was 13.17% in Kattampatti and 70.44% in Kodangi-palayam watershed (Palanisami and Suresh Kumar 2004). Increase in per capita income and household income helps the rural households to enhance their asset position. The asset position of the households increased significantly from 13 to 50% (AFC 2001). The increased income helps the households to ensure quality food and achieve nutritional security.

Any development program is expected to generate adequate employment to the local people. Casual employment was created during the implementation of works such as bunding, leveling, construction of check-dams, percolation ponds, summer plowing, crop demonstration, retaining wall, and plantation. Also, the watershed development program reduces out-migration. As sufficient employment opportunities are created due to watershed intervention through expansion in cropped area, the landless rural labor households and other marginal and small farmers get adequate employment to earn their livelihoods. This helps reduce out-migration. Evidences show that out-migration has been reduced by 20–50% in many watersheds (Sastry *et al.*, 2002). In some watersheds, the reduction is up to 43%.

Like all other development programs, watershed development program banks heavily on the participatory approach. Though watershed development program envisages an integrated and comprehensive plan of action for the rural areas, peoples' participation at all levels of its implementation is of critical importance. For this to happen, it is necessary that every farmer having land in the watershed accepts and implements the recommended watershed development plan. As the issue of sustainable NRM becomes more and more crucial, it has also become clear that sustainability is closely linked to the participation of the community. This requires sustained efforts

(i) to inform and educate the rural community, demonstrate to them the benefits of watershed development, and that the project can be planned and implemented by the rural community with expert help from government and non-government sources; and (ii) to critically analyze the various institutional and policy aspects of watershed development programs in relation to participatory watershed management.

Experience on the evaluation study of 15 DPAP watersheds conducted in Coimbatore district of Tamil Nadu showed that the overall community participation was 42%. The participation was found to be 55, 44, and 27%, respectively at planning, implementation, and maintenance stages. This suggests that the community participation in watershed development program has to be greater. Similarly, overall contribution for works on private land was found to be 14% and varied from a minimum of 7% for fodder plots to a maximum of 22% for horticulture and farm pond. However, the contribution in terms of cash or kind towards development of structures in common lands such as percolation ponds, and check-dams was found to be nil. Level of adoption of various soil and moisture conservation measures and their maintenance indicate that there is a wide variation in the level of adoption, with a minimum of 2.4% in farm pond, 30.40% in summer plowing, 36.80% in land leveling, and 44% in contour bunding. Follow-up by farmers was also found to be poor in most of the technologies and it accounted for 5.23% in farm ponds and plowing, 21.58% for contour bunding (Sikka *et al.*, 2000).

The Water Technology Centre at the Tamil Nadu Agricultural University carried out mid-term evaluation of 18 watersheds under Integrated Wasteland Development Program in Pongalur Block of Coimbatore district, Tamil Nadu. The results revealed that Peoples' Participation Index at the planning stage was 52.69%, followed by implementation stage (39.28%). This shows low peoples' participation at both the stages of the project (Palanisami *et al.*, 2002b). In several watersheds, the structures are not maintained due to lack of funds as well as lack of coordination among beneficiaries. Also, many of the presidents of the Watershed Association were not reelected in the local panchayat elections, resulting in lack of coordination particularly during the post-project management. There is a decline in interest in watershed structures during the post-implementation phase attributable to (i) failure or collapse of the new institutions set up to manage watersheds; and (ii) lack of clear norms on how to operate WDFs. Thus ensuring peoples' participation in different stages of watershed implementation and management is crucial for achieving the objectives of watershed development in a sustained manner.

11.3.4.4 Environmental impacts

The watershed development activities generate significant positive externalities including improving agricultural production, productivity, and socioeconomic status of the people who directly or indirectly depend upon the watershed for their livelihoods. The environmental indicators include water level in the wells, changes in irrigated area, duration of water availability, water table of wells, surface water storage capacity, differences in number of wells, number of wells recharged/defunct, differences in irrigation intensity, and Watershed Eco Index (WEI).

The impact assessment studies conducted by different agencies and scientists across regions imply that watershed development activities generated significant positive

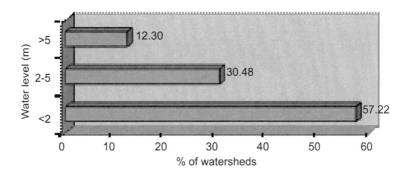

Figure 11.8 Distribution (%) of watersheds by water level in the wells

impacts on the environment. One of the important objectives of watershed development is in-situ water and soil conservation. Water resources development in the watershed village and the treatment activities helped in conservation and enhancement of water resources. Most of the studies report that water level in the wells increased leading to expansion in irrigated area in the watershed. In practice, only a few studies actually measured the water level in the wells. The increase in water level varied across seasons from 0.1 to 3.5 m. Similarly, the expansion in irrigated area due to watershed development activities varied from 5.6 to 68% across region and season. Experiences show that the increase in water level in the wells is observed to be less than 2 m (57.22% of watersheds). About 30.48% of watersheds witnessed an increase of 2–5 m and only 12.3% of watersheds had an increase of more than 5 m increase in water level in the wells (Figure 11.8).

Watershed development activities produced significant positive impact on water table, perenniality of water in the wells, and pumping hours that resulted in an increased irrigated area and crop diversification (Sikka *et al.*, 2000, 2001). Conservation and water harvesting measures in the watershed helped in improving the groundwater recharge, water availability for cattle and other domestic uses, increasing perenniality of water in the streams, increasing water table in the wells, sediment trapping behind the conservation measures/structures, and stabilization of gully beds (Madhu *et al.*, 2004). The productivity of crops increased from 6.65 to 16.59% in the watershed village.

Planting trees in private and common lands is also being undertaken as part of the watershed development. This has created additional green cover, improving the environment. The WEI, which reflects the additional green cover created, varied from 1.8 to 43% (Sikka *et al.*, 2000, 2001; Ramaswamy and Palanisami 2002; Palanisami and Suresh Kumar 2004; Ramakrishna *et al.*, 2006).

11.3.4.5 Overall economic impacts

Experiences show that watershed development activities have overall positive impacts on the village economy. Thus, it is essential to assess the impact of these activities using key indicators such as NPV, BCR, and IRR (Figures 11.9 and 11.10). Though these

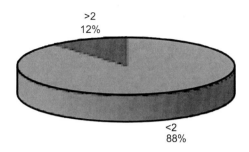

Figure 11.9 Distribution (%) of watersheds by BCR category

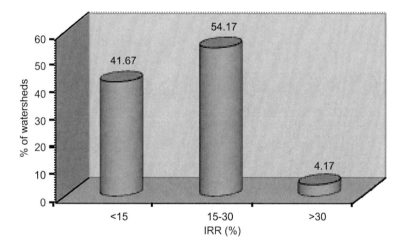

Figure 11.10 Distribution (%) of watersheds by IRR category

indicators show the overall impact of watershed development activities, only very few studies have quantified the benefits and actually estimated the NPV, BCR, and IRR. The reasons for this are: (i) Most of the evaluating agencies are not familiar with the techniques; (ii) Inadequate data availability for quantifying benefits and costs; and (iii) Non-familiarity with computer software used. The overall impact of watershed development activities in terms of NPV, BCR, and IRR are reviewed and discussed hereunder.

A few studies (Palanisami *et al.*, 2002a; Ramaswamy and Palanisami 2002; Milkesha Wakjira 2003; Palanisami and Suresh Kumar 2004, 2006; Lokesh *et al.*, 2006) made attempts to assess the overall impact of watershed development activities through BCR and NPV. The BCR, which is the return per rupee of investment, ranged from 1.27 to 3.7. The size of BCR also depends on the magnitude of the benefits accrued due to the watershed development activities, which in turn critically depend upon the rainfall. The watersheds which have high BCR (>2) received an annual rainfall of 700 to 900 mm. Similarly, the watersheds that receive rainfall <700 mm and

700–900 mm had relatively higher IRR. The analysis also revealed that the BCR is worked out to more than 2 in around 12% of watersheds. About 88% of watersheds have the BCR < 2. Similarly, 41.67% of watersheds showed IRR of 41.67%; 54.17% of the watersheds have IRR ranging between 15 and 30%; and only 4.17% of the watersheds have IRR > 30%.

The BCR varies across regions and depends upon the agroclimatic conditions. For instance, financial analysis of the impact of watershed development indicates that the BCR varied from 1.43 to 1.51, implying that the returns to public investment such as watershed development activities are feasible. Similarly, the IRR was 26 and 24%, respectively for Kattampatti and Kodangipalayam watersheds, which is higher than the long-term loan interest rate by commercial banks (12.75%), indicating the worthiness of the government investment on watershed development (Palanisami and Suresh Kumar 2004). The studies proved that the watershed development activities have high benefit-cost ratio of 3.5 (Lokesh *et al.*, 2006) and a fairly high IRR of 38% (Ramaswamy and Palanisami 2002).

Another viable indicator, viz., net returns per rupee of irrigation cost, shows the overall impact of watershed development activities. The net returns per rupee of irrigation cost ranged from 1.4 to 16.32 and varied with type of watershed and season. The watershed development activities have increased the net returns per rupee of irrigation cost. The net returns have increased from 6.52 to 16.32 after the implementation of watershed development activities. Similarly, the watershed development activities have had differential impacts and varied across size groups. The net return per unit of water (i.e., acre inch of groundwater applied) increased by 3% and 30%, respectively for small and large farmers under watershed development program implementation. Water use and net returns per acre of gross irrigated area (GIA) for farmers in the upstream increased by 68% and 66% respectively and in downstream by 48% and 110% respectively (Mengesha 2000).

The Net Present Worth (NPW) indicates that the watershed development activities produced desired results as evidenced from positive NPV. The NPV of the benefits derived from various watershed treatment activities was ₹1.24 million (Milkesha Wakjira 2003). From these indicators [NPV (positive), BCR (>1), and IRR (>the opportunity cost of capital)], one can speculate that the watershed development activities are financially feasible and economically viable.

11.3.5 Comparison of the methods

The methods discussed above have their strengths and weaknesses (Table 11.9). The major constraints are the non-availability of reliable data and the expertise to analyze and interpret the data. However, depending on the situation and data availability, the method of evaluation can be targeted. For example, in situations where no detailed data is available, simple BCR and IRR will give some idea about the impact of the watershed investment; whereas in regions where detailed information on biophysical aspects are available and the aim of the evaluation is mainly to evaluate the impact of the different biophysical factors, then biophysical modeling will be an appropriate choice. Once some key data on biophysical aspects are available from the model watersheds in each region, the bioeconomic models can be easily targeted. In the case of the economic analysis where only the total benefits of the watersheds should be analyzed,

Table 11.9 Comparison of different watershed evaluation methods

Method	Major advantage	Major limitation[a]
Conventional Analysis	Quick to estimate	Sensitive to i and n*
Econometric Models	All sectors included	εd and εs sensitive**
Bio economic Models	Whole system included; optimization	Too much experimental details
Meta Analysis	Macro picture given	Aggregation bias

[a]i = discount rate; n = life period; εd = price elasticity of demand; εs = price elasticity of supply

the economic surplus methods will be appropriate as it takes into account both the producers and consumers surpluses.

11.4 CONCLUSIONS AND POLICY RECOMMENDATIONS

With the large investment of financial resources in the watershed program, it is important that the program becomes successful. Hence the challenges in watershed impact assessment should be given due importance in the future planning and development programs. Realizing the potential and importance of watershed development and their likely impact on the economy, enough efforts have been made to identify and develop indicators for proper monitoring and evaluation of watershed development projects. This will be useful for the researchers, government agencies, and other agencies involved in the monitoring and evaluation of watershed development projects.

This chapter has thus demonstrated the importance of different watershed evaluation methods with adequate explanation with the field data and derived results from the analysis. The results had indicated that watershed development activities have been found to have significant impact on groundwater recharge, access to groundwater and hence the expansion in irrigated area. In addition to these public investments, private investments through construction of farm ponds may be encouraged as these structures help in a big way to harvest the available rainwater and hence groundwater recharge. Thus, the combination of public and private investment will enhance the return to investment in watershed programs. Therefore, the policy focus must be on the development of these water harvesting structures, particularly percolation ponds wherever feasible.

Once the groundwater is available, high water intensive crops are introduced. Hence, appropriate water saving technologies like drip is introduced without affecting farmers' choice of crops. The creation and implementation of regulations in relation to depth of wells and spacing between wells will reduce the well failure, which could be possible through Watershed Association. The existing NABARD norms such as 150 m spacing between two wells should be strictly followed.

People's participation, involvement of the PRIs, local user groups, and NGOs alongside institutional support from different levels, viz., the central and state government, district and block levels should be ensured to make the program more participatory, interactive, and cost effective.

The watershed development technologies benefit not only the participating farm households but also the non-participating farm and other rural households in the watershed villages. The economic surplus method has emphasized the need for enhancing the farm income through the adoption of alternative farming system combining agricultural crops, trees, and livestock components with comparable profit as both consumers and producers will have enhanced benefits.

In order to strengthen the applicability of the watershed evaluation methods, strong data base is important. The data generation through model watersheds in each region will help strengthen the evaluation mechanism in an effective manner. Also in each implementing department, separate data bank should be maintained starting from the benchmark data on the watersheds. The details should cover all aspects of costs and benefits of the watershed development programs. The staff can be given the needed training on data collection, data storage, and basic analysis. The officials from the government departments, evaluation departments, and research institutions should be sensitized about the use of different watershed evaluation methods including handling of the data from the fields.

REFERENCES

AFC. 2001. Report on evaluation study of the scheme of soil conservation in the catchment of river valley projects and flood prone rivers, Kundah Catchment, Kerala. Agricultural Finance Corporation Ltd.

Baker, J.L. 2000. *Evaluating the impacts of development projects on poverty.* Washington, DC, USA: World Bank.

Bantilan, M.C.S., K.V. Anupama, and P.K. Joshi. 2005. Assessing economic and environmental impacts of NRM technologies: an empirical application using the economic surplus approach. In *Natural resource management in agriculture: methods for assessing economic and environmental impacts,* ed. B. Shiferaw, H.A. Freeman, and S.M. Swinton, 245–268. Wallingford, UK: CAB International.

Barbier, B. 1998. Induced innovation and land degradation: results from a bioeconomic model of a village in West Africa. *Agricultural Economics* 19:15–25.

Barbier, B., and G. Bergerson. 2001. Natural resource management in the hillsides of Honduras: bioeconomic modeling at the micro watershed level. *Research Report no. 123.* Washington, DC, USA: International Food Policy Research Institute.

Barrett, C.B., J. Lynam, F. Place, T. Reardon, and A.A. Aboud. 2002. Towards improved natural resource management in African agriculture. In *Natural resource management in African agriculture: understanding and improving current practices,* ed. C.B. Barrett, F. Place, and A.A. Aboud, 287–296. Wallingford, UK: CAB International.

Deshpande, R.S., and N. Rajasekaran. 1997. Impact of watershed development programme: Experiences and issues. *Artha Vijnana* 34(3):374–390.

Deshpande, R.S., and V. Ratna Reddy. 1991. Differential impact of watershed based technology: analytical study of Maharashtra. *Indian Journal of Agricultural Economics* 46(3):261–269.

Erappa, S. 1998. Sustainability of watershed development programmes (WDPs) to dryland Agriculture in Karnataka: A study of two sub-watersheds. *Project Report.* Bangalore, India: Institute for Social and Economic Change.

Evaluation and Applied Research Department. 1981. *An evaluation report on soil conservation scheme under the DPAP in Ramanathapuram District.* Chennai, India: Evaluation and Applied Research Department.

Evaluation and Applied Research Department. 1991. *Report on the evaluation of soil conservation works executed in Sholur Micro Watersheds in Nilgris District under HADP.* Chennai, India: Evaluation and Applied Research Department.

Fan, S., P. Hazell, and D. Thorat. 1999. Linkages between government spending, growth, and poverty in rural India. *Research Report no. 110.* Washington, DC, USA: International Food Policy Research Institute.

Farrington, J., and C. Lobo. 1997. *Scaling up participatory watershed development in India: lessons from the Indo-German Watershed Development Programme.* Natural Resource Perspective 17. London, UK: Overseas Development Institute.

Farrington, J., C. Turton, and A.J. James. 1999. *Participatory watershed development: challenges for the 21st Century.* New Delhi, India: Oxford University Press.

Fernandez, A. 1994. The MYRADA experience: The interventions of a voluntary agency in the emergence and growth of peoples' institutions of sustained and equitable management of micro watershed. Research Report. Bangalore, India: MYRADA. (http:/www.myrada.org/publications.htm)

Freeman, H.A., B. Shiferaw, and S.M. Swinton. 2005. Assessing the impact of natural resource management interventions in agriculture: concepts, issues and challenges. In *Natural resource management in agriculture: methods for assessing economic and environmental impacts,* ed. B. Shiferaw, H.A. Freeman, and S.M. Swinton, 4–16. Wallingford, UK: CAB International.

Government of India. 1990. *WARASA: National Watershed Development Project for Rainfed Areas (NWDPRA) Guidelines.* 1st edition. New Delhi, India: Ministry of Agriculture.

Government of India. 1994a. *Reports of the Technical Committee on Drought Prone Areas Programme and Desert Development Programme.* New Delhi, India: Ministry of Rural Development.

Government of India. 1994b. *Guidelines for Watershed Development.* New Delhi, India: Ministry of Rural Development.

Government of India. 2000. *Common approach for watershed development.* New Delhi, India: Ministry of Agriculture and Cooperation. (http:/ www.nic.in/agricoop/guide.htm)

Government of India. 2001a. *Agricultural statistics at a glance.* New Delhi, India: Ministry of Agriculture.

Government of India. 2001b. *Guidelines for Watershed Development.* (Revised.) New Delhi, India: Department of Land Resources, Ministry of Rural Development.

Government of India. 2003. Guidelines for Hariyali, New Delhi, India: Department of Land Resources, Ministry of Rural Development.

Gregerson, H., and Contreras, A. 1992. Economic assessment of forestry project impacts. *FAO Forestry Paper No. 106.* Rome, Italy: FAO.

Hanumantha Rao, C.H. 2000. Watershed development in India: Recent experiences and emerging issues. *Economic and Political Weekly* 35(45):3943–3947.

Hinchcliffe, F., J. Thompson, J. Pretty, I. Guijit, and P. Shah (ed.) 1999. *Fertile ground: the impacts of participatory watershed management.* London, UK: Intermediate Technology Publications.

Holden, S., and B. Shiferaw. 2004. Land degradation, drought and food security in a less-favoured area in the Ethiopian highlands: a bioeconomic model with market imperfections. *Agricultural Economics* 30:31–49.

Holden, S., B. Shiferaw, and J. Pender. 2004. Non-farm income, household welfare, and sustainable land management in a less-favoured area in the Ethiopian highlands. *Food Policy* 29:369–392.

Holden, S.T. 2005. Bioeconomic modeling for natural resource management impact assessment. In *Natural resource management in agriculture: methods for assessing economic and environmental impacts,* ed. B. Shiferaw, H.A. Freeman, and S.M. Swinton, 295–318. Wallingford, UK: CAB International.

IJAE. 1991. Subject 1. Watershed development. *Indian Journal of Agricultural Economics* 46(3):241–327.

Janssen, S., and M.K. van Ittersum. 2007. Assessing farm innovations and responses to policies: A review of bio-economic farm models. *Agricultural Systems* 94(3):622–636.

Joshi, P.K., A.K Jha, S.P. Wani, P. Laxmi Joshi, and R.L. Shiyani. 2005. *Meta-analysis to assess impact of watershed program and people's participation.* Comprehensive Assessment Research Report 8. Colombo, Sri Lanka: Comprehensive Assessment Secretariat.

Joshi, P.K., V. Pangare, B. Shiferaw, S.P. Wani, J. Bouma, and C. Scott. 2004. Socioeconomic and policy research on watershed management in India: Synthesis of past experience and needs for future research. *Global Theme on Agroecosystems Report No.7.* Patancheru, Andhra Pradesh, India: International Crops Research Institute for the Semi-Arid Tropics.

Kakade, B.K., H.C. Kulkarni, K.J. Petare, G.S. Neelam, A. Marathe, and P.R. Nagargoje. 2001. *Integrated drinking water resource management.* Pune, India:BAIF Development Foundation.

Kerr, J.M., and K.R. Chung. 2005. Evaluating the impact of watershed management projects: a practical econometric approach. In *Natural resource management in agriculture: methods for assessing economic and environmental impacts,* ed. B. Shiferaw, H.A. Freeman, and S.M. Swinton, 223–243. Wallingford, UK: CAB International.

Kerr, J., G. Pangare, V.L. Pangare, and P.J. George. 2000. *An evaluation of dryland watershed development projects in India.* EPTD Discussion Paper No. 68. Washington, DC, USA: International Food Policy Research Institute.

Libardo, R.R., J.A. García, C. Seré, S.J. Lovell, L.R. Sanint, and D. Pachico. 1999. *Manual on Economic Surplus Analysis Model (MODEXC).* Colombia: International Center for Tropical Agriculture.

Lokesh, G.B., Chandrakanth, M.G., and Chinnappa Reddy, B.V. 2006. Total economic valuation of watershed development programme. In *Impact assessment of watershed development – issues, methods and experiences,* ed. K. Palanisami, and D. Suresh Kumar, 251–267. New Delhi, India: Associated Publishing Company Ltd.

Madhu, M., Subhash Chand, P. Sundarambal, and A.K. Sikka. 2004. *Report on impact evaluation of DPAP watersheds in Coimbatore District (IV Batch).* Uthagamandalam, Tamil Nadu, India: Central Soil and Water Conservation Research and Training Institute, Research Centre.

Maluccio, A.J., and R. Flores. 2005. Impact evaluation of a conditional cash transfer program: The Nicaraguan Red de Social. *Research Report 141.* Washington, DC, USA: International Food Policy Research Institute.

McCarl, B.A., and J. Apland. 1986. Validation of linear programming models. *Southern Journal of Agricultural Economics,* December, 155–164.

Mengesha, B.A. 2000. *Access to water resource for irrigation: economics of watershed development in a drought prone area of Karnataka.* MSc (Agri.) thesis, University of Agricultural Sciences, Bangalore, India.

Milkesha Wakjira. 2003. *Economic analysis of watershed development study of Rajanakunte Micro-watershed, Karnataka.* PhD thesis, University of Agricultural Sciences, Bangalore, India.

Mohr, L.B. 2000. *Impact analysis for programme evaluation.* 2nd edition. California, USA: Sage Publications.

Nanan, K.N. 1998. An assessment of European-aided watershed development projects in India from the perspective of poverty reduction and the poor. *CDR Working Paper 98.3.* Copenhagen, Denmark: Centre for Development Research. (http:/www.cdr.dk/working_papers)

Nedumaran, S. 2009. Inter-temporal impacts of technological interventions of watershed development programme on household welfare, soil erosion and nutrient flow in semi-arid India: an integrated bioeconomic modeling approach. Selected paper at the Tri-annual Conference of the International Association of Agricultural Economists, Beijing, China, 16–22 August 2009.

Nedumaran, S. 2007. *Assessing the impact of technological and policy interventions for microwatershed management in semi-arid India: a bioeconomic modeling approach.* PhD thesis, Tamil Nadu Agricultural University, Coimbatore, India.

Okumu, B., M. Jabbar, D. Coleman, N. Russell, M. Saleem, and J. Pender. 2000. Technology and policy impacts on the nutrient flows, soil erosion and economic performance at watershed level: the case of Ginchi in Ethiopia. Presented in conference on Beyond Economics – Multidisciplinary Approaches to Development, December 11–14, 2000, Tokyo, Japan.

Okumu, B.N., M. Jabbar, D. Coleman, and N. Russel. 2002. A bioeconomic model of integrated crop-livestock farming systems: the case of the Ginchi Watershed in Ethiopia. In *Natural resource management in African Agriculture: understanding and improving current practices*, ed. C.B. Barrett, F. Place, and A.A. Aboud. Wallingford, UK: CAB International.

Palanisami, K., S. Devarajan, M. Chellamuthu, and D. Suresh Kumar. 2002a. *Mid-term evaluation of IWDP watersheds in Coimbatore District of Tamil Nadu.* Coimbatore, Tamil Nadu, India: Tamil Nadu Agricultural University.

Palanisami, K., and D. Suresh Kumar. 2004. Participatory watershed development: institutional and policy issues. *Indian Journal of Agricultural Economics* 59(3):376.

Palanisami, K., and D. Suresh Kumar. 2006. Challenges in impact assessment of watershed development. In *Impact assessment of watershed development: methodological issues and experiences*, ed. K. Palanisami, and D. Suresh Kumar. New Delhi, India: Associated Publishing Company Ltd.

Palanisami, K., and D. Suresh Kumar. 2007. Watershed development and augmentation of groundwater resources: Evidence from Southern India. Presented at Third International Groundwater Conference, February 7–10, 2007, Tamil Nadu Agricultural University, Coimbatore, India.

Palanisami, K., D. Suresh Kumar, and B. Chandrasekaran. (ed.) 2002b. *Watershed management: issues and policies* for 21st century. New Delhi, India: Associated Publishing Company Ltd.

Palanisami, K., D. Suresh Kumar, S.P. Wani, and Mark Giordano. 2009. Evaluation of watershed development programmes in India using economic surplus method. *Agricultural Economics Research Review,* Vol. 22, July-December, pp. 197–207.

Pender, J. 2005. Econometric methods for measuring natural resource management impacts: theoretical issues and illustration from Uganda. In *Natural resource management in agriculture: methods for assessing economic and environmental impacts*, ed. B. Shiferaw, H.A. Freeman, and S.M. Swinton, 127–154. Wallingford, UK: CAB International.

Rajagopalan, V. 1991. Integrated watershed development in India: some problems and perspective. *Indian Journal of Agricultural Economics* 46(3):242–260.

Ramakrishna, Y.S., Y.V.R. Reddy, and B.M.K. Reddy. 2006. Impact assessment of watershed development programme in India. In *Impact assessment of watershed development – issues, methods and experiences*, ed. K. Palanisami, and D. Suresh Kumar, 223–238. New Delhi, India: Associated Publishing Company Ltd.

Ramaswamy, K., and K. Palanisami. 2002. Some impact indicators and experiences of watershed development in drought prone areas of Tamil Nadu. In *Watershed management: issues and policies for 21st century*, ed. K. Palanisami, D. Suresh Kumar, and B Chandrasekaran, 182–191. New Delhi, India: Associated Publishing Company Ltd.

Reddy, R.V. 2000. Sustainable watershed management: institutional approach. *Economic and Political Weekly* 35(40):3943–3947.

Reddy, R.V., M. Gopinath Reddy, S. Galab, and Oliver Springate-Baginski. 2001. Watershed development and livelihood security: an assessment of linkages and impact in Andhra Pradesh, India. Draft Report. Hyderabad, India: Centre for Economic and Social Studies; and Leeds, UK: School of Geography, University of Leeds.

Sankhayan, P.L., and O. Hofstad. 2001. A village-level economic model of land clearing, grazing, and wood harvesting for Sub-Saharan Africa: with a case study of southern Senegal. *Ecological Economics* 38:423–440.

Sastry, G., Y.V.R. Reddy, and H.P. Singh. 2002. Appropriate policy and institutional arrangements for efficient management of rain-fed watersheds in 21st century. In *Watershed
management: issues and policies for 21st century*, ed. K. Palanisami, D. Suresh Kumar, and
B. Chandrasekaran, 228–234. New Delhi, India: Associated Publishing Company Ltd.

Schipper, R.A. 1996. *Farming in a fragile future: economics of land use with applications in
the Atlantic of Costa Rica*. PhD thesis, Wageningen Agricultural University, Wageningen,
Netherlands.

Shah, A. 2001. Who benefits from participatory watershed development? Lesson from Gujarat,
India. *SARLP Gatekeeper Series No. 97*. London, UK: International Institute for Environment
and Development.

Shah, A. 2000. Watershed programmes: a long way to go. *Economic and Political Weekly*,
35(35&36), August 26, pp. 3155–3164.

Sharma, R. 2001. Foreword. In *Watershed at work*, ed. M.M. Joshi, S.K. Dalal, and
C.K. Haridas. New Delhi, India: Ministry of Agriculture, Department of Agriculture and
Cooperation, Government of India.

Shiferaw, B., R.V. Reddy, S.P. Wani, and G.D. Nageswara Rao. 2003. Watershed management
and farmers conservation investments in the semi-arid tropics of India : analysis of determinants of resource use decisions and land productivity benefits. *Working Paper No. 16*.
Patancheru, Andhra Pradesh, India: International Crops Research Institute for the Semi-Arid
Tropics.

Shiferaw, B., and H.A. Freeman. (ed.) 2003. *Methods for assessing the impact of natural resource
management research. Summary of the Proceedings of an International Workshop, 6–7
December 2002*. Patancheru, Andhra Pradesh, India: International Crops Research Institute
for the Semi-Arid Tropics.

Shobha Rani, S. 2001. *Economics of groundwater recharge in Huthur Watershed in Southern
Dry Zone of Karnataka*. MSc thesis, University of Agricultural Sciences, Bangalore, India.

Sikka, A.K., Subhash Chand, M. Madhu, and J.S. Samra. 2000. *Report on evaluation study
of DPAP watersheds in Coimbatore District*. Uthagamandalam, Tamil Nadu, India: Central
Soil and Water Conservation Research and Training Institute.

Sikka, A.K., N. Narayanasamy, B.J. Pandian, V. Selvi, Subhash Chand, and Ayyapalam.
2001. *Report on participatory impact evaluation of Comprehensive Watershed Development Project, Tirunelveli*. Uthagamandalam, Tamil Nadu, India: Central Soil and Water
Conservation Research and Training Institute.

Sreedharan, C.K. 2002. Joint forest management and watershed development programme in
Tamil Nadu: An experience in TAP. In *Watershed management – issues and policies for 21st
century*, ed. K. Palanisami, D. Suresh Kumar, and B. Chandrasekaran, 265–274. New Delhi,
India: Associated Publishing Company Ltd.

Wani, S.P., P.K. Joshi, K.V. Raju *et al.* 2008. Community watershed as a growth engine for devel-
opment of dryland areas. A comprehensive assessment of watershed programs in India. *Global
Theme on Agroecosystems Report No. 47*. Patancheru, Andhra Pradesh, India: International
Crops Research Institute for the Semi-Arid Tropics.

Wani, S.P., T.K. Sreedevi, H.P. Singh, P. Pathak, and T.J. Rego. 2002. *Innovative farmers participatory integrated watershed management model: Adarsha Watershed, Kothapally, India
(A success story)*. Patancheru, Andhra Pradesh, India: International Crops Research Institute
for the Semi-Arid Tropics.

Woelcke, J. 2006. Technological and policy options for sustainable agricultural intensification
in eastern Uganda. *Agricultural Economics* 34:129–139.

World Bank. 1990. *Staff Appraisal Report, India: Integrated Watershed Development (Plains)
Projects*. Washington, DC, USA: World Bank.

Chapter 12

Watershed management through a resilience lens

Jennie Barron and Patrick Keys

12.1 WATERSHED MANAGEMENT IN SMALLHOLDER RAINFED AGROECOSYSTEMS

All farming systems, from rainfed to irrigated, are embedded in a landscape that has an inherent capacity to provide a range of ecosystem services with various functions. By managing these landscapes for agricultural production, different ecosystem services may take precedence over others. These human-made shifts in agricultural landscapes put emphasis on particular ecosystem services, and this emphasis may alter the natural balances of ecosystems and linkages to adjacent ecosystems at lower or higher spatial scales. The emphasis in agricultural landscapes is usually on the provisioning ecosystem services, in particular increased biomass from certain crops or produce from livestock.

In this chapter, we focus on smallholder farming agroecosystems, predominantly rainfed, subject to intensification, from a watershed perspective. Active watershed management, with reference to India in particular, has been successful in transforming rural livelihoods and enhancing agricultural productivity (Sreedevi *et al.*, 2004; Joshi *et al.*, 2008; Wani *et al.*, 2009). Emerging meta-analyses from other large-scale adoption and adaptation of agriculture (Noble *et al.*, 2006; Barron *et al.*, 2010) suggest that landscape scale impacts are detectable in both natural and social capital, with potentially significant changes in coupled socio-ecological systems. It is important, however, to approach successes analytically. In this chapter, we first ask whether these successes have also benefited or impacted the ecosystem services in the landscape, further affecting poverty alleviation and increased yields (two key targets in watershed management). Secondly, we explore whether more holistic approaches for assessing benefits, dis-benefits, and impacts are needed to address various scales and dimensions of coupled livelihood-landscape systems especially in cases of accelerated development. Thirdly, we consider whether a resilience framing can be a constructive approach to analyze key socio-ecological system processes or characteristics for more effective and sustainable management entry points (MEPs), and more socially and environmentally beneficial outcomes. The dynamics of rainfed agroecosystems are particularly important in some of the most underprivileged regions in the world. These regions are characterized by sensitive ecosystems which often face a broad range of development pressures. The development agendas often aim to achieve not only environmental sustainability, but also multiple social and human development goals, such as alleviating poverty, increasing local empowerment, and addressing social inequity, including gender inequality. Taking a systems perspective on agroecosystems, embedded in a landscape as well as socioeconomic context, will show that agroecosystems are both determined by, as well as impacted by, various critical ecosystem services generated in the landscape. Here we will focus on the water-related ecosystem services, and discuss them in relation to several development interventions to increase agricultural production and productivity, many of which are examined at length by Sahrawat *et al.* and Singh *et al.* in this volume.

Whilst achieving crop yield or livestock gains, rainfed agriculture, as with other types of agriculture, has increasingly been associated with negative externalities, which have adversely affected other users of landscape ecosystem services (Foley *et al.*, 2005; Falkenmark *et al.*, 2007). For smallholder rainfed agroecosystems, particular concerns have been associated with soil degradation, in particular decreased water-holding capacity, soil erosion and soil health, loss of vegetation, and loss of pollinators (MEA 2005; World Resources Institute 2005; Raudsepp-Hearne *et al.*, 2010). At the same time, spatial expansion of farm land has affected landscape ecosystem services (loss of habitats, loss of biodiversity), and the increased use of agrochemicals has affected both native fauna and water resources. In some parts of the world, the increased withdrawal of surface and shallow groundwater has caused impacts on both provisioning and regulating support services of water in landscapes (World Resources Institute 2005).

Despite these drawbacks of agricultural development, are there cases and situations where active consideration and management for improving smallholder farming systems has been achieved, whilst maintaining or enhancing other provisioning and/or supporting ecosystem services? Furthermore, have these achievements been sustainable and made these agroecosystems less vulnerable to shocks and stresses that commonly affect smallholder farmers' livelihoods?

In this chapter, we discuss whether the application of resilience framing can shed additional insights on these poverty affected and sometimes fast changing agroecosystems. The justification for this academic exercise is to systematically examine development interventions in smallholder agroecosystems management and to target the activities that will yield the most impact for improving livelihoods with less (or least) ecosystem service depletion. The resilience frame helps focus on the identification of key system variables that indicate both the ability of the system to remain stable despite shocks and stresses, and identify the drivers that can flip the system out of a negative/undesired stable state to a desired/positive state or development trajectory (Folke *et al.*, 2004).

Section 12.2 introduces case studies chosen deliberately to illustrate various examples of watershed (landscape) management in smallholder farming systems. The cases describe both top-down development interventions and bottom-up, farmer-led efforts that have been interpreted by both the research and practitioner communities as "successes". Next, we examine how these "successes" in rainfed agroecosystem management have impacted the ecosystem services (positively and/or negatively) in the watershed (landscape) scale.

Section 12.3 discusses in what ways the ecosystem services in the cases are related to system stability; specifically, whether the positive and negative impacts to ecosystem services have affected system vulnerability to common barriers in development of smallholder rainfed agroecosystems. This section also discusses how potential climate change may influence these common barriers for progress in socioeconomic and environmental sustainability.

Section 12.4 examines how the barriers for progress are related to system stability (including relationships to drivers, positive feedbacks, thresholds, etc.). The case studies are then analyzed further to understand whether the barriers for progress are related to key system processes or characteristics, which multiply benefits/co-benefits and/or avoid multiple negative impacts.

Section 12.5 discusses how resilience framing may be useful for identifying the watershed management activities that enhance ecosystem services to subsequently enhance key system processes, and, as a result, reduce vulnerability to shocks and stresses. The section then concludes with a list of recommendations, future research needs, and a list of resources that can aid the incorporation of resilience framing into rainfed agricultural research and management.

12.2 EMBEDDING SMALLHOLDER FARMING IN LANDSCAPE ECOSYSTEM SERVICES

12.2.1 Introduction to management successes and failures

Within the past several decades, effective watershed management has been a designated development vehicle target at smallholder farmers in rainfed agroecosystems. As such, there is a growing body of work that helps identify common characteristics and patterns, and perhaps even MEPs, for smallholder rainfed agroecosystems (Rockström *et al.*, 2007; Joshi *et al.*, 2008; Wani *et al.*, 2009). The ultimate goal is often to close the current yield gap to the benefit of smallholder farmers, increasing their income and improving livelihood security, sometimes even enabling individuals, households, and communities to move out of poverty. In a recent study by Rockström and Karlberg (2009), 'hotspot' areas for reduction of yield gap and high poverty incidence were mapped, and three areas, Sub-Saharan Africa (SSA), South Asia and East Asia were identified as areas where more than 100 million people may be affected (Figure 12.1). However, the goals of increasing agricultural production as means to decrease poverty and hunger by closing yield gaps in these 'hotspot' areas are not necessarily environmentally sustainable when taken to scale. In the past such agricultural water management interventions at watershed scale rarely addressed both social and environmental issues holistically (Barron *et al.*, 2010; de Bruin *et al.*, 2010). Thus, often unexpected negative effects that emerge as some gains for the smallholder farmer are achieved. An increasing number of recorded cases can be found discussing negative impacts of agricultural development, both environmental and socioeconomic (World Resources Institute 2005; Enfors and Gordon 2008). Can these emerging unexpected impacts be addressed through a more holistic management approach, including explicitly considering ecosystem services when development actions are initiated? We start by discussing the smallholder farmer as embedded in a landscape with inherent ecosystem services. A key question is: What do healthy ecosystem services provide farmers beyond crop production, with particular reference to semi-arid and subhumid rainfed systems?

12.2.2 Smallholder agroecosystems and ecosystem services

Typically, smallholder rainfed agroecosystems in tropical and subtropical environments are closely coupled with the surrounding landscapes (World Resources Institute

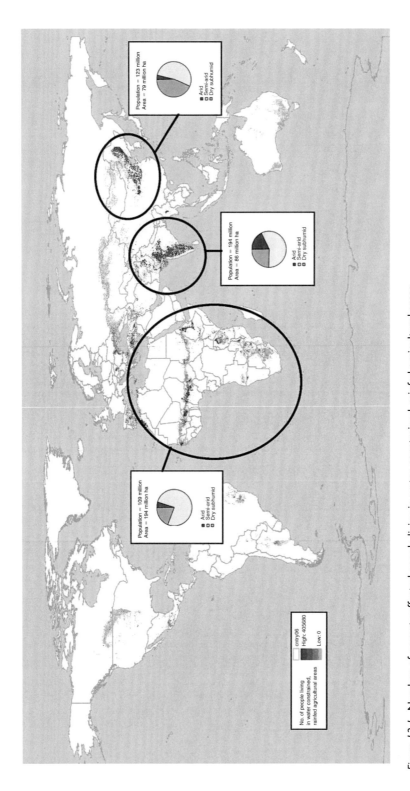

Figure 12.1 Number of poverty affected people living in water-constrained, rainfed agricultural areas.
Note: The three circles indicate the occurrence of global hotspots where more than 100 million people may be affected. Source: Rockström and Karlberg (2009) (See color plate section)

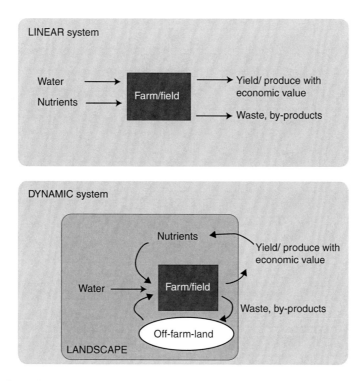

Figure 12.2 The conceptual differences between linear and dynamic systems

2005). Their farms range in size from 5 ha in SSA and Asia to 10 ha in Latin America (Kulecho and Weatherhead 2006; Glover 2007). In these smallholder rainfed agroecosystems the ecosystems on-farm and in the surrounding landscape provide a range of ecosystem services, which provide complementary support and sometimes necessary support to livelihoods and income. Researchers and practitioners alike have demonstrated the close linkages that exist between vulnerability of smallholder livelihoods and the reliance of smallholders on the provisioning and supporting capacity of the landscape (e.g., Mortimore and Adams 2001; World Resources Institute 2005; Kosoy *et al.*, 2007; Enfors and Gordon 2008). Thus, as researchers and development agents shift from viewing the farm systems as being an almost isolated entity of the landscape with only input and output connectivity, more and more it is being recognized that farms are impacted and impacting the surrounding landscape and its various ecosystem, in turn affecting livelihoods, well-being, and economic benefits (e.g., Kosoy *et al.*, 2007). Figure 12.2 displays a simplified conceptual diagram for understanding linear and dynamic farming systems.

The dynamic system illustrated in Figure 12.2 includes the feedbacks between the 'off-farm land' as well as the 'landscape' with the farm. These additional relationships between the different spatial and temporal scales are especially important for including ecosystem services into the assessment of management success. The importance of the dynamic system is that it characterizes the "embeddedness" of the smallholder

Table 12.1 Principal ecosystem services in smallholder rainfed agroecosystems taken from the case studies in the text[a]

Type of ecosystem service	Example valued service for people in landscape
Provisioning – Products obtained from the ecosystem	Cash crops Subsistence crops Livestock Fodder Charcoal Timber NTFPs (non-timber forest products) including fire wood Bricks Wild fruits, vegetables, and herbs Fiber
Regulating – Benefits obtained from the regulation of ecosystem processes	Nitrogen fixation Water purification Water regulation Sediment transport and/or trap Carbon sequestration Pollination Pest regulation Air quality
Cultural – Cultural or other benefits obtained from the functioning of ecosystems	Sacred sites, places of worship Recreational/ecotourism (in upper watershed for non-smallholders) Knowledge systems Social systems
Supporting – Services which are preconditions for all other ecosystem services	Primary productivity Nutrient cycling Water cycling

[a]Source: MEA (2005); Enfors and Gordon (2008).

farming system. Ecosystem services are divided into four categories: provisioning, regulating, cultural/ spiritual, and supporting (MEA 2005). Table 12.1 explains these four categories, with relevant examples from smallholder rainfed agroecosystems.

When communities manage different ecosystem services for specific purposes, trade-offs may emerge as various ecosystem services depend on the same land, nutrients, and/or water for their generation (Gordon *et al.*, 2008). This is particularly so for provisioning services, which are directly harvested and often valued in economic terms (see Table 12.1).

Regulating and supporting ecosystem services are more difficult to directly value in economical terms, due to the often larger and/or cross-scale spatial and temporal scales at which they operate (MEA 2005). For example, the regulating function of carbon sequestration performed by the biosphere occurs at much larger spatial and temporal scales than the farm to landscape scale which a single farmer or community of farmers can impact through their soil and plant actions. Thus market rules are challenging to put into place an operational transfer of funds for actions taken on the ground.

In order for watershed management activities to be considered successful, especially in the long-term, it is important to understand: (a) which ecosystem services the agroecosystem is dependent upon; (b) which ecosystem services the management activities impact; and (c) how these two aspects (a and b) are connected to livelihoods, human well-being, and economies.

12.2.3 Watershed (landscape) management "successes"

The evidence for successful watershed management activities is primarily identified by case studies from: (1) specific development interventions, often performed by an NGO (non-government organization), and/or government entity, and (2) academic analyses of farmers' internal management, without explicit NGO and/or government intervention. Complementary evidence was found from summary reports and articles published by government agencies, NGOs, and academic institutions, or some combination thereof.

The case studies that are described below were identified based on a set of search criteria. These search criteria, broadly defined, required the case studies to examine water-related issues in smallholder rainfed agroecosystems. The sources of the case studies were a combination of academic research, government, and NGO reports. The target spatial scope was the 'intermediate catchment' (approximately 1 to $<1000\,\text{km}^2$). The timeframe of the case studies was for the most part within the past two decades. It should be noted that the collection of case studies presented below is by no means comprehensive, and should rather be considered as a sample of the different types of case studies, from which conclusions may be drawn for identifying the key entry points for watershed management in smallholder rainfed agroecosystems.

The relevant characteristics of some case studies are summarized in Table 12.2. The additional materials of government/NGO reports and "state-of-the-art" articles on management activities related to smallholder rainfed agroecosystems are introduced. These additional materials are also used later in this chapter, in the categorization and analysis of the "successes" identified by the case studies.

12.2.4 Summary of selected cases

Sreedevi *et al.* (2004) explored the Adarsha watershed in Kothapally, Andhra Pradesh, India and reported the results of a comprehensive watershed management project for Kothapally's smallholder rainfed agroecosystem. The scope of the management activities included soil and water conservation, wasteland development and tree planting, integrated pest management, integrated nutrient management, nutrient budgeting, generation of nitrogen-rich manure, worm farming, and HNPV (*Helicoverpa* nuclear polyhedrosis virus, natural pesticide) production. This project is considered a strong success, with specific successes including reduced runoff and soil loss, improved groundwater levels, improved land cover and vegetation, increased productivity of the agroecosystem, and and improved livelihoods. This project also serves as the example for similar projects around the world as well as the baseline from which things may be scaled-up. A later evaluation of impacts on water flows of the watershed shows that the enhancement of yields and subsequent incomes has affected landscape water flows both positively through retaining soil water and recharging groundwater,

Table 12.2 Characteristics of selected cases of watershed management interventions

Case	Agroecosystems issues	Response to issues	Impact on ecosystem services	Short-term success	Long-term success	Reference
Kothapally watershed management, Andhra Pradesh, India	Droughts and dry spells Soil erosion, degraded soils Low crop production and low farm income	Adoption of soil, and water conservation (SWC); Wasteland development and tree planting Integrated pest management Integrated nutrient management Nutrient budgeting Generation of N-rich manure Worm farming HNPV production	Increased infiltration and increasing groundwater table, reduced soil loss, increased soil health and nutrient status, re-vegetation	Yes	Yes; but impacts beyond watershed boundary are not measured (only modeled)	Sreedevi et al. 2004
Darewadi, India	Poor drinking water accessibility Degraded vegetation/ loss of tree cover Degraded soils	Phased grazing bans and regeneration of landscape	Increased vegetation and improved soil health	Yes	Yes	WRI 2005

Case	Threat	Response	Ecosystem service outcome	Adaptive?	Sustainable?	Reference
Shinyanga, Tanzania	Overgrazing; loss of habitat for wildlife Loss of indigenous trees (miombo woodland)	Restore *ngitili*, or *sukuna* for enclosures, of trees for livestock fodder	Increased forest provisioning services (for fodder)	Maybe, loss of livelihood due to wildlife could negate improvements	Unknown	WRI 2005
Soil and water conservation, Machakos, Kenya	Drought Decreased access to land for fallow Soil erosion	Adoption of SWC Increased cropping density	Improved soil health/ increased infiltration, vegetation growth trap sediments, increased value of land	Yes	Yes, value of land increased after SWC adoption gave incentive to manage land better	Mortimore and Tiffen 1994
Adaptation strategies in rainfed farming systems, Old Peanut Basin, Senegal	Drought and dry spells Degraded/low nutrients in soils	Sold livestock to buy food	Possible increase in soil structure and organics with decrease in livestock pressure	Yes, in terms of coping	No, since buffer (livestock) are gone now	Tshakert 2007
Farming systems improvement and land revegetation, Burkina Faso and Niger	Low yields Falling water table	Rehabilitate land and forests	Improved water table, forest resources	Yes, in terms of increased yields	Unknown, but likely, given widespread acceptance of methods	Reij *et al.* 2009
Rainfed farming and landscape transition, Makanya, Tanzania	Drought and dry spell Low yields Low food security and low income	Harvest resources from ecosystem Supplement food from market	Reduced provisioning Ecosystem services for future drought	Yes	Unknown, but doubtful due to reduced ecosystem service capacity	Enfors and Gordon 2008

and reducing sediment flows and also reducing downstream water flows out of the catchment.

World Resources Institute (2005) describes a case study of watershed management in Darewadi, India [conducted by IWMI (International Water Management Institute) in 2002], that focused specifically on relieving village dependence on water tanker truck visits, and regenerating local water resources, soils, and trees. The primary activities were phased grazing bans and regeneration of the landscape. This project is described as yielding an enormous amount of benefits, including elimination of water tanker visits, improved groundwater table, increased fodder available for livestock, increase in cropped area, increased summer milk production, increase in average livelihood, increase in the number of crops that were cultivated, and many more benefits. A key characteristic of this case study is that it explicitly enhanced the local provisioning ecosystem services, notably soil and forest biomass as well as regulating services, such as increased stream and groundwater flow. A key driver that was fundamental to project success was the consensus of the community to work together to improve private and public gains.

World Resources Institute (2005) describe a case study in Shinyanga, Tanzania that aimed to restore local acacia and miombo woodlands to the landscape, thus improving the traditional *ngitili*, or 'natural enclosures' for livestock grazing. The project achieved improvements in average livelihoods from harvesting the livestock fodder in the *ngitili* as well as harvesting other plants for both consumption and sale. However, a significant issue existed in terms of increased damages by wildlife to *ngitili*, which effectively canceled or reversed any benefits accrued from the *ngitili*. Additional problems included a growing human and livestock population, scarcity of land and lack of land tenure, and a growing and unregulated problem of individually owned *ngitili*. Although this project did represent a management "success" for some communities, and certainly a success for local ecosystem services (wildlife, native vegetation, etc.) overall there were significant negative impacts to the stability of the social system, which question the long-term sustainability of this development intervention.

12.2.5 Academic reviews of unmanaged case studies

Reij *et al.* (2009) explore the agro-environmental transformation of agricultural landscapes that has taken place in Burkina Faso and Niger. The farmer-managed land rehabilitation efforts in Burkina Faso have improved crop yields (in terms of area planted and volume), increased the number of on-farm trees, and increased the height of the groundwater table. The farmer-managed forest regeneration efforts in Niger increased the number of on-farm trees, improved crop yields, and increased the above-ground biomass. The effectiveness of the farmer's efforts in Niger are specifically important for understanding the relationship between regeneration of peripheral (off-farm) ecosystems, and the acquisition of tangible benefits (e.g., increased crop yields). The impacts of watershed water balances have been modeled for conventional and improved vegetation cover, showing no significant differences in water flows for the two landscape types with/without intercropped trees (Barron et al., 2010).

Enfors and Gordon (2007, 2008) explored the strategies that smallholder farmers in Tanzania use to cope with drought, with specific emphasis on the local *Ndiva* system of community irrigation. The authors examined what strategies the residents of the target community utilized, specifically the relative importance of food sources after a drought period, and the relative importance of livelihood generating activities. They found that after drought the primary source of food became the market, and that a very significant portion of community livelihoods originated from provisioning ecosystem services off-farm especially in times of shocks (i.e., consecutive dry spells or droughts resulting in complete crop failure). It was also interesting that the local *Ndiva* system of irrigation provided much less in terms of coping capacity for the communities as the population has grown over the last 30 years, and the water allocation through the *Ndiva* was too small for effective dry spell mitigation strategy. Further, the resilience of smallholder rainfed agroecosystems in SSA was explored using two aggregate variables of ecosystem health, the ecosystem insurance capacity (EIC) related to ecosystem services and the soil water index (SWI), related to soil moisture and productivity. A framework was developed for evaluating the resilience of a smallholder agroecosystem, and a community in the Makanya catchment in Tanzania was evaluated. The smallholder agroecosystem had moved towards a state of degradation but the agroecosystem could be lifted out of degradation by first improving the SWI by technological innovation, and then improving the EIC.

Mortimore and Tiffen (1994) provide an environmental history of the Machakos district, encompassing several catchments in Kenya, and describe in detail the transformation of a semi-arid/dry subhumid agricultural environment severely degraded during colonial and post-colonial times, which due to local and external investments has been rehabilitated from the 1980s and onwards. The main investment has been in soil water conservation measures reducing soil erosion and increasing productivity in smallholder farming systems. Challenging the dominant "victim" paradigm, which characterizes smallholder rainfed farmers in SSA as helpless, the authors suggest that farmers are the best-suited and best-equipped to identify the best strategies to adapt to changing physical conditions.

Tschakert (2007) explored the dynamics of smallholder farmer vulnerability in the old Peanut Basin in Senegal, specifically looking at coping capacity and strategies for dealing with long-term drought. Tschakert (2007) discovered that most of the smallholder farmers were "utterly unprepared with no strategies to moderate damage or take advantage of the opportunity (of flood waters)". This conclusion conflicts with the conclusions of the "smart farmers" identified by Mortimore and Tiffen (1994), Mortimore and Adams (2001), and others as it suggests that not all smallholder rainfed agroecosystems can "choose the best technology". It appears that if the internal agroecosystem knowledge-base lacks the capacity to develop coping mechanisms in response to drought, then the only option is for those coping mechanisms to originate outside the system.

12.2.6 Additional case studies

Kosoy *et al.* (2007) explored several projects that utilized payments for water-related environmental services (PES) between upstream providers of environmental services and downstream buyers of those services in Central and Latin America. They also explored the impacts of the PES schemes in terms of management goals, types of ecosystem services that were protected, change in livelihood for providers and buyers, and overall success of the project. Kosoy *et al.* (2007) found that in general the opportunity costs of foregone landuse (e.g., timber harvest) payments to providers of pristine forests were higher than the amounts that were ultimately paid to the providers.

Mortimore and Adams (2001) have provided examples of 'smart-farmers' in Nigeria who, in response to a shifting external climate, develop strong coping strategies to overcome the unexpected problems. They also identified a four-stage process of increasing livelihood diversification, with the key first step being acquiring livestock, which can generate short-term capital as a means to develop other income diversification pathways.

12.2.7 Landscape limits and trade-offs between ecosystem services

For any given landscape unit there is an inherent limit to biomass production and productivity given the soil conditions and prevailing climatic conditions. With undisturbed natural vegetation, this would constitute a net primary production capacity. When humans impact the landscape resources, for example through farming, extraction of produce tends to travel beyond the source of the landscape unit. Thus, there is a loss of nutrients and fibers, which need to be compensated for, through, for example, additional fertilizers. In low-yielding agroecosystems, the production and productivity (per unit land or unit water) is often far from achievable, or even close to potential equivalent net primary production. A degraded landscape, as in depleted soil nutrients or poor soil health or reduction in flora and fauna, would further move the landscape unit from the inherent potential. Thus, we hypothesize that for degraded landscapes (watersheds) with low-yielding agroecosystems, there may be a large potential with marginal trade-offs for enhancing, provisioning, and various supporting and regulating ecosystem services while enhancing the agroecosystems into higher production (Figure 12.3). This enhancement of agroecosystem production would be closer to the equivalent net primary production level, where the agroecosystem may mimic a natural system and its various ecosystem service capacities; whereas when a landscape unit (watershed) moves into a high production agroecosystem, often relying on substantial inputs of water, nutrients, and energy to sustain high crop yields, there may be adverse impacts on the supporting and regulating capacity of both local and external ecosystems.

Additional challenges in smallholder agroecosystems are the potential trade-offs that may occur between the different types of ecosystem services. For example, choosing to grow maize on a plot of land or using it for grazing livestock rather than as a habitat for non-cultivated flora and fauna may have implications beyond the change in provisioning ecosystem services. Very often development of smallholder farming systems are targeted at maximizing one ecosystem service, such as crop (including cultivated trees and bushes) production and/or livestock, without explicitly considering what the change in landscape or social resource use will imply on other provisioning and/or supporting-regulating services. Thus negative externalities may emerge. Understanding ecosystem service dynamics requires comprehending whether ecosystem service trade-offs exist when development occurs. If so, are there alternatives with less negative impacts and more desirable outcomes for stakeholders? The challenge is to identify the magnitudes and spatio-temporal characteristics of these trade-offs. A further challenge is to possibly offer strategies for managing these potential trade-offs

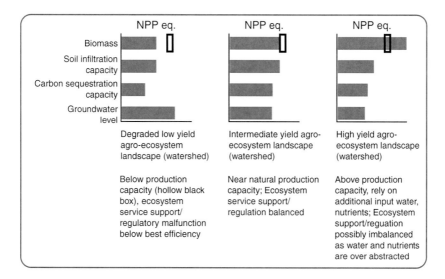

Figure 12.3 Principal levels and possible trade-offs between various specified provisioning and supporting-regulating services for degraded, healthy, and high-intensity farming watersheds

among the agricultural production in the landscape and the resources that are needed in land, water, and nutrients for this production, and its potential impact on other landscape ecosystem services.

Recent analysis has suggested the benefits of analyzing bundles of ecosystem services as indicative of ecosystem services that reinforce one another (Raudsepp-Hearne *et al.*, 2010). However, as this is a relatively new field of research, there are currently few case studies available. A theoretical approach has been presented by Foley *et al.*, (2005), further adapted by Falkenmark *et al.* (2007) for smallholder farming systems under agricultural water management strategies for increasing yields, food security, and income (Figure 12.4).

The importance of ecosystem services as both a source of livelihoods and agro-ecosystem stability is supported by the above case studies. Furthermore, ecosystem service analysis, as demonstrated by Enfors and Gordon (2008), can provide an empirical framework for prioritizing the management of specific ecosystem services, specifically regulating services, to increase system resilience, and the continuity of provisioning and cultural services (Bennett *et al.*, 2009; Raudsepp-Hearne *et al.*, 2010).

12.2.8 Long-term sustainability and the hidden impacts of management successes and failures

Based on the above successes, and in some cases failures, in smallholder rainfed agricultural management, there are some indications that "successful" management activities seem to correspond to positive impacts to the local ecosystem services (Table 12.2)

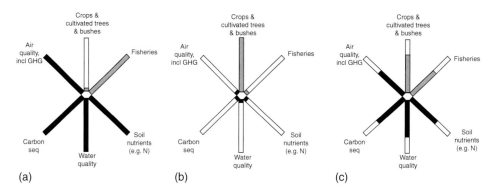

Figure 12.4 Ecosystem service bundles in smallholder rainfed agroecosystems; frame (a) indicates a non-agricultural landscape with the naturally occurring ecosystem services fully intact, (b) indicates a landscape that has maximized agricultural production, and (c) indicates a landscape that has balanced agricultural production with the other ecosystem services in the landscape [Adapted from Foley *et al.* (2005) and Falkenmark *et al.* (2007)]

(Mortimore and Tiffen 1994; Sreedevi *et al.*, 2004; Reij *et al.*, 2009). Likewise, the "failed" management activities seem to correspond to negative impacts to local ecosystem services (Tschakert 2007; Enfors and Gordon 2008). However, there are anomalies, where "successful" management activities have negative impacts to ecosystem services and vice versa (World Resources Institute 2005; Kosoy *et al.*, 2007; Enfors and Gordon 2008).

The reason this finding is important is that actions that are perceived as successes, i.e., improved livelihoods and income for smallholder farmers, may undermine long-term sustainability, and create unexpected, or hidden, barriers of progress. These barriers may be understood as a general term to capture anything that reduces the sustainability of both the social and ecological system. Examples of barriers include negative impacts from soil degradation or positive feedbacks of drought induced livestock losses.

Conversely, actions that are perceived as failures may hold hidden opportunities for management to make significant improvements to livelihoods. An example of a short-term coping strategy undermining long-term sustainability is explained by Enfors and Gordon (2008), where farmers sold their investments (i.e., bricks, livestock) as a means to cope with crop failure, and a need to pay for food at market. Bricks and livestock represent capital investments, which are steps on the ladder of livelihood improvement. As the smallholders sold their bricks and livestock they reset their livelihood to the bottom of the ladder, and without their investments, made themselves more susceptible to barriers in the future.

Interpreting the positive and negative impacts of development activities requires understanding the relevant development context. The next section will categorize short-term and long-term impacts on ecosystem services as barriers and/or 'opportunities' for management, and then connect these to specific goals in smallholder agroecosystem sustainability.

12.3 IMPACTS ON ECOSYSTEM SERVICES AND THE RELATIONSHIP TO BARRIERS TO DEVELOPMENT OF SUSTAINABLE ECOSYSTEM SERVICES

12.3.1 Impacts on system stability as obstacles in smallholder rainfed agroecosystems

Embedding smallholder farming systems in a landscape enables the accounting of various ecosystem services beyond the provisioning capacity of crop or livestock production. However, the complexity of the embedded farming system is multiplied by the biophysical landscape's ecosystem services and the socioeconomic landscape's varying management scales. In this context resilience framing can aid in the understanding of smallholder agroecosystem characteristics and behavior. It can also aid in understanding what may be desirable and undesirable states of the studied agroecosystem, and whether these states are more or less resilient to changing into more desirable states. We note here that resilience is used to describe a characteristic of the system, i.e., the agroecosystem sensitivity to a systemic change. Thus, resilience does not refer to a particular state of the smallholder agroecosystem, but can be a feature of an undesired state such as high poverty caused by frequent dry spell/drought affecting a subsistence farm (Tschakert 2007; Enfors and Gordon 2008), or a desired state, such as smallholder farmers with continuous market access due to rehabilitated landscape ecosystem services (Sreedevi *et al.*, 2004; World Resources Institute 2005).

There are certain features in the agroecosystems that may be more sensitive or less sensitive to drivers of change. These drivers of change are key to the system's resilience, in particular the drivers that disturb the stability of the agroecosystem. The aggregate impact of multiple obstacles is that they reduce the stability (or resilience) of an agroecosystem, and increase the overall vulnerability. As such, it is useful to understand the constituent obstacles that aggregate to form "vulnerability". A specific barrier to change is more easily quantifiable, thus allowing measurement, comparison, and evaluation. Not all potential barriers can be quantified, but attempting to do so can clarify the varying level of risk associated with different barriers, and therefore help to prioritize the order in which obstacles should be addressed.

This section includes a simplified framework for categorizing barriers specific to smallholder agroecosystems in developing context. These categories are: external vs. internal and predictable vs. short term. It is especially useful to distinguish between shocks and stresses. Shocks and stresses both threaten the stability of agroecosystem, but shocks have low predictability, whereas stresses are impacts on the system that have high predictability. Separating barriers by temporal scale seems natural, but may be inappropriate since droughts can be unexpected, occurring in either a single year or multiple years.

Depending how the agroecosystem is defined in a spatial and temporal scale, an obstacle can fall into more than one category. For example, climatic drought would be external and unpredictable; and farm-scale over-irrigation, would be internal and predictable.

12.3.2 Understanding barriers as parts of ecosystem processes

These barriers are components of ecosystem process loops, and can be sorted according to which part of the process in which they occur. Several authors have applied their own

Table 12.3 Proposed categorization of potential barriers to sustainability of smallholder rainfed agroecosystems sorted according to (a) whether the barrier is a shock (unpredictable) or a stress (predictable), (b) whether it originates outside the agroecosystem (external) or within it (internal)

Origin	Shock	Stress
External	Drought	National/International regulations
		National and sub-national institutional capacity to support smallholder farmers
		Access to input and output markets
		Access to new information/knowledge
	Dryspell	N pollution
	Flood	Reduced stream flow
	Pests and diseases on crop/livestock	Reduced groundwater reach
Internal	Pests and disease	Pests and diseases
	Crop disease (e.g., potato blight)	Crop disease (e.g., potato blight)
	Soil erosion	Failing/inappropriate land tenure
	Lack of labor	Land scarcity or low poor farm productivity
		Overstocking of cattle
		Soil erosion
		Salinization
		N pollution/water quality issues
		Poor soil quality
		Mono crop agriculture
		Inadequate access to capital (to purchase fertilizer, monocrop equipment, labor)
		No community coordination (food banking, micro-lending)

frameworks for categorizing and understanding barriers (not always so called), and therefore we will combine elements of their categorization (Barron 2004; Tschakert 2007; Gordon *et al.*, 2008) (see Table 12.3).

The utility of categorizing the obstacles (Table 12.3) is that it identifies the obstacles that are within the "reach" of management of the local smallholder rainfed agroecosystem. At first glance it appears many of the obstacles are local stresses, occurring at the internal scale. Thus, such relatively simple categorization may aid in efforts to address multiple constraints, in particular issues under 'stress' that will inhibit actual change unless addressed simultaneously.

Beyond simply categorizing the obstacles (as in Table 12.3), it is possible to identify how and where these obstacles are situated within the broader management framework. The process loop in Figure 12.5 identifies how these obstacles may relate to one another. These shocks and stresses can then be integrated with the available management options; an example is illustrated using the development interventions discussed by Sreedevi *et al.* (2004) (Figure 12.6). Based on the success of Sreedevi *et al.* (2004), the development interventions at the internal scale with community endorsements interrupted the cycle of lower crop yields, livestock losses, soil erosion, and soil degradation. This case study also illustrates the fact that the management activities are not focused

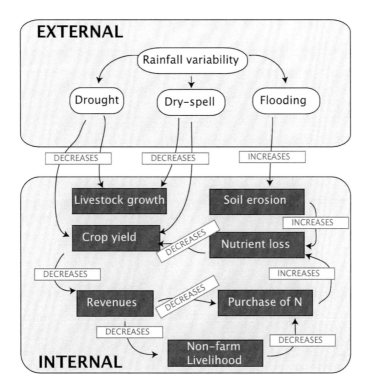

Figure 12.5 A simplified feedback loop among different barriers, with ovals and rectangles representing shocks and stresses respectively, and light and dark objects representing physical and social obstacles respectively

on mitigating the external shocks and stresses (e.g., drought and/or dry spells), but rather the internal shocks and stresses, specifically the predictable stresses of human origin (referring to those obstacles that are a result of human actions or are in the control of human societies). This idea of intervening at the scale of the farmer is reiterated by many authors (Rockström *et al.*, 2002; Barron 2004; Wani *et al.*, 2009). However, there remains a gap where these internal shocks and stresses are addressed without external knowledge or capital input.

Thus far, resilience science is at an early stage to enhance our understanding how to improve management in smallholder rainfed farming systems. However, we will now zoom into the specific obstacle of soil degradation, to understand how the specific management activities identified by Sreedevi *et al.* (2004) impacted the local ecosystem services. This is illustrated in Figure 12.7 using the "Dynamic farming system" from Figure 12.2 and the management activities from Sreedevi *et al.* (2004).

An important point that is clearly illustrated above is that the successes described in the management case studies specifically included activities that enhanced local ecosystem services, namely reforestation, wasteland development, and improved water infiltration (Sreedevi *et al.*, 2004; World Resources Institute 2005). Furthermore,

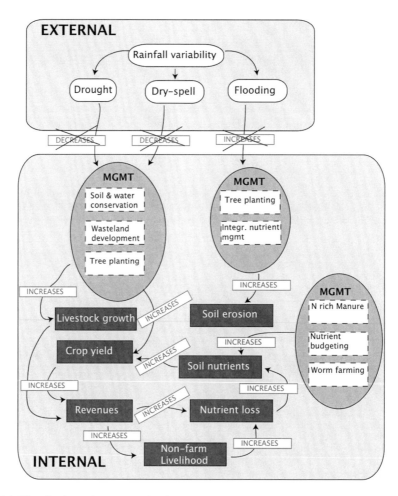

Figure 12.6 The shocks and stresses of a degraded Indian watershed and its management activities, which interrupt the cycle of soil degradation, and encourage system stability/resilience (Source: Sreedevi *et al.*, 2004)

the dynamic positive feedback of the development interventions improves the overall stability of the system.

12.3.3   Climate change as an over-arching pressure

In many smallholder rainfed agroecosystems, climate variability is an everyday reality. Expected climate change over these areas is currently highly uncertain (Pauchauri and Reisinger 2007) but is broadly expected to affect precipitation variability, and certainly increase temperatures and evapotranspiration rates. However, climate change should not be thought of as an additional obstacle to rainfed agricultural systems, but rather that it will affect the entire smallholder farming system, since the farming system is

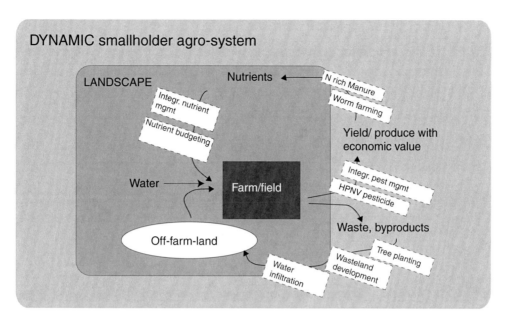

Figure 12.7 The dynamic smallholder rainfed agroecosystem with management activities from Sreedevi *et al.* (2004).

Note: The activities specially influencing the off-farm provisioning ecosystem services include: Tree planting, Wasteland development, and Water infiltration; these appear at the bottom of the figure as white boxes with broken outlines

embedded in the landscape. For example, if climate change increases the variability of dry spells in the Sahel, a management response will not be focused on climate change specifically, but rather on increasing the flexibility of the system through, perhaps, complementary income-generation activities. This section will not go into the details of specific threats from climate change to rainfed agriculture (as this is covered by Craufurd *et al.*, in this volume), but we will emphasize the role of resilience framing in addressing the unique problems of climate change, namely the twin issues of uncertainty and variability (Cooper *et al.*, 2008).

The Intergovernmental Panel on Climate Change (IPCC) defines 'uncertainty' as "An expression of the degree to which a value (e.g., the future state of the climate system) is unknown. Uncertainty can result from lack of information or from disagreement about what is known or even knowable . . ." (Pauchauri and Reisinger 2007). Similarly, variability is defined as a change in the statistical distribution of events, or, in other words, the range of possible temperatures and rainfall events may increase. From a resilience perspective, the management responses to increased variability and uncertainty will be similar to current measures; resilience-based management responses will increase the stability of the system through increased flexibility of the smallholder farming system to respond to various impacts. A relevant example of an uncertain impact of climate change is whether or not annual precipitation will change in the future in a rainfed agricultural system in India. The management regime cannot base plans on historical climate, because the future is not guaranteed to behave like the past. However, there are management responses that can be taken that increase the overall stability of the system, and are independent of a wet or dry climate. Examples include:

- Enhance non-food provisioning ecosystem services, such as timber harvesting;
- Encourage diversification of livelihood activities to include raising livestock;
- Increase access to agricultural input and output markets;
- Increase access to new knowledge in relation to the agroecosystem;
- Increase non-agricultural adaptive capacity, including basic health improvements, education, and increased gender equity.

Although this volume is about rainfed agricultural management, livelihood improvement is not limited to agricultural activities. For example, Tschakert (2007) discovers that, the most significant concern among all members of the surveyed community is 'poor health'. The dominant livelihood activity may be rainfed agriculture, but the primary concern is perhaps not agriculturally related. This simply illustrates that development interventions that have 'livelihood improvement' as the central goal should consider what is a priority to the community, and understand how interventions can be complementary to the community's non-agricultural concerns.

Tschakert (2007) recommends activities that "enhance generic adaptive capacity," and that "too narrow focus on augmenting the adaptive capacity of the agricultural sector through reforestation, anti-salinization measures, and other management practices ... is unlikely to produce a poverty-reducing and resilience-enhancing trickle down effect for poor farmers." This concept is important to understand for more fully integrating cross-disciplinary livelihood improvement activities. The work done by Sreedevi et al. (2004) is an excellent example of a fully integrated cross-disciplinary effort, improving both agricultural sector adaptive capacity as well as increasing gender equality, improving access to markets, and enhancing off-farm, non-food provisioning ecosystem services.

The coping strategies currently utilized by farmers in smallholder rainfed agroecosystems will remain absolutely essential for coping with future climate variability (Cooper et al., 2008). However, it is worth considering the fact that there are case studies wherein farmers were unable to cope with either a specific drought event (Tschakert 2007) or the coping strategy for a single event weakened the long-term overall coping capacity (Enfors and Gordon 2008). These findings are important since they provide a reason for utilizing resilience-based management strategies that explicitly address long-term sustainability, specifically in terms of resistance to potential barriers in the existing agroecosystem. Climate change is predicted to affect so many different aspects of the physical systems on which smallholder farmers depend (evapotranspiration, precipitation, and temperature) with secondary impacts on both on-farm and off-farm ecosystem services in the landscape, that understanding which activities affect which obstacles becomes even more important.

12.3.4 Barriers for development reinterpreted as management opportunities

The next step in utilizing a resilience frame to better understand the smallholder rainfed agroecosystem is to consider whether certain obstacles impact agroecosystem stability more than others, and which activities influence key system characteristics or processes.

12.4 UNDERSTANDING SUCCESSES AND LONG-TERM AGROECOSYSTEM STABILITY

12.4.1 Introduction to agroecosystem stability

The authors of the case studies described in Section 12.2 provide evidence that relationships exist between ecosystem services and the stability of smallholder rainfed agroecosystems. There is also ample evidence to suggest that some watershed management activities enhance system stability whereas others weaken stability. This "weakening" can also be understood as vulnerability. However, since vulnerability is only one part of system stability, it is useful to first have a framework for understanding stability as a whole, and second, for understanding where vulnerability lies within that framework.

12.4.2 Resilience defined

Resilience does not have a positive or negative connotation (Box 1): a system that is resilient simply means that the system state is resistant to change. The "system-state" in a smallholder rainfed agroecosystem depends on the system specific physical conditions as well as its social characteristics to sustain agricultural production. This varies based on inherent landscape and climatic conditions as well as the man-made alterations in the past in the landscape such as crop type, growing season, and number of growing seasons. The social preconditions to sustain agricultural production include functioning institutions, community networks, knowledge (and access to knowledge), and access to markets to acquire improved technologies and inputs as well as to sell produce.

Box 1.   Glossary of Resilience Terms

Stability = the ability of a system to remain in a given state

Resilience = the ability of a system to return to a given state, after being disturbed away from that stable state

Vulnerability = the ability of a system to be disturbed, potentially out of a stable state

Driver = a force internal or external to the given system, which can push the system towards or away from a stable state; this is generally *not* self-reinforcing

Positive feedback = a self-reinforcing cycle that pushes a system towards or away from a stable state.

Regime shift = the result of a transition from one stable state to another; could include a different climate, different food-web, etc.

Some readers may ask: "Why do we need resilience science? What is the added-value that resilience science provides to the management of smallholder rainfed agro-ecosystems?" Our response is that the added-value of resilience science is that it can help smallholder watershed managers and farmers to understand system dynamics more clearly, and target the management activities that are most important to system sustainability. By understanding the relevant drivers and positive feedbacks that encourage system sustainability (or that encourage system instability), it is possible to prioritize management activities and investments in the areas that will yield the most results.

This discussion of resilience science may seem too theoretical, so here is an example that clarifies why and how this is useful. Enfors and Gordon (2008) expected the traditional *Ndiva* irrigation system to make a significant contribution to the drought-based coping strategies of the smallholder farmers in the Makanya catchment. However, their results indicate that the smallholder communities depended more on the surrounding provisioning ecosystem services to provide complementary food and income especially in times of external shocks such as droughts or loss of yield due to pests and diseases. The major finding here is that provisioning ecosystem services, particularly the services that provide supplementary food and income (timber, charcoal, and wild plants), are a key system characteristic. Although, the traditional supplemental *Ndiva* irrigation systems make some contribution to smallholder coping strategies, it is the provisioning ecosystem services that maintained the stability of the smallholder farmers' livelihoods.

It is clear from the case studies reviewed above that many practitioners and researchers are utilizing several aspects of resilience science, such as identifying key drivers of stability, without necessarily applying an over-arching framework. A brief description of the different aspects of the resilience-science approach is given.

12.4.3   Drivers of system stability

System stability is often maintained with a variety of different key variables known as drivers, which allow a given state to persist. Within a rainfed agricultural system there are several drivers. However, only a few of these drivers are likely to be key drivers, i.e., drivers that are the dominant influences of the system's stability. These key drivers may act and interact at different spatial and temporal scales. In rainfed agro-ecosystems, a powerful driver of system stability or instability is the 'meteorological' driver, which can drive the physical and social system towards stability or instability. For example, meteorological drought acts on a multi-month to multi-year temporal scale, and from very local to regional or continental spatial scale (Kasei *et al.*, 2010). The meteorological drivers are powerful, external drivers of a system and can set in motion events that have long-term impacts to the stability of a given smallholder rainfed agroecosystems (Mortimore and Tiffen 1994; Enfors and Gordon 2008). Access to local irrigation to supplement or fully replace absent rainfall is also a driver of stability in rainfed agroecosystems. These two drivers act on the same system, but occur at completely different scales, presenting a number of challenges, for example the scale of institutional response (local, regional, national, or international).

Furthermore, interactions between drivers can lead to cascading effects or positive feedbacks, which can push a system entirely out of its stability domain. A physical-social positive feedback is described by the following example from Enfors and Gordon (2008). Meteorological drought leads to farmers lacking food sources, so they sell off livestock to generate income to purchase food at market. However, this reduces their coping capacity, and if the drought persists, the farmer may reach a point where they will run out of livestock to sell, and consequently have a much lower ability to purchase food.

12.4.4 Tipping points and regime shifts

Folke *et al.* (2004) suggest that most systems can exhibit multiple stability domains. Within rainfed agroecosystems this often is simplified into productive and degraded stability domains (Enfors and Gordon 2007), with 'productive' implying that sustainable crop production is possible, and 'degraded' implying that sustainable crop production is not possible. A more nuanced distinction of productivity is characterized by low actual yields, large yield gaps, and/or low water, land, or labor productivity. Additional indicators should also capture other ecosystem services such as pollination, biodiversity and indices on capacity of supporting and regulating services needed to sustain the provisioning capacity. Within smallholder rainfed agroecosystems, stability is not limited to on-farm crop productivity, but includes all activities which contribute to the livelihoods, including off-farm activities (e.g., timber harvesting).

Understanding that systems can move between different stable states is fundamental to applying resilience framing to watershed management. Gordon *et al.* (2008) identifies several specific regime shifts that can take place in water-related agroecosystems. Water is both a key input and a resource being affected in various ways by agricultural production systems, even low input sites. Because of this embedded characteristic of water, several regime shifts associated with water resources in landscapes can be associated with various agricultural practices. The most important aspect of the regime shifts concept is that these shifts can be unexpected, cross-scale, and in some cases, irreversible. Later in this section, Figure 12.8 will examine how several of the case studies in this chapter have shifted from stable to unstable states (or vice versa), and from undesirable to desirable states (or vice versa).

12.4.5 Defining key system components and processes

By analyzing the shocks and stresses that can drive (sometimes referred to as push or pull) an agroecosystem out of a stability domain, patterns emerge, and entry points for management activities become apparent. Figure 12.5 from the previous section illustrated an example of how different shocks and stresses and thus opportunities for management interventions occur at different parts of the system. The next step is to examine the key system components that, when acted upon, produce the largest benefits to smallholder farmers as well as ecosystems to attain sustainability and improved income. Several case studies identified key system components that are critical to the functioning of the system, and that when acted upon, yield multiple benefits. However, key system components vary among different landscapes and smallholder communities. In some cases, labor might be a key component (Mortimore and Adams 2001) whereas functioning off-farm provisioning ecosystem services are key in other situations (Enfors and Gordon 2008).

It is impossible to identify a generic key agroecosystem component for all smallholder farms and even less so for all farming systems. However, several authors have

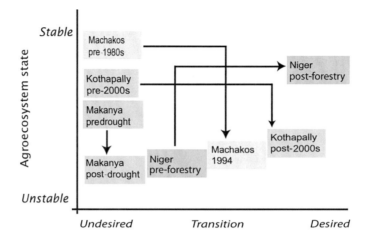

State of agro–ecosystem embedded in landscape

Figure 12.8 Principal change of cases studies by Mortimore and Tiffen (1994), Sreedevi *et al.* (2004), Enfors and Gordon (2008), and Reij *et al.* (2009), in terms of desired system state on the x-axis, and system stability on the y-axis

suggested that in the many water-constrained regions of the world, a common characteristic of smallholder rainfed farmers is a pronounced yield gap that can be eliminated (in some cases) with access to supplemental irrigation and improved soil nutrients (e.g., Rockström *et al.*, 2002; Barron 2004; Singh *et al.*, 2009). These authors recommend that by interrupting the key drivers of drought and soil degradation, crop yields may be improved.

12.4.6 Interpreting successes in terms of overall agroecosystem stability

Based on the discussion of stable and unstable system states, and the discussion of desired and undesired system states, it is useful to compare several of the case studies in a conceptual diagram. While interpreting the diagram it is important to understand that a given agroecosystem could be stable (often perceived as "good"), but in an undesirable state (often perceived as "bad"). Likewise, a system could be unstable (perceived as "bad") and in a desirable state (perceived as "good"). These system state "paradoxes" are important (Figure 12.8), because for any development intervention within smallholder farming communities to be considered sustainable in the long-term, that activity must produce not just stability, but desired stability.

The Kothapally project, described by Sreedevi *et al.* (2004) and Wani *et al.* (2009), indicates a system that moves from a mostly-stable, undesired state towards a less stable, but desired state. This project has been very successful by most measures, and its aspects are expected to continue to be successful. However, it is considered somewhat unstable now because the system is not yet self-reinforcing. To date, active, external development interventions were required for the watershed and its livelihoods to remain sustainable. The measure of whether the Kothapally community has truly moved to a *stable* and *desired* system state will be if the community is able to continue the various management activities on its own, i.e., without external support. This has happened as the project was completed in 2003; however, as researchers continue to study, the villagers do get benefits of their visits but without their inputs and have the system in place.

Evaluating watershed management in terms of its impact on system stability and desirability is at the core of what resilience framing can contribute to sustainable development of smallholder rainfed farmers. In order to achieve both stability and desirability, it is critical not only to act upon the key system processes that will yield the most benefits but also to address the issues that are of greatest concern to the target community.

12.5   IDENTIFYING MANAGEMENT ENTRY POINTS USING A RESILIENCE FRAME

Resilience framing may increase the strength of existing watershed management in a development context by equipping current managers, stakeholders, researchers, and farmers to better prepare and respond to system-wide surprises. This may be an especially effective tool for addressing increased uncertainty, variability, and other constraints expected in future scenarios of climate change.

For a smallholder farmer who is heavily reliant on rainfed agriculture for his livelihood, income generation is fraught with complexity, partly due to unpredictable climate regimes, pre-existing poverty, and various spatial and temporal scales of governance. This chapter argues that it is precisely because of these aspects of uncertainty and variability that resilience framing, as a complementary method of watershed planning and management can only serve to improve the effectiveness of poverty alleviation in smallholder farming systems by emphasizing the embedded landscapes that provide multiple types of ecosystem services for human well-being and income. Although resilience framing is still in its early stages for strongly manipulated agroecosystems with high incidence of poverty, a number of existing 'lessons learned' can serve as a first step towards more generic conclusions.

In this chapter we have used a set of well-known and fairly well documented cases to discuss how resilience framing may help understand opportunities and constraints in moving these coupled livelihood-agroecosystems into a more positive state with the double aim of improving human well-being and ensuring/enhancing ecosystem services in productive landscapes. Although tentative, we propose some emerging conclusions and ways ahead.

12.5.1   Management entry points – Learning from case studies

Based on the case studies reviewed in this chapter, a set of general conclusions can be made regarding the characteristics of smallholder management entry points (MEPs). These general conclusions originate from the use of resilience-based analysis, namely identification of drivers, positive feedbacks, tipping points, and regime shifts. To clarify this term, Wani *et al.* (2009) described the watershed scale as the "entry point" for effective management of smallholder agroecosystems. Tschakert (2007) refers to entry points, as suitable points of policy intervention. We use "entry point" to refer to "a specific point of entry for managers or farmers to actively intervene in the dynamic smallholder rainfed agroecosystem."

- *Recognize agroecosystem produce as a result of ecosystems services, and agricultural activities as potentially affecting other regulating and/or supporting ecosystem services:*

This requires a systems approach recognizing that smallholder farmer communities are often dependent on various ecosystem services being generated for livelihood, well-being, and income, in addition to crop yield/livestock.

- *Act on key drivers and feedbacks that can leverage change into more desirable agroecosystem states:*

There is often a set of key drivers within an embedded smallholder agroecosystem. It may be that the target system drivers are either functioning too poorly, or function too well (keeping the system in an undesirable state). Through the process of identification, a better targeted action/activity strategy may be implemented, which may be more effective in addressing unintended negative impacts on ecosystem services and on human well-being. Acting on a key driver serves the overall effectiveness of the given MEP. The MEPs should also impede or obstruct destabilizing positive feedbacks and/or enhance stabilizing positive feedbacks within the system. This is particularly important for smallholder agroecosystems, where unpredictable shocks (such as meteorological drought) can have long-term, self-reinforcing impacts for multiple growing seasons.

- *Activity at local scale, but origin of activity may be external:*

The case studies examined above almost all dealt with local-scale activities. However, they are nearly all enabled by external pressure (investment, actors, knowledge sometimes access to infrastructure and markets). Since smallholder rainfed agroecosystems are implicitly tied to the local landscape, it follows that MEPs are likely to be most effective if they occur at local spatial and social scales. There are exceptions to this rule in terms of government subsidies for groundwater pumps, regional electrical grids, and other non-local policies; however in general, successful activities occur at the scale of the local watershed. This may seem obvious, but it is an important component, often overlooked by governments or even international NGOs. However, an important nuance is that not all communities have the internal, or traditional, knowledge to effectively cope with, let alone thrive, during prolonged stress and/or shock events such as recurrent drought events (Tschakert and Tappan 2004; Tschakert 2007). The important MEP here is to understand that although actions must be taken at the local scale, the origin of the management activity may need to come from beyond the local scale.

- *Act on stresses, rather than shocks:*

Based on unpredictability of shocks (such as droughts, dry spells, etc.), and the relative predictability of stresses (for example, soil degradation, deforestation, poor health affecting labor, gender inequality, etc.), it may be more rewarding in the long-term to act on the stresses. The categorization of different obstacles may at first seem arbitrary, but it is important for identifying which obstacles are worth the time, effort, and money of local communities to try and address.

- *Enhancing and building off-farm ecosystem service capacity may have several direct and indirect benefits to smallholder farmer's livelihoods, benefits, and income:*

Many case studies illustrated the value of rehabilitating off-farm natural processes to enhance ecosystem services. In most cases these rehabilitations ultimately enhanced provisioning ecosystem services, which subsequently provided additional and alternative livelihood strategies. Enhancing regulating ecosystem services (at the local watershed scale) will encourage long-term stability of a desired state. Enhancing non-crop provisioning ecosystem services is a particularly useful coping strategy for increasing generic coping capacity, and improving overall livelihoods, especially with regard to the uncertain and variable impacts of climate change (Tschakert 2007; Enfors and Gordon 2008).

12.5.2   Recommendations for future research

Resilience framing is critical to understanding these longer time-scale dynamics, especially for farmers operating at the margins of productivity. As suggested earlier in the chapter (Sections 12.2, 12.3, and 12.4), resilience framing provides practical methods for individual farmers and groups of farmers to identify the best parts of their systems to manage. As yet, resilience framing is still in its early stages in application and generic conclusions for smallholder farming systems in development context. The issues of development on account of environment is becoming less acceptable as side effects or negative externalities of agricultural development quite often are immediately experienced by other communities, societies and/or resource users in the landscape. An improved documentation of development assessing both on-farm and off-farm indicators of potential impacts, coupled with social and human impacts of development would assist such analysis further.

The issue of how much more effective resilience framing is compared to alternatives in directing investments for development is unclear, and further evidence is needed for cost-effectiveness and of multiple goals of poverty alleviation and sustainable watershed management. The rapid development of various tools for data collection and management (also discussed in detail by Wani *et al.*, in this volume) will further help the application of resilience framing to define systems and thresholds. There is a range of participatory approaches that may be increasingly relevant to address these issues more systematically at different scales: What is desirable? What is stable/unstable? What are the system boundaries and relevant key drivers? It is likely that these answers should be supplied by stakeholders rather than researchers.

12.5.3   Resources to assist practitioners

Despite increasing interest in resilience science, further case studies are needed to draw more robust and generic conclusions. In order to translate new, resilience-based lessons learned to policies and investments for the smallholder farmers to improve livelihoods, income, and environmental sustainability, a set of tools that is freely available is included.

First, it is worthwhile to consult the Resilience Alliance website, specifically the handbooks for practitioners and for scientists. Depending on the needs of the specific farming community, one or the other may be more useful. Second, there are many online databases that are freely accessible, with hundreds of case studies for various ecosystems and social systems. The amount of content available on the topic of smallholder farming is increasing, with specific content available on soil erosion, salinization, semi-arid crop management, and social measures. Four databases of notable value are: the social-ecological systems E-library (CSID 2010), the Resilience Alliance bibliography, with nearly 800 documents (Resilience Alliance 2010a), the Resilience Alliance researchers database (Resilience Alliance 2010b), and the Resilience Alliance thresholds database (Resilience Alliance 2010c). Third, it is important that knowledge sharing takes place between researchers, practitioners, and farmers around the world. For this reason, it is important that the databases identified above are both open source and updated regularly.

We hope that the concepts of resilience science and the real-world case studies will provide ample evidence for the utility of resilience science as a complementary tool for evaluating smallholder watershed management activities. We encourage the reader to consider how to connect the concepts, tools, and strategies of resilience framing to the current "state-of-the-art" as explained in other chapters in this volume. Only through combining the collective knowledge and experience of researchers and practitioners in smallholder agroecosystems can the necessary paradigm shift towards long-term sustainability take place.

ACKNOWLEDGMENTS

This chapter was developed building on a range of ideas and discussions with colleagues at the Stockholm Resilience Centre (SRC), and various partners in research projects coordinated by the Stockholm Environment Institute (SEI) and SRC. We especially thank Louise Karlberg for her support and ideas at an early stage. The funding to support the development of this chapter was provided by FORMAS "Formel Exec" grant to SEI, Stockholm University, and Beijer.

REFERENCES

Barron, J. 2004. *Dry spell mitigation to upgrade semi-arid rainfed agriculture: Water harvesting and soil nutrient management for smallholder maize cultivation in Machakos, Kenya.* Stockholm, Sweden: Stockholm University.

Barron, J., E. Enfors, H. Cambridge, M. Adamou. 2010. Coping with rainfall variability: Dryspell mitigation and implication on landscape water balances in smallscale farming systems in semi-arid Niger. *Int. J. Water. Res. Dev.* 26:523–542

Barron, J., S. Noel, and M. Mihkail. 2010. Review of agricultural water management intervention impacts at the watershed scale: a synthesis using the sustainable livelihoods framework.
Project Report. Stockholm/York: Stockholm Environment Institute.

Bennett, E.M., G.D. Peterson, and L.J. Gordon. 2009. Understanding relationships among
multiple ecosystem services. *Ecology Letters* 12:11.

Cooper, P.J.M., J. Dimes, K.P.C. Rao, B. Shapiro, B. Shiferaw, and S. Twomlow. 2008. Coping
better with current climatic variability in the rain-fed farming systems of sub-Saharan Africa:
An essential first step in adapting to future climate change? *Agriculture, Ecosystems &
Environment* 126(1–2):24–35.

CSID. 2010. Social Ecological System E-Library. (http://dev.commons.asu.edu/csid/) (Accessed
9 July, 2010.)

de Bruin, A., M. Mihkail, S. Noel, and J. Barron. 2010. AWM interventions and monitoring
and evaluation at the watershed scale: Potential approaches at the watershed level. *Project
Report*. Stockholm/York: Stockholm Environment Institute..

Enfors, E.I., and L.J. Gordon. 2007. Analyzing resilience in dryland agroecosystems: a case
study of the Makanya catchment in Tanzania over the past 50 years. *Land Degradation and
Development* 18:17.

Enfors, E.I., and L.J. Gordon. 2008. Dealing with drought: The challenge of using water system
technologies to break dryland poverty traps. *Global Environmental Change: Human and
Policy Dimensions* 18(4):607.

Falkenmark, M., M. Finlayson, L. Gordon *et al.* 2007. Agriculture, water, and ecosystems: avoiding the costs of going too far. (http://library.wur.nl/WebQuery/wurpubs/360019)
(Accessed 24 June, 2010.)

Foley, J.A., R. DeFries, G.P. Asner *et al.* 2005. Global consequences of land use. *Science*
309(5734):570.

Folke, C., S. Carpenter, B. Walker, M. Scheffer, T. Elmqvist, L. Gunderson *et al.* 2004. Regime
shifts, resilience, and biodiversity in ecosystem management. *Annual Review of Ecology,
Evolution and Systematics* 5:557–581.

Glover, D. 2007. Monsanto and smallholder farmers: a case study in csr. *Third World Quarterly*
28(4):851–867.

Gordon, L.J., G.D. Peterson, and E.M. Bennett. 2008. Agricultural modifications of
hydrological flows create ecological surprises. *Trends in Ecology and Evolution* 23(4):
211–219.

Joshi, P.K., A.K. Jha, S.P. Wani, T.K. Sreedevi, and F.A. Shaheen. 2008. Impact of watershed program and conditions for success: a meta-analysis approach. *Global Theme on Agroecosystems
Report No. 46*. Patancheru, Andhra Pradesh, India: International Crops Research Institute
for the Semi-Arid Tropics.

Kasei, R., B. Diekkrüger, and C. Leemhuis. 2010. Drought frequency in the Volta Basin of West
Africa. *Sustainability Science* 5(1):89–97.

Kosoy, N., M. Martinez-Tuna, R. Muradian, and J. Martinez-Alier. 2007. Payments for environmental services in watersheds: Insights from a comparative study of three cases in Central
America. *Ecological Economics* 61(2–3):446–455.

Kulecho, I.K., and K. Weatherhead. 2006. Issues of irrigation of horticultural crops by
smallholder farmers in Kenya. *Irrigation and Drainage Systems* 20(2):259–266.

MEA. 2005. *Ecosystems and human well-being*. Washington, DC, USA: World Resources
Institute.

Mortimore, M.J., and W.M. Adams. 2001. Farmer adaptation, change and 'crisis' in the Sahel.
Global Environmental Change 11:9.

Mortimore, M.J., and M. Tiffen. 1994. Population growth and a sustainable environment.
Environment 36(8):16.

Noble, A.D., D.A. Bossio, F.W.T. Penning de Vries, J. Pretty, and T.M. Thiyagarajan. 2006. Intensifying agricultural sustainability: An analysis of impacts and drivers in the development of 'bright spots'. *Comprehensive Assessment Research Report 13*. Colombo, Sri Lanka: Comprehensive Assessment Secretariat.

Pauchauri, R., and A. Reisinger. 2007. *Climate Change 2007: Synthesis report. Contribution of Working Groups I, II, III to the Fourth Assessment Report of the Intergovernmental Panel on Climate Change*. Geneva, Switzerland: IPCC Secretariat.

Raudsepp-Hearne, C., G.D. Peterson, and E.M. Bennett. 2010. Ecosystem service bundles for analyzing tradeoffs in diverse landscapes. *Proceedings of the National Academy of Sciences* 107(11):6.

Reij, C., G. Tappan, and M. Smale. 2009. *Agroenvironmental transformation in the Sahel: Another kind of "Green Revolution"*. USA: International Food Policy Research Institute.

Resilience Alliance. 2010a. Bibliography. (http://www.resalliance.org/2.php) (Accessed 8 July 2010.)

Resilience Alliance. 2010b. Researcher database. http://www.resalliance.org/2986.php (Accessed 8 July 2010.)

Resilience Alliance. 2010c. Thresholds database. Organization. (http://www.resalliance.org/183.php) (Accessed 8 July 2010.)

Rockström, J., J. Barron, and P. Fox. 2002. Rainwater management for increased productivity among small-holder farmers in drought prone environments. *Physics and Chemistry of the Earth, Parts A/B/C* 27(11–22):949–959.

Rockström, J., N. Hatibu, T. Oweis *et al.* 2007. Managing water in rainfed agriculture. In *Water for food, water for life: a comprehensive assessment of water management in agriculture*, ed. D. Molden London, UK: Earthscan; and Colombo, Sri Lanka: International Water Management Institute.

Rockström, J., and L. Karlberg. 2009. Zooming in on the global hotspots of rainfed agriculture in water-constrained environments. In *Rainfed agriculture: unlocking the potential*, ed. S.P. Wani, J. Rockström, and T. Oweis, 36–43. Wallingford, UK: CAB International.

Singh, P., P.K. Aggarwal, V.S. Bhatia *et al.* 2009. Yield gap analysis: modelling of achievable yields at farm level. In *Rainfed agriculture: unlocking the potential* ed. S.P. Wani, J. Rockström, and T. Oweis, 81–123. Wallingford, UK: CAB International.

Sreedevi, T.K., B. Shiferaw, and S.P. Wani. 2004. *Adarsha watershed in Kothapally: understanding the drivers of higher impact*. Patancheru, Andhra Pradesh, India: International Crops Research Institute for the Semi-Arid Tropics.

Tschakert, P. 2007. Views from the vulnerable: Understanding climatic and other stressors in the Sahel. *Global Environmental Change* 17(3–4):381–396.

Tschakert, P., and G. Tappan. 2004. The social context of carbon sequestration: Considerations from a multi-scale environmental history of the Old Peanut Basin of Senegal. *Journal of Arid Environments* 59(3):535–564.

Wani, S.P., J. Rockström and T. Oweis. 2009. *Rainfed agriculture: unlocking the potential*. Wallingford, UK: CAB International.

World Resources Institute. 2005. The wealth of the poor: managing ecosystems to fight poverty. *World Resources Report*. World Resources Institute.

Chapter 13

Impacts of climate change on rainfed agriculture and adaptation strategies to improve livelihoods

Peter Q. Craufurd, S.V.K. Jagadish, and Jon Padgham

13.1 INTRODUCTION

Farmers living and working in the semi-arid tropics (SAT) of Africa and Asia are acutely vulnerable to climate variability and change due to their limited natural and financial resources coupled with poor infrastructure, institutional support, and governance (World Bank 2008). Coping with variability is nonetheless a way of life for many of these farmers, and farmers in many different regions of the world have adopted or adapted strategies to manage variability. In this chapter we first describe the impacts of climate change on crop and livestock production, water resources, and prices, poverty, and malnutrition in South Asia and sub-Saharan Africa (SSA). Secondly, we examine adaptation strategies, focusing on the social/institutional aspects needed to support farmers' adaptation strategies as well as describing briefly strategies used by farmers.

13.2 CLIMATE CHANGE IMPACTS

Climate change impacts on agriculture in the near to medium term (next one or two decades) are more likely to arise from increased climate variability, and increased frequency and intensity of extreme events, rather than from changes in mean or average climatic conditions. Rising temperatures and changes in rainfall patterns, including increased seasonal and inter-annual rainfall variability, can directly reduce crop yields, and indirectly affect irrigation water availability and increase the water requirement of the crops (Nelson *et al.*, 2009). In addition, there are a number of secondary effects of climate change, such as increased pest and disease pressure (Anderson *et al.*, 2004) and heightened risk of soil erosion and other land degradation processes (Boardman 2006) that can negatively impact food production. These factors are usually not accounted for in crop loss models but their effects could be quite significant. The most vulnerable agricultural systems occur in arid, semi-arid, and dry subhumid regions in the developing world, where extreme rainfall variability results in recurrent droughts and floods regularly disrupting food production leading to pervasive poverty (Hyman *et al.*, 2008).

The long-term impacts of climate change on agricultural productivity are not expected to be geographically uniform. Small increases in yield and production could occur in certain high latitude locations, e.g., parts of Europe, northern China, and northern North America, while yields and production in much of Africa, South and Central Asia, the Mediterranean Basin, the Andes, and parts of Central America are likely to be greatly reduced (Baettig *et al.*, 2007; Easterling and Aggarwal 2007), with the maximum impact predicted to be in SSA and South Asia (Nelson *et al.*, 2009) (Figure 13.1). These discrepancies arise in part because at higher latitudes future warming, up to about 2°C, will be favorable for crop development and growth in these cold limited zones (see Box 1). In contrast, at lower latitudes temperatures are already close to the optimum for crop production and dryland conditions are widespread, so any further increase in temperatures and adverse changes in rainfall patterns are damaging. However, more favorable temperatures at high latitude zones would not automatically sustain production at existing levels as crops, cropping systems, and appropriate management practices will still need to be modified and adapted to future conditions, which could include more extreme events as exemplified by record-setting high temperatures and drought in Russia and elsewhere in northern Europe in the summer of 2010.

Box 1. Response to Temperature

The figure below shows a typical rate response to temperature – in this case rate of development – but also applicable to other processes such as dry matter production. As temperature increases, the rate of development increases till an optimum value – approximately 20 to 25°C in temperate species (gray line) and 27 to 32°C in tropical species (black line). Above the optimum, rate decreases and flowering is delayed or dry matter production is reduced. Impacts of climate change, both positive and negative, are strongly linked to how close current ambient temperatures are to the optimum temperature of different crop species. It should also be noted that extreme hot and cold temperatures at certain stages of crop development, notably flowering, cause sterility and hence very poor yields (Matsui *et al.*, 1997; Wheeler *et al.*, 2000; Gunawardena *et al.*, 2003; Prasad *et al.*, 2006; Jagadish *et al.*, 2007).

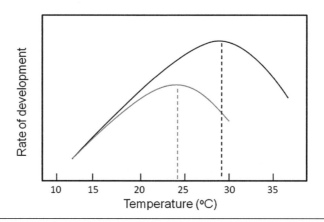

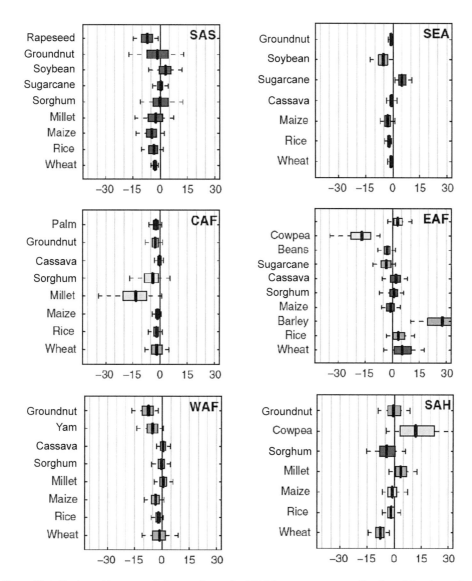

Figure 13.1 Projected impacts of climate change by 2030 for major crops in South and Southeast Asia and most of Africa.

[Note: Probabilistic projections of production impacts in 2030 from climate change (expressed as a percentage of 1998 to 2002 average yields). Broken lines extend from 5th to 95th percentile of projections, boxes extend from 25th to 75th percentile, and the middle vertical line within each box indicates the median projection. Region codes SAS, SEA, CAF, EAF, WAF, and SAH are for South Asia, Southeast Asia, Central Africa, Eastern Africa, West Africa, and Sahel, respectively. Modified and adapted from Lobell *et al.* (2008a, 2008b).]

13.2.1 Crop and livestock production

A recent global modeling study using the outputs from two global climate change models [NCAR (wetter) and CSIRO (drier)], suggests that production of major crops growing in the developing countries will predominantly decline while those in developed countries will be less affected (Nelson *et al.*, 2009). For example, 14% decline is predicted in rice production relative to the no climate change scenario, 44 to 49% decline in wheat production, and 9 to 19% fall in maize production (Table 13.1). Even with CO_2 fertilization effect (on C3 species only; see Long *et al.*, 2006, 2007 for a fuller discussion), yield will still be substantially reduced (Nelson *et al.*, 2009). Apart from SSA and South Asia, other regions predicted to suffer major yield losses are semi-arid northeastern Brazil and areas in Central America (Magrin *et al.*, 2007; Lobell *et al.*, 2008a). A separate study of percent yield change among major crops across most of Africa and South and Southeast Asia compared with the baseline (1980–2000) and projections for 2020–40, assuming an approximate 1°C increase in temperature between 1980 and 2000 (Lobell *et al.*, 2008b), is presented in Figure 13.1. This study also predicts significant negative impacts of climate change on food security that could occur as early as 2030 for several crops in these regions.

More than 600 million people depend on livestock for their livelihoods (Thornton *et al.*, 2009) and hence impacts on this sector are also important though frequently overlooked and not well researched. Livestock will be impacted by climate change directly (heat, diseases) and indirectly (feed quality and quantity, water resources). The Intergovernmental Panel on Climate Change (IPCC) predicts negative impacts of climate change on livestock in arid and semi-arid regions, but positive effects in humid temperature regions, in line with the principles governing crop species adaptation (Christensen *et al.*, 2007; IPCC 2007). Animals, like plants, also grow (and produce milk) best at certain temperatures and are negatively impacted by high temperatures. The ideal range of ambient environmental temperatures for animals is termed as the 'thermoneutral zone'. High temperature stress is defined as a point at which the animal cannot dissipate an adequate quantity of heat to maintain body temperature balance, which is normally calculated as temperature humidity index (THI) based on ambient temperature and relative humidity. Heat stress begins to occur in dairy cattle, beef cattle, swine, and poultry when the THI is above 72, resulting in reduced intake and milk yield, and higher milk temperature in dairy cows (West *et al.*, 2003). In Georgia, for example, cool periods with temperatures of 18 (minimum) and 30°C (maximum) have THI of

Table 13.1 Recent extreme climate events and their impacts on agriculture in sub-Saharan Africa[a]

Country/Region	Period	Climatic event	Impact
Kenya	1997–2000	Severe flooding followed by drought	10% loss of national GDP
Malawi	1991–92	Drought	60% maize yield loss
	2000–01	Floods	30% maize yield loss
Zimbabwe and Zambia	1992	Drought	8–9% loss of GDP from agriculture
Mozambique	2000	Floods	2 million people affected
	2002–06	Drought	800,000 people affected

[a]Source: Padgham (2009).

about 70 compared with a THI of 78 during hot periods with temperatures of 23 (minimum) and 34°C (maximum) (West *et al.*, 2003). St-Pierre *et al.* (2003) have estimated the yearly loss from heat stress without abatement to be US$2.4 billion in the US alone which could be reduced to US$1.7 billion if heat stress abatement practices (e.g., shading) are implemented. The largest proportion of losses was reported for dairy cattle (52%) compared to losses of 21% for beef cattle, 17% for swine, and 10% for poultry.

Water availability, both for direct consumption and for fodder/forage, is also likely to be impacted by climate change. While there are uncertainties in the predictions of water availability for pasture and fodder, the effect of temperature on demand for water is well known (Thornton *et al.*, 2009). For *Bos indica*, water demand increases from about 3 kg dry matter intake at 10°C to 5 kg at 30°C and 10 kg at 35°C. *Bos taurus* requires 3, 8, and 14 kg at the same temperatures. In Australia, water demand for beef cattle is predicted to be 13% higher under predicted climate change. During the severe El Niño year of 1980 countries as widespread as Botswana, Niger, and Ethiopia suffered 20 to 62% cattle deaths.

13.2.2 Water resources

Droughts or floods that last a few months can be highly destructive but when they last for decades the effects can be devastating or even irreversible (Conway 2008). Although significant disagreement among climate models still exists regarding long-term precipitation changes, warmer air holds more moisture; thus rainfall is likely to become increasingly aggregated, with a shift towards fewer but more intense storms and longer periods between rainfall events, as has already been observed across several land areas (Trenberth *et al.*, 2007). Although the total percentage global land area affected by drought has been quite stable from 1950 to 1980s, there has been a significant increase in the area subjected to water deficit stress from 1990 to 2000 (Figure 13.2) and some key cropping systems for food security are highly vulnerable (Hyman *et al.*, 2008).

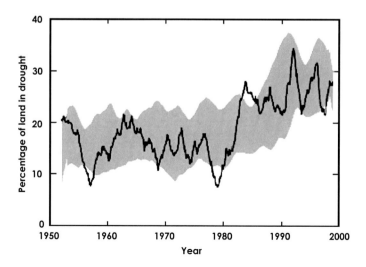

Figure 13.2 Global land area under drought between 1950 and 2000. (http://www.metoffice.gov.uk/research/hadleycentre/pubs/brochures/COP12.pdf)

13.3   REGIONAL IMPACTS

In the following section we examine impacts of climate change on crop production, food prices, poverty, and malnutrition in the highly vulnerable regions of SSA and South Asia.

13.3.1   Sub-Saharan Africa

Northern and Southern Africa are projected to have hotter and drier conditions by the end of this century, potentially resulting in a much greater risk of drought in what are already highly drought-prone sub-regions. The direction of mean annual precipitation change in West Africa is uncertain while East Africa could experience increased precipitation (Christensen *et al.*, 2007), though other analyses (e.g., Funk *et al.*, 2008) indicate a potential drying trend in that sub-region. Median annual temperature changes across Africa by the end of the century are projected to exceed 3°C, assuming a mid-range scenario of greenhouse gas (GHG) emissions (Christensen *et al.*, 2007).

The majority of African countries are highly dependent on natural resources and their agricultural sector for food, employment, income, tax revenue, and exports. Changes in the weather conditions which can damage the agricultural sector will have a major impact on people's incomes and livelihoods. Moreover with weak Government and institutions which are poorly resourced, people are mostly left to cope on their own. For example, in Northeastern Ethiopia, between 1998 and 2000 drought-induced crop and livestock losses were estimated at US$266 per household, which is significantly greater than the annual average cash income for more than 75% of rural households (Carter *et al.*, 2004).

Rainfed agriculture currently constitutes about 90% of Africa's staple food production, making it highly vulnerable to reduced quantity, distribution, and timing of rainfall; in addition growing season length will likely decrease due to higher temperatures (Conway 2008). It is estimated that large areas of the semi-arid and dry subhumid regions could lose 5 to 20% of their growing season length, with the Sahel potentially experiencing >20% loss by 2050 (Thornton *et al.*, 2006; Cooper *et al.*, 2008). There will also be an increased percentage of failed seasons throughout the continent. Moreover, increased climate variability within climate change poses a significant risk to food production in Africa in the near- to medium-term. Africa currently experiences a variety of weather-related disasters on a regular basis (Cornford 2003) that combined with widespread poverty, land degradation, and poor governance reduce its capacity to effectively cope with current climate risks and adapt to future climate change.

An account of the recent extreme water-related disasters dominated either by excess or shortage of water and their impact on agriculture and people in SSA is presented in Table 13.1. Further, the IPCC has estimated that 75 to 250 million more people in Africa will face increased water shortage by 2020, and a 10% drop in precipitation in semi-arid areas of SSA could decrease surface drainage volumes by 50%, according to de Wit and Stankiewicz (2006). Moreover large increases (5 to 8%) in the proportion of arid and semi-arid lands by 2080s, in addition with depleted water resources, will result in more prominent chronic hunger. In some countries the projected yield decline could be as much as 50% by 2020, and crop net revenues could fall by as much as 90% by 2100, with small-scale farmers being the most vulnerable (Carter *et al.*, 2004).

At the ground level, in southern Africa and across western and north-central Africa lower rainfall may also lead to shorter crop growing season, threatening the probability of getting a second crop in some areas and even the viability of a single crop in others (ILRI 2006).

Africa is less likely to be damaged by rising sea levels than Asia. The most extensive inundation is likely to be in the Nile delta. A one-meter rise is predicted to affect nearly 6 million people and inundate lagoons and the low-lying reclaimed lands (http//www.grida.no/climate/vitalafrica/english/16.htm). This in turn would affect one-third of Egypt's fish catches made predominantly in the lagoons and by changing the water quality, the fishing community could be badly affected.

13.3.2 South Asia

Food production in South Asia also faces significant risks from climate change. South Asia's agriculture critically depends on the June-September southwest monsoon, which generates 70% of the subcontinent's total annual precipitation. However, the distribution and timing of monsoon precipitation can be highly variable. For example, under extreme cases, a significant percentage of seasonal rainfall can occur within a period of several days resulting in severe flooding (Mall et al., 2006a). At the other end of the spectrum, failures of the Indian monsoon, which have historically had a strong positive relationship with El Niño events, create widespread drought (Mall et al., 2006b).

The Indian monsoon is expected to intensify with climate change, potentially producing a slight increase in overall precipitation for the subcontinent in the long-term (Christensen et al., 2007). However, greater regional variations in rainfall are possible, with dry regions potentially becoming drier and wet regions wetter, and increase in the number of additional years of record or near-record precipitation (Baettig et al., 2007). These hydrologic changes will occur against a backdrop of rising temperatures, with the region projected to experience an annual median temperature rise of around 3°C by the end of this century, under a mid-range of greenhouse gas emission scenarios (Christensen et al., 2007). Temperature rise will also produce fundamental changes in the dry-season supply of glacial meltwater, an important water source for irrigated agriculture especially in the Indo-Gangetic Plain of South Asia.

Climate change is likely to magnify the adverse effects of existing pressures on agricultural systems in South Asia. For example, more intense rainfall and runoff could reduce groundwater recharge in areas where the unsustainable extraction of groundwater for irrigation has resulted in rapidly declining water tables. The region's two major cereal crops are quite vulnerable to increases in temperature. Wheat is currently near its maximum temperature range, with high temperatures during reproductive growth and grain filling, representing a critical yield-limiting factor for wheat in significant portions of the Indo-Gangetic Plain. Incremental increases in temperature with climate change could thus have a large impact. Ortiz et al. (2008) estimate that by 2050 approximately half of the highly productive wheat areas of the Indo-Gangetic Plain could be reclassified as a heat-stressed, short-season production mega-environment. The other major cereal crop in the region, rice, is also quite susceptible to temperature rise, particularly warmer night temperatures, which increase respiration losses (Peng et al., 2004).

Widespread flooding is also expected to increase in Asia. Many small islands and delta regions, for example the Mekong delta, are highly vulnerable to flooding. In

Myanmar, floods caused by the tropical cyclone Nargis (during May 2008) devastated 1.75 million ha of rice land (USDA/FAS 2008) while in Bangladesh, cyclone Sidr caused production losses in the range of 800,000 t of rice during 2007 (IRIN 2008).

13.4   PRICES, POVERTY, AND MALNUTRITION

The direct and indirect effects of climate change on agriculture can be tracked through an economic system wherein climatic change will bring greater volatility to production costs and consumption prices of production and consumption, productivity investments, food demand, and ultimately human well-being (Parry et al., 2009). With no climate change, world prices for rice, wheat, and maize could increase between 2000 and 2050, mainly driven by population and income growth along with a declining productivity. The price of rice could rise by 62% and maize by 63% (Nelson et al., 2009). However, with climate change (note that CO_2 fertilization effects on price are not large) an additional increase in prices by 32 to 37% for rice and 52 to 55% for maize is predicted (Table 13.2). Among livestock products, beef prices are predicted to be 33% higher by 2050 with no climate change and 60% higher with climate change. Similarly prices of all other livestock products including pork, lamb, and poultry were predicted to increase with the same magnitude with both the drier (CSIRO) and the wetter (NCAR) model.

By analyzing the diminishing consumption of cereals, Nelson et al. (2009) showed that without climate change caloric availability would increase throughout the world between 2000 and 2050, except for a small decline in Latin America and the Caribbean. The largest increase would be in SSA (12.6%). Even by including climate change in the model, caloric availability not only was lower than the no climate change scenario in 2050 but also declined relative to 2000 levels throughout the world. However, with the beneficial effect of CO_2 fertilization, the decline was predicted to be 3 to 6% less severe though still a considerable decline relative to the no climate change scenario (Table 13.3). In terms of number of malnourished children, only SSA is projected to have an increase in the number of malnourished children between 2000 and 2050 even without climate change, with the other developing countries recording greater reductions in numbers.

Burney et al. (2010) provide evidence against the prevailing assumption that higher prices lead to increased poverty in the world given that poor people tend to spend a larger share of their income on food. They indicated that poor people who own their own land could actually benefit from higher crop prices while rural wage laborers and people living in cities will definitely be negatively affected. Hence the study revealed a surprising mix of winners and losers depending on the projected global temperature and the scenario considered. In Thailand, for example, the poverty rate for people in the non-agricultural sector was projected to rise 5%, while the rate for self-employed farmers dropped more than 30%. With the most likely scenario of crop production meeting expectations, a 1°C increase by 2030 in crop yields, food prices, and poverty rates could be relatively small. But under the "low-yield" scenario (crop production towards the low end of expectations), with 1.5°C increase would result in 10 to 20% drop in agricultural productivity and 10 to 60% rise in the price of rice, wheat, and maize, in turn increasing the overall poverty rate by 3% in the 15 countries surveyed.

Table 13.2 Production and price of rice, maize, millet, and sorghum in 2000, 2050 with no climate change (CC), and percent change with CC (range from CSIRO and NCAR models) in 2050 relative to 2050 without CC[a]

Agriculture product	Production (million t^{-1})			World price (US$ t^{-1})
	South Asia	Sub-Saharan Africa	World	
Rice				
2000	119.8	7.4	390.7	190
2050 No CC	168.9	18.3	455.2	307
2050 CC (% change)	−14.3 & −14.5	−14.5 & −15.2	−11.9 & −13.5	32.0 & 36.8 (−15.1 & −17.0)[b]
Maize				
2000	16.2	37.1	619.2	95
2050 No CC	18.7	53.9	1061.3	155
2050 CC (% change)	−18.5 & −8.9	−9.6 & −7.1	0.2 & −0.4	55.1 & 51.9 (−12.6 & −11.2)[b]
Millet				
2000	10.5	13.1	27.8	—
2050 No CC	12.3	48.1	67.0	—
2050 CC (% change)	−19.0 & −9.5	−6.9 & −7.6	−8.4 & −7.0	—
Sorghum				
2000	8.4	19.0	59.9	—
2050 No CC	9.6	60.1	123.5	—
2050 CC (% change)	−19.6 & −12.2	−2.3 & −3.0	−2.6 & −2.5	—

[a]Source: Nelson *et al.* (2009).
[b]Values in parentheses show price changes with CO_2 fertilization (i.e., % change from no CO_2 fertilization).

Table 13.3 Projected number ('000) of malnourished children below the age of 5 in 2000, and in 2050 with no climate change (No CC) and with climate change excluding a CO_2 fertilization effect (+CC) averaged from CSIRO and NCAR predictions[a]

Region	2000	2050		CF effect[b] (%)
		No CC	+ CC	
South Asia	75621	52374	58168	−3
East Asia and Pacific	23810	12018	16537	−8
Europe and Central Asia	4112	2962	3909	−4
Latin America and Caribbean	7687	5433	6728	−4.5
Middle East and North Africa	3459	1148	2016	−10
Sub-Saharan Africa	32669	38780	48875	−5
All developing countries	147357	112714	136232	−4.5

[a]Source: Nelson *et al.* (2009); adapted and modified from Parry *et al.* (2009).
[b]Percentage difference between the number of malnourished children in 2050 with and without the CO_2 fertilization (CF) effect taking the average of CSIRO and NCAR predictions.

13.5 ADAPTATION

Many of the impacts outlined in the previous section are now regarded as inevitable, given the lag in the climate system that ensures continued warming for several decades even if GHG emissions were to somehow immediately cease, and the fact that efforts to reach a global agreement on limiting GHG emissions has to date failed (Parry *et al.*, 2008). Thus adaptation is now essential rather than optional. The profound changes to the climate system that are projected to occur within this century will have a pronounced effect on crop and livestock production and livelihoods of the poor, resulting in more intense poverty, malnutrition, and conflict. In this section key concepts of coping, adaptation, and resilience are examined, framing a discussion about links between development and adaptation, and the need to enable or 'adapt' to adaptation. Lastly, technological options for adaptation are described briefly, as many of these have been covered elsewhere in this book.

Long-term investments in agriculture not only enhance the capacity of agriculture to better manage risks from climate change but also produce double dividends with respect to slowing the growth of greenhouse gases in the atmosphere. For example, a recent study by Burney *et al.* (2010) demonstrates the beneficial effects of investment in agricultural research: they estimated that from 1961 forward, emissions of three major greenhouse gases (methane, nitrous oxide, and CO_2) were reduced by a quarter ton for every dollar invested in agricultural research. Although, GHG emissions have increased with agricultural intensification, those emissions are far outstripped by the emissions that would have been generated in converting additional forest and grassland to farm land. Considering the total amount of agricultural research funding related to yield improvements since 1961 through 2005, a very nominal price ranging between approximately US$4 and US$7.50 has been invested for each ton of CO_2 that was not emitted. Hence this study clearly demonstrated the huge potential that can be presently achieved, and subsequently reaped by the future generations, by investing in agricultural research, as well as the opportunity costs associated with under-investment in agriculture as has been the case over the past couple of decades.

13.5.1   Coping, adaptation, and resilience

Most poor smallholder farmers are vulnerable to climate variability and change, being highly dependent on agriculture, and especially on natural resources/assets, for their livelihoods (Conway 2008). These natural resources of land, soil, water, and biodiversity are often degraded or overexploited; a situation that is exacerbated by widespread poverty, weak institutions, poor support mechanisms and governance, and lack of infrastructure. Hence smallholder farmers are acutely vulnerable to shocks and stresses, both from climate variability and other factors. Farmers have developed coping strategies over time that allow them to cope with the vagaries of climate and other factors, but these are short-term strategies that respond to expected and observed seasonal variation, and are usually risk averse strategies designed for below-average seasons. While many of these coping strategies can contribute to adaptation, such strategies are essentially internal and are not sufficient for adaptation.

Adaptation is defined as an 'adjustment in natural or human systems in response to actual or expected climatic stimuli or their effects, which moderates harm or exploits beneficial opportunities' [Christensen *et al.* (2007); for other definitions of adaptation, see Levina and Tirpak (2006)]. Adaptation thus includes both responses to threats and opportunities, the latter being frequently overlooked. Indeed, in future, making use of opportunities to maximize production and profit could become important as a means of ameliorating the impact of poor years, assuming forecasts of seasonal climate conditions are sufficiently robust, and other production factors adequate, to allow for opportunistic farming. Adaptation includes dimensions of biophysical, social, and economic change and considers coupled human-natural systems and not just biophysical impacts. Adaptation is a key strategy for building resilience, which broadly describes the ability of systems or individuals to cope with sudden (shock) or gradual (stress) changes (Conway 2008).

Adaptation and adaptive capacity need to be understood in the context of sustainable livelihoods and development in general, and not viewed as, or indeed implemented as, a separate package of largely technical fixes (Mortimore 2010). The livelihoods approach has been found to be useful for understanding food insecurity as it emphasizes the importance of looking at an individual's capacity for managing risks as well as external threats to livelihood security such as droughts (Chambers *et al.*, 1989; Scoones 1998). Adaptive capacity at its core comprises the major elements of sustainable livelihoods (Carney 1998); natural or biophysical assets (soil, water, land, biodiversity), human or socioeconomic assets (literacy, gender equality, social networks), and financial and technological assets. However, long-term risks from climate change require that additional measures beyond sustainable livelihoods frameworks be considered. Such measures should foster 'climate aware' development, and may include, inter alia, building capacity for: appropriately interpreting and applying output from regional downscaled climate models; conducting integrated assessments on vulnerability, impacts, and adaptation; and developing climate risk communication strategies and tools appropriate to the needs of vulnerable groups.

Many studies have shown how important sustainable livelihoods factors are in the ability of farmers to effectively manage and cope with risks under current conditions (Chambers *et al.*, 1989; Scoones 1998; Mortimore and Adams 1999). It is essential to understand how any technology that putatively contributes to adaptation will affect such livelihoods-based coping strategies and especially the sustainable use of natural resources. Similarly, in targeting the most vulnerable, who are frequently women, children, and the landless, these factors have to be carefully considered. For example, women are more likely to do natural resource management related livelihood diversification (market gardens, production) while men are more likely to do wage-related diversification as a strategy for coping with risks, both climatic and potentially for adapting to longer term climate change.

It is also important in discussing adaptation technologies, especially in relation to climate change, to recognize that many technologies used by farmers as part of their coping strategies are regarded by them not as means for managing climate risks but for productivity and profitability. This is important to note for at the core of all agricultural development is the need for 'incentive' in order for farmers to adopt technology. Part of this incentive may include technology or technology adoption approaches that have sufficient flexibility so as to allow potential adopters to reconfigure the technology to most effectively meet needs for coping with risk (Nederlof and Dangbégnon 2007). Thus, adaptation strategies need to be devised around understanding constraints and hence entry points linked to incentives or tangible benefits. While stating that coping strategies are short term, we nonetheless fully concur with the sentiments of Cooper *et al.* (2009), who said that first stage in adaptation is to support farmers to cope better with current variability. We add that a parallel tract is needed to identify entry points for developing policies, promoting communication between decision makers at multiple levels and the research community, and building individual and institutional capacity for generating and disseminating new knowledge, which would allow societies to begin to prepare for longer term manifestations of climate change, including that of food production.

It is increasingly recognized that many small farmers have not benefited from technologies available today (Cooper *et al.*, 2009), due to failure either of the technology or more commonly of the delivery mechanism (Renkow and Byerlee 2010). As discussed elsewhere, this is often rooted in either a failure to understand livelihoods and assets or the impact pathway and other actors and organizations needed to deliver technology (Hall *et al.*, 2005). Technology may be 'necessary', but it is rarely if ever 'sufficient' for impact. Collective action, including participatory approaches, that use or build stronger social networks, has been shown to be important in all societies, rich and poor, for technology adoption (Pretty 2008) and for strategies that more broadly reduce impacts from climate change (Adger 2003). Equally important are favorable enabling environments in terms of government and other sectors' support and policies, both national and local (Mortimore 2010). Indeed, for adaptation the role and importance of local organizations and their capacity for supporting adaptation is frequently overlooked, despite the fact that these organizations will be the ones supporting farmers directly. Where farmers perceive weak support for adaptation interventions they are less likely to try what they perceive to be riskier technologies (Pedzisa *et al.*, 2010). As Kandlikar and Risbey (2000) note in a review of adaptation challenges for agriculture, "[F]armers in low income countries face high downside risks from failure of new technologies, especially if information and government support is limited or lacking. In such cases, they are likely to choose options that have been well tested in the past. Studies of [climate change] adaptation need to pay greater attention to these issues to be truly relevant in a global sense."

Another factor frequently overlooked in technology transfer is knowledge transfer, two-way knowledge exchange, and the adaptation or modification of technology to suit local needs and environments. Natural resource technologies are knowledge intensive, especially in comparison with seed-based technology, and not easily adapted without knowledge transfer and exchange and capacity building, as well as technology adaptation in many cases (Pound 2008). As such natural resource based interventions are often local rather than global, including responding to the local policy environment, also limiting their impact (Renkow and Byerlee 2010).

Larger scale natural resource interventions, such as watershed management (Wani *et al.*, 2008), also require community action and may also involve processes around property rights and common property resources (Meinzen-Dick *et al.*, 2004). Again, these interventions require a better understanding of farmer and community livelihood strategies and a 'toolbox' of appropriate skills to facilitate the process. Roncoli *et al.* (2001) also indicate that farmers need more than information to be able to respond optimally to a forecasted climate shock. There is a need for integrating science and development interventions in ways that help improve livelihood options and the productive capacity of farming households, especially those with limited resources. Access to labor saving technologies that accelerate land preparation and planting, and timely availability of locally adapted seed varieties were some of the key elements to more effectively manage risks associated with climate variability, identified by Roncoli *et al.* (2001) in semi-arid Burkina Faso (see Box 2).

Box 2.   Case Study – Adaptation in Burkina Faso

The Sahel has long suffered from climate variability and farmers' coping and adaptation strategies to drought have been studied by many (e.g., Mortimore and Adams 1999, Roncoli *et al.*, 2001, Barbier *et al.*, 2009). In a recent study of Tougou in Burkina Faso, adaptation strategies were studied in contrasting seasons in 2004 and 2006. Tougou is an area of high population density (170 km² in 1998) and intensive land use and farmers operate in a fairly typical Sahelian context of increasing population, poor policy and enabling environment, declining soil fertility, and poorly functioning markets. The average farm size is 5 ha supporting 12 people and sorghum and millet being the main cereal crops. Farmers' strategies are aimed at increasing yield but reducing variability. Farmers adopted a wide range of low-cost strategies both for crops and animals (see below).

Strategy	Adoption (%)	Strategy	Adoption (%)
Crop management		**Animal management**	
Stone bunds	60	Bull fattening	47
Micro-water harvesting (Zai)	49	Purchased feed	4
Water harvesting (demi-lune)	6	Sorghum stover	54
Soil restoration	49	More animals	13
Row planting	30	Hay	48
Improved seed	49	More milk production	4
Plow	46		
Draft animals	25	**Preferred adaptation**	
Weeder	10	Animal sale	82
Mineral fertilizer	21	Less meals	56
Coralling	42	Diversification, improved seed	32
Manure	41	Change of grazing areas for cattle herds	15
Compost	56	Other activities (gold mining, trade …)	10
Lowland production	51	Less food	70
Vegetable production	61	Waiting for irrigation during the dry season	44
Fertilization of vegetables	59	Migrate to other regions	20
Crop insurance	0	Temporary migration	12
		More fertilization (organic matter, inorganic fertilizers)	6

Many of these strategies contributed to intensification of production, especially crop/livestock systems, as well as reducing variability. Diversification into vegetable production was also important where access to irrigation water was possible. When asked about future strategies in the event of another drought, selling animals would be the most important strategy, followed by eating less meals and consuming less food.

13.5.2 Adaptation strategies

Adaptation strategies often contain both social and technical elements that sometimes act independent of each other and at other times interact. Among social adaptation strategies are maximization of family labor use, including generating remittances from temporary or permanent migration; diversification into non-agricultural enterprises; deployment of social protection schemes and employment schemes; crop and livestock insurance; and realization of collective action and community-based empowerment efforts.

Resilience, in the context of the social elements mentioned above, is strongly associated with diversification of income-generating opportunities that reduce exposure to livelihoods shocks from climatic and non-climatic stressors. Long-term village-level studies in India (Walker and Ryan 1990) have shown that incomes have been diversified over time in response to long-term changes in climate and other changes in agricultural policies and markets, and that agricultural production constitutes a smaller proportion of livelihood than previously (see Box 3). Off-farm and non-farm income sources,

---

Box 3. Case Study – Adaptation and Coping in India's SAT

ICRISAT initiated a series of long-term village-level studies (VLS) in 1975 in Andhra Pradesh and Maharashtra which provide many insights into coping and adaptation (Walker and Ryan 1990; Bantilan and Anupama 2006). Farmers report that rainfall has become more uneven with more frequent drought years and declining groundwater levels. Mean temperatures have indeed increased slightly (by about 0.7°C) and number of rainy days decreased. The incidence of extreme temperature events has not changed significantly. Over time, there have been adaptations at:

- Farm level: change in cropping patterns, adoption of shorter duration cultivars, diversification away from staple cereals (millet, sorghum) to higher value non-cereal crops
- Institutional level: diversification of agricultural income sources (livestock and dairy, vegetables), more formal credit/lending institutions, rural employment schemes, food security systems
- Technological level: micro-irrigation, rainwater harvesting
- Social level: increase self help groups (SHGs), diversification to non-agricultural sources of income, seasonal and permanent outmigration

Among social wealth classes, adaptation responses also vary:

Household	Adaptation strategy
Landless	Seasonal migration, Government employment scheme (in some states)
Marginal	Work as laborer, lending money, some seasonal migration
Medium	Lending money, selling of limited stocks
Large	Using savings, reducing expenses, selling of stock, investing in dairy, irrigation

The most preferred short-term strategies are reducing household expenditure and food intake, selling some assets, and changing planting dates. Selling livestock, changing cropping patterns, or introducing new crops are less preferred.

including migration are now much more important than was previously the case. Likewise village-level studies in SSA have shown that in severe droughts farm families increase the number and type of off-farm income-generating activities (Mortimore and Adams 1989). Rural livelihood programs [e.g., Andhra Pradesh Rural Livelihoods Programme (APRLP), India] also promote non-agricultural livelihoods as a core part of their strategy to cope in drought-prone environments.

Diversification may include greater crop and livestock integration and in many cases intensification or specialization, including dairy production. Small ruminants commonly replace large ruminants. Diversification also takes place into market gardens and vegetable and fruit production, and where livestock are important into fodder production. High-value fruits and vegetable production are associated strongly with market access and demand, and this may be facilitated where small-scale irrigation is required (Wani *et al.*, 2008). Small-scale market gardens are an important entry point for women in particular and this approach has been successfully used in West Africa.

Increasingly important for adaptive capacity will be efforts to bolster support systems provided by government and civil society organizations (CSOs) and to nurture community-based efforts to develop or strengthen local-level support systems. These may include national schemes such as employment guarantee schemes or social protection schemes, or local schemes implemented by CSOs through, for example, drought relief programs. For these schemes to be implemented effectively, and better linked to agriculture, greater capacity building of local organizations is required. Often 'products' are delivered to meet targets rather than in the best interests of the target populations. A counterbalance to these top-down schemes are the myriad autonomous forms of support organized at the local level, as described by Agrawal and Perrin (2008).

A major area for investment to help farmers adapt (and cope) with climate variability is seasonal forecasting. The science and delivery of seasonal forecasting is still in its infancy but has considerable potential, especially for taking advantage of better than average years and not just ameliorating poorer than average years (Meza *et al.*, 2008). While considerable uncertainties remain in the forecast itself, more attention is needed on how to deliver these forecasts to farmers and indeed to local government and CSOs that interface with farmers. This is because a forecast, however accurate, is useless without the options being understood and available to farmers, e.g., the availability of seeds of a shorter or longer duration cultivar for a below or above average season, respectively. The primary constraints to realizing the full potential of seasonal climate forecasts include: lack of specificity of the forecasts with respect to end-user needs and inadequate coordination between forecasters and end-users; poor communication and interpretation of forecasts; and inability of farmers to act on forecasts (Vogel and O'Brien 2006; Archer *et al.*, 2007; Patt *et al.*, 2007).

Lastly, all of the above, and many of the technical options in Table 13.4, require much greater investment in capacity building among communities, individuals, and supporting institutions, and a greater orientation of organizations involved in technology delivery towards participatory and collective or community action programs. Participatory extension has begun to take hold over the last several years, and has led to more responsive service delivery by introducing new technologies and the means to empower technology uptake and innovation by farmers (reviewed by Padgham 2009). Supporting expansion of the participatory extension model could aid adaptation efforts by promoting joint learning and the communication and sharing of knowledge among farmers. For example, Thomas *et al.* (2005) found that support for group visits and

Table 13.4 Some examples of technology-based adaptation options

Intervention	Example
Change resource allocation between fields	Fallow, abandon outer fields; concentrate effort on inner fields
Rainwater harvesting	Zai pits/planting basins; demi-lunes; bunds (rock, earth), small tanks/pits; small dams
Supplementary irrigation	Drip irrigation
Conservation-effective practices	Minimum tillage; mulching; semi-permanent ground cover
Sowing dates	Earlier, staggered
Crop and livestock species or cultivars	Drought tolerant species and cultivars
Cropping system diversification and agroforestry	Intercropping and within farm diversification; greater use of tree products
Good agricultural practice	Integrated soil fertility management; integrated pest management; weeding strategies; fertilizer strategy
Livestock management	Number; type; grazing and feeding strategies/feeding (kraaling); crop/livestock integration
Agricultural enterprise diversification	Market gardening; fruit and other trees; dairy; payment for ecosystem services
Seasonal forecasting	Change crop management practices such as cultivar, sowing time, and plant density
Crop insurance	Compensation for drought failure of crops

farmer-to-farmer exchange networks were an effective and low cost means for relaying adaptation-relevant knowledge and information.

There are many technical options that can enhance climate risk management and promote adaptation; some examples are listed in Table 13.4. Cooper *et al.* (2006) suggested that such interventions could be grouped by the timing of the decision, namely, prior to the season (ex-ante), within the season, and after the season (ex-post). Pre-season options may include investing in water conservation technologies (e.g., digging zai pits) or choosing drought tolerant (short season) crop varieties. In-season options (response farming) include adjustments to crop and livestock management in response to weather, and may include abandoning outer fields and concentrating on the home field, or not applying fertilizer to conserve cash or avoid debt, or the converse, applying fertilizer when seasonal forecasts or other decision parameters are favorable. At the end of the season, farmers may make decisions that attempt to either reduce the negative effects of, or in some cases exploit, production outcomes. These post-season actions including such actions as sale of assets and temporary migration for wage labor are often used to protect livelihoods and compensate for insufficient food production. Understanding the complexities of household decision-making at different time periods in agricultural cycles is therefore critical when developing adaptation strategies.

13.6 CONCLUSIONS

Farmers have evolved many coping and adaptation strategies in the face of climate variability and other factors affecting their livelihoods. Indeed, the role of non-climatic factors such as policy, markets, and other external drivers of change should not be underestimated. Understanding the role of these factors, and promoting good enabling policies and a support for organizations that help farmers to adapt is a key component of any adaptation strategy. At the end of the day, farmers need a range of options and in many cases support to utilize those options in order to adapt.

REFERENCES

Adger, W.N. 2003. Social capital, collective action, and adaptation to climate change. *Economic Geography* 79:387–404.

Agrawal, A., and N. Perrin. 2008. Climate adaptation, local institutions, and rural livelihoods. *IFRI Working Paper No. W081-6*. Ann Arbor, Michigan, USA: University of Michigan.

Anderson, P.K., A.A. Cunningham, N.G. Patel, F.J. Morales, P.R. Epstein, and P. Daszak. 2004. Emerging infectious diseases of plants: pathogen pollution, climate change and agrotechnology drivers. *Trends in Ecology and Evolution* 19:535–544.

Archer, E., E. Mukhala, S. Walker, M. Dilley, and K. Masamvu. 2007. Sustaining agricultural production and food security in Southern Africa: an improved role for climate prediction? *Climatic Change* 83(3):287–300.

Baettig, M.B., M. Wild, and D.M. Imboden. 2007. A climate change index: where climate change may be most prominent in the 21st century. *Geophysical Research Letters* 34:L01705 (6 pages).

Bantilan, M.C.S., and K.V. Anupama. 2006. Vulnerability and adaptation in dryland agriculture in India's SAT: Experiences from ICRISAT's village-level studies. *Journal of SAT Agricultural Research*, Vol. 2. (ejournal.icrisat.org)

Boardman, J. 2006. Soil erosion science: Reflections on the limitations of current approaches. *Catena* 68:73–86.

Barbier, B., Hamma Yacouba, Harouna Karambiri, Malick Zorome, and Blaise Some. 2009. Human vulnerability to climate variability in the Sahel: Farmers' adaptation strategies in Northern Burkina Faso. *Environmental Management* 43:790–803.

Burney, J.A., J.D. Steven, and B.L. David. 2010. Greenhouse gas mitigation by agricultural intensification. *PNAS* 107:12052–12057.

Carney, D. 1998. Sustainable rural livelihoods. What contribution can we make? Presented at the Department for International Developments Natural Resources Advisers Conference, July 1998. (http://hdl.handle.net/10068/545709)

Carter, M.R., P.D. Little, T. Mogues, and W. Negatu. 2004. *Shock, sensitivity and resilience: tracking the economic impacts of environmental disaster on assets in Ethiopia and Honduras.* Wisconsin, USA: Basis.

Chambers, R., A. Pacey, and L.A. Thrupp. 1989. *Farmer first: farmer innovation and agricultural research.* London, UK: Intermediate Technology Publications Ltd.

Christensen, J.H., B. Hewitson, A. Busuioc *et al.* 2007. Regional climate projections. In *Climate change 2007: the physical science basis.* Contribution of Working Group I to the Fourth Assessment Report of the Intergovernmental Panel on Climate Change, ed. S. Solomon, D. Qin, M. Manning *et al.* Cambridge, UK: Cambridge University Press.

Conway G. 2008. *The science of climate change in Africa: impacts and adaptation.* UK: Department of International Development.

Cooper, P.J.M, J. Dimes, K.P.C. Rao *et al.* 2008. Coping better with current climate variability in the rain-fed farming systems of sub-Saharan Africa: an essential first step in adapting to future climate change. *Agriculture, Ecosystems and Environment* 126:24–35.

Cooper, P.J.M., J. Dimes, K.P.C. Rao, B. Shapiro, B. Shiferaw, and S.J. Twomlow. 2006. Coping better with current climatic variability in the rainfed farming systems of sub-Saharan Africa:

A dress rehearsal for adapting to future climate change? *Global Theme on Agroecosystems Report No. 27.* Nairobi, Kenya: International Crops Research Institute for the Semi-Arid Tropics.

Cooper, P., K.P.C. Rao, P. Singh *et al.* 2009. Farming with current and future climate risk: Advancing a "Hypothesis of Hope" for rainfed agriculture in the semi-arid tropics. *Journal of SAT Agricultural Research,* Vol. 7. (ejournal.icrisat.org)

Cornford, S.G. 2003. The socio-economic impacts of weather events in 2002. *WMO Bulletin* 52:269–290.

de Wit, M., and J. Stankiewicz. 2006. Changes in surface water supply across Africa with predicted climate change. *Science* 311(5769):1917–1921.

Easterling, W., and P.K. Aggarwal. 2007. Food, fibre and forest products. In *Climate change 2007: impacts, adaptation and vulnerability.* Contribution of Working Group II to the Fourth Assessment Report of the Intergovernmental Panel on Climate Change, ed. M.L. Parry, O.F. Canzianai, J.P. Palutikof, P.J. van der Linden, and C.E. Hanson, 273–313. Cambridge, UK: Cambridge University Press.

Funk, C., M.D. Dettinger, J.C. Michaelsen *et al.* 2008. Warming of the Indian Ocean threatens eastern and southern African food security but could be mitigated by agricultural development. *Proceedings of the National Academy of Sciences* 105(32):11081–11086.

Gunawardena, T.A., S. Fukai, and F.P.C. Blamey. 2003. Low temperature induced spikelet sterility in rice. I. Nitrogen fertilisation and sensitive reproductive period. *Australian Journal of Experimental Agriculture* 54:937–946.

Hall, A., L. Mytelka, and B. Oyeyinka. 2005. Innovation systems: implications for agricultural policy and practice. *ILCA Brief 2.* Addis Ababa: ILCA.

Hyman, G., S. Fujisaka, P. Jones *et al.* 2008. Strategic approaches to targeting technology generation: Assessing the coincidence of poverty and drought-prone crop production. *Agricultural Systems* 98:50–61.

ILRI. 2006. *Mapping climate vulnerability and poverty in Africa.* Report to Department for International Development. Nairobi, Kenya: International Livestock Research Institute.

IPCC. 2007. Summary for policymakers. In *Climate change 2007: impacts, adaptation and vulnerability.* Contribution of Working Group II to the 4th assessment report of the Intergovernmental Panel on Climate Change (IPCC), ed. M.L. Parry, O.F. Canziani, J.P. Palutikof, P.J. van der Linden, and C.E. Hanson. Cambridge, UK: Cambridge University Press.

Jagadish, S.V.K., P.Q. Craufurd, and T.R. Wheeler. 2007. High temperature stress and spikelet fertility in rice (*Oryza sativa* L.). *Journal of Exprimental Botany* 58:1627–1635.

Kandlikar, M., and J. Risbey. 2000. Agricultural impacts of climate change: if adaptation is the answer, what is the question? *Climatic Change* 45:529–539.

Levina, E., and D. Tirpak. 2006. *Adaptation to climate change: key terms.* Paris, France: Organisation for Economic Cooperation and Development.

Lobell, D.B., M. Burke, C. Tebaldi, M. Mastrandrea, W. Falcon, and R. Naylor. 2008a. Policy brief: Prioritizing climate change adaptation needs for food security to 2030. Stanford University, Program on Food Security and the Environment.

Lobell, D.B., M. Burke, C. Tebaldi, M. Mastrandrea, W. Falcon, and R. Naylor. 2008b. Prioritizing climate change adaptation needs for food security in 2030. *Science* 319:607–610.

Long, S.P., E.A. Ainsworth, A.D.B. Leakey, J. Nosberger, and D.R. Ort. 2006. Food for thought: Lower-than-expected crop yield stimulation with rising CO_2 concentrations. *Science* 312:1918–1921.

Long, S.P., E.A. Ainsworth, A.D.B. Leakey, D.R. Ort, J. Nosberger, and D. Schimel. 2007. Crop models, CO_2, and climate change - Response. *Science* 315:459–460.

Magrin, G., C. Gay García, D. Cruz Choque *et al.* 2007. Latin America. In *Climate change 2007: impacts, adaptation and vulnerability.* Contribution of Working Group II to the Fourth Assessment Report of the Intergovernmental Panel on Climate Change, ed. M.L. Parry,

O.F. Canziani, J.P. Palutikof, P.J. van der Linden, and C.E. Hanson, 581–615. Cambridge, UK: Cambridge University Press.

Mall, R.K., A. Gupta, R. Singh, R.S. Singh and L.S. Rathore. 2006a. Water resources and climate change: an Indian perspective. *Current Science* 90(12):1610–1626.

Mall, R.K., R. Singh, A. Gupta, G. Srinivasan, and L.S. Rathore. 2006b. Impact of climate change on Indian Agriculture: a review. *Climatic Change* 78:445–478.

Matsui, T., O.S. Namuco, L.H. Ziska, and T. Horie. 1997. Effect of high temperature and CO_2 concentration on spikelet fertility in Indica rice. *Field Crops Research* 51:213–219.

Meinzen-Dick, R., M. DiGregorio, and N. McCarthy. 2004. Methods for studying collective action in rural development. *Agricultural Systems* 82:197–214.

Meza, J. Francisco., J.W. Hansen, and D.Osgood. 2008. Economic value of seasonal climate forecasts for agriculture: review of ex-ante assessments and recommendations for future research. *Journal of Applied Meteorology* 47:1269–1286.

Mortimore, M. 2010. Adapting to drought in the Sahel: lessons for climate change. *WIREs Climate Change* 1:134–143.

Mortimore, M., and W.M. Adams. 1999. *Working the Sahel: environment and society in northern Nigeria*. London, UK: Routledge.

Nederlof, E.S., and C. Dangbégnon. 2007. Lessons for farmer-oriented research: Experiences from a West African soil fertility management project. *Agriculture and Human Values* 24:369–387.

Nelson, G., M. Rosegrant, J. Koo *et al.* 2009. *Climate change: impact on agriculture and costs of adaptation*. USA: International Food Policy Research Institute.

Ortiz, R., K.D. Sayre, B. Govaerts *et al.* 2008. Climate change: Can wheat beat the heat? *Agriculture, Ecosystems and Environment* 126:46–58.

Padgham, J. 2009. *Agricultural development under a changing climate: Opportunities and challenges for adaptation*. Washington, DC, USA: World Bank.

Parry, M., A. Evans, M.W. Rosegrant, and T. Wheeler. 2009. *Climate change and hunger responding to the challenge*. World Food Programme. Rome, Italy: C.G. Viola; 68–70.

Parry, M., J. Palutikof, C. Hanson, and J. Lowe. 2008. Squaring up to reality. *Nature Reports Climate Change* 2:1–3.

Patt, A.G., L. Ogallo, and M. Hellmuth. 2007. Learning from 10 years of Climate Outlook Forums in Africa. *Science* 318:49–50.

Pedzisa, T., I. Minde, and S. Twomlow. 2010. An evaluation of the use of participatory processes in wide-scale dissemination of research in micro dosing and conservation agriculture in Zimbabwe. *Research Evaluation* 302:1912–1914.

Peng, S.B, J.L. Huang, J.E. Sheehy *et al.* 2004. Rice yields decline with higher night temperature from global warming. *Proceedings of the National Academy of Sciences* 101:9971–9975.

Prasad, P.V.V., K.J. Boote, and L.H. Allen Jr. 2006. Adverse high temperature effects on pollen viability, seed-set, seed yield and harvest index of grain-sorghum [*Sorghum bicolor* (L.) Moench] are more severe at elevated carbon dioxide due to higher tissue temperatures. *Agricultural and Forest Meteorology* 139:237–251.

Pretty, J. 2008. Social capital and the collective management of resources. *Journal of Applied Meteorology* 47(5).

Pound, B. 2008. Livelihoods and rural innovation. In *Agricultural systems: agroecology and rural innovation for development*, ed. S. Snapp, and B. Pound, 27–52. London, UK: Academic Press.

Renkow, M., and D. Byerlee. 2010. The impacts of CGIAR research: A review of recent evidence. Food Policy. doi:10.1016/j.foodpol.2010.04.006

Roncoli, C., K. Ingram, and P. Kirshen. 2001. The costs and risks of coping with drought: livelihood impacts and farmers' responses in Burkina Faso. *Climate Research* 19:119–132.

Scoones. 1998. Sustainable rural livelihoods: A framework for analysis. *Working Paper 72.* Scoones-Brighton, Institute of Development Studies.

St-Pierre, N.R., B. Cobanov, and G. Schnitkey. 2003. Economic losses from heat-stress by US livestock industries. *Journal of Dairy Science* 86(E. Suppl.):E52–E77.

Thomas, D., H. Osbahr, C. Twyman, N. Adger, and B. Hewitson. 2005. Adaptive: Adaptations to climate change amongst natural resource-dependant societies in the developing world: across the Southern African climate gradient. *Tyndall Centre for Climate Change Research Technical Report 35.*

Thornton, P.K., P.G. Jones, T.M. Owiyo, R.L. Krusha *et al.* 2006. *Mapping climate vulnerability and poverty in Africa.* Report to the Department for International Development. Nairobi, Kenya: International Livestock Research Institute.

Thornton, P.K., J. van de Steeg, A. Notenbaert, and M. Herrero. 2009. The impacts of climate change on livestock and livestock systems in developing countries: a review of what we know and what we need to know. *Agricultural Systems* 101(3):113–127.

Trenberth, K.E., P.D. Jones, P. Ambenje *et al.* 2007. Observations: surface and atmospheric climate change. In *Climate change 2007: The physical science basis.* Contribution of Working Group I to the Fourth Assessment Report of the Intergovernmental Panel on Climate Change, ed. S.D. Solomon, M. Qin, Z. Manning *et al.* Cambridge, UK: Cambridge University Press.

USDA/FAS. 2008. http://www.pecad.fas.usda.gov/highlights/2008/05/Burma_Cyclone_Nargis_Rice_Impact.html (Accessed 20 October 2010.)

Vogel, C., and K. O'Brien. 2006. Who can eat information? Examining the effectiveness of seasonal climate forecasts and regional climate-risk management strategies. *Climate Research* 33:111–122.

Walker, T.S., and J.G. Ryan. 1990. *Village and household economies in India's semi-arid tropics.* Baltimore, USA: The Johns Hopkins University Press.

Wani, S.P., T.K. Sreedevi, T.S. Vamsidhar Reddy, B. Venkateshvarlu, and C. Shambhu Prasad. 2008. Community watersheds for improved livelihoods through consortium approach in drought prone rainfed areas. *Journal of Hydrological Research and Development* 23:55–77.

West, J.W., B.G. Mullinix, and J.K. Berrnard. 2003. Effects of hot, humid weather on milk temperature, dry matter intake, and milk yield of lactating dairy cows. *Journal of Dairy Science* 86:232–242.

Wheeler, T.R., P.Q. Craufurd, R.H. Ellis, J.R. Porter, and P.V. Vara Prasad. 2000. Temperature variability and the yield of annual crops. *Agriculture, Ecosystems and Environment* 82:159–167.

World Bank. 2008. *Agriculture for development.* Washington, DC, USA: World Bank.

Index

Page numbers in **bold** refer to figures and tables

Agricultural productivity 3, 27, 47, 74, 82, 132, 147, 186, **194**, 200, 205, 207, 249, 252, 254, 256, 273, 281–284, 297, 301, 306, 319, 332, **360**, 391, 421, 428

Agroclimatic zone 163, **250**, **317–318**, 342, 355, 361, 384

Agroecosystems **38**, 47, 283, 297, 320, 326, 391–396, **398**, 401, 402, **404**, 405, **406**, 408, 410–413, 416, 418

Agroforestry 12, 30, 187, **360**, **436**

Animal husbandry 62, 72, **119**, 142–3

Aquaculture 17–18, 26, 27, 29, 72

Aroecological characteristics 56

Balanced nutrient management 175, 281–2, 285, 292, **293–297**, 300, 316

Bio-diesel 12, 20–21, 59, 76, 77, 80–1

Biodiversity 2, 23–4, 37, 53, 79, 81, 130, 392, 413, 430–1

Biodiversity conservation 24

Bioeconometric models 57

Biofertilizers 69, 73

Biogas plants 17, 71–2

Biopesticides 59, 73

Biophysical interventions 82

Blue water 42, **43**, 44, 46–7, 53, 265, 276

Capacity building 2, 13, 48, 57, 69, 72–73, 78, 98, 114, 132, 136, 141, 146, 150–2, 154, 240, 242, 271, 275, 303, 306, **320**, 342–3, 352, 432, 435

Carbon sequestration 1, 24, 81, 163, 198, 396, **403**

Catchment 46–47, 53–54, **134**, 137, 234, **235**, 239, 249, 255, **258**, 259, 261, 263, **271**, 386, 397, 400–1, 412

Client oriented breeding 18

Climate change 3, 6, 10, 21–22, 24, 28, 37–39, 44, 56, 60, 82, 152, 163–164, 251, 320, 342, 392, 408, 416, 421, **422**, **423**, 424, 425–8, **429**, 430, 431, 432

  adaptation 2, 10, 20, 26, 211, 273, 341, 364, 391, **399**, 421, 430–7

  impacts on agriculture 421, **424**

    crop and livestock production 424, 430

    malnutrition 25, 39, 53, 283, 421, 426, 428, 430

    poverty 4, 12, 74, 80, 82, 88, 263, 349, 426, 430

    prices 131, 144–146, 232, 337, 365–366, 421, 426, 428

    water resources 10, 12, 22, 35, 42, 44, 46, 56, 62, 87, 89, 95, 96, 102–3, 111, 114, 118, 129–130–131, 161, 163, 175, 193, 197, 205–206, 249, 251, 274, 283, 315, 341, 378, 382, 392, 400, 413, 421, 424–5

  regional impacts 426

    South Asia 35, 38–40, 42, 44, 172–173, **174**, 315, 421

    sub-Saharan Africa 35, **37–40**, 42–45, 249, 283, 315, 393, 421, 426, 429, **430**

  risk management 2, 175, 341, 436

Climate variability 183, 408, 410, 421, 426, 430, 432, **433**, 435–6

Common property resource 15, 102, 110, 357

Community empowerment 21, 56, 62

Community participation 33, 56, 62–63, 78, 82, 92, 98, 110, 130, **134**, 145, 147, 353, 355, 381
Community resource centers 11
Comprehensive assessment 28, 37, 40, 47, 129, 136, 145, 152, 207, 315, 318
Conservation agriculture 29, 132, 253, 266–267, **319–20**, 324, 339–340
Crop breeding 18, 25, 28, 340
Crop diversification 70, 80, **319**, 330, **331**, 337, **360**, 367, **368**, 379, 382
Crop intensification 70, 80, 173, 198, 316, **319**, 336
Crop pest control 20, 69
Crop productivity 35, 45, 61, 64–5, 68–9, 79, 88, 136, 197, 205–206, 237, 243, 253, 281, 283, 292, 298, 315, 317, **320**, **323**, 343, 379, 413
Crop quality 295, 297, 306
Crop yield 39, **40**, 45, **66**, 70, **72**, 131, 164, 185, 211–212, 216, 218, 221, 223, 228–9, **236**, 240, 253, 264, 267–8, 284, 302, 317–320, 327, 336, 339, 354, 377, 400, 402, 414, 421, 428
Crop-growth simulation model 163
Cropping intensity 70, 136, 168, 173, 186, 198, 222, 231, 338, **360**, 367, **368**, 378–9

Decision-making 9, 22, 87, 98, 110, 113, **119**, 120–21, 130, 141, 147, 165, 179, 199, 271, **360**, 372, 436
Drip irrigation systems 17, 328
Drought 1, 3–4, 6, 22, 28, 39, 53, 60–61, 102, 131, 134, **138**, 144, 159, 163–4, 166, 183, 185, 189, **190**, 207, **208**, 234, 251, 255, 262, 269–270, 298, 315, 319, **320**, 334, 338, 341, 350, 363, 376, 399, 401, 404, 405, **406**, 407, 410, 412–414, 416, 422, 424–**425**, 426, 433, 434–435, **436**

Ecosystem services 82, 90, 205, 391–393, **396**, 397, 400–3, **404**, 405, **409**, 410–13, 417, **436**
Environment 1–3, 6, 14, 19, 23, 27, 30, 35, 44, 46, 69, 79, 81, 111, 130, 132, 134–135, 144, 154, 198, 324, 331, 335, 339, 354, 371
Equity 3, 7, 39, 60, 62, 72, 77, 79, 87–93, 97–100, 102–103, 106, 108–111,

122–4, 130, **133**, 136, **140**, 144, 150, 153, 325, 341, 353, 356, 410
Evapotranspiration 43, 57, 180, 183–**184**, 201, 253, 273, 316, 408, 410

Farming systems 26, 40, **44**, 59, 60, 64, 79, 87, 153, 205, 241, **325**, 339, 391–2, 395, 401–3, 405, 407, 413, 415, 417
Food security 2, 35, 44, 47–8, 53, 62, 129, 132, 173, 175, 200, 252, 261, 267, 271, 328, 330, 342, 343, **360**, 372, 378, **399**, 403, 424–25, **434**
Forestry 3, 37, 57, 101, **119**, 122, 131, **134**, 179, 370

Geographical information system (GIS) 29, 57, 159, 160, 303
Green water 42, 44, 46–47, 53, 255, 261–263, 272
Greenhouse gas 426–427
Groundwater 2, 6, 15, 21–22, 29, 42, 53, 57, **58**, 61, 68, **73**, 79, 81, 87–89, 94, 97, 116, 131–2, 135, 144, 177, 180, 184, 189, 192, 197, 198, 207, **208**, 231, **232**, **238**, 239–240, 273, **319**, **324–6**, 342, 350, **360**, 364, 367, 373, 385, 392, 397, **398**, 400, **403**, **406**, 417, 427, 434

Horticulture 12, 27, 62, 72, 101, **119**, 142, 187, 189, 192, 381
Human rights 6

Impact assessment 108, 115, 139, 152, 163, 186, 188, 355–358, 362, 365, 371, 377, 380, 385
  approaches 356, 359, 361, 372, 378, 418
  discount rate 240, 356, 358, 370
  methods 356
  selection of indicators 356–357
Improved cultivars 45, 59, 69–70, 240, 296
Improved livelihoods 46, 80, 150, 201, 242, 272, 397, 404
Income-generating activities 9, 11, 16, 72, 80, 82, 87–88, 91, 98, 108, 115, 146–7, 435
Integrated nutrient management 69, 80, **319**, 372, 397, **398**
Integrated pest management 27, 59, 69, 335, 372, 397, **436**

Integrated Watershed Management 1, 32–33, 36, 46, 48, 50–51, 56, 75, 79, 82, 98, 101, 129, 141, 147, 151, 160, 207, 243, 246, 248, 270, 273, 282, 284, 346, 372

Land degradation 24, 36–37, 45, 48, 53, 57, 62, 68, 82, 129, 159, 198, 206, 242, 244, 340, 342, 349, 421, 426
Landscape ecosystem services 392–393, 403, 405
Livelihood system 18
Livestock 2, 5, 9, 11–12, 18–19, 21, 25–26, 56, **58**, 71, 75, 78, 80, 91, **105**, 108, 120, 122–3, 130–31, **134**, 146, 149, 152, 219, 249–50, 255–6, 258, 269, **270**, 272, 354, 367, **368**, 373, 377, 386, 391–92, **396–99**, 400, 402, 406, 410, 416, 421, 424, 428, 430, **434**, **436**

Macronutrients 45, 67, 68
Major nutrients 20, 25, 281–282, 284, 292, 299, 305
Micronutrients–12, 20, 25, 45, 59, 67, 69, 77, 175, 281–282, 285, 292, 299, 302, 305, 332–34
Micro-watershed 54, 88, 104, 108, 119, 138, 162–163, **194**
Migration 2, 4–5, 7–10, 14, **58**, 59, 105, 118, 354, 378, 380, **433**, 434–36
Multi-nutrient deficiencies 334

Natural resource 3, 30, 38, 47, 57, 60, 62, 78, 98, 110, 132, 137, **138**, 148, 150, 159, 206, 207, 243, 253, 282, 319, 354, 358, 371–72, 376, 432
Natural resource management 62, 98, 110, 129, 137, 282, 319, 353, 372, 431
New Science Tools 78–79, 132, 149, 159–160, 200–201
    automatic weather station 166, 372
    data communication devices 165
    digital elevation models 162, 177
    field sensors 165, 199
    geographical information system (GIS) 29, 57, 159, 160, 303
    global positioning system- 166
    mobile devices 166–167

remote sensing 24, **58**, 60, 64, 78, 160, 161, **162**, 168, 171, 173, 179, 186–187
    satellite images 162, 174
    spatial simulation modeling 197
    spatial water balance modeling 180
Nutrients
    deficiencies 282, 284, 292, 297, 299, 306, 334
    diagnosis 282, 284, 306
    management 62, 68–70, 80, 175, 281, 285, 292, **293–96**, 300–306, 316, **319**, 338, 340
    Status 282–283, 376

On-farm research 235, 282, 302
Organic fertilizers 17, 306

Participatory crop selection 4
Participatory research and development 68–69, 172, 302
Participatory rural appraisal 14, 110, 137, 302
Participatory varietal selection 18
Population growth 3, 6, 38, 152, 205, 373
Production systems 79–80, 119, 163, 167, 169, 242, 281–284, 292, 300, 303, 306, 342, 413

Rainfall management **319**, 320
    broad-bed and furrow 68, 172, 215, **216**, 217, **218**, 321
    conservation agriculture 29, 132, 253, 266–267, **319–320**, 324, 339–340
    groundwater recharge 21, 68, 131, 184, **209**, **238**, 324, 367, 382, 385, 427
    land surface 214, 319, 321, **328**
    soil conservation -10, 14, 18, 59, 61, 80, 99, 101, 130, 146–147, 160, 207, 221, 239–40, 241, 264, 350, 378, 382
    water conservation 2, 18, 21, 24, 56, 59, 62, 68–69, 71–73, 80, 95, 99–101, **119**, 130, **134**, 144–145, 147, 152, 179, 205, 207, **208**, 209, 212–213, 221–222, 231, **238**, 239, 241-43, 255, **263**, 268, 270, 282, 300, 301, **319**, 321, 335, 350–351, 353, **360**, **372**, 376, 378, 397, **398**, 401, 436

Rainfall management (*continued*)
  water harvesting 2, 18, 29, 79–81, 96,
    98–99, 104, 106, 116, 132, 180, 184,
    197, **208**, **232**, 233, **234**, 235, **238**,
    240, 243, 255, 256, **257**, 261–263,
    271–272, 316–318, 321, 324, **325**,
    326, 335, 350, 364, 382, 385, **433**
Rainfall use efficiency 206, 298, **299, 300,**
    316, **318, 322, 333, 337**
Rainfall variability 2, **38**, 421
Rainfed agriculture 1, 2, 36, 38–40, 42, 45,
    47–48, 78, 129, 148, 167, 201, 207,
    231, 234, 239, 242, 253, 287, 315,
    318, **319**, 335, 336, **340**, 342, 349,
    392, 421, 426
Rainfed farming 2–3, 7, 129, 205, 235, 399,
    407
Rainwater harvest 61, 71, 130, **133**, 147,
    183, **238**, 240, 249, **252**, 271, 273,
    357, **436**
  in-situ conservation 79, 81, 240, 249, 324
Rainwater harvesting 61, 71, 130, **133**, 147,
    183, **238**, 240, 249, **252**, 271,
    273–74, 357, **436**
  blue water 42, **43**, 44, 46–47, 53, 265,
    276
  earth dams 255, 259, 270
  green water 42, 44, 46–7, 53, 255,
    261–263, 272
  micro irrigation 255, 270, **434**
  rooftop 255, 258–259
  runoff storage **238**
  sand dams 255, 274
  socioeconomic issues 255, 270
  surface runoff 137, 206, 234, 253–255,
    262, 273
  weirs 22, 255, 259, 274
Remote sensing 24, **58**, 60, 64, 78, 160,
    161, **162**, 168, 171, 173, 179, 186–87
Renewable energy 10–11, 18, 20, 24, 29
Resilience framing 391–93, 405, 409, 413,
    415, 417–18
  ecosystem services 82, 90, 205, 391–393,
    396, 397, 400–403, **404**, 405, **409**,
    410–11, 413, 417, **436**
  management entry points 391, 415
  provisioning services 396, **399**
  regulating services 400, **402–403**, 413
  supporting services
Resource management 46, 62, 98, 110, 118,
    129, 132, 135, 137, 146, 160, 163,

166, 195, 282, 316, 319, 353, 372,
    431
Risk management 2, 175, 341, 436
Runoff 19, 22, 46, 53, 57, 60, 68, 71, 73,
    79, 81, 130, 134, 137, 144, 169, 177,
    181, 183, 199, 205, 206, **208**,
    212–214, **215**, **218**, 219, 221–223,
    224, 225, **227–29**, 230–231, 234,
    238, 240, 249, 254–55, **256**,
    259–160, **261**, 262, 265, **271**,
    273–274, 276, 334–335, 367, 378,
    397
Rural development 1, 4, 16, 27, 30, 56, 57,
    91, 94, 101, 114, 135, 138, 141, 148,
    240, 352–353
Rural livelihood 198, 435
Rural poverty 4, 12, 74, 80, 82, 88, 263,
    349

Secondary nutrients 67, 80, 218, 281–282,
    285, 305, 332
Seed priming 17, 19, 21, 174–75, 200
Semi-arid tropics 1–2, 31–32, 39, 45, 49, 59,
    80, 98, 132, 159, 213, 216, 221, 228,
    282, 327, 372, 421
Silt yield index 57
Simulation models 57, 163–64, 180,
    198–99, 212
Smallholder farming 44, 153, 391–93, 403,
    405, 408–10, 414–18
Soil conservation 10, 14, 18, 59, 61, 80, 99,
    101, 130, 146–147, 160, 207, 221,
    239–40, 241, 264, 350, 378, 382
Soil degradation 45, 130, 132, 238, 283,
    350, 392, 404, 407, **408**, 414, 417
Soil erosion 2, 3, 23, 25, 56, 71, 87, **105**,
    130, **133–134**, 136, 159, 179, 205,
    209, 221–22, 225, 239, 242,
    263–265, 273, 321, 324, 340, 364,
    372–73, 376, 378, 392, **398**, 401,
    406, 418, 421
Soil fertility 2, 7, 11, 18, 20, 23, 25–26, 45,
    59, 69, 72, 152, 159, 235, 239, 242,
    251, 261, 266, 268–269, 281,
    283–84, 292, 297–98, 301, 305–6
Soil fertility management 2, 235, 261, 268,
    282, 292, 298, 306, 323, 332, **436**,
    372, 397, **398**
Soil health 45, 68, 272, 295, 298, 301, 392,
    399, 402

Soil organic matter 45, 283–84, 297, 324, 333, 340

Soil testing 12, 20, 282, 301, 303, 305–306

Soil water conservation 144, 147, 401

broad-bed and furrows -68, 172, 215, **216**, 217, **218**, 321

compartmental bunding 224–25, **238**, 242

conservation furrows 208, 209, **211–12**, **238**, 242, 322, **323**

contour bunding 61, **107**, 130, 132, 189, 221–22, 242, 322, 381

contour cultivation 208, 209, **210–12**, 224–225, 242

field bunding 224, **238**, 242

graded bunding 137, 223, 242

indigenous practices 238

in-situ conservation 79, 81, 240, 249, 324

pitting 213, 238, **239**

runoff harvesting 231, 255, 260–262, 273, 276

scoops 68, **208**, **213–215**, 242

tied ridges 208, 212–215, 242, **267**

vegetative barriers 189, 225, 227, 242, 263

zero tillage 229, **230**, 243, 267

Soil water uptake 331

crop agronomy 319, 331, 337

crop diversification 70, 80, **319**, 330, **331**, 337, **360**, 367, **368**, 379, 382

crop intensification 70, 80, 173, 198, 316, **319**, 336

crop protection 319, 323, 334, 335

crop varieties 18, 27, 62, 153, 187, 243, 334, 340, 436

water conservation 2, 18, 21, 24, 56, 59, 62, 68–69, 71–73, 80, 95, 99–101, **119**, 130, **134**, 144–145, 147, 152, 179, 205, 207, **208**, 209, 212–213, 221–222, 231, **238**, 239, 241-243, 255, **263**, 268, 270, 282, 300, 301, **319**, **321**, 335, 350–351, 353, **360**, **372**, 376, 378, 397, **398**, 401, 436

Spatial resolution 162, 171, 186

Spatial technologies 159, 167

Spatial water balance 180

Stream order 54, **55**

Subsidies 7, 17, 92, 98, 103, 107–8, 121, 144–46, 148, 220, 242, 268, 354, 416

Supplemental irrigation 29, 70, 79, 81, 231, 233–34, **235–37**, 239, 243, 249, 254–56, **257**

262, 265, **266**, 272, 298, 325–26, **327–29**, 330–**31**, 414

Supplementary irrigation 2, 21, 256, 273, 274, **436**

Sustainability 4, 16, 17–18, 28, 35, 37, 39, 54, 56, 60, 62, 75, 89, 91, 106, 109, 111, 120, 123, 130, 141, 143–145, 147, 150–151, **194**, 198, 240, 243, 252, 283–284, 324, 338, 340, 342, 353–54, 356, 361, 371–72, 376, 380, 391–92, 400, 403–04, **406**, 410, 412, 413, 418

Water availability 42, 45, 54, 71, 81, 94, 129, 133, 136, 183, 185, 226, 240, 272, 318, **319**, 320, 331, 340–341, **360**, 367, 381–382, 421, 425

Water conservation 2, 18, 21, 24, 56, 59, 62, 68–9, 71–3, 80, 95, 99–101, **119**, 130, **134**, 144–45, 147, 152, 179, 205, 207, **208**, 209, 212–213, 221–22, 231, **238**, 239, 241–43, 255, **263**, 268, 270, 282, 300, 301, **319**, **321**, 335, 350–51, 353, **360**, **372**, 376, 378, 397, **398**, 401, 436

Water harvesting 2, 18, 29, 79–81, 96, 98–9, 104, 106, 116, 132, 180, 184, 197, **208**, **232**, 233, **234**, 235, **238**, **240**, 243, 255, 256, **257**, 261–263, 271–72, 316–18, 321, 324, **325**, **326**, 335, 350, 364, 382, 385, **433**

Water harvesting structures 80–81, 98, 104, 106–107, 114, 116, **208**, 239–40, 271, 324–25, **326**, 385

Water productivity 1, 29, 42–44, 47, 234, 235, 265, 324, 328, 335, 339–41

conservation agriculture 29, 132, 253, 266–67, **319–20**, 324, 339–40

crop breeding 18, 25, 28, 340

evaporation management 339

microclimate 320, 339

mulches 22, 231, 339

water use efficiency 44–45, 123, **206**, 218, 243, 249, 315, 319, 338

Water Resources 10, 12, 22, 35, 42, 44, 46, 56, 62, 87, 89, 95, 96, 102–3, 111, 114, 118, 129–31, 161, 163, 175, 193, 197, 205–06, 249, 251, 274, 283, 315, 341, 378, 382, 392, 400, 413, 421, 424–25

Water scarcity 2, 6, 42, 48, 53, 62, 82, 89, 94, 129, 144, 206, 215, 253, 266, 320, 322, 342
Water table 3, 6, 59, **105**, 240, **360**, 367, 378, 381, 382, **399**
Water use 21, 42, 100, 116, 123, 206, 218, 243, 249, 315–17, 319–20, 340, 342, 364, 367, 384
Watershed
  approach 1, 20, 30, 60, **61**, 64, 99, 131, 132, 135, 242, 301
  comprehensive assessment 28, 37, 40, 47, 129, 136, 145, 152, 207, 315, 318
  consortium approach 63, 64, 74–8, 82, 132, 149–150, 152
  development 17, 35, 55–7, 60–64, 69, 72, 74, 78, 82, 87, 89–92, 95–100, 103, 109, 117, **119**, 120–22, 130–38, 141–48, 150–53, 159, 187, **194**, 200, 240, 243, 349–60, **362**, 363, 365, 366, 367, **369**, 370–371, 373, 378–386
  entry point activity 65, 146, 207, 240, 305
  exit strategy 17
  implementing agencies 17, 60, 63, 92, 98, 122, 140, 243, 319, 320, 342, 349, 353
  interventions 1, 16, 27, 54, 96, 365, **366**, 369
  management 21, 53, 56, 58, 63, 68–70, 74, 79, 98, 130–132, 135, 137–138, 141, 147, 149–151, 159, 165, 177, 186, 198, **199**, 276, 284, 353, 372, 391, 393, 397, 400, 411, 413, 415, 417, 418, 432
Watershed
  capacity building 2, 13, 48, 57, 69, 72–73, 78, 92, 98, 114, 132, 136, 141–42, 146, 150, 52–154, 240, 242, 272, 275, 303, 306, 320, 342–343, 352, 432, 435
  community-based organizations 62, 114, 143, 320
  concept 53, 135
  convergence 63–64, 78–79, 80, 90, 98, 102, 120, 122, 124, 135, 137, 142, 149, 167, 240
  development 17, 35, 55–7, 60–64, 69, 72, 74, 78, 82, 87, 89–92, 95–100, 103, 109, 117, **119**, 120–22, 130–38, 141–48, 150–53, 159, 187, **194**, 200,

240, 243, 349–60, **362**, 363, 365–67, **369**, 370–71, 373, 378–86
  entry point 65, 146, 207, 240, 305
  evaluation 349, 364, 377, 385–86
  multiple benefits 80, 82, 130, 262, 413
  partnership 1, 18, 60, 74, 77–8, 136, 159, 180–181, 201, 240, 251, 341, 356
  post-program intervention 60
  scaling-up/out 75
  self-help groups 9, 58, 69, 88, 143, 198
  sustainability 4, 8, 16–17, 28, 35, 37, 39, 54, 56, 60, 62, 75, 90–91, 106, 109–111, 121, 23, 130, 141, 143–45, 147, 150–53, 174, 193–194, 240, 252, 283–284, 324, 338, 340, 42, 354, 357, 361, 370–72, 380, 391–92, 400, 403–04, 406, 410, 412–13, 418
  team building 74, 78
Watershed development 17, 35, 55–57, 60–64, 69, 72, 74, 78, 82, 87, 89–92, 95–100, 103, 109, 117, **119**, 120–22, 130–38, 141–48, 150–53, 159, 187, **194**, 200, 240, 243, 349–60, **362**–63, 365–67, **369**, 370–71, 373, 378–386
  beneficiaries 13, 91, 105, **106**, 107–09, 130, 146, 150, 265, 272, 349, 354–55, 381
  biophysical characteristics 92, 96–98
  capacity building 2, 13, 48, 57, 69, 72–73, 78, 98, 114, 132, 136, 141, 146, 150–52, 154, 240, 242, 271, 275, 303, 306, **320**, 342–343, 352, 432, 435
  climate risk management 175, 436
  climatic water balance 183
  common guidelines 92, 111, 119, 132, 135–137, **138–140**, 144, 147–153, 352, 370
  common lands 45, 80, 89, 99–100, 111, 114, 357, 381, 382
  common property resources -15, 102, 110, 357
  conservation agriculture 29, 132, 253, 266–267, **319–320**, 324, 339–340
  decision-making 9, 22, 87, 98, 110, 113, **119**, 120–121, 130, 141, 147, 165, 179, 199, 271, **360**, 372, 436
  entry point activity 65, 146, 207, 240, 305
  funds allocation 91, 103
  gender equity 62, 77, 117, 123, 153, 410
  groundwater and equity 89, 385

information and communication
technology (ICT) 160, **199**
information and management systems 79,
149
institutional arrangements 75, 89, 93,
102–103, 110–11, 113, 117, 137,
141, 143, 148, 149, 151, 241, 351,
354, 370
institutional constraints 145
institutional links 143, 145
land ownership 15, 55, 200, 316, 343
leaf area index 198
micro-enterprises 79, 118, 120, 149–150
project implementing agency **141**, 142
property relations in land and water 92
public–private partnership 77, 276
rainfall forecast 175
satellite data 162–63, 168, 170, 172, **178**,
186–88, 190, 193, 198, 200
self-help groups 9, 58, 69, 88, 143, 198
sharing biomass 102
sharing of water 101, 109
spatial variability 175
subsidies 7, 17, 92, 98, 103, 107–08, 121,
144–46, 148, 220, 242, 268, 354, 416
user groups 91, 111–12, **141**, 143, 342,
387
user rights 93, 114–115, 117, 121, 123,
140
water harvesting 2, 18, 29, 79–81, 96,
98–99, 104, 106, 116, 132, 180, 184,
197, **208**, **232**, 233, **234**, 235, **238**,
240, 243, 255–57, 261–63, 271–72,
316–18, 321, 324–**26**, 335, 350, 364,
382, 385, **433**
water use efficiency 44–45, 123, **206**, 218,
243, 249, 315, 319, 338
watershed policies 109, 132, 154
watershed treatments 89, 91, **104**, **107**,
380
weather forecasting 28, 185
women self-help groups 118
women's participation 118, **360**
Watershed evaluation 349, 364, 377, **385**,
386
benefit-cost analysis 361

bioeconomic modeling 370–72, 376
econometric methods 362
meta analysis 6, 136, 377, **385**
Watershed guidelines 56, 91–92, 121, **138**,
143, 147
Watershed management 21, 53, 56, 58, 63,
68–69, 74, 79, 98, 130, 132–**33**, 141,
149, 151, 159, 165, 186, **199**, 353,
372, 432
consortium approach 63, 64, 74–78, 82,
132, 149–150, 152
convergence- 64, 79, 80, 90, 98, 102,
120–122, 124, 135, 137, 142, 149,
167, 240, 252
equity 3, 7, 39, 60, 62, 72, 77, 79, 87–93,
97–100, 102–103, 106, 108–111,
112–124, 130, **33**, 136, **140**, 144,
150, 153, 325, 341, 353, 356, 410
evaluation 349, 364, 377, 385–86
impact assessment 108, 115, 139, 152,
163, 186, 188, 355–358, 362, 365,
371, 377–78, 380, 385
information and communication
technology (ICT) 160, **199**
information system 56, 159, 180, 198–99,
303
post-project sustainability 143, 151
project management 13, 151
water budgeting 180
watershed monitoring 186
Watershed Management Information System
56
Water use efficiency 44–45, 123, **206**, 218,
243, 249, 315, 319, 338
concepts 3, 103, 183, 196, 316
definitions 316–17, 430
irrigation schedule 319
supplemental irrigation 29, 70, 79, 81,
231, 233–34, **235**–37, 239, 243, 249,
254–56, **257**, 261–62, 265, **266**, 272,
298, 325–26, **327**–29, 330, **331**, 414
water balance 46, 180, 183–85, 315, **317**,
335
yield gap 40, **42**, 60, 81, 164, **165**, 200,
207, 253, 318, 393, 414